GIScience and Geo-environmental Modelling

The “GIScience and Geo-environmental Modelling” book series seeks to publish a broad portfolio of scientific books addressing the interface between geography and the environment. The aim of the book series is to present geospatial technology approaches to data mining techniques, data analytics, modeling, risk assessment, visualization, and management strategies. The series includes peer-reviewed monographs, edited volumes, textbooks, and conference proceedings. The focus of Geo-environmental is on geospatial modelling in the frontier area of earth-environment related fields, such as urban and peri-urban environmental issues, ecology, hydrology, basin management, geohazards, estuarine-ecology, groundwater, agriculture, climate change, land-water, and forest resources, and related topics.

Geo-environmental modelling techniques have enjoyed an overwhelming interest in recent decades among the earth environmental and social sciences research communities for their powerful ability to solve and understand various complex problems and develop novel approaches toward sustainable earth and human society. Geo-environmental modelling using data mining, machine learning, and cloud computing technology is focused on spatiotemporal data analysis and modeling for sustainability in our environment.

Jayanta Das • Sudip Kumar Bhattacharya
Editors

Monitoring and Managing Multi-hazards

A Multidisciplinary Approach

Springer

Editors
Jayanta Das
Department of Geography
Rampurhat College
Rampurhat, West Bengal, India

Sudip Kumar Bhattacharya
Department of Geography and Applied Geography
University of North Bengal
Siliguri, West Bengal, India

ISSN 2730-7506 ISSN 2730-7514 (electronic)
GIScience and Geo-environmental Modelling
ISBN 978-3-031-15376-1 ISBN 978-3-031-15377-8 (eBook)
https://doi.org/10.1007/978-3-031-15377-8

This Springer imprint is published by the registered company Springer Nature Switzerland AG
The registered company address is: Gewerbestrasse 11, 6330 Cham, Switzerland

Dedicated to beloved teachers, parents and daughters

Foreword

The book *Monitoring and Managing Multi-hazards: A Multidisciplinary Approach* edited by Dr. Jayanta Das and Dr. Sudip Kumar Bhattacharya, is a noble effort to assemble the research works on multi-hazards and related issues, as well as their monitoring and management, contributed by the eminent academicians and researchers from the different corners of the globe. In the last few decades, human civilization has been confronted with a variety of environmental hazards due to an exorbitant growth in population, unplanned anthropogenic activities, as well as climate change. Numerous scientific researches have also shown that the frequency of these hazards is expected to rise over time. Therefore, understanding the nature, magnitude, and systemic risk posed by multi-hazards and their management has evoked the interest of researchers and the scientific community on a global and national scale. Given the preceding importance, the decision to publish the current book and its full implementation will open a new horizon in the scientific community.

This book has been divided into two parts—the first part describes the nature, extent, and characteristics of various natural hazards using different statistical and MCDM techniques in remote sensing and GIS platform, and second part deals with the various mitigation measures to reduce threats posed by the different hazards. The entire book consists of 22 chapters on multi-hazards and allied issues related to sea level upsurge and aridity, soil erosion, flood, landslides, climate change, climatic hazards, groundwater vulnerability, groundwater depletion, chlorophyll and sea surface temperature changes, social issues like arsenic contamination, the mitigation measures such as terrain sensitivity and people's perception-based risk area management, wastewater treatment, adaptation to climate change, and impacts of COVID-19 on air quality by the investigators across the world. Many of the investigators are well known in their fields. The book has been edited with the motive to cover up both the review works and regional case studies to focus on the dynamicity of climate change, the change in natural resources, landscape, water, river, and agricultural ecosystems; some social issues at various spatio-temporal scales, including both theoretical and applied aspects, will thus help as guideline information for further research. This book includes conventional and advanced technologies commonly employed in environmental and socio-economic circumstances.

I wholeheartedly admire the efforts of Dr. Jayanta Das, the Department of Geography in Rampurhat College, Burdwan University, West Bengal, India and Dr. Sudip Kumar Bhattacharya, the Department of Geography and Applied Geography, North Bengal University, West Bengal, India, being the editors of this book, to create a platform for young and expert researchers from different academic fields to play their responsible role in unveiling the nature of multi-hazards and their best possible mitigation strategies. In addition, I would also like to thank all the staff of Springer Publishing Group for their full support and cooperation at all the stages of publication. I firmly believe that this book volume will depict new advancements in assessing various hazards and their management strategies that will benefit planners, scientists, and researchers in their research work. I wish all the success of this book.

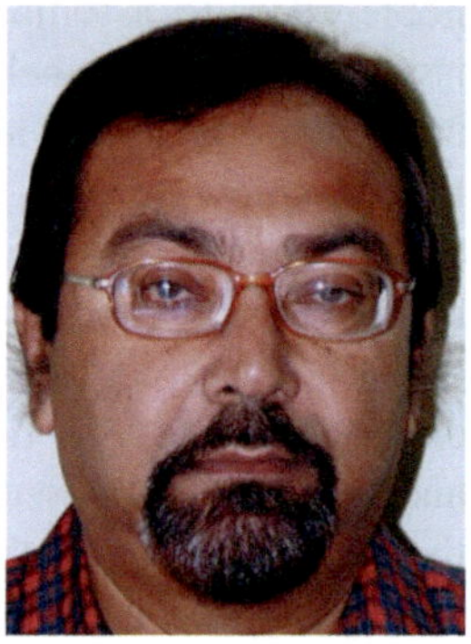

December 2021

Dr. Subir Sarkar Ph.D.
Professor
Department of Geography and Applied Geography
University of North Bengal
Siliguri, West Bengal, India

Preface

Nowadays, most part of the world frequently faces natural and anthropogenic hazards, even sometimes in the form of disasters, like drought, flood, landslides, deforestation, and cultivation-related sever soil erosion and many more which impend many people into catastrophic destruction and damage. Since natural hazards cannot be eliminated, quantifying these events and reliable forecasts can alleviate detrimental effects, building a more resilient and safe society. This book is the first detailed assessment of available knowledge of the multi-hazards in response to monitoring and management. It intends to fulfill the gap between science, policy, and the community concerned. It encompasses both the thematic and regional case studies to highlight dynamicity of climate changes, imbalance in natural resources, landscape, water, river, and agricultural ecosystems at various spatio-temporal scales, including theoretical and applied aspects, and helps as guideline information for future research. In a nutshell, this book provides traditional techniques and advanced geospatial technologies used in atmospheric, lithospheric, hydrospheric, biospheric, and socio-economic contexts on all spatial and temporal scales regarding hazards and disasters and sustainable development and management for the future. This book will be an essential product for researchers, scientists, NGOs, policymakers, and academic personnel in hazards and disaster resilience, climate change studies, environmental sciences, applied geomorphology, remote sensing, natural resources management, GIS, hydrology, and soil sciences. It also focuses on the application of precision techniques of remote sensing and GIS, quantification of various natural and environmental hazards along with the capacity building and sustainable mitigation strategies toward resilient societies.

This book is divided into two parts which consist of 22 chapters. Part I *Multi-hazards Monitoring* contains 16 chapters dealing with the subject of natural resources, aridity and sea level upsurge, active tectonics and channel shifting, soil loss rate and sediment yield, flood frequency, susceptibility and risk, effects of climatic hazards on agriculture, review of articles published on landslides in Mizoram, groundwater quality vulnerability assessment, review of the impact of COVID-19 pandemic on land surface temperature and air quality, variation of chlorophyll and sea surface temperature, landslide susceptibility mapping, the social impact of arsenicosis, and groundwater depletion zonation.

Chapter 1, entitled "Evaluating the Multi Hazards Threats Due to Aridity, Sea Level Upsurge in the Coastal Areas of North Tamil Nadu, South India", authored by Dhanya, focuses on land degradation in terms of changing climate, aridity. Pinna combinative index (IPmi) and De Martonne aridity index (IDmi) were used to calculate the potential impacts on land degradation with respect to aridity. RegCM data under IPCCs RCP 4.5 trajectories were used for simulating 127-year period of aridity in the erstwhile Chengalpet district. During these years, the maximum and minimum mean AI values observed in the study area ranged between 10.11 and 29.88 mm/°C. The mean IPmi value for the reference period was found to be 13.6 mm/°C and likely to be reduced by 0.55 mm/°C in the near century. Similarly, the rise in sea level for the coasts of Thiruvallur and Kancheepuram may be 73.2 cm by the end of 2100 and 50.3 cm by 2100, respectively. It indicates possibilities of saline water intrusions to the coastal wetlands and the need for CZM.

Chapter 2, "Active Tectonics and Associated Channel Shifting Pattern of Neora River Basin, Darjeeling Himalaya", contributed by Barman et al. states that most of the Himalayan Rivers are affected by the tectonic evolution of Himalaya. The study employed two morphotectonic indices: basin asymmetry and transverse topographic symmetry. Both factors point out to the river tilting toward the east. The results were validated by examining the regional geology and channel shifting pattern. The river Neora runs through Neora national park. Thus, there is a need to implement sustainable land and water management measures in the catchment.

Chapter 3, "Estimating Soil Loss Rate and Sediment Yield of the Proposed Ngololweni Earth Dam, Kingdom of Eswatini", contributed by Hlanze et al. states that soil loss and sediments yield affect most of the reservoirs, which reduce reservoir useful life through loss of storage. The study employed the Revised Universal Soil Loss Equation (RUSLE), runoff plots, and satellite images for the analysis. The study results show that the annual average sediment yield was 82.04 tons/year and sediment delivery ratio (SDR) was 0.46; thus, there is a need to implement sustainable land and water management measures in the catchment.

Chapter 4, "Flood Susceptibility Mapping Using GIS and Multi-criteria Decision Analysis in Dibrugarh District of Assam, North-East India", contributed by Bora et al. is the first study of flood susceptibility mapping using GIS and AHP in the study region to the best of the author's knowledge. The results of the study show that 434.56 km^2 (12.86%) of Dibrugarh district is very high susceptible to flooding, and the most effective flood causing factor is elevation. The flood susceptibility model is validated using the ROC curve and found to be good enough with an AUC value of 74%.

Chapter 5, Mamun et al. showed the effects of climatic hazards on agriculture in the Teesta basin of Bangladesh using both primary and secondary data. They found that inundation is the main problem of *Aman* rice at the sowing stage, which causes huge losses for the farmers. Hailstorms affect jute cultivation in Kursamari and Azam khan villages. Cold waves caused serious damage to *Rabi* crops. Diseases associated with cold and fog also lessen farmer's winter crops.

Chapter 6, "Mizoram, the Capital of Landslide: A Review of Articles Published on Landslides in Mizoram, India", contributed by Barman et al. states that northeast India is under the direct impact of SE monsoon. The high rainfall intensity throughout the year, coupled with rapid unscientific road construction and urbanization, is the main cause of landslides in hilly areas like Mizoram. The study does a review of articles that were published based on Mizoram as a study area. The review results show that most of the landslide occurs due to heavy rainfall and along the roadside. Thus, the authors hope this review will help future research in town planning, settlement development, road construction, etc., in the hilly regions, especially northeast India.

Chapter 7, deals with applying Geographical Information System (GIS) and geostatistics to evaluate groundwater vulnerability in the National Capital Territory of Delhi. Hydrogeochemical data from 50 sample points were analyzed to estimate the vulnerability, and 10 thematic layers were prepared, such as pH, TDS, bicarbonate, chloride, sulfate, nitrate, fluoride, calcium, magnesium, and total hardness. Integration of these datasets was done using multi-criteria decision making (MCDM). The results are extremely beneficial for the proper planning and management of groundwater resources and delineating the zone of groundwater vulnerability.

Chapter 8, "A Literature Review of the Impact of COVID-19 Pandemic on Land Surface Temperature and Air Quality of India", contributed by Vanlalchhuanga et al. states that the unprecedented upsurge of the COVID-19 has disrupted human life in many ways. The study employed a review on various research literatures in land surface temperature (LST) and air quality monitoring during India's pandemic lockdown. In addition to air quality, the LST also decreased during the lockdown, with major cities experiencing a decrease in the mean surface temperature during the lockdown in India. All the polluted Indian cities, within just four days of implementing lockdown, saw improvement in air quality by 40 to 50% with a massive reduction by 60 to 30% of the major air pollutants.

Chapter 9, "Seasonal and Inter-Annual Variation of Chlorophyll and Sea Surface Temperature in Northern and Southern Arabian Sea, India", contributed by Abhilash et al. states the annual mean chl-a values in NAS is 1.89 mg/m^3, which is a large difference compared to 0.44 mg/m^3 in SAS. From 2003 to 2019, chl-a density and SST were analyzed by using MODIS data and SeaDAS software. Authors found that chlorophyll and SST are not stable and change with the seasons. Increased evaporation, reduced solar radiation, convection, and surface cooling in the northern region make NAS more productive in the winter and also in summer compared to SAS.

Chapter 10, "Application of a Geospatial-Based Subjective MCDM Method for Flood Susceptibility Modeling in Teesta River Basin, West Bengal, India", contributed by Poddar et al. states the development of a flood susceptibility map in the Teesta River basin of West Bengal, India. The study used the geospatial-based AHP method for developing a flood susceptibility map. The results of the study showed that 11.99%, 13.90%, 21.48%, 23.93%, and 28.70% of study areas as very high, high, moderate, low, and very low susceptible areas, respectively. The accuracy of the flood susceptibility map was tested using ROC and AUC analysis. The obtained AUC value was 0.739, confirming the reliability of the applied method in flood modeling. The finding of this study will be useful in preparing effective flood management strategies.

Chapter 11, entitled "Flood Frequency Analysis of Baitarani River Using Three Probability Distributions", by Sinam, deals with the method of finding the most suitable probability distribution model for the river basin, which will help in estimating the design flood. The author examined three distributions, namely Gumbel, Generalized Extreme Value, and Log Pearson Type 3 distribution, for the Baitarani River of Odisha using time series data for 43 years. The result of the study showed that out of the three models, Log Pearson Type 3 is the best-fit model for the sub-basin.

The study of Chap. 12 employed a GIS-based MCDA technique for flood vulnerability and hazard to estimate the risk in Sub-Himalayan Alipurduar district, West Bengal. Authors found that Falakata, Alipurduar I, and Alipurduar II block reveal a high risk of inundation problems. It is seen that 19.15% and 5.42% area of Alipurduar district represent high and very high flood risk conditions. Therefore, there is a need to implement sustainable flood management, control, and mitigation.

Chapter 13, Guhananth et al. studied landslide susceptibility mapping over Kegalle district, Sri Lanka, using the AHP method integrated with GIS. They used a few trigger parameters of landslides such as slope, aspect, soil class, lithology, rainfall, land use, distance to roads, and distance to streams. They concluded that about 6% and 12% of the entire Kegalle district is at a very high and high landslide-prone zone.

Chapter 14, "Landslide Susceptibility Evaluation and Analysis: A Review on Articles Published During 2000 to 2020", contributed by Barman et al. states that landslide is one of the significant hazards in hilly areas worldwide. They reviewed different methodologies applied in landslide susceptibility mapping and zonation. They concluded that the logistic regression model has the highest accuracy in landslide susceptibility mapping. Review articles covered 46 countries, with the maximum work coming from countries in the Southeast and East Asian countries. The authors hope this review work will help future research in landslide susceptibility and hill area development programs.

Chapter 15, Poddar et al. assessed the social impact of arsenicosis through groundwater arsenic poisoning in the Maldah district using primary data. They concluded that arsenicosis creates social hazards among the poor, especially for rural women. The most prominent social hazards such as difficulties in getting married to arsenicosis victims' women, the problem of social instability, the dominance of the dowry system, preferences to remain unidentified, and the mental stress of women were identified.

Chapter 16, Gupta et al. studied groundwater depletion zonation using TOPSIS-MCDM, MK, Senslope, and RAI tests in Raipur District, Chhattisgarh, India. They summarized that the one-third area of this district is in the critical stage; the proper sustainable groundwater management plan is needed to replenish this precious natural resource.

Part II *Multi-hazards Management* contains six chapters relating to terrain sensitivity guided and people's perception-based risk area management of the hills, wastewater treatment, adaptation to climate change in agriculture, land use/land cover change detection, groundwater potentiality mapping, and climate-induced alterations in sugarcane yield distribution and its agronomic adaptation strategies.

Chapter 17, "Terrain Sensitivity Guided and People's Perception Based Risk Area Management of the Hills of Darjeeling District, India", contributed by Sudip Kumar Bhattacharya, is an attempt to identify the 'Risk Areas' for human habitation in the hilly terrain of the Darjeeling district, West Bengal, India. The risk areas have been identified based on terrain sensitivity assessment as well as people's perception study on the built environment. Various degrees of terrain sensitivity (MCDA, GIS, and field-based) and the local people's perception based on a close-ended questionnaire have identified different degrees of 'risk areas' based on which appropriate management strategies have been proposed.

Chapter 18, "Wastewater Treatment in India—A New Perspective", contributed by Mandal et al. described how the innovative biotechnological tools could be used to treat domestic and industrial wastewater in a cost-effective and cost-effective way and eco-friendly manner. They suggested that microbial population, plants, and fungi could be used for sequestration and removal of toxic heavy metals, degradation of complex organic matter, and biotransformation and biovolatilization of harmful substances in the wastewater. They also concluded that bioremediation, phytoremediation, and mycoremediation techniques could be independently used or amalgamated with existing technologies for wastewater treatment.

Chapter 19, Mamun et al. investigated the indigenous and institutional adaptation measures of the farming households using survey data collected from three districts of Bangladesh. They suggested that institutional support such as technical and financial support is needed to promote farmers' adaptation to climate change.

Chapter 20, "Land Use/Land Cover Change Detection Through the Spatial Approach: A Case Study of the Badiadka Panchayath, Kerala", contributed by Nirmala et al. investigated the changes in LULC pattern in the Badiadka Panchayath from 2013 to 2020 by using satellite images. The changes have been made by the Krishi project, Kerala, after 2017, which has attracted more people to their homes and surrounding farming. The results of the LCR and LAC-based study reveal that the agricultural land and settlements area increased from 25% to 47% and 14% to 22%, respectively, in 2020 due to rapid growth of population and declining vegetation and wasteland.

Chapter 21, "Delineation of Groundwater Potential Zones Through AHP: A Case Study from Tamil Nadu, India", contributed by Rahaman et al. employed the AHP method using seven causative factors for delineating the potential groundwater zones. The result divided the study area into five groundwater potential zones from very high zone (covering 17.79 km^2), high (1987.22 km^2), moderate (595.86 km^2), low (731.82 km^2), and very low zone (covering 6.746 km^2), respectively. The high drainage density (high rate of infiltration) and low lineament density enhance the area marginally only in the very poor to poor prospective zones.

Chapter 22, entitled "Expected Climate-Induced Alterations in Sugarcane Yield and Its Agronomic Adaptation Strategies", authored by Dhanya et al. evaluates implications of climate change impacts on sugarcane crop using DSSATv 4.7 crop simulation model. The study used dynamically downscaled climate data from the RegCM model for 128 years for Kancheepuram and Thiruvallur. The projections showed a decline in yield for sugarcane. The yield change was just −1.8, −2.6, and −2.8% for the near, mid, and end of the twenty-first century, respectively. Findings show sugarcane planted on March 15th can augment yield rates by 3.5% under the RCP 4.5. Implementation of planned agronomic adaptation is necessary.

We are thankful to all the authors who have meticulously completed their documents on a short announcement and co-operated in building this a very edifying and beneficial publication. We hope that the information provided in this volume will encourage the research community to use this synthesis and develop spatial and temporal models to monitor and manage multi-hazards worldwide.

Rampurhat, West Bengal, India — Jayanta Das
Siliguri, West Bengal, India — Sudip Kumar Bhattacharya

Acknowledgements

We would like to express our deep appreciation to all the contributing authors for their time, effort, and research ideas presented in the individual chapters. We would also express our sincere thanks to the series editors, Biswajeet Pradhan, Pravat Kumar Shit, Gouri Sankar Bhunia, Partha Pratim Adhikary, Hamid Reza Pourghasemi and publishing editor Doris Bleier of 'GIScience and Geo-environmental Modelling' for their kind patience and generosity to undertake the editorial work for such diverse topics on *Multi-hazards Monitoring and Management*.

All the chapters have been peer-reviewed, and the authors updated the research in response to the reviewers' comments. We want to express our gratitude to all of the anonymous reviewers who made the effort to submit comments on the chapters.

We are deeply indebted to Prof. Subir Sakar (Department of Geography and Applied Geography, University of North Bengal) for agreeing to write the Foreword and to continuously keep patience in revising it as we kept adding new chapters and/or changing chapter titles in the book.

We would like to express our special thanks to Prof. Ranjan Roy, HoD, Department of Geography and Applied Geography, the University of North Bengal, for his very positive attitude and valuable suggestions regarding the preparation of this book.

Dr. Jayanta Das wishes thanks to Dr. Asish Banerjee, President of Rampurhat College and also Deputy Speaker of West Bengal Legislative Assembly and Dr. Buddhadeb Mukherjee, Teacher-in-Charge, for their encouragement and support. He also expresses his due acknowledgment to the Department of Geography, Rampurhat College, for providing infrastructure facilities. Dr. Das wishes again to thank his colleagues and students, especially Rajib Mitra, for helping him with some of the copy-editing parts.

Dr. Sudip Kumar Bhattacharya wishes to thank all of his colleagues, well-wishers, and students from whom he has got encouragement to take such a venture in his hand. He too expresses his due acknowledgment to the Department of Geography and Applied Geography, University of North Bengal, for getting right ambience and infrastructure.

Finally, we also thank the Springer Management team for accommodating us and providing us such a platform for this book for its successful completion. It was a pleasure to work with them throughout the process.

Rampurhat, West Bengal, India Jayanta Das
Siliguri, West Bengal, India Sudip Kumar Bhattacharya

About This Book

Over the years, natural and social hazards have increased from local, regional, to global scales mainly due to ever-increasing population and anthropogenic activities. To monitor multi-hazards, remote sensing and GIS-based multi-criteria decision-making (MCDM) techniques have been extensively used in recent years worldwide. Since natural hazards cannot be eliminated, only quantification of these events and reliable forecasting can alleviate their detrimental effects, which can build a more resilient and safe society. Social hazards and disasters are also equally catastrophic where scientific know-how and management can hold back the society in a stable state. Therefore, cultivating the proper knowledge of the multi-hazards and their monitoring and management can fill the gap between science, policy, and the community concerned. In an endeavor to understand and characterize the various hazards, *Monitoring and Managing Multi-hazards: A Multidisciplinary Approach* presents a synthesis of what cross-disciplinary researchers know about these hazards and indigenous adaptation strategies.

This book intends to focus on the use of precision techniques, remote sensing, and GIS technologies to quantify various natural, environmental, and social hazards along with the capacity building and sustainable mitigation strategies toward resilient societies. It encompasses both the thematic and regional case studies to highlight the dynamicity of climate change, change of natural resources, landscape, water, river, agricultural, and social ecosystems at various spatio-temporal scales, including theoretical and applied aspects and helps as guideline information for future research. Combining reviews, methods, and case studies, the book gives readers research information and analysis of traditional and advanced geospatial technologies on atmospheric, lithospheric, hydrosphere, biospheric, and socio-economic contexts, on all spatial and temporal scales regarding hazards and disasters and sustainable development and management for the future. Overall, this book will be an essential product for researchers, scientists, NGOs, academic personnel, policymakers, and advanced learners in hazards and disaster resilience, climate change, environmental sciences, geomorphology, remote sensing, natural resource management, GIS, hydrology, and soil sciences.

Key Features

- Manifests a comprehensive nature and characteristics of multi-hazards by integrating several numerical models and geoinformatics technologies.
- Provides a complete overview of the role of remote sensing and GIS-based multi-criteria decision-making (MCDM) techniques in hazards monitoring.
- Unveils how various hazards and climate change affect people and natural resources.
- Demonstrates how data analysis can be used for managing hazards.
- International and cross-disciplinary contributions by globally renowned researchers.

Contents

Part II Multi-hazards Management

Editors and Contributors

About the Editors

Dr. Jayanta Das is an Assistant Professor at the Department of Geography in Rampurhat College, The University of Burdwan, West Bengal, India. He has completed his master's and Ph.D. degrees at the Department of Geography and Applied Geography, University of North Bengal, India. Dr. Das's field of interest includes multi-hazards modeling such as landslide, groundwater, flood, drought analysis, climate change, watershed management, hydrological modeling, water quality, geospatial data analysis, data mining, and GIS applications with more than 12 academic years of experience. Dr. Jayanta Das has published more than 30 research articles in peer-reviewed journals, focusing mainly on: climate change, agricultural suitability analysis, natural and man-made hazards analysis, risk management, and spatial data analysis.

Dr. Sudip Kumar Bhattacharya is a permanent Associate Professor in the Department of Geography and Applied Geography, North Bengal University, West Bengal, India. He started his academic carrier in the fields of geomorphology, environment, and climatology. He has 60 research publications on the topics of landslide disaster and management, landslide vulnerability zonation, impact of deforestation and management, soil erosion and conservation, water scarcity in the hill areas and management, terrain evaluation and land use planning, terrain sensitivity assessment, climate-related issues, population and urbanization-related issues, etc. in the various reputed international and national journals like *Transactions: Japanese Geomorphological Union*; *Environmental Earth Science*: Springer, *Environment, Development and Sustainability*: Springer, *Modeling Earth Systems and Environment*: Springer, *Spatial Information Research*: Springer, *SN Applied Sciences*: Springer, *International Journal of Geomatics and Geosciences, Indian Journal of Soil Conservation*; *Geographical Review of India, MAUSAM, Indian Journal of Spatial Science*, and many other standard journals and edited books and monographs. Dr. Bhattacharya did his graduation in Geography Honors from Vidyasagar College, Kolkata, under Calcutta University and completed his masters in Geography from Burdwan University, West Bengal. He was awarded Ph.D. degree by North Bengal University. His Ph.D. research work was on 'River Basin Management'. Six research scholars have been awarded Ph.D. degrees under his supervision. Some research scholars are currently carrying out Ph.D. research work under his supervision.

Contributors

H. D. Abeysiriwardana Department of Civil Engineering, Faculty of Engineering, Sri Lanka Institute of Information Technology, Malabe, Sri Lanka

S. Abhilash Department of Marine Geology, Mangalore University, Karnataka, India, India

Md Nashim Akhtar Department of Civil Engineering, Aliah University, New Town, Kolkata, India

Nadeem Akhtar Department of Civil Engineering, Aliah University, New Town, Kolkata, India

Md. Abdullah Al Mamun Institute of Bangladesh Studies, University of Rajshahi, Rajshahi, Bangladesh

G. M. Monirul Alam Department of Agribusiness, Bangabandhu Sheikh Mujibur Rahman Agricultural University, Ghazipur, Bangladesh

Jiarul Alam Department of Geography and Applied Geography, University of North Bengal, Darjeeling, India

Suparna Bagchi Department of Molecular Biology and Biotechnology, Sripat Singh College (Under University of Kalyani), Jiaganj, Murshidabad, West Bengal, India

Jonmenjoy Barman Department of Geography and Resource Management, Mizoram University (Central), Aizawl, India

Amiya Basak Department of Geography, Cooch Behar Panchanan Barma University, University in Cooch Behar, Cooch Behar, West Bengal, India

Abhishek Basu Department of Molecular Biology and Biotechnology, Sripat Singh College (Under University of Kalyani), Jiaganj, Murshidabad, West Bengal, India

Sudip Kumar Bhattacharya Department of Geography and Applied Geography, University of North Bengal, Siliguri, West Bengal, India

Kalyan Bhuyan Department of Physics, Dibrugarh University, Dibrugarh, Assam, India

Brototi Biswas Department of Geography and Resource Management, Mizoram University (Central), Aizawl, India

Sayan Biswas Department of Molecular Biology and Biotechnology, Sripat Singh College (Under University of Kalyani), Jiaganj, Murshidabad, West Bengal, India

Shyam Lochan Bora Department of Physics, D.D.R. College, Chabua, Assam, India;
Department of Physics, Dibrugarh University, Dibrugarh, Assam, India

Subhashis Choudhury Centre for Himalayan Studies, University of North Bengal, Darjeeling, West Bengal, India

Jayanta Das Department of Geography, Rampurhat College, Rampurhat, Birbhum, West Bengal, India

Satyajit Das Department of Geography and Applied Geography, University of North Bengal, Siliguri, West Bengal, India

Sayan Das Department of Molecular Biology and Biotechnology, Sripat Singh College (Under University of Kalyani), Jiaganj, Murshidabad, West Bengal, India

Rajalakshmi Dhandapani Tamil Nadu Agricultural University, Coimbatore, Tamil Nadu, India

P. Dhanya Tamil Nadu Agricultural University, Coimbatore, Tamil Nadu, India

W. M. Dlamini Geography, Environmental Science and Planning, University of Eswatini, Matsapha, Eswatini

Sk. Fulchand Department of Molecular Biology and Biotechnology, Sripat Singh College (Under University of Kalyani), Jiaganj, Murshidabad, West Bengal, India

Pattiyage I. A. Gomes Department of Civil Engineering, Faculty of Engineering, Sri Lanka Institute of Information Technology, Malabe, Sri Lanka

Kupendrathas Guhananth Department of Civil Engineering, Faculty of Engineering, Sri Lanka Institute of Information Technology, Malabe, Sri Lanka;
Department of Civil Engineering, Faculty of Engineering, University of Jaffna, Jaffna, Sri Lanka

W. Gumindoga Department of Civil Engineering, University of Zimbabwe, Harare, Zimbabwe

Pooja Gupta Department of Geology, Government Science College, Jabalpur, Madhya Pradesh, India

Mohd Sayeed Ul Hasan Department of Civil Engineering, Aliah University, New Town, Kolkata, India;
Centre for Oceans, Rivers, Atmosphere and Land Sciences, Indian Institute of Technology Kharagpur, Kharagpur, West Bengal, India

Partha Jyoti Hazarika Department of Statistics, Dibrugarh University, Dibrugarh, Assam, India

D. K. Hlanze Department of Water Resources Engineering, University of Dar es Salaam, Dar es Salaam, Tanzania

Md. Aminul Islam Department of Media Studies and Journalism, University of Liberal Arts Bangladesh, Dhaka, Bangladesh

R. Jagannathan Tamil Nadu Agricultural University, Coimbatore, Tamil Nadu, India

Md. Kamruzzaman Institute of Bangladesh Studies, University of Rajshahi, Rajshahi, Bangladesh

Dhananjay Kathal Plant Pathology, College of Agriculture, Pawerkheda, India

Braj Kishor St. John's College, Agra, India

E. Krishnaveni Prabhu Center for Climate Change and Adaptation Research, Anna University, Guindy, Chennai, Tamil Nadu, India

S. Mabaso Geography, Environmental Science and Planning, University of Eswatini, Matsapha, Eswatini

Md. Abdullah Al Mamun Institute of Bangladesh Studies, University of Rajshahi, Rajshahi, Bangladesh

Debjani Mandal Department of Molecular Biology and Biotechnology, Sripat Singh College (Under University of Kalyani), Jiaganj, Murshidabad, West Bengal, India

Rudrajit Mandal Department of Molecular Biology and Biotechnology, Sripat Singh College (Under University of Kalyani), Jiaganj, Murshidabad, West Bengal, India

Rajib Mitra Department of Geography and Applied Geography, University of North Bengal, Darjeeling, India

Atanu Mondal Department of Molecular Biology and Biotechnology, Sripat Singh College (Under University of Kalyani), Jiaganj, Murshidabad, West Bengal, India

Subhankar Mondal Department of Molecular Biology and Biotechnology, Sripat Singh College (Under University of Kalyani), Jiaganj, Murshidabad, West Bengal, India

Koyel Mukherjee Department of Geography, Rampurhat College, Birbhum, West Bengal, India

Sarbari Mukhopadhyay Department of Geography, Maynaguri College, Maynaguri, Jalpaiguri, India

D. M. M. Mulungu Department of Water Resources Engineering, University of Dar es Salaam, Dar es Salaam, Tanzania

K. S. Harikrishna Naik Department of Marine Geology, Mangalore University, Mangaluru, Karnataka, India

P. Ndomba Department of Water Resources Engineering, University of Dar es Salaam, Dar es Salaam, Tanzania

R. Nirmala Department of Marine Geology, Mangalore University, Mangaluru, Karnataka, India

Surajit Paul Department of Geography and Applied Geography, University of North Bengal, Siliguri, West Bengal, India

Surjapada Paul Department of Geography and Applied Geography, University of North Bengal, Siliguri, West Bengal, India

Debapriya Poddar Department of Geography and Applied Geography, University of North Bengal, Darjeeling, India

Indrajit Poddar Department of Geography and Applied Geography, University of North Bengal, Darjeeling, India

Aneesah Rahaman CNDHS, University of Madras, Chennai, India

A. T. M. Sakiur Rahman Hydrology Lab, Department of Earth and Environmental Science, Kumamoto University, Kumamoto, Japan

M. Mizanur Rahman Department of Geography and Environmental Studies, University of Rajshahi, Rajshahi, Bangladesh

Abhishek Kumar Rai Centre for Oceans, Rivers, Atmosphere and Land Sciences, Indian Institute of Technology Kharagpur, Kharagpur, West Bengal, India

A. Ramachandran Center for Climate Change and Adaptation Research, Anna University, Guindy, Chennai, Tamil Nadu, India

Sufia Rehman Department of Civil Engineering, National Institute of Technology Karnataka, Surathkal, India

Debasish Roy Department of Geography and Applied Geography, University of North Bengal, Siliguri, West Bengal, India

Aravind Sahay Space Application Centre (SAC), ISRO, Ahmedabad, Gujarat, India

Souradip Seal Department of Molecular Biology and Biotechnology, Sripat Singh College (Under University of Kalyani), Jiaganj, Murshidabad, West Bengal, India

Most. Shayda Shamsea Aziz Shabee Kobe City, Hyogo, Japan

Rebati Sinam Centre for the Study of Regional Development, School of Social Sciences, Jawaharlal Nehru University, New Delhi, India

David Durjoy Lal Soren Department of Geography and RM, Mizoram University, Aizawl, India

V. Madha Suresh CNDHS, University of Madras, Chennai, India

Shams Tabrez King Fahd Medical Research Center, King Abdulaziz University, Jeddah, Saudi Arabia;
Department of Medical Laboratory Sciences, Faculty of Applied Medical Sciences, King Abdulaziz University, Jeddah, Saudi Arabia

S. Tfwala Geography, Environmental Science and Planning, University of Eswatini, Matsapha, Eswatini

Sanjay Tignath Department of Geology, Government Science College, Jabalpur, Madhya Pradesh, India

Vanlalchhuanga Department of Geography and RM, Mizoram (Central) University, Aizawl, Mizoram, India

Samina Wasi Department of Biochemistry, College of Medicine, Imam Abdulrahman Bin Faisal University, Dammam, Saudi Arabia

Abbreviations

AD	Anderson-Darling
AHP	Analytical hierarchy process
AM	Annual maximum
ANCOVA	Analysis of covariance
ANFIS	Adaptive neuro-fuzzy inference system
AQI	Air quality index
As	Arsenic
AUC	Area under curve
C/I	Capacity–inflow ratio
Ca^{2+}	Calcium
CDF	Cumulative distribution function
CDPs	Chiefdom development plans
CF	Certainty function
CGWB	Central Ground Water Board
Chl-a	Chlorophyll-a
CI	Consistency index
Cl^{-}	Chloride
CO	Carbon monoxide
CP	Child population
CPCB	Central Pollution Control Board
CR	Consistency ratio
CSM	Crop simulation model
DD	Drainage density
DEM	Digital elevation model
DENFIS	Dynamic evolving neural-fuzzy inference system
DFS	Distance from flood shelter
DH	Distance from hospital
DR	Distance to river
DS	Divisional Secretariat
DSSAT	Decision support system for agrotechnology transfer
DTR	Diurnal temperature ranges
EBF	Evidential belief function
EL	Elevation
ER	Employment rate
EV	Extreme value
F^{-}	Fluoride
FAO	Food and Agriculture Organization

FAO-PM	FAO Penman–Monteith method
FR	Frequency ratio
FRI	Flood risk index
FSC	Full supply capacity
FSI	Flood susceptibility index
FSZ	Flood susceptibility zonation
FVI	Flood vulnerability index
G	Geomorphology
GDEM	Geographical digital elevation model
GDZ	Groundwater depletion zonation
Ge	Geology
GEV	Generalized extreme value
GGWB	Central Groundwater Board
GIS	Geographic Information System
GL	Geology
GLO	Generalized logistics
GMIS	Global Man-made Impervious Surface and Settlement Extent
GOF	Goodness of fit
GPA	Generalized Pareto
GPZ	Groundwater potential zone
GWD	Groundwater depth
GWPI	Groundwater potential index
HCO_3^-	Bicarbonate
HD	Household density
IDmi	De Martonne aridity index
IDW	Inverse distance weighted
IDW	Inverse distance weighting
IFAD	International Fund for Agricultural Development
IMD	India Meteorological Department
IPCC	Intergovernmental Panel on Climate Change
IPmi	Pinna combinative index
KAP	Kappa
KS	Kolmogorov–Smirnov
L	Lineament density
LAC	Land absorption coefficient
LANDSAT	Land satellite
LCR	Land consumption rate
LN3	Log Normal Type 3
LP3	Log Pearson Type 3
LPM	Liter per minute
LR	Logistic regression
LS	Least squares
LSE	Land surface emissivity
LSM	Landslide susceptibility mapping
LST	Land surface temperature
LULC	Land use land cover
MARS	Multivariate adaptive regression splines
MBT	Main boundary thrust

MCDA	Multi-criteria decision analysis
MCDM	Multi-criteria decision making
MCT	Main central thrust
MFI	Modified Fourier index
Mg^{2+}	Magnesium
MK	Mann-Kendall
ML	Maximum likelihood
mMK	Modified Mann-Kendall
MNDWI	Modified normalized difference water index
MODIS	Moderate resolution imaging spectroradiometer
MOM	Method of moments
MRSA	Map removal sensitivity analysis
MSL	Mean sea level
NAAQS	National Ambient Air Quality Standards
NAS	Northern Arabian Sea
NASA	National Aeronautics and Space Administration
NBSS	National Bureau of Soil Science
NCT	National Capital Territory
NDBI	Normalized difference built-up index
NDVI	Normalized difference vegetation index
NEI	Northeast India
NO_2	Nitrogen dioxide
NO_3	Nitrate
NRSC	National Remote Sensing Center
Pc	Undivided Precambrian rock
PD	Partial duration
PD	Population density
PDF	Probability density function
PE3	Pearson Type 3
pH	Potential of hydrogen
PM10	Particulate matter 10
PM2.5	Particulate matter 2.5
PWM	Probability weighted moments
Pz	Undivided Paleozoic rock
Q	Quaternary sediment
R	Rainfall
RAI	Rainfall anomaly index
RCI	Random consistency index
RegCM	Regional climate model
RI	Rainfall intensity
RI	Random consistency index
RNN	Recurrent neural network
ROC	Receiver operating characteristic
RS	Remote sensing
RUSLE	Revised Universal Soil Loss Equation
RUSLEFAC	Revised Universal Soil Loss Equation for Application in Canada
S	Soil

SAS	Southern Arabian Sea
SDR	Sediment delivery ratio
SeaDAS	SeaWiFS data analysis system
SHI	Social hazards index
SI	Sensitivity index
SL	Slope
SLR	Sea level rise
SO_2	Sulfur dioxide
SO_4^{2-}	Sulfate
SPI	Stream power index
SPSA	Single-parameter sensitivity analysis
SRTM DEM	Shuttle Radar Topography Mission digital elevation model
SST	Sea surface temperature
STI	Sediment transport index
SWIR	Short-wave infrared
SWOT	Strengths, weaknesses, opportunities, and threats
TDS	Total dissolved solids
TH	Total hardness
TOPSIS	Technique for Order of Preference by Similarity to Ideal Solution
TWI	Topographic wetness index
USDA	United States Department of Agriculture
USGS	United States Geological Survey
VCI	Vegetation condition index
VIF	Variance inflation factor
WAK	Wakeby
WASPAS	Weighted aggregated sum product assessment

Part I

Multi-hazards Monitoring

1 Evaluating the Multi-Hazards Threats Due to Aridity, Sea Level Upsurge in the Coastal Areas of North Tamil Nadu, South India

P. Dhanya

Abstract

Climate change and improper soil management cause an increased rate of aridity and decline in soil quality in water-stressed areas across the globe. The world's arid zones have the lowest annual precipitation and the highest intensities of heat. The southern arid regions of India are in the rain shadow zones of the Western Ghats, which cover many parts of Tamil Nadu. This paper focuses on land degradation impacts on changing climate, aridity, and soil quality. Pinna combinative index *(IPmi)* and De Martonne Aridity Index *(IDmi)* were used to calculate the potential impacts on land degradation with respect to aridity. Simulated climate data from RegCM model under IPCCs Representative Concentration Pathway 4.5 (RCPs) trajectories were used for projecting 127 year period of aridity from 1971 to 2098 for 14 grids in the erstwhile Chengalpet district. The mean precipitation and air temperature for the 14 grids in the study area are utilized to compute aridity for the yearly time scale. For the coldest and driest months of winter and summer, the anticipated aridity conditions were computed for the near (2040s), mid (2070s), and end century (2100). During these years, the maximum and minimum mean AI values observed in the study area ranged between 10.11 and 29.88 mm/°C. The mean Pinna combinative index value for the reference period was found to be 13.6 mm/°C for the study area; however, it was found to be reduced to 13.05 (by 0.55 mm/°C) during the near-century period under RCP 4.5. Similarly, the upsurge in sea level for the coast of Thiruvallur and Kancheepuram district is likely to be 44.6 cm by mid, 73.2 cm by the end of 2100, 32.00 cm by mid, and 50.3 cm by 2100, respectively. It indicates possibilities of saline water intrusions to the coastal wetlands. Combined exposure to these hazards may exacerbate land degradation in these areas. Proactive evaluation and planning using geospatial technologies are essential for sustainable land management.

Keywords

Land degradation · Pinna combinative index *(IPmi)* · De Martonne aridity index *(IDmi)* · Climate change · Sea level rise · South India

1.1 Introduction

Climate change is transforming the context in which living organisms live. The speed and intensity of environmental and climatic change

P. Dhanya (✉)
Tamil Nadu Agriculture University, Coimbatore, India
e-mail: dhanya.eptri@gmail.com

J. Das and S. K. Bhattacharya (eds.), *Monitoring and Managing Multi-hazards*,
GIScience and Geo-environmental Modelling, https://doi.org/10.1007/978-3-031-15377-8_1

are accelerating day by day (Das et al. 2019). It has become a global need to assess the vulnerabilities and understand the local changes in climate in developing countries (Das and Bhattacharya 2018). Building community resilience through proactive management measures for the susceptible coastal areas is critical at the moment (Brown et al. 2010; Abiy et al. 2014; Martínez et al. 2017). Global mean sea level rise and extreme weather events are anticipated to intensify in many parts of the world (IPCC 2014, 2018). Strong undesirable impacts are reported on tropical agriculture production due to climatic changes (Naresh Kumar et al. 2012; Hatfield et al. 2011).

According to IPCC, inter-annual, seasonal variations in precipitations, and resultant moisture stress will be increased in the future. Desertification may affect a billion sources of income in over a hundred nations (UNCCD 2016). It is known that any changes in the hydrological cycle may force far-reaching consequences on rainfall availability, soil moisture conditions, and land quality in an area. A regional climate simulation study reported by Chaturvedi et al. (2012) indicated a rise in temperature for India in the range of 2.9–3.3 °C by the end of the century under the moderate emission pathway RCP 4.5. There will be more break days, less monsoon days for India. Increases in the frequency of extreme climate events like floods and droughts will decline crop production and negatively impact food security, cost of living, and ultimately worsen rural poverty in developing countries (Marengo et al. 2016). The changes in rainfall patterns and evaporation and transpiration also impact crop yield positively or negatively depending on its intensity (Młyński et al. 2018; Das et al. 2020). Alterations in the hydrological cycle could enhance flooding risk and soil erosion in coastal wetlands (Guhathakurta et al. 2011; Dai 2011). Unexpected climatic perturbations enhance the susceptibility of lands degradation.

Interior and western Tamil Nadu have a tropical semi-arid climate. Drier conditions in arid and semi-arid regions are highly probable to cause severe water scarcity in the future. (Dutta et al. 2015; Shifteh Some'e et al. 2013). A coastline of 7500 km and a share of 16.7% of the world's population make India vulnerable to ecological challenges Teri (2016). Indian coastal districts share 17% of our national population, making these areas more and more sensitive to climate aberrations (Shukla et al. 2003; Brown et al. 2018). The topographical settings of Tamil Nadu make it vulnerable to natural disasters such as cyclones, droughts, floods, and earthquake-induced tsunami. Research publications are revealing that the sea level of Chennai and adjacent coasts, including the study area, may rise by 0.92 m above MSL under IPCCs moderately strong emission reduction scenario RCP 4.5 (Dhanya and Ramachandran 2018). A study published by Varadan et al. (2015) states that Chengalpet is the most exposed district to climate change with an Index of 2.43 due to declining South West Monsoon rainfall in combination with a stagnant North-East Monsoon. The present study area is also frequently prone to natural disasters. Karikattukuppam and Sathankuppam, villages located in the Pulicat lake region of Kancheepuram and Thiruvallur district approximately 50–60 km from Chennai, were seriously impacted by the 2004 Tsunami and associated coastal erosions. The study area is hit severely by tropical cyclones during the North-East monsoon seasons, the recent ones being the Vardah cyclone (December 2016) and the Nivar cyclone (November 2020).

1.2 Materials and Methods

1.2.1 General Profile of the Study Area

The present study region covers two districts in the north coastal Tamil Nadu, namely Thiruvallur and Kancheepuram, erstwhile known as Chengalpet (Fig. 1.1). The study area forms part of the northeast agro-ecological zones of Tamil Nadu. This low-lying coastal belt has an average elevation of 100–200 m above MSL. Unlike other parts of the nation, retreating monsoon during October to December months provides a

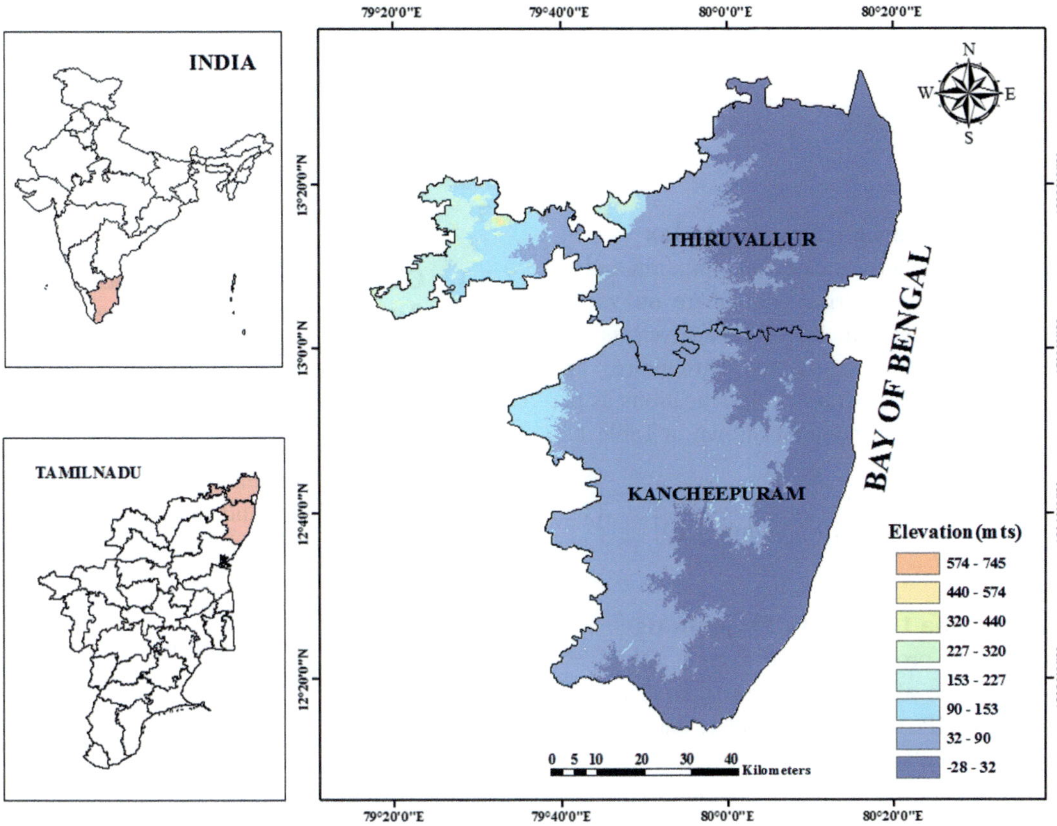

Fig. 1.1 Location map of the study area with Digital Elevation Model (DEM) of erstwhile Chengalpet (the present, Kancheepuram and Thiruvallur Districts)

major share of rainfall to Tamil Nadu. The present study area is frequently hit by tropical cyclones almost every year. The recent ones being Gajja, Okhi, Vardah and Amphan and Nivar have always created havoc on these areas' natural and cultural resources. This year the storm surge during the recent Nivar cyclone was in the range of 2–5 m above the high tide level as per IMD. It has impacted severe coastal erosion in this coastal belt.

1.2.2 Methods

1.2.2.1 Climate Data Analysis and Models

In order to get data pertaining to future climate conditions, dynamical downscaling of the regional climate model is used. Regional Climate Model-RegCM 4.4 output under Representative Concentration Pathway 4.5 (RCP4.5) of CMIP5 has been used for analysis purposes (Bhattacharjee and Zaitchik 2015; Murari et al. 2015).

1.2.2.2 De Martonne Aridity Index

Rain-fed farming systems can be seen in arid and semi-arid areas of our country. The De Martonne Aridity Index was calculated using the regional climate model (RegCM) output based on the RCP 4.5 emission trajectory. The average precipitation and air temperature for the 14 grids in the area of interest were used to obtain aridity values on an annual time frame (Shifteh Some'e et al. 2013; Pravalie 2013). Aridity prognoses were made for the near (2010–2040) and mid-term (2041–2070) periods. Arc GIS tools were employed for mapping.

The formula for calculating De Martonne aridity index (DM_{Ia}) is given below (Table 1.1).

$$DMI_a = \frac{P}{T + 10} \quad (1.1)$$

1.2.2.3 Pinna Combinative Index

Multiannual mean values of precipitation and air temperature of the driest month are utilized to derive this index Eq. 1.2. This analysis is significant for water resources, agriculture, and irrigation planning. Climate classifications as per Pinna combinative index are shown in Table 1.2.

$$IPi = 1/2\left[\frac{Pyi}{Tyi + 10} + \frac{(12Pdmi)}{Tdmi + 10}\right] \quad (1.2)$$

1.2.2.4 Sea Level Rise Projections

Agricultural production and community livelihood may be affected by the rise in sea levels worldwide (IPCC 2014). A climate and sea level simulator model is used to generate sea level projections for the twenty-first century based on low, medium, and high emission representative concentration pathways. The global SLR data were downloaded and processed to understand future sea levels (Source: http://www.psmsl.org/data and https://tidesandcurrents.noaa.gov/sltrends/) by the end of the twenty-first century. Data are represented using box and whisker plots.

1.3 Results and Discussion

1.3.1 Observed and Simulated Anomaly in Temperature and Rainfall Over Chengalpet

Tropical semi-arid to dry sub-humid climate prevail in the study area, in that coastal stretch comes under the latter and interior parts comes under the former category. Rainfall distribution is mainly under the influence of both southwest and northeast monsoons. Most of the precipitation is obtained during northeast monsoon and caused due to cyclonic disturbances formed in the Bay of Bengal, during the months from October to December. The rainfall is highly erratic.

The annual mean temperature anomaly shows a significant increase in warming from 1990s onwards. Rising trend was observed in both day and night-time temperatures. All seasons and

Table 1.1 De' Martonne index based aridity index categories

Climate	Values of DMI
Dry	$DM_I < 10$
Semi-dry	$10 \leq DM_I \leq 20$
Mediterranean	$20 \leq DM_I < 24$
Semi-humid	$24 \leq DM_I < 28$
Humid	$28 \leq DM_I < 35$
Very humid	a. $35 \leq DM_I \leq 55$ b. $DM_I > 55$

Table 1.2 Pinna combinative index

Climate	Values of PCi
Dry	$IP_I < 10$
Semi-dry and Mediterranean	$10 \leq IP_I \leq 20$
Humid	$20 \leq IP_I < 30$
Very humid	$30 \leq IP_I$

annual mean minimum and maximum temperatures showed a significant increasing trend except for the south west monsoon season. The study area receives 462.7 mmrainfall during SW season, 697.2 mm during NEM season, 32.1 mm during winter, and 60.1 mm during summer months. The coldest months, January and February, receive only 22 mm and 7 mm of rainfall. March and April are the hottest and driest periods, receiving only 4.7 and 14.2 mm, respectively.

Post-monsoon season, i.e., the northeast monsoon season was found to be contributing more 54.91% to the overall rainfall. The largest contribution was from the month of November, contributing around 20.55% followed by October by about 16.6%. The southwest monsoon months were contributing about 38.77%, the major portion of it is coming from September and August of about 13 and 12%, respectively, and the remaining rainfall was obtained from summer (6.38%) and winter (2.15%) months. During the summer season, May month contributes 4.5%, and between the winter months, January and February contribute about 1% each.

The model output shows significant warming in the mean annual temperatures for the study area. Winter season temperature may rise more during the latter part of the twenty-first century. The results revealed that May month was the hottest followed by June and April. The coolest month on record was January followed by December. The monthly MMaxT varied in the range of 28.20–37.43 °C. The mean daytime and night-time temperature of the hottest month May were 37.31 °C and 26.73 oC, respectively. The majority of the coastal areas may experience a rise in warming of around 2.4–2.6 °C by 2100 (Fig. 1.2). Bapuji Rao et al. (2014) and Chaturvedi et al. (2012) have also estimated a substantial rise in mean temperature over India. Seasonal analysis of the simulated climate parameters shows that daytime temperature may have an average annual increase of 2.3 °C and I.8 °C during the end and mid of twenty-first century, respectively. The rise in maximum temperature is more pronounced during winter months, i.e., January and February. It is projected to be 2.8 and 1.8 °C during the mid and end centuries. However, the night-time temperature showed a higher increase during summer and southwest monsoon season. The night-time temperature seemed to increase more annually than the daytime temperature. The projected night-time warming is likely to be in the range of 2.4 °C during the end century period. In the summer and monsoon periods, the predicted night-time temperature may increase by 2.7 °C and 2.6 °C, respectively. High-temperature stress during the sensitive growth stages, especially at flowering and 4 days before flowering, is highly critical for heat stress in groundnut crops. Winter season rainfall is expected to decline by −31.6% during winter season in the mid-century period, followed by (−9.6%) end century and an insignificant decrease of 4.7% during end century. Rainfall during summer shows a maximum increase of 28.5% during the end century. Southwest monsoon rainfall is projected to decline by 4% in the mid-century. The annual rainfall increases by 8.9% by the end century. Changes in Mean annual temperature in Kancheepuram District Tamil Nadu for near, mid and end century period as per RCP 4.5.

1.3.2 Projected Aridity Status Over Chengalpet

Worsening aridity status in an area put additional stress on agriculture production (Asha Latha et al. 2012). Climate change, especially increasing temperature, severely affects the soil moisture regime and the overall quality of the soil strata. A quantitative assessment of the possible changes in the aridity was undertaken to determine the status of aridification and understand future directions of land degradation. The long-term trend in the observed annual mean aridity for the baseline period and projected aridity for 14 grid points is shown in Figs. 1.3 and 1.4. It shows the downward slope in aridity values under RCP 4.5 Pathway. It reaffirms the area to be semi-arid in the future, and the same conditions may worsen in the future.

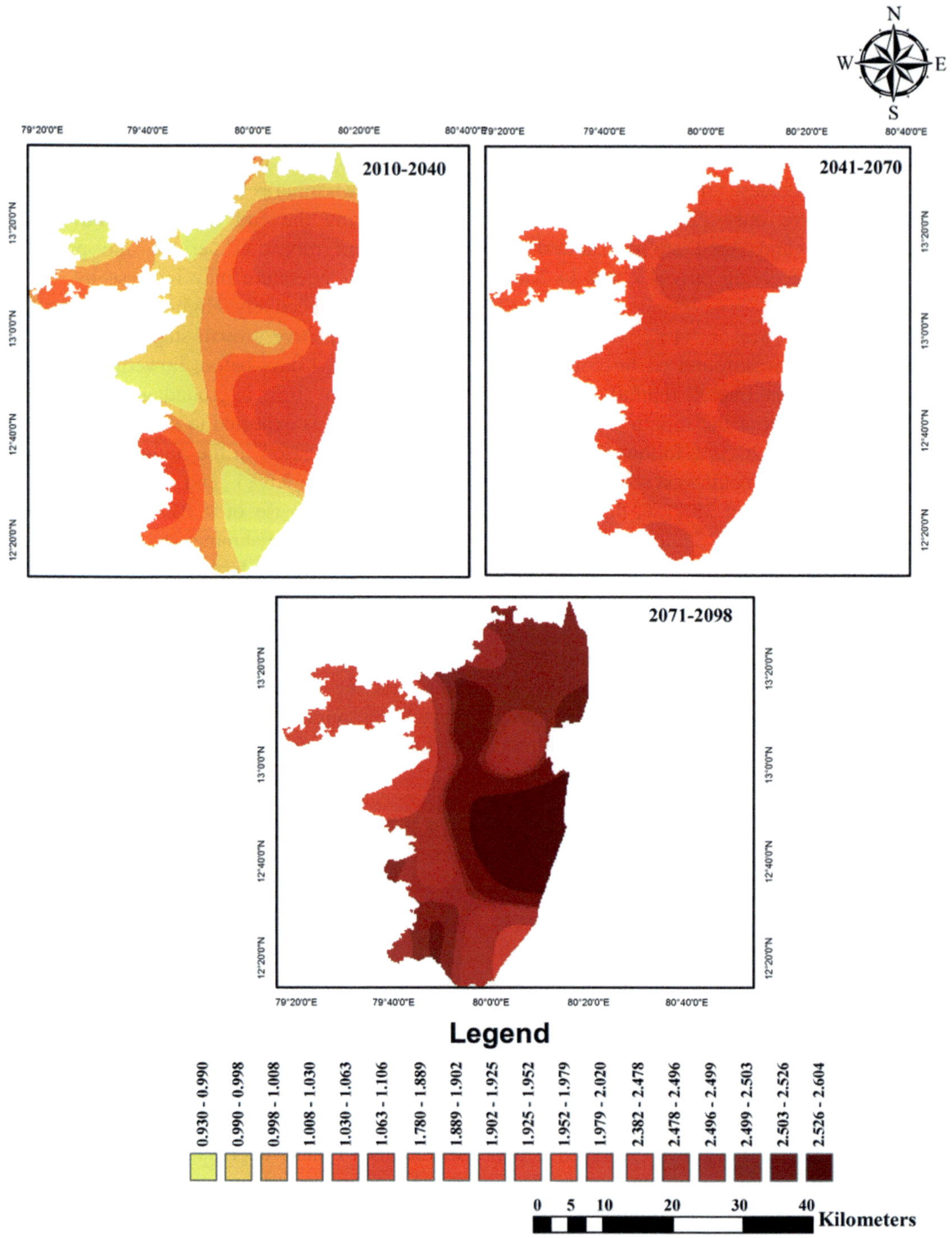

Fig. 1.2 Projected change in mean annual temperatures warming (in °C) for the study area

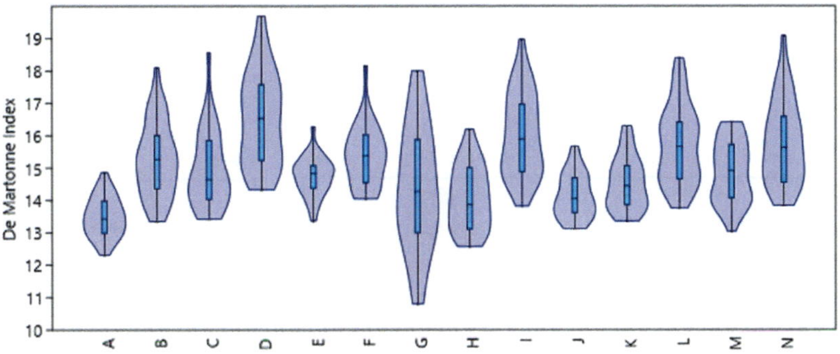

Fig. 1.3 Violin and box plot showing mean De Martonne index values from 1970 to 2000 for 14 grid points

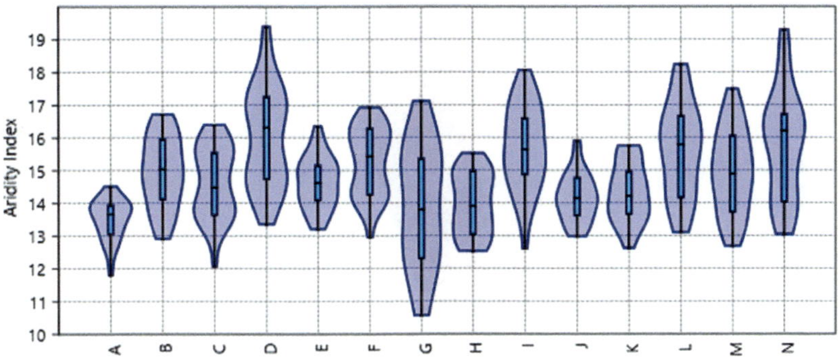

Fig. 1.4 Violin and box plot showing the projected changes in monthly mean De Martonne aridity values for 14 grid locations for the period 2010 to 2040

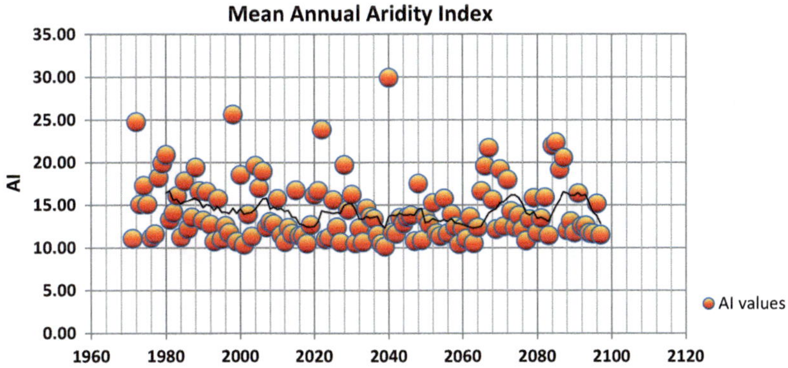

Fig. 1.5 Simulated annual mean aridity index values from 1970 to 2100s

The average annual aridity index values for the 127 year period are 14.3 mm/°C, portraying this area to own a semidry climate. The highest and lowest mean AI values recorded ranged from 10.1 to 29.9 mm/°C over these years (Fig. 1.5). The drying trend is from northwestern parts to southeastern parts of Chengalpet. An exponential rise in the saturation vapor pressure may be caused due to global warming. It may escalate the vapor pressure deficit (VPD), (Willett et al. 2007). Increasing aridity enhances the water demand in crops. Water stress is extremely detrimental during the critical growth stages of cereal crops (Hatfield et al. 2011; Barbosa and Lakshmi Kumar 2016; Ramachandran et al. 2017; Evans and Geerken 2004). Prevailing aridity condition is a critical sign of diminishing land quality. The driest months add to the land degradation, and soil becomes compact in these areas during these months (Fig. 1.6a and b).

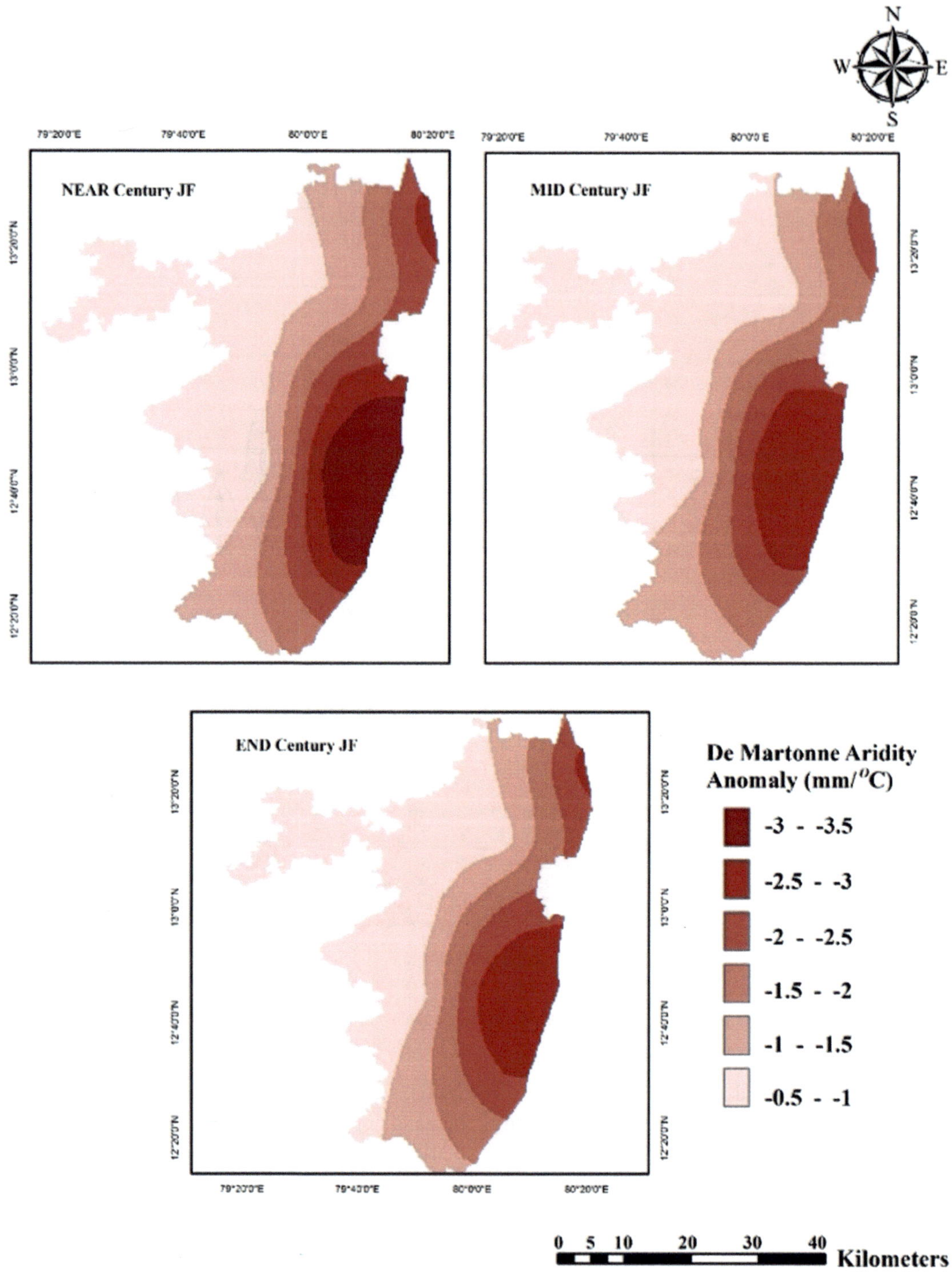

Fig. 1.6 **a** and **b** Projected variations in the aridity conditions in the driest and wettest months

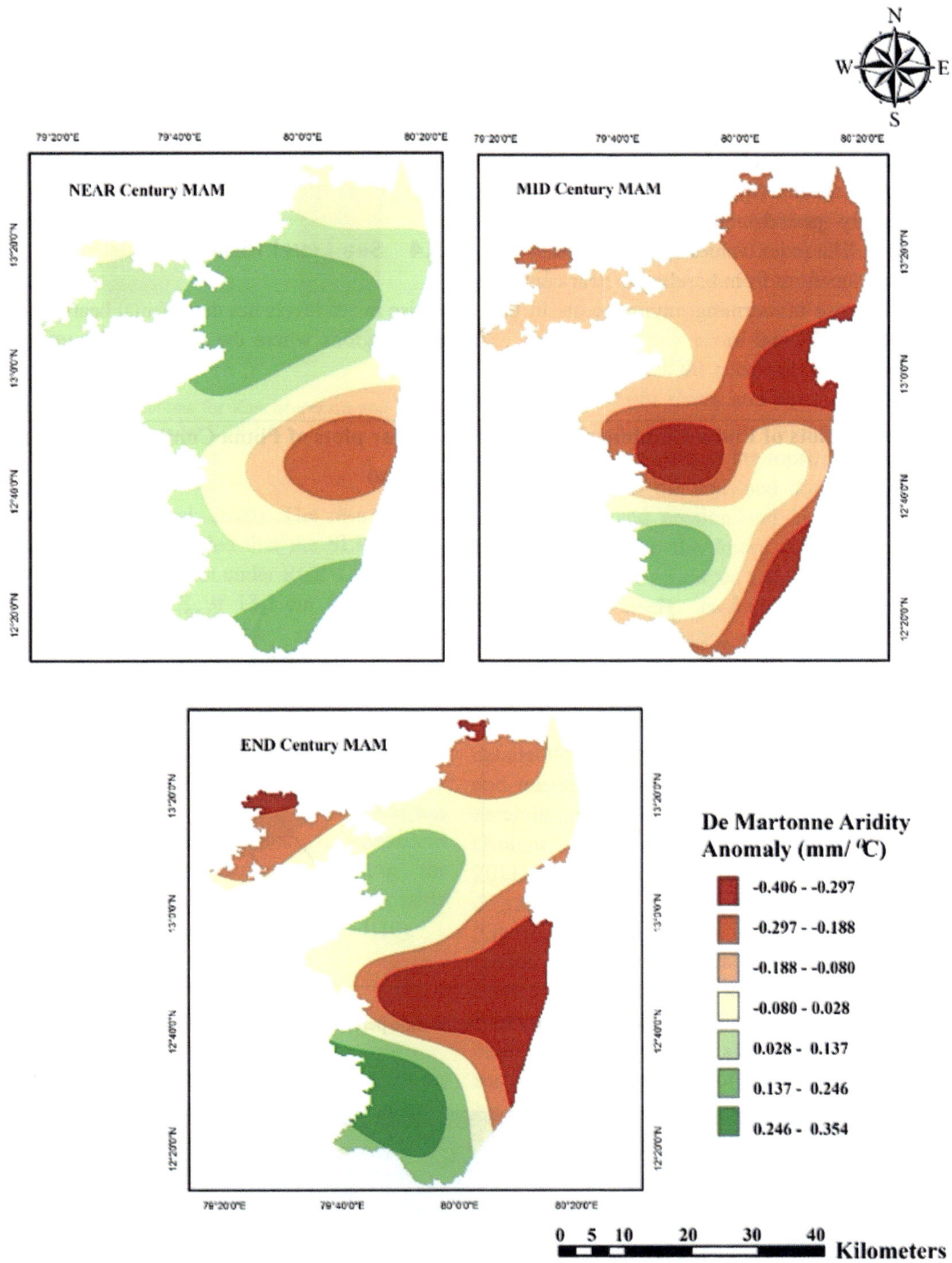

Fig. 1.6 (continued)

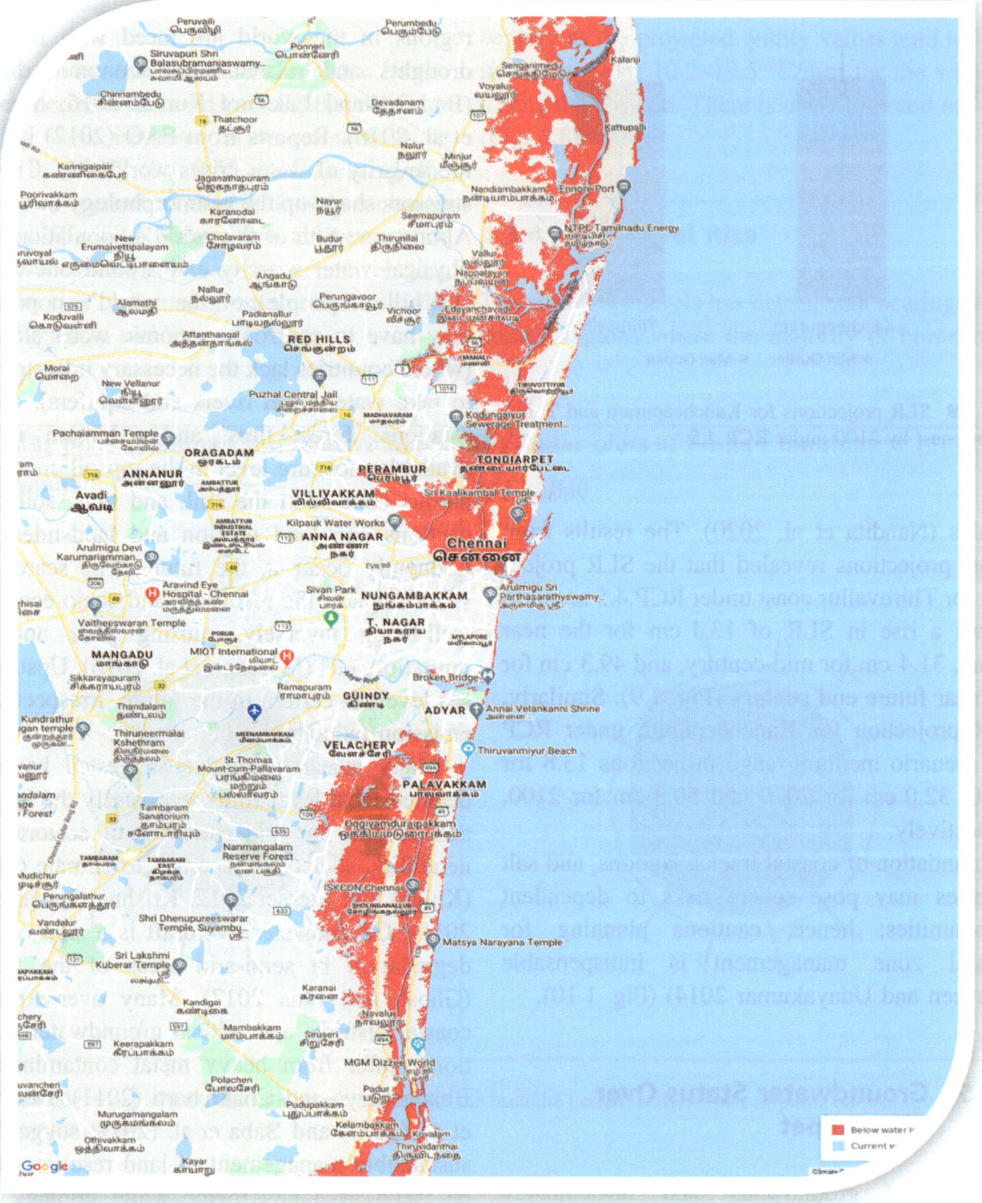

Fig. 1.10 SLR-induced coastal inundation for Chengalpet areas. *Data Source* NOAA 2019

aridification status. According to both the indices, the aridity is going to be aggravated in the future for grid points in the study area as the index values tend to decline from baseline to near-century period. The regional climate model data have represented the spatial heterogeneity of land degradation in northeast Tamil Nadu and projects aridity more reliably. The results of climate simulations show that majority of the coastal areas may experience more rise in warming than the interior areas of the district. AI values recorded ranged from 10.11 to

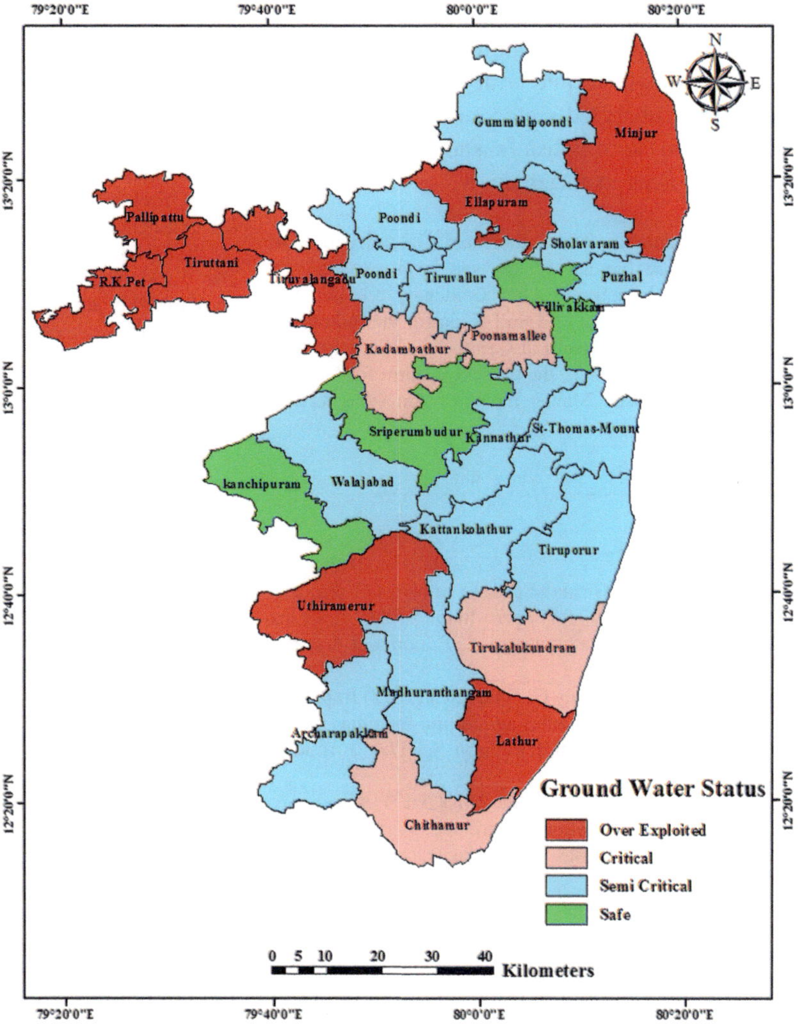

Fig. 1.11 Groundwater status of erstwhile Chengalpet district

29.88 mm/°C over the analyzed years. Even under moderate emission, the mean Pinna combinative index value for the reference period was found to be 13.6 mm/°C for the study area. However, it was found to reduce to 13.05 (by 0.55 mm/°C) during the near-century period under RCP 4.5 emission trajectory Poor SOC values in the study area is yet another indicator of poor soil quality and characterizing this area to have degraded land resources. In order to draw comprehensive conclusions on dryland degradation, sea level rise data are also used. Sea level rise for the coast of Thiruvallur and Kancheepuram district even under the medium emission range projection with emission reduction shows a rise more than half a meter by 2100, indicating possibilities of saline water intrusions to the vast stretch of low-lying wetlands in the study area. Combined exposure to these hazards may exacerbate the land degradation in these areas. Proactive evaluation and planning are essential for sustainable land management. Depleted groundwater resources also indicate that low-lying coastal areas of Minjur, Lathur, Chithamur, and Thirukalikundram may fall as highly vulnerable to sea level rise and associated land degradations. Micro-level studies are the essential prerequisites for any developmental planning process. Robust action plans envisioned to tackle multi-hazard risk reductions strategies are the need of the hour to tackle the degradation threat. It is significant to have coordinated actions and inclusive growth in line with INDCs and SDGs for all interconnected sectors and departments in the state to accomplish sustainable land and water resources management.

Acknowledgements The authors are grateful and wish to acknowledge the use of the sea level rise data of NOAA. Special thanks to Anna Centenary research fellowship of Government of Tamil Nadu for the fellowship support and the entire team of Centre for Climate change and Adaptation Research, the present (Centre for Climate Change and Disaster Management) of Anna University, Guindy, Chennai.

References

Abiy G, Quraishi S, Girma M (2014) Analysis of seasonal rainfall variability for agricultural water resource management in southern region, Ethiopia. J Nat Sci Res 4(11):56–79

Asha latha KV, Munisamy G, Bhat ARS (2012) Impact of climate change on rainfed agriculture in India: a case study of Dharwad. Int J Environ Sci Dev 3(4):368–371

Baba AN, Tifwa YH, Achoba IM (2017) A synergy between urban planning and the public towards achieving sustainable development goals (SDGs). Int J Tech Res Sci

Bapuji Rao B, Santhibhushan Chowdary P, Sandeep VM, Rao VUM, Venkateswarlu B (2014) Rising minimum temperature trends over India in recent decades: implications for agricultural production. Global Planet Change 117:1–8. https://doi.org/10.1016/j.gloplacha.2014.03.001

Barbosa HA, Lakshmi Kumar TV (2016) Influence of rainfall variability on the vegetation dynamics over Northeastern Brazil. J Arid Env 124:377–387. https://doi.org/10.1016/j.jaridenv.2015.08.015

Bhattacharjee PS, Zaitchik BF (2015) Perspectives on CMIP5 model performance in the Nile river headwaters regions. Int J Climatol 35(14):4262–4275. https://doi.org/10.1002/joc.4284

Bhattacharya HN, Chakrabarti S (2011) Incidence of fluoride in the groundwater of Purulia district, West Bengal: a geo-environmental appraisal. Curr Sci 101 (2):152–155

Brown S, Nicholls RJ, Lázár AN et al (2018) What are the implications of sea-level rise for a 1.5, 2 and 3 °C rise in global mean temperatures in the Ganges-Brahmaputra-Meghna and other vulnerable deltas? Reg Environ Change 18:1829–1842. https://doi.org/10.1007/s10113-018-1311-0

Brown HCP, Nkem JN, Sonwa DJ, Bele Y (2010) Institutional adaptive capacity and climate change response in the Congo Basin forests of Cameroon. Mitig Adapt Strat Glob Change 15(3):263–282. https://doi.org/10.1007/s11027-010-9216-3

Chaturvedi RK, Joshi J, Jayaraman M, Bala G, Ravindranath NH (2012) Multi model climate change projections for India under representative concentration pathways (RCPs): a preliminary analysis. Curr Sci 103:791–802

Dai A (2011) Drought under global warming: a review. Wiley Interdisc Rev Clim Change 2:45–65

Das J, Mandal T, Saha P, Bhattacharya SK (2020) Variability and trends of rainfall using non-parametric approaches: a case study of semi-arid area. MAUSAM 71(1):33–44

Das J, Bhattacharya SK (2018) Trend analysis of long-term climatic parameters in Dinhata of Koch Bihar district, West Bengal. Spat Inf Res 26(3):271–280. https://doi.org/10.1007/s41324-018-0173-3

Das J, Mandal T, Saha P (2019) Spatio-temporal trend and change point detection of winter temperature of North Bengal, India. Spat Inf Res 27(4):411–424. https://doi.org/10.1007/s41324-019-00241-9

De Vente J, Chotte JL, Bernoux M, Kust G, Sanz MJ, Ruiz I, Almagro M, Alloza JA, Vallejo R, Castillo V, Hebel A, Akhtar-Schuster M (2017) Sustainable land management contribution to successful land-based climate change adaptation and mitigation. A report of the science-policy interface. United Nations Convention to Combat Desertification (UNCCD), Bonn, Germany

Dhanya P, Ramachandran A (2018) Constructing local sea level rise scenarios for assessing possible impacts and adaptation needs: insights from coasts of India. In: Sea level rise and coastal infrastructure. Intech. https://doi.org/10.5772/intechopen.74325

Dutta D, Kundu A, Patel NR, Saha SK, Siddiqui AR (2015). Assessment of agricultural drought in Rajasthan (India) using remote sensing derived vegetation condition index (VCI) and standardized precipitation index (SPI). Egypt J Remote Sens Space Sci 18(1):53–63

Evans J, Geerken R (2004) Discrimination between climate and human-induced dryland degradation. J Arid Environ 57(4):535–554

Ghosh PK, Jana NC (2017) Groundwater potentiality of the Kumari river basin in drought-prone Purulia upland, Eastern India: a combined approach using quantitative geomorphology and GIS. Sustain Water Resour Manag 1–17

Guhathakurta P, Sreejith OP, Menon PA (2011) Impact of climate change on extreme rainfall events and flood risk in India. J Earth Syst Sci 120(3):359–373

Hatfield JL, Boote KJ, Kimball BA, Ziska LH, Izaurralde RC, Ort D, Thomson AM, Wolfe D (2011) Climate impacts on agriculture: implications for crop production. Agron J 103:351–370

IPCC (2014) Climate change 2014—impacts, adaptation, and vulnerability. Part A: global and sectoral aspects. Contribution of working group II to the fifth assessment report of the international panel of climate change

IPCC (2018) Special report on global warming of 1.5 °C (Masson-Delmotte V et al. (eds)) WMO

Knight KW, Messer BL (2012) Environmental concern in cross-national perspective: the effects of affluence, environmental degradation, and world society. Soc Sci Q 93(2):521–537

Krishna Kumar K, Patwardhan SK, Kulkarni A, Kamala K, Rao K, Jones R (2011) Simulated projections for summer monsoon climate over India by a high-resolution regional climate model (PRECIS). Curr Sci 101(3):312–326

Marengo JA, Torres RR, Alves LM (2016) Drought in Northeast Brazil—past, present, and future. Theor App Clim 20:1–12. https://doi.org/10.1007/s00704-016-1840-8

Martinez et al (2017), Stone P (2001) The effects of heat stress on cereal yield and quality. In: Basra AS (ed) Crop responses and adaptations to temperature stress. Food Products Press, Binghamton, NY, pp 243–291

Mlynski D, Cebulska M, Walega A (2018) Trends, variability, and seasonality of maximum annual daily precipitation in the Upper Vistula Basin, Poland. Atmosphere 9:313. https://doi.org/10.3390/atmos9080313

Murari KK, Ghosh S, Patwardhan A, Daly E, Salvi K (2015) Intensification of future severe heat waves in India and their effect on heat stress and mortality. Reg Environ Change 15:569–579. https://doi.org/10.1007/s10113-014-0660-6

Nanditha JS, van der Wiel K, Bhatia U, Stone D, Selton F, Mishra V (2020) A seven-fold rise in the probability of exceeding the observed hottest summer in India in a 2 °C warmer world. Environ Res Lett 15:044028

Naresh Kumar S, Singh AK, Aggarwal PK, Rao VUM, Venkateswaru B (2012) Climate Change and Indian agriculture salient achievements from ICAR network project. Indian Agricultural Research Institute publications, New Delhi

Pravalie R (2013) Climate issues on aridity trends of southern Oltenia in the last five decades. Geographia Technica 1:70–79

Praveen K, Udayakumar P (2014) Effects of sea level change on vulnerable east coast of India. Res J Mar Sci 2321–1296. 2(1):1–5. International Science Congress Association 1

Ramachandran A, Dhanya P, Jagannathan R, Rajalakshmi D, Palanivelu K (2017) Spatiotemporal analysis of projected impacts of climate change on the major C_3 and C_4 crop yield under representative concentration pathway 4.5: insight from the coasts of Tamil Nadu, South India. PLoS ONE 12(7):e0180706. https://doi.org/10.1371/journal.pone.0180706

Shifteh Some'e B, Ezani A, Tabari H (2013) Spatiotemporal trends of aridity index in arid and semi-arid regions of Iran. Theoret Appl Climatol 111(1–2):149–160

Shukla PR, Sharma SK, Ravindranath NH, Bhattacharya S, Garg A (eds) (2003) Climate change and India: vulnerability assessment and adaptation. Universities Press, Hyderabad

Steduto P, Hsiao TC, Fereres E, Raes D (2012) Crop yield response to water. Food and Agriculture Organization of the United Nations (FAO), Rome, Italy

Teri (2016) Executive summary economics of desertification, land degradation and drought (DLDD) in India. In: Sharma JV IFS, Chopra K, PiaSethi (eds) Macroeconomic assessment of the costs of degradation in India, vol I. Lodhi Road, New Delhi

UNCCD, United Nations Convention to Combat Desertification, UNCCD-SPI (2016) Land in balance. The scientific conceptual framework for land degradation neutrality. Science-policy brief 02. United nations convention to combat desertification, science-policy interface (UNCCD-SPI). (http://www.unccd.int/Lists/SiteDocumentLibrary/Publications/10_2016_spi_pb_multipage_eng.pdf)

Willett KM, Gillett NP, Jones PD, Thorne PW (2007) Attribution of observed surface humidity changes to human influence. Nature 449:710–712

Active Tectonics and Associated Channel Shifting Pattern of Neora River Basin, Darjeeling Himalaya

2

Jonmenjoy Barman, Brototi Biswas, and Jayanta Das

Abstract

The river Neora flows through its upper course in the Darjeeling Himalaya and lowers course in the Dooars plain. The Himalaya mountain is separated from the Indian great plain by the Himalayan front thrust, which has a significant role in the evolution of drainage patterns and tectonic activities. Most of the Himalayan foot plain is covered by the Precambrian rock structure. The elongation basin shape of the Neora river expresses the tectonic activities of the area. ALOS PALSAR digital elevation model has been taken for morphometric analysis. It was found that Neora is a 3rd order stream and has created a dendritic drainage pattern. Two significant tectonics indices, namely the Asymmetry factor and Transverse topographic symmetry factors, are applied to quantify the basin tilting. Asymmetry factors indicate the basin tilting direction whereas Transverse topographic symmetry factors represent the tilting magnitude. Both indices show the basin as a tectonically active region and tilting toward the right side except for sub-basin 3. Tilting of the basin has, in turn, influenced the channel shifting of Neora river and the same was confirmed after doing temporal analysis of Google earth images. The river Neora is the lifeline of Neora wildlife sanctuary. It is full of biodiversity and natural resources. Rapid change of the river course causes soil erosion, deduction of forest area. It is important to proper assessment of tectonic activities which is a major cause behind landslide. The authors hope the present study will be helpful for the geologist, regional planner, zoologist, etc.

Keywords

Active tectonics · Asymmetry factor · T vector · Darjeeling Himalaya

2.1 Introduction

A river basin is a fundamental hydrologic and geomorphic unit that develops by the interaction of hydrological processes, geology, climate, and configuration of the earth's surface (Saha et al. 2022). Neora river is a small river running through Neora national park. The term asymmetry means the ratio of the proportion of area from both sides of the river thalweg (Dhanya and Tiruvanatpuram 2014). Asymmetry indicates the nature of basin

J. Barman · B. Biswas (✉)
Department of Geography and Resource Management, Mizoram University (Central), Aizawl 796004, India
e-mail: brototibiswas@gmail.com

J. Das
Department of Geography, Rampurhat College, Rampurhat, Birbhum 731224, India

J. Das and S. K. Bhattacharya (eds.), *Monitoring and Managing Multi-hazards*, GIScience and Geo-environmental Modelling, https://doi.org/10.1007/978-3-031-15377-8_2

tilting and simultaneously plays a significant role in drainage development. The Himalaya mountain range is the result of continental collapse between the Eurasian and Indian plates during the tertiary era (Luirei et al. 2014). It has a significant footprint on regional geology since the N-S compression has resulted in the formation of several lineaments, fault, and fold. The Himalaya mountain ranges are geologically very young fold mountains that have been divided at several places by innumerous fault-like structures—MCT, MBT, etc. Different geomorphic indices have been used by scholars for reconnaissance studies of active tectonics like Asymmetry factor (Cox 1994; Hare and Gardner 1985), Valley floor width- Valley height ratio (Bull and McFadden 1977), Mountain front sinuosity (Bull and McFadden 1977), Hypsometric integral (Strahler 1952), Stream length-gradient index (Hack 1973) and Transverse topographic symmetric factor (Keller and Pinter 1996). Schumm et al. (1986) and Holbrook and Schumm (1999) have done a detailed review on active tectonics on the alluvial river. Neotectonics activities have given birth to topographic roughness, diversion of channels, and block tilting (Virdi et al. 2006). Starkel et al. (2008) reported channel patterns to vary by the influence of hydrology and active tectonic movement in the Ganga–Brahmaputra alluvial plain. It shows the uplift or subsidence and expresses the uplift tendency for aggradation or downcutting.

Among the above-mentioned indices Asymmetry factor (A_f) and the Transverse Topographic symmetry factor (T_{sf}) are the two qualitative indices for basin tilting analyzing channel shifting. The present study is aimed toward the qualitative analysis of the Neora river basin area. The basin area is fractured by a number of thrusts like Goubathan thrust, Matiali thrust, and Chalsa thrust. According to Ouchi (1985), alluvial rivers respond to vertical deformation in the basin by changing the above-mentioned fluvial features. The river basin is covered by natural resources from both sides, and hence it is important to monitor and analyze the channel shifting pattern and the causes behind it. The authors are hopeful that the study will be useful for the hydrologist, rural planners for better watershed management.

2.2 Materials and Methods

2.2.1 Study Area

The river Neora is a small tributary of the Teesta river that originates from Rechila Hill Reserve Forest at nearly 3200 m altitude from MSL. It covers 291 km^2 and lies between 26°40′N and 27°6′N latitudes and 88°41′E to 88°43′E longitudes and flows through Darjeeling and Jalpaiguri districts of West Bengal (Fig. 2.1). The study area is filled with natural resources and it is an attractive place for tourists. The river Neora is a 3rd order stream according to Strahler's method of stream order. It has an elongated basin shape similar to most of the Himalayan foothill rivers indicating the prevalence of tectonics activities. Most of the areas are covered with undivided Precambrian rock with quartzite, mica schist, gneiss, calcgranulite being the most abundant lithological units.

2.2.2 Database

The link between drainage pattern and evolution with tectonic context is shown in morphotectonic indices. With the help of Arc Hydro Tool, the stream features have been extracted from the Alos Palsar digital elevation model (12.5 m resolution). The stream is ordered using the Strahler stream ordering method. Considering the stream order above 2, the basin was divided into four sub-basins. Google earth historical image is used for digitization of the historical river courses.

2.2.3 Methods

2.2.3.1 Asymmetry Factor

It is a quantitative method to check basin asymmetry, first developed by Hera and Gardner in 1985 (Chen et al. 2016). The Asymmetry factor is measurement of tectonic existence. It is formulated as the ratio of basin area on the right to the trunk stream and the total basin area as represented in Eq. 2.1 (Keller and Pinter 1996).

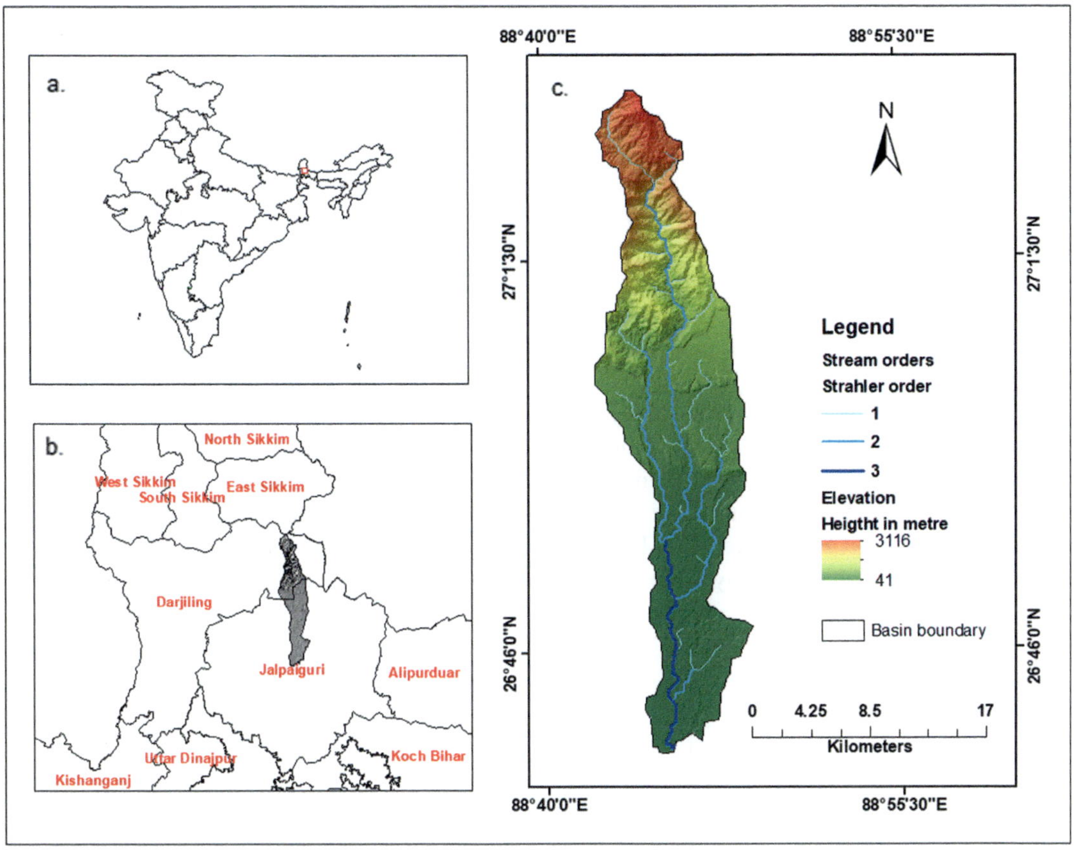

Fig. 2.1 Location of the study area **a** India, **b** Darjeeling and Jalpaiguri District, West Bengal, and **c** Neora river basin

$$A_f = (Ar/At) * 100 \quad (2.1)$$

where Ar is the area on the right to the trunk stream and at is the total basin area. A basin is called symmetric when the As value is equal to 50 (Salvany 2004). A basin is considered to be titled when the A_f value is greater than or less than 50 (Bahrami 2013). If the As value is less than 50 it is called right-facing tilting; otherwise, vice-versa (Clement and Brook 2008). An asymmetric basin indicates the differential soil erosion and lithologic control over the flow. In tectonically active topography, the landforms are characterized by relatively steep, mountainous sides and flat floors. The steep sides are created by displacement on faults such that the valley floor moves down relative to the surrounding margins, or, conversely, the margins move up relative to the floor. This movement results in basin tilting and causes the river to migrate laterly and deviate from the basin midline.

2.2.3.2 Transverse Topographic Symmetric Factor (T Vector)

T vector is a method that is used to evaluate the amount of asymmetry and how the symmetry varies with the length within a basin (Cox 1994). T is a vector with both magnitude and bearing (Virdi et al. 2006). It is represented as Eq. 2.2

$$T = Da/Dd \quad (2.2)$$

where Da is the distance of stream from the basin midline and Dd is the distance of basin midline from the watershed boundary. The imaginary midline represents the symmetrical drainage location which is drawn according to the maximum axis between river source to confluence (Mathew et al. 2016). It ranges from 0 to 1. *T* value 1 indicates the stream has shifted further from midline to basin boundary.

2.2.4 Regional Geomorphology and Geology

The origin of the Himalayan ranges is a long history. About 50–55 million years ago, Indian and Eurasian plates collided, which slowly started producing the world's highest and youngest mountain range. Tectonically, the Himalayan mountain belt is subdivided into five tectonic regions—(a) Trans- Himalayas (b) Himadri-Himalayas (c) Himachal Himalayas (d) Siwalik-the Himalayas, and (e) Quaternary sediment belt. Each of these zones is separated from the other by major thrust zones. Main Central Thrust lies between the Greater and Himadri Himalayas, main boundary thrust lies between Himachal and Siwalik Himalayas, while the Himalayan frontal thrust lies between Siwalik and quaternary sediments. Darjeeling Himalayas are mainly covered by Q (Quaternary sediment), Pz (Undivided Paleozoic Rock), and Pc (Undivided Precambrian rock) (Fig. 2.2). The three major thrusts of the present study are Goubathan, matali, and Chalsa thrust. Goubathan thrust is located north of Matiali fault, an extension of MCT (Guha et al. 2007). Mataili thrust is an expression of MBT and Chalsa thrust is a footprint of HFT (Fig. 2.3). Both the Chalsa and Matiali faults are located in an NNE-SSW direction (13–193°),

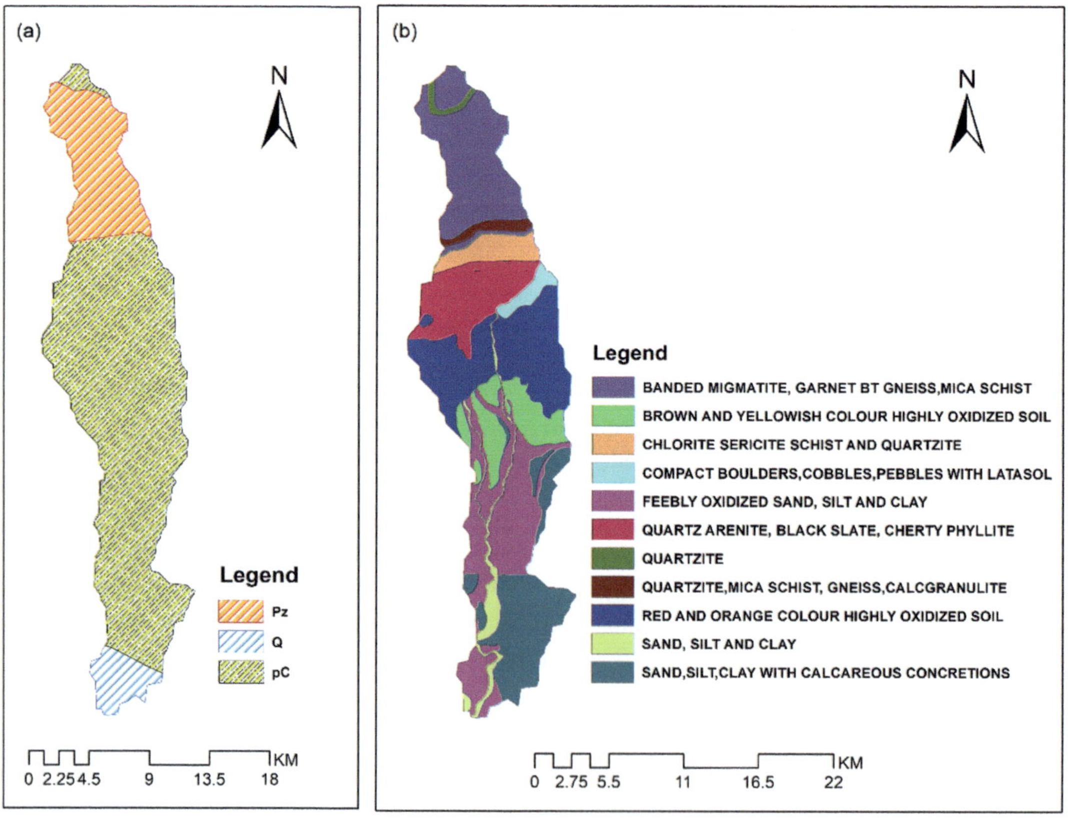

Fig. 2.2 Physical setting of the study area, **a** Geology, **b** Lithology

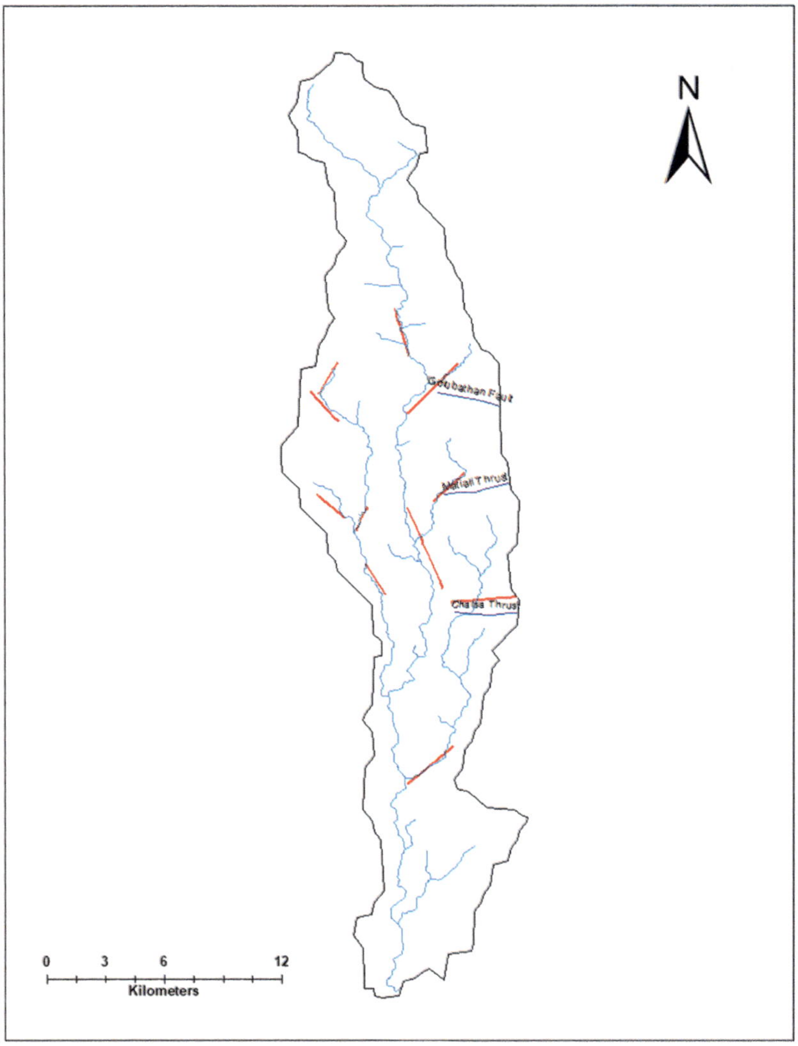

Fig. 2.3 Major lineament of the study area

lying in the present course of Neora and Murti rivers, respectively. Banded migmatite, gneiss, mica schist, quartzite, quartz arenite, black slate, cherty phyllite of Proterozoic age, sand, silt, clay with calcareous concretions, compact boulders, cobbles, pebbles with latasol, red and orange color highly oxidized soil of Pleistocene age, sand, silt and clay of Meghalaya age and feebly oxidized sand, silt and clay of holocene are the prominent lithological units of Neora basin area (Table 2.1). The Chalsa thrust works as a boundary between the northern boulder and gravel bed from the southern alluvium bed.

2.2.5 Results

The Neora river basin area is divided into four sub-basins to understand the micro-level tectonics. Two asymmetric indices, namely the Asymmetry factor and Transverse Topographic symmetry factor, are applied to each sub-basins to assign the basin tilting. As usual lithological and geological factors control Neora river course. After examining the A_f factor, it ranges between 61.47 and 29.98 (Table 2.2). The A_f value of sub-basins 1, 2, and 4 have harmonious low-value tendency, whereas sub-basin 3 has the highest

Table 2.1 Geological composition of Neora basin

Age	Group name	Formation	Lithologic
Proterozoic	Central crystalline gneissic complex	Kanchenjunga gneiss	Banded migmatite, garnet bt gneiss, mica schist
Proterozoic	Daling	Gorubathan	Chlorite sericite schist and quartzite
Proterozoic	Daling	Reyang	Quartz arenite, black slate, cherty phyllite
Proterozoic	Central crystalline gneissic complex	Chungthang	Quartzite
Proterozoic	Central crystalline gneissic complex	Chungthang	Quartzite, mica schist, gneiss, calcgranulite
Pleistocene–Holocene	Older alluvium	Baikunthapur	Sand, silt, clay with calcareous concretions
Pleistocene	Older alluvium	Duars	Brown and yellowish color highly oxidized soil
Pleistocene	Older alluvium	Duars	Compact boulders, cobbles, and pebbles with latasol
Pleistocene	Older alluvium	Duars	Red and orange color highly oxidized soil
Meghalayan	Newer alluvium	Present day deposits	Sand, silt, and clay
Holocene	Newer alluvium	Jalpaiguri	Feebly oxidized sand, silt, and clay

Table 2.2 Asymmetry factor of sub-basins

Sub basin	Ar	At	A_f
1	20.8651	42.979	48.55
2	50.2382	135.253	37.14
3	30.2243	49.1655	61.47
4	19.18	63.9865	29.98

value. The result indicates that except for sub-basin 3 all the other sub-basins are tilting toward the left side (Fig. 2.4). The box plot in Fig. 2.5 shows that the average asymmetry factor of the basin is 44.29. The value indicates eastward tilting of the river basin. On the other hand, T vector has been applied to the Neora river basin, which ranges between an average of 0.30–0.41 (Tables 2.3, 2.4, 2.5 and 2.6). The T vector displays both the tilting direction and magnitude of tilting to each sub-basins. Sub-basin- 1 has dissimilarity in both tilting magnitude and direction where the T vector ranges between 0.61 and 0.13. Figure 2.6 shows that the channel shifting tendency has occurred from the west to the east. A similar scenario is found in cross-sectional studies in different sections of the river basin. All other sub-basins have the same tendency except sub-basin 3. Sub-basin 3 has the same dissimilarity in tilting magnitude. However, the direction is opposite to the sub-basin 1, which is in the westward direction with a magnitude ranging between 0.07 and 0.68.

2.2.6 Discussion

The study reveals that the river Neora is affected by neotectonics activities which are induced by the regional geological structure and the

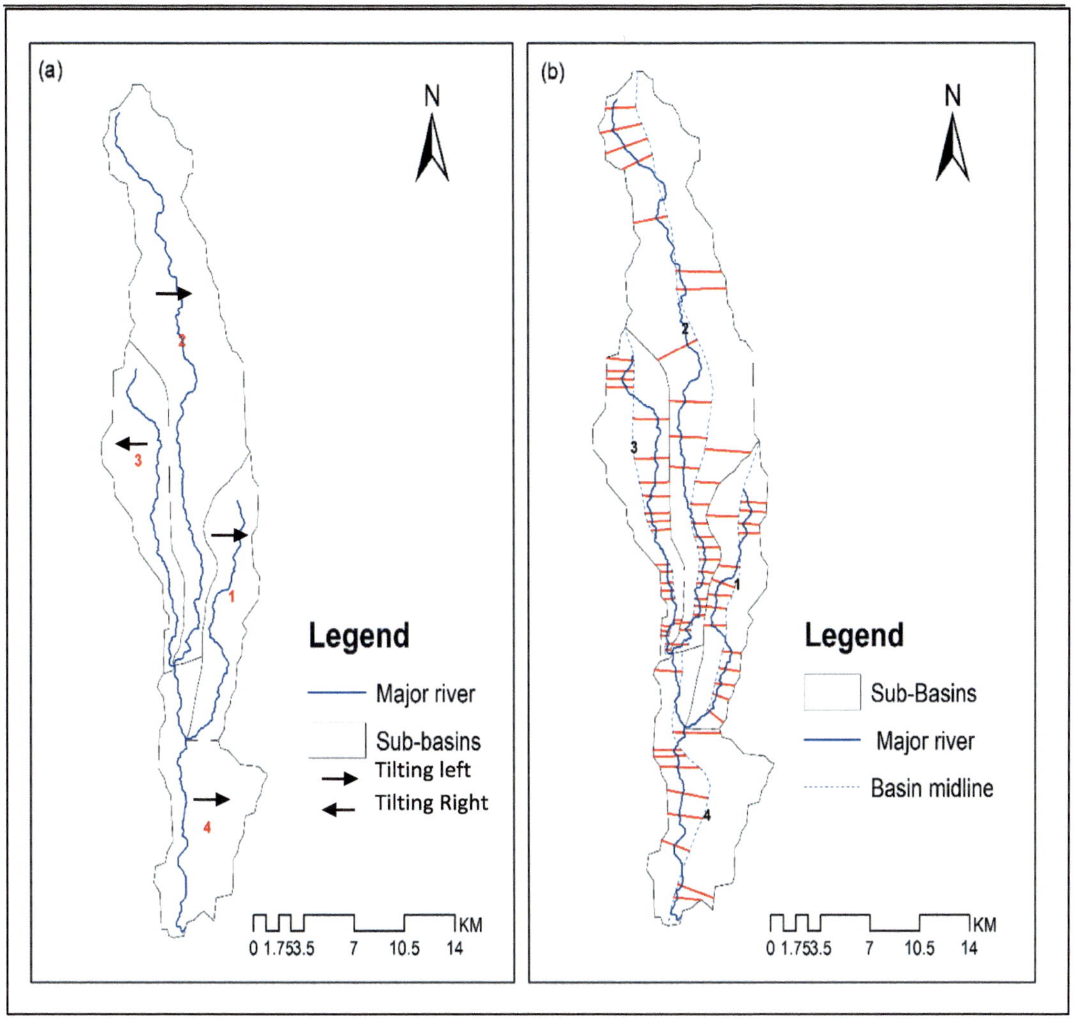

Fig. 2.4 **a** Asymmetry factor showing sub-basins tilting direction and **b** distribution of T vector

Fig. 2.5 Box plot showing the average asymmetry factor

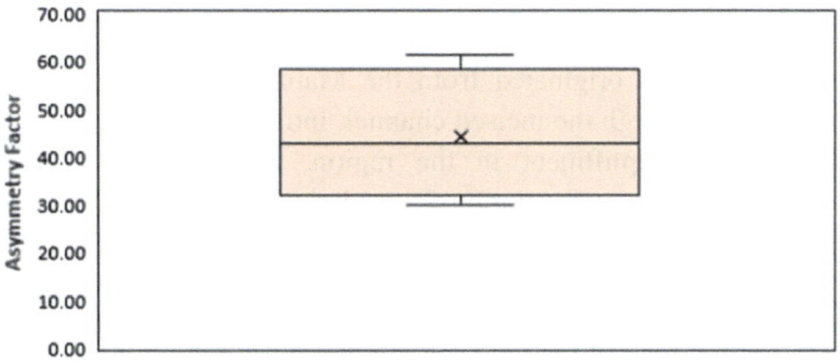

Table 2.3 T vector of sub-basin 1

Sub basin-01				
Sl. no.	Da	Dd	T vector	Avg
1	0.17	1.25	0.13	0.41
2	0.27	1.06	0.26	
3	0.38	1.21	0.31	
4	0.41	1.33	0.31	
5	0.59	1.17	0.50	
6	0.92	1.41	0.65	
7	0.69	1.27	0.54	
8	1.01	1.51	0.67	
9	0.92	1.50	0.61	
10	0.22	1.35	0.16	
11	0.51	1.33	0.38	
12	0.32	1.32	0.24	
13	0.75	1.65	0.45	
14	0.40	1.59	0.25	
15	1.38	2.04	0.67	

lineament of the area. Goswami et al. (2013) have done a case study on the relationship between geomorphology and tectonic activity and the study revealed the existence of two major scarps—the Matiali and Chalsa scarp, which are oriented in an E-W direction owing to the movement of north dipping thrust. A 24 m displacement of Chalsa fault was recorded due to the earthquake before 1100 A.D. in Nepal (Kumar et al. 2011). Guha et al. (2007) has explained the evolution of Matiali fan based on ratio carbonate dating and have concluded that both the Matiali and Chalsa scarps extend over the major river channel. The erosional work by the stream has originated from the Matiali fan and flows through the incised channel, indicating the tectonic upliftment in the region. Radial drainage pattern to the north of the Matiali scarp of the Neora and Murti rivers changes after crossing the scarp. This is a significant footprint of the presence of tectonism due to transverse faulting along the Murti and Neora lineament. Grayish soil color indicates a newly formed surface, while two Knick points at Matiali and Chalsa scarp along the Neora river show structural development due to tectonic activity (Fig. 2.7).

2.3 Conclusions

The present study attempts to understand the tectonic nature of the Neora river basin, which encompasses 291 km^2. Geological analysis reveals that the basin is affected by neotectonics activities which are induced by the regional geological structure and the lineament of the area. The link between drainage pattern and evolution with tectonic context is shown through morpho-tectonic indices. The region has three major thrusts- Goubathan, Matali, and Chalsa thrust. The study area has been divided into four sub-basins for geological evaluation. Two asymmetric indices, namely the Asymmetry

Table 2.4 T vector of sub-basin 2

Sub basin 02				
Sl. no.	Da	Dd	T vector	Avg
1	0.38	1.00	0.39	0.39
2	0.33	0.71	0.46	
3	0.12	0.68	0.18	
4	0.35	0.72	0.48	
5	0.35	0.77	0.45	
6	0.17	0.78	0.21	
7	0.30	0.83	0.35	
8	0.21	0.98	0.22	
9	0.35	1.32	0.27	
10	0.30	1.24	0.24	
11	0.35	1.39	0.25	
12	0.20	1.01	0.20	
13	0.63	1.19	0.53	
14	0.59	1.58	0.37	
15	1.47	2.87	0.51	
16	0.95	1.85	0.51	
17	1.59	2.34	0.68	
18	1.66	2.65	0.62	
19	0.65	2.68	0.24	
20	0.56	3.11	0.18	
21	0.23	2.86	0.08	
22	0.68	2.11	0.32	
23	0.95	2.03	0.47	
24	1.39	1.85	0.75	
25	1.81	2.58	0.70	
26	1.49	2.72	0.55	

factor and Transverse Topographic symmetry factor, were applied to each sub-basins to assign the basin tilting. It was found that lithological and geological factors control the Neora river course and the river basin. The result indicates that except for sub-basin 3 all the other sub-basins are tilting toward the right side. The region has two major scarps, namely Matiali and Chalsa extending over the major river channel. The channel morphology of the basin and soil color indicates tectonic upliftment. The nature of the river basin and the underlying tectonic factors are the pre-requisites for river basin planning and integrated watershed development. Hence the present study will help the geomorphologists and planners to develop the region keeping in mind the neo-tectonic nature of the Neora river basin.

Table 2.5 T vector of sub-basin 3

Sub basin 03				
Sl. no.	Da	Dd	T vector	Avg
1	0.10708	0.34151	0.31354	0.3039
2	0.1777	0.40744	0.43614	
3	0.12415	0.46504	0.26697	
4	0.15271	0.53358	0.28619	
5	0.08533	0.60806	0.14032	
6	0.08818	0.5106	0.1727	
7	0.0562	0.49979	0.11245	
8	0.20803	0.71458	0.29113	
9	0.13436	0.79023	0.17003	
10	0.27523	0.78427	0.35093	
11	0.2866	0.75999	0.37711	
12	0.14759	0.89581	0.16476	
13	0.31182	1.19863	0.26015	
14	0.28685	1.40579	0.20405	
15	0.38473	1.54435	0.24912	
16	0.56505	1.67779	0.33678	
17	0.98898	1.85143	0.53417	
18	1.27456	2.1354	0.59687	
19	1.52883	2.24546	0.68085	
20	0.43787	1.68891	0.25926	
21	0.73958	1.68995	0.43764	
22	0.46614	1.68507	0.27663	
23	0.11874	1.64884	0.07202	

Table 2.6 T vector of sub-basin 4

Sub basin-04				
Sl. no.	Da	Dd	T vector	Avg
1	0.20255	1.70412	0.11886	0.36209
2	0.28153	2.40666	0.11698	
3	0.98496	1.82448	0.53986	
4	1.58308	2.60323	0.60812	
5	1.41407	2.45952	0.57494	
6	1.45394	2.17644	0.66804	
7	0.53079	2.02539	0.26207	
8	0.47598	1.7564	0.27099	
9	0.67842	3.00394	0.22584	
10	0.40651	1.72816	0.23523	

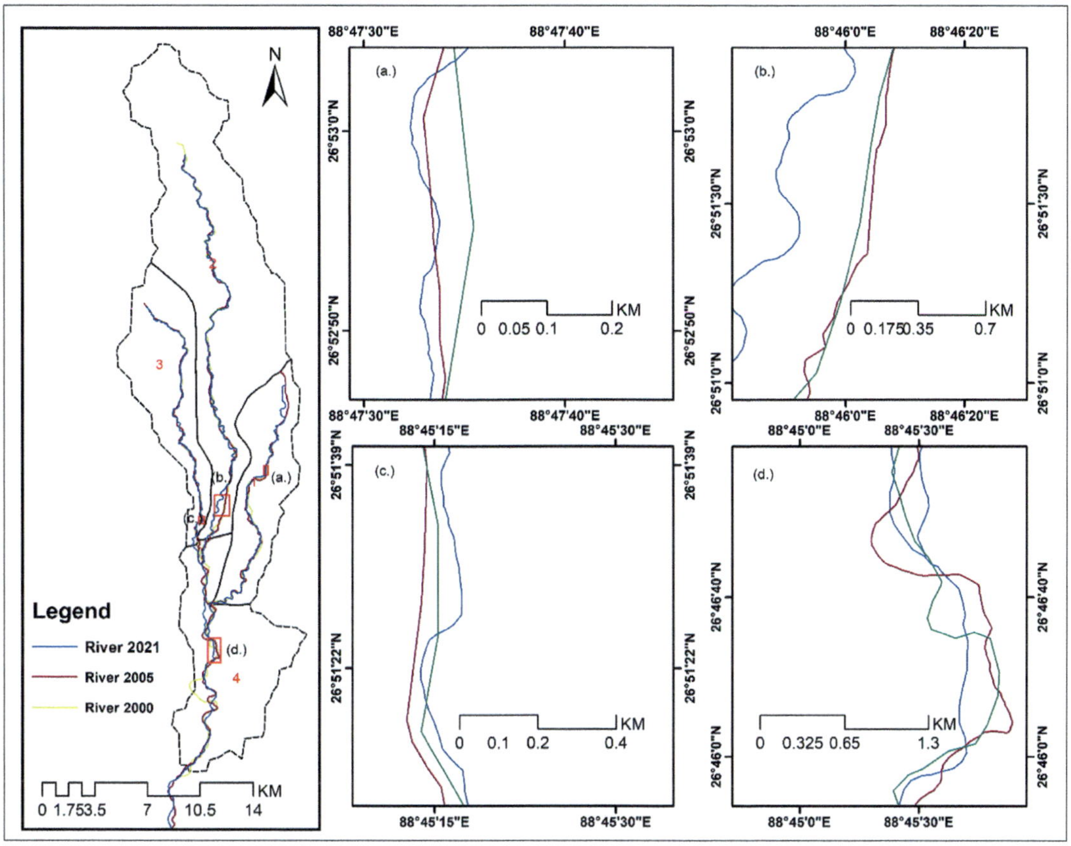

Fig. 2.6 Channel shifting pattern **a** Sub-basin 1 toward east, **b** Sub-basin 2 toward east, **c** Sub-basin 3 toward west, **d** Sub-basin 4 toward east

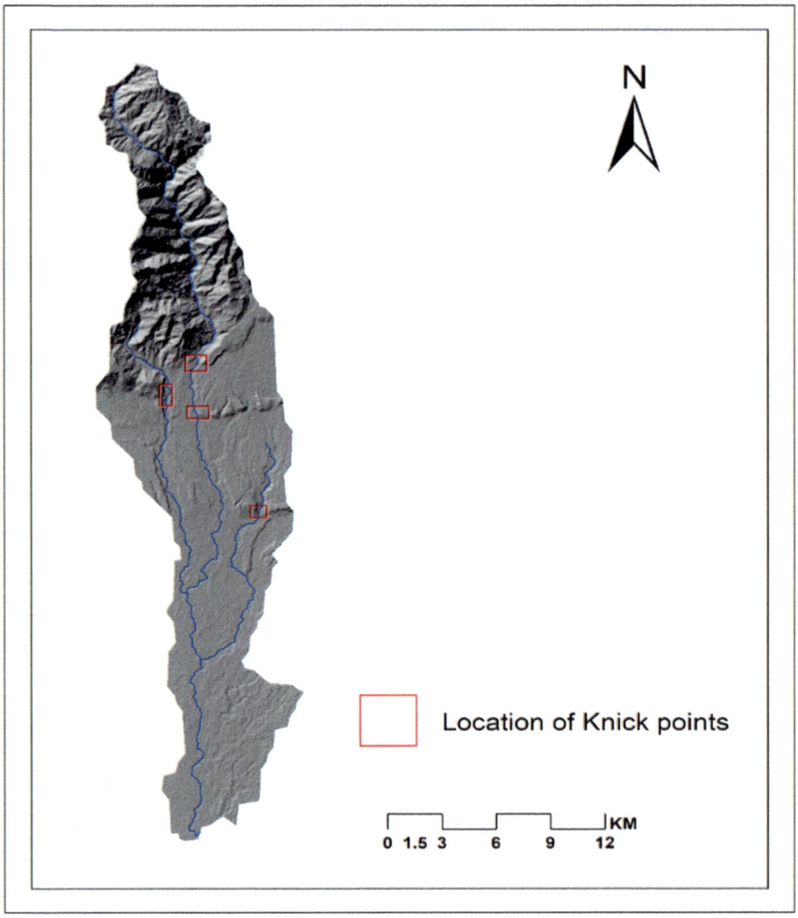

Fig. 2.7 Knick point showing the area of tectonic activities

References

Bahrami S (2013) Analyzing the drainage system anomaly of Zagros basins: implications for active tectonics. Tectonophysics 608:914–928

Bull W, McFadden L (1977) Tectonic geomorphology north and south of the Garlock fault, California. In: Doehring DO (ed) Geomorphology in Arid regions. Publications in Geomorphology, State University of New York at Bingamton, pp 115–138

Chen Y, Zhang Z, Li K, Yu H, Wu T (2016) Geochemistry and zircon U–Pb–Hf isotopes of early Paleozoic arc-related volcanic rocks in Sonid Zuoqi, Inner Mongolia: implications for the tectonic evolution of the southeastern Central Asian Orogenic Belt. Lithos 264:392–404

Clement AJ, Brook MS (2008) Tilting of active folds and drainage asymmetry on the Manawatu anticlines, New Zealand: a preliminary investigation. Earth Surf Proc Land Br Geomorphol Res Group J 33(11):1787–1795. https://doi.org/10.1002/esp.1632

Cox RT (1994) Analysis of drainage-basin symmetry as a rapid technique to identify areas of possible quaternary tilt-block tectonics: an example from the Mississippi Embayment. Geol Soc Am Bull 106(5):571–581

Dhanya V, Tiruvanatpuram K (2014) Basin asymmetry and associated tectonics: a case study of Achankovil river basin, Kerala. Transactions 36(2)

Goswami CC, Mukhopadhyay D, Poddar BC (2013) Geomorphology in relation to tectonics: a case study from the eastern Himalayan foothills of West Bengal, India. Quatern Int 298:80–92

Guha D, Bardhan S, Basir SR, De AK, Sarkar A (2007) Imprints of Himalayan thrust tectonics on the quaternary piedmont sediments of the Neora–Jaldhaka valley, Darjeeling–Sikkim Sub-Himalayas, India. J Asian Earth Sci 30(3–4):464–473

Hack JT (1973) Stream-profile analysis and stream-gradient index. J Res US Geol Surv 1(4):421–429

Hare PW, Gardner TW (1985) Geomorphic indicators of vertical neotectonism along converging plate margins, Nicoya Peninsula, Costa Rica. In: Morisawa M, Hack JT (eds) Tectonic geomorphology. Proceedings

of the 15th annual Binghantom geomorphology symposium, Sep 1984. Allen & Unwin, Boston
Holbrook J, Schumm SA (1999) Geomorphic and sedimentary response of rivers to tectonic deformation: a brief review and critique of a tool for recognizing subtle epeirogenic deformation in modern and ancient settings. Tectonophysics 305(1–3):287–306
Keller EA, Pinter N (1996) Active tectonics, vol 338. Prentice Hall, Upper Saddle River, NJ
Luirei K, Bhakuni SS, Suresh N, Kothyari GC, Pant PD (2014) Tectonic geomorphology and morphometry of the frontal part of Kumaun Sub-Himalaya: appraisal of tectonic activity. Zeitschriftfür Geomorphologie 58:435–458
Mathew MJ, Menier D, Siddiqui N, Ramkumar M, Santosh M, Kumar S, Hassaan M (2016) Drainage basin and topographic analysis of a tropical landscape: insights into surface and tectonic processes in northern Borneo. J Asian Earth Sci 124:14–27. https://doi.org/10.1016/j.jseaes.2016.04.016
Ouchi S (1985) Response of alluvial rivers to slow active tectonic movement. Geol Soc Am Bull 96:504–515
Saha S, Das J, Mandal T (2022) Investigation of the watershed hydro-morphologic characteristics through the morphometric analysis: a study on Rayeng basin in Darjeeling Himalaya. Environ Challenges 100463. https://doi.org/10.1016/j.envc.2022.100463
Salvany JM (2004) Tilting neotectonics of the Guadiamar drainage basin, SW Spain. Earth Surf Proc Land 29 (2):145–160
Schumm SA (1986) Alluvial river response to active tectonics. Act Tectonics 80–94
Starkel L, Sarkar S, Soja R, Prokop P (2008) Present-day evolution of the Sikkimese-Bhutanese Himalayan piedmont. IGiPZ PAN 219
Strahler AN (1952) Hypsometric (area-altitude) analysis of erosional topography. Geol Soc Am Bull 63 (11):1117–1142
Virdi NS, Philip G, Bhattacharya S (2006) Neotectonic activity in the Markanda and Bata river basins, Himachal Pradesh, NW Himalaya: a morphotectonic approach. Int J Remote Sens 27(10):2093–2099. https://doi.org/10.1080/01431160500445316

Estimating Soil Loss Rate and Sediment Yield of the Proposed Ngololweni Earth Dam, Kingdom of Eswatini

3

D. K. Hlanze, D. M. M. Mulungu, P. Ndomba, S. Tfwala, S. Mabaso, W. M. Dlamini, W. Gumindoga, Rajib Mitra, and Jayanta Das

Abstract

Soil loss and sediment yield affect most of our reservoirs globally. Sediments deposited in water bodies, especially reservoirs can reduce reservoir useful life through loss of storage. The main objective of the study was to estimate soil loss and sediment yield of the proposed earth dam in *Mntjuzalala* Catchment, in Ngololweni chiefdom, Eswatini. The study employed the Revised Universal Soil Loss Equation (RUSLE), runoff plots, and satellite images for the analysis. Using ArcGIS, land use/land cover (LULC) results indicated that there was a significant change between 1988 and 2018. Grassland cover was observed to have decreased from 39.8 to 18.5%, while bushland increased from 13.3 to 37.1%. Cultivation also experienced an increase in recent years, from 3.2 to 21%. This was due to an increase in settlements from 11.5 to 19.7%, which resulted in increased human activities, thus reducing wetland from 32 to 4.5% in between the study period. From estimated soil loss, it was observed that the annual average sediment yield was 82.04 tons/year and Sediment Delivery Ratio (SDR) was 0.46; thus, sediment deposit on reservoir site was 30.19 tons/year. The sedimentation rate/annual storage loss was estimated at 1.55% per annum, which would lead to 50% of the reservoir storage having been lost after 32 years when the annual sediment deposition is 1946.75 m^3. In terms of sediments sources, it was observed that this earth dam site is adjacent to gullies

D. K. Hlanze (✉) · D. M. M. Mulungu · P. Ndomba
Department of Water Resources Engineering, University of Dar es Salaam, Dar es Salaam, Tanzania
e-mail: kennyhlanze@gmail.com

D. M. M. Mulungu
e-mail: deorgm@yahoo.com

P. Ndomba
e-mail: preksedis.ndomba@dit.ac.tz

S. Tfwala · S. Mabaso · W. M. Dlamini
Geography, Environmental Science and Planning, University of Eswatini, Matsapha, Eswatini
e-mail: samkelet@gmail.com

S. Mabaso
e-mail: mabasosd@yahoo.com

W. M. Dlamini
e-mail: mwdlamini@gmail.com

W. Gumindoga
Department of Civil Engineering, University of Zimbabwe, Harare, Zimbabwe
e-mail: wgumindoga@gmail.com

R. Mitra
Department of Geography and Applied Geography, University of North Bengal, Darjeeling 734013, India
e-mail: rajibmitrageo@gmail.com

J. Das
Department of Geography, Rampurhat College, Rampurhat, Birbhum 731224, India
e-mail: jayanta.daas@gmail.com

J. Das and S. K. Bhattacharya (eds.), *Monitoring and Managing Multi-hazards*, GIScience and Geo-environmental Modelling, https://doi.org/10.1007/978-3-031-15377-8_3

and cattle tracks which are active sources of sediments during the rainy season. From the soil erosion and sedimentation outcomes, there is a need to implement sustainable land and water management measures in the catchment. This should include activities on soil conservation and sediment retention structures, development and implementation of chiefdom development plan as well as improved institutional arrangements.

Keywords

RUSLE · LULC · Sedimentation rate · Multi-criteria analysis · SWOT analysis · Mntjuzalala catchment

3.1 Introduction

Soil is a vital global livelihood resource for human beings (Patil et al. 2014). However, soil erosion is a global challenge affecting the whole world today with possible threats to the natural environment, agriculture, and water resources. Soil erosion is defined as the washing away of the land surface by water, especially rainfall and wind, including processes such as gravitational creep (Jeje et al. 1997). The effect of soil erosion is negatively affecting reservoir storage, increasing sedimentation rate, and reducing their life span in the process. The erosion process results in soil loss from a watershed, and it is difficult to estimate soil loss as it arises from a complex interaction of various hydro-geological processes. Through the process of soil erosion, the soil particles are dislodged, mainly by rainfall. During rainfall, there is a concentration of water droplets, leading to the formation of overland flows. According to, Mavhima et al. (2011), the force of the rain droplets and the accumulated flows has the ability to break soil particles through a process known as sediment generation. The erodibility of the soil influences the amount of sediment produced, the amount of energy formed by the intensity of the rainfall event and the land use, land cover, slope of catchment, and its length. Senzanje et al. (2008) state that small surface reservoirs (earth dams) are crucial for supporting rural communities' livelihoods development by providing water for different purposes to safeguard against dry and drought spells. Water from small reservoirs is used for beneficial purposes in small-scale communal gardens, providing water for livestock, making bricks, construction, dip-tank watering, and fisheries. In as much as such water resources improve the lives of people in rural semi-arid communities, there are challenges in the sustainable management of the water resources so to ensure long-term benefits.

The government of Eswatini is currently in the process of constructing such small reservoirs/earth dams across the country to provide rural farmers with water for commercial irrigation, livestock, dipping tanks, washing, and possibly other environmental and socio-economic services. This proposed reservoir, just like all other water bodies, is subject to sedimentation due to soil erosion within the Mntjuzalala Catchment on which it is located. Sedimentation, which is defined as the deposition of sediments on the bottom of water bodies, especially reservoirs as a result of sediment entering waterways, is a major challenge for small and large reservoirs. It is also one of the major challenges for water resources management across the globe and may have serious economic implications in water infrastructure and as a result of such a challenge. It is estimated that most reservoirs in Southern Africa are losing their storage by 0.3% annually, while globally, storage loss of all reservoirs due to sedimentation is 0.8% per year (Morris and Fan 1998). As a management measure, there is a need to conduct sediment yield research that may play a significant role in different soil and water conservation planning, including reservoir sedimentation analysis, studies on river morphology changes, and siltation of river bed planning of agricultural projects. The government of Eswatini, through the Ministry of Tinkhundla Administration and Development, is currently facilitating the development of chiefdom development plans (CDPs) at the chiefdom level. These plans are integrated with nature, they focus on three main pillars: social, economic, and environmental aspects of development, and it is envisaged that their implementation will play a significant role in sustainable land and water management.

Estimating the soil loss risk and its spatial distribution is, therefore, one of the key factors for successful erosion assessment. Thus, governments can develop, implement and monitor policies that will reduce the effect of soil loss and water bodies siltation under different geographical conditions. According to a design report by IFAD (2014) conducted on behalf of the government of the Kingdom of Eswatini, the country has areas prone to soil erosion, where it was discovered that the Mntjuzalala Catchment, in the Ngololweni community falls within the land degradation "hotspots" in Eswatini. The government, through a project known as Smallholder Market-Led, is currently assisting rural farmers in improving their income through developing irrigated communal command areas and creating access to markets. This will be achieved through the construction of small earth dams for irrigation in different rural communities after feasibility studies have been conducted.

The Ngololweni community is one of such communities; however, due to the persistence of soil erosion and land use and land cover changes within this catchment, the proposed project can be negatively impacted (IFAD 2014). Mntjuzalala Catchment, though small, it experiences varying land use practices that have an influence on soil stability. It has mainly sparse vegetation, shrubs, grassland to bare areas, and sloppy. These characteristics have made the catchment vulnerable to soil erosion. However, it is not known how much soil is lost per annum within this area and what is its potential impact on the proposed reservoir for smallholder irrigation in the catchment. Therefore, the novelty of the study is estimating how much annual soil loss and potential sedimentation rate of the proposed irrigation reservoir in this catchment.

This study focuses on assessing soil erosion and potential sediment yield and concentration in the proposed earth dam to be constructed in the Mntjuzalala Catchment under the Ngololweni Chiefdom. It used data for the past 30-year period, including rainfall and land use/land cover data.

3.2 Materials and Methods

3.2.1 Study Area

The Mntjuzalala Catchment is found in the Ingwavuma Basin, in the Highveld of Eswatini, and is located in the Shiselweni region lies at coordinates 31° 06′ 19.05″ S and 31° 25′ 25.21″ E. The proposed earth dam will be constructed about 300 m from Ingwavuma River on Mntjuzalala stream. The area has smallholder irrigation scheme below the outlet of the catchment. This irrigation scheme has approximately thirty members who formed an association to produce and supply vegetables to entrepreneurs, including larger supermarkets in the country. A conveyance system where the irrigation line will be laid has already been demarcated. The study area is located in the Shiselweni region, the southern part of the country, and is shown in Fig. 3.1.

The catchment is characterized by about five classes of soil; loamy, sandy loamy, clay, gravel loamy, and rock outcrop. The red loamy soil has depths of up to 1.2 m deep. Vegetation cover is mainly grassland, shrubs, and sparsely spaced bushes. The main land uses are settlements and cultivation. The topography of the area is 7% downward steep slope and 1% steep slope across the catchment. The elevation in the Mntjuzalala Catchment ranges from 559 to 791 m above sea level, with the highest areas observed on the north-western part. Areas of low altitude are those closer to the proposed reservoir site toward the outlet of the catchment. The conceptual framework of the study is shown in Fig. 3.2.

3.2.2 Data Acquisition and Processing

The data needed for the study was obtained from primary and secondary data, including questionnaires, and it is summarized in Table 3.1.

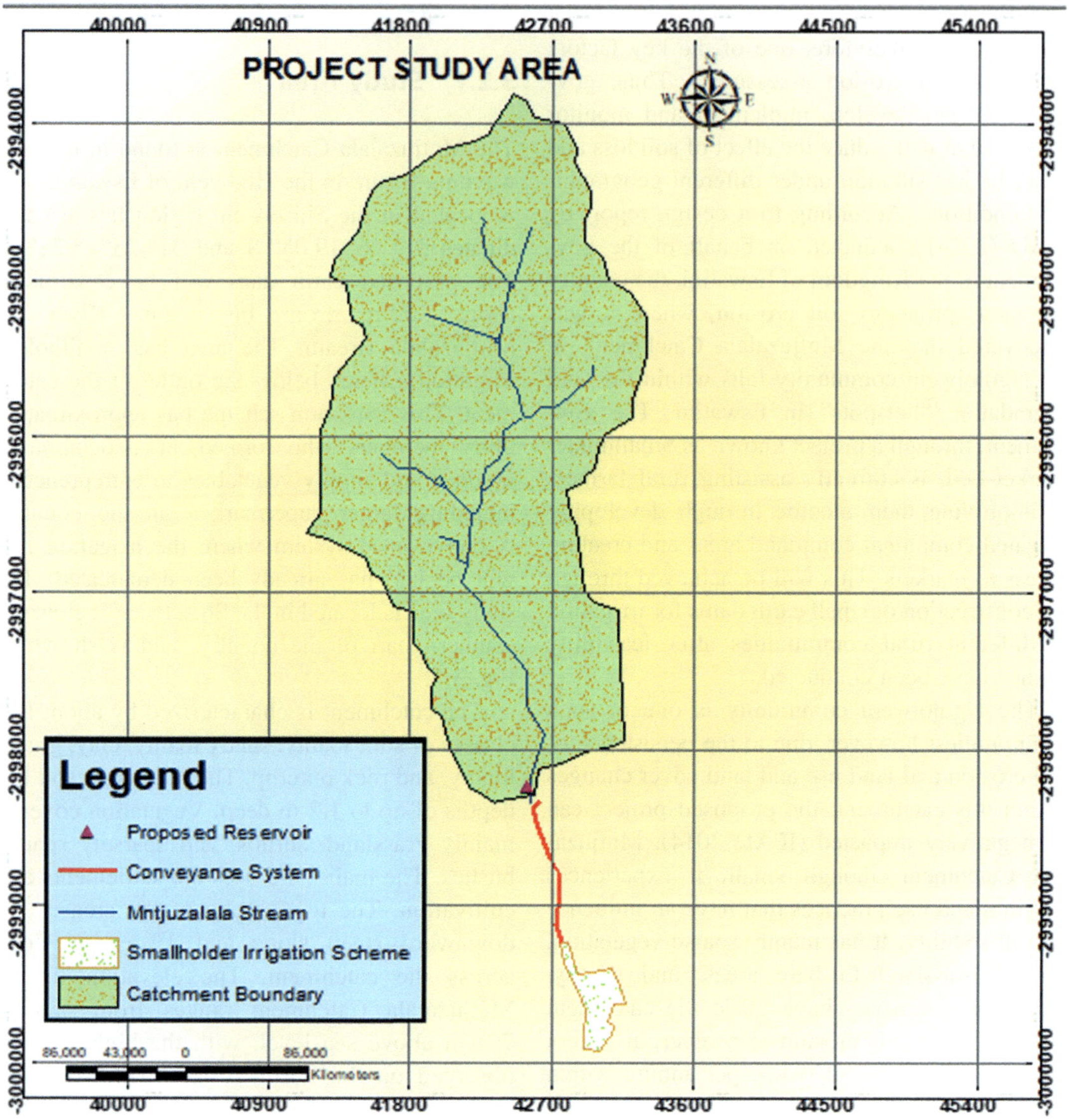

Fig. 3.1 Location map of the study area

3.2.3 Meteorological and Hydrological Data Acquisition and Processing

Rainfall

The rainfall data were acquired from the Swaziland Meteorological Services. It covered the period from 1988 to 2018 and was obtained from three stations; Kubuta, Sandleni (SAN), and Nhlangano (NHO). The annual rainfall from the three neighboring stations is shown in Fig. 3.3.

Streamflow measurement

Streamflow was measured and estimated using the area-velocity method by using a floating object over a known distance and time on the stream channel. In order to get an average/ accurate figure, three sites were chosen where the flow measurements were done. At each site, the

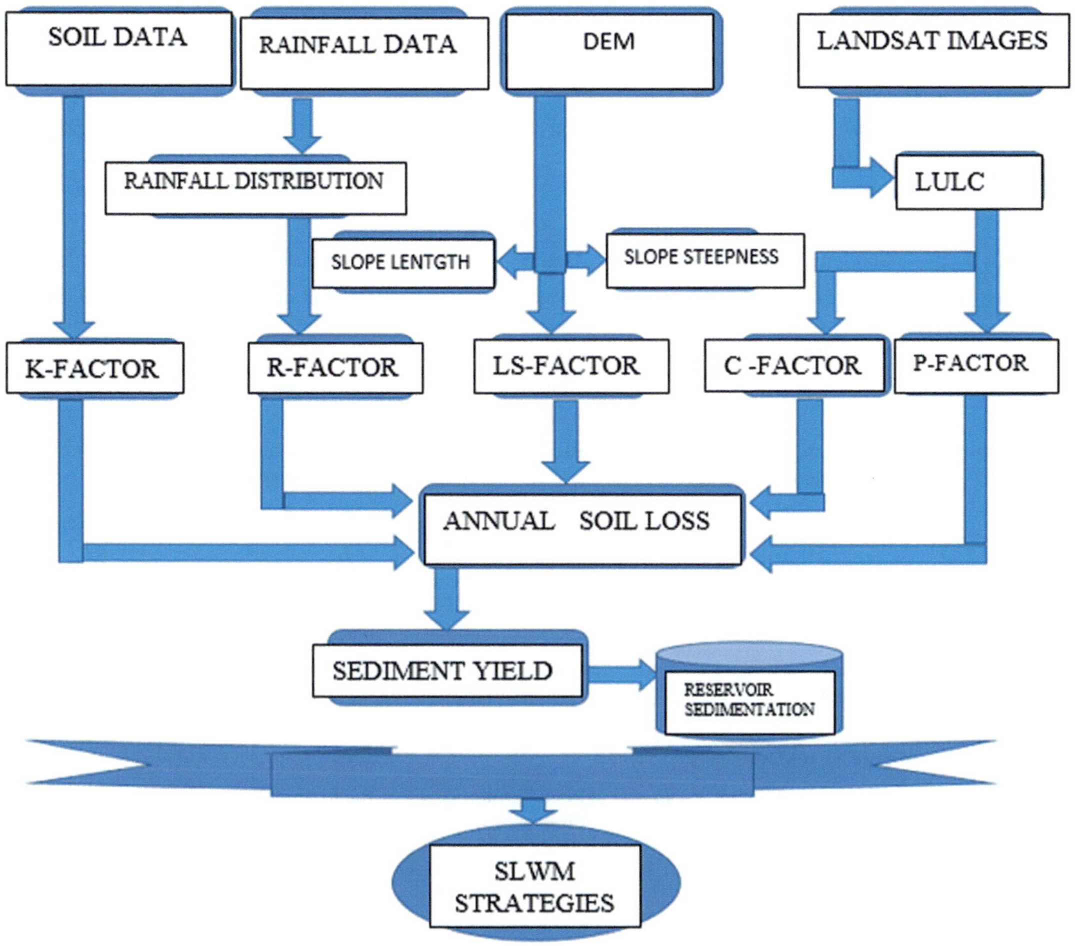

Fig. 3.2 Conceptual framework using the RUSLE model in GIS. Adopted from Breetzke et al. (2013) and modified by the Researchers

floating method was repeated three times, and an average of time and distance taken were recorded. The discharge was then calculated using Eq. 3.1.

$$Q = AV \tag{3.1}$$

where, Q = Discharge (m^3/s), A = Cross sectional area (m^2), V = Stream Velocity (m/s).

Figure 3.4 is an illustration of the area-velocity method. A floating material was used. Intervals of 0.2 m on channels width of 1 m were estimated. Due to the unevenness of the channel, distances of 10 m were chosen, and the float material covered that distance over a marked period using a stopwatch.

3.2.4 Land Use and Land Cover Images

For the year 1988, Landsat 4–5 TM images were used for mapping land cover changes, while for 1998 and 2008, Landsat 7 images were used, and Landsat 8 for the 2018 image. The 30 years was basically chosen so to observe significant changes in land use and land cover over time which will improve the accuracy of the results. After the images were downloaded from USGS, they were imported into ArcMap, combining bands 3, 2, 1 for Landsat 4–5 TM and 4, 3, 2 for Landsat 8 images. The researcher was familiar with the study area, but due to the limitation of its size, and low resolution of Landsat images, unsupervised

Table 3.1 Study data availability

Data type	Data description	Data source	Data purpose
Objective 1: to estimate past and current LULC changes in the catchment			
Landsat images (30 m * 30 m)	1988, 1998, 2008 and 2018 images	USGS website	Assess LULC changes, input in RUSLE
Objective 2: to estimate potential soil loss in the catchment			
Soil data	Soil types of Swaziland	Ministry of Agric	Input in RUSLE
Meteorological	Monthly rainfall (1988–2018), daily rainfall during research period	Swaziland MET services, rainguage on site	Rainfall distribution, input in RUSLE
DEM	30 m * 30 m	ASTER-GDEM website	Catchment delineation, slope calculation
Soil sediments	Sediments from runoff plots and stream	Runoff plots and stream in the catchment	Particle size analysis, estimation of soil loss during the experimentation period
Objective 3: to estimate the potential sedimentation of the proposed reservoir			
Streamflow Data	Streamflow during grab sampling annual discharge	Field Ministry of Agric/SWADE	To estimate sediment load with discharge area-velocity method estimate reservoir capacity
Reservoir design characteristics	Capacity, and other design parameters	Ministry of Agric/SWADE	Estimation of sediment yield
Objective 4: to propose SLWM strategies			
Questionnaire data	Responses to questions in relation SLWM practices	Focus group discussions	Ranking and prioritizing strategies

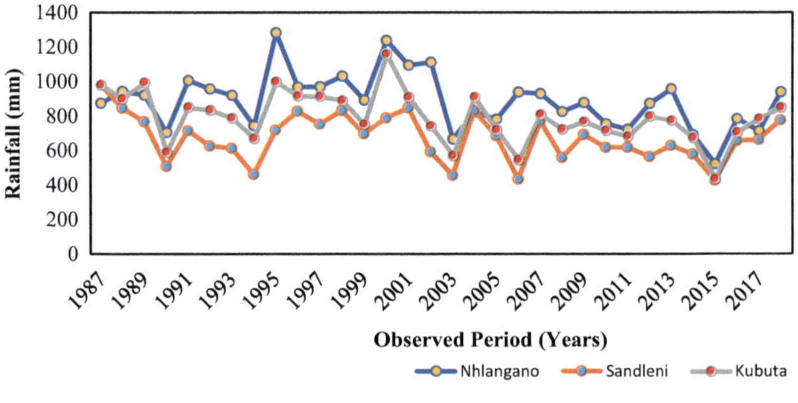

Fig. 3.3 Yearly average rainfall (1987–2018) of Kubuta, Sandleni (SAN), and Nhlangano (NHO) stations

classification was used and then reclassified function to limit the classes to the desired number of dominant classes in the catchment. Five land cover/land use classes were selected: riverine vegetation (stream), grassland, cultivation, bushes/forests, and settlements. In order to explain the extent of changes experienced between the past and present, percent of change (Ebrahim 2017) and rate of change (Abate 2011; Temesgen 2014) were calculated using the following equations:

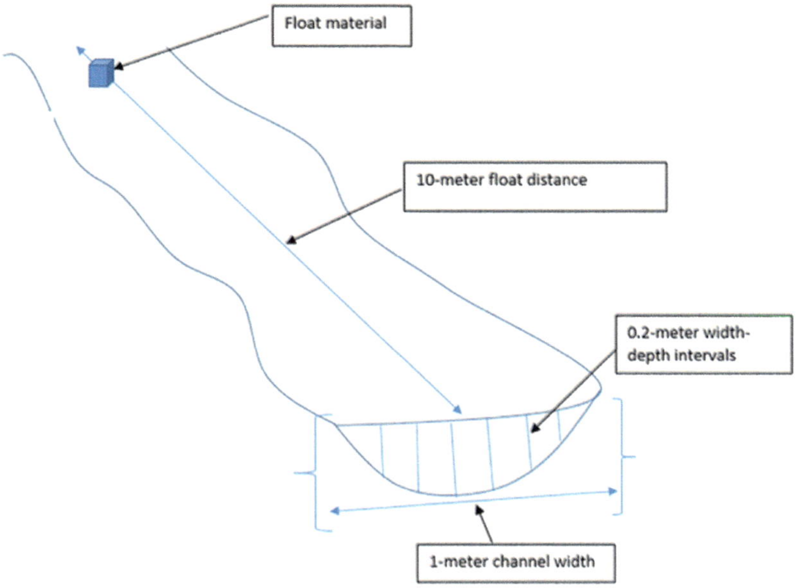

Fig. 3.4 The above is an illustration of the area-velocity method

$$\text{Percent of Change} = ((X - Y))/Y \quad (3.2)$$

$$\text{Rate of Change(ha/year)} = ((X - Y))/Z \quad (3.3)$$

where X, is area of LULC (ha) in time 2, Y, is area of LULC (ha) in time 1, and Z is time interval between X and Y in years.

3.2.5 Validation of Land Use/land Cover Classification

Since there is no classification that can be accurately correct, an accuracy assessment of the classified images was performed using an error matrix in ArcMap 10.4. The Kappa coefficient was also performed (Saha et al. 2022). To come up with the error matrix, a total of 244 control points were established for the five classes in the catchment. This was established by following the thumb rule, which says that the number of control points should be ten times the number of classes. In this case, it means there were supposed to be fifty (50) control points per class. The following equation for accuracy assessment was used:

$$\text{Overall Accuracy} = \frac{\text{Sum of Correctly Classified Cells}}{\text{Total Number of Cells}} \quad (3.4)$$

During classification and using the control points, you realize that some cells were not correctly classified. So, to determine the accuracy of your classification, you have to add all the correctly classified cells and divide them by the total number of cells in the catchment, then multiply by 100% to get the accuracy assessment. For the Kappa Coefficient, Eq. 3.5 below was used:

$$\text{K} = \frac{(\text{Total Cells} * \text{Sum(Correctly Classified)}) - \text{Sum(Row Total} * \text{Column Total)}}{\text{Total Cells Squared} - \text{Sum(Row Total} * \text{Column Total)}} \quad (3.5)$$

3.2.6 Application of the RUSLE Model for Estimation of Sediment Yield in the Catchment

The rationale behind choosing the RUSLE model was the fact that it has been used successfully to estimate soil loss and, subsequently, sediment yield by many researchers around the world, including the southern African region (Breetzke et al. 2013). Therefore, there was enough information available on literature to guide the researcher on how to estimate soil loss using the RUSLE combined with GIS, especially in Swaziland catchments where there have been limited studies on soil loss with which an already locally approved model could be selected from. The RUSLE model is expressed as shown in Eq. 3.6.

$$A = R \times K \times LS \times C \times P \tag{3.6}$$

where, A = annual soil loss (tons/ha/year), R = rainfall erosivity factor (MJ mm/ha/h), K = soil erodibility factor (ton/ha/MJ/mm), LS = topographic factor (dimensionless), C = cover and management factor (dimensionless), P = support practice factor (dimensionless).

Rainfall erosivity factor/*R*-Factor

According to Lal (1990), rain erosivity is defined as the rain compressional power during the incidence of erosion. To determine the R-factor, rainfall stations in the study area were identified, then monthly and annual rainfall was reconstructed using these stations and the years of interest (1988, 1998, 2008, and 2018). The rainfall erosivity factor was calculated using the equation below:

$$R = (4.17 * MFI) - 152 \tag{3.7}$$

where MFI is given by:

$$MFI = \frac{\sum_{i-1}^{n} Pi^2}{P} \tag{3.8}$$

where Pi is monthly rainfall and P is annual rainfall.

Soil erodibility factor/*K*-Factor

According to (Parysow et al. 2003), the soil erodibility factor (K) is soil erodibility rate per rain erosion index unit that is measured in the catchment and often determined by using soil characteristics. These characteristics include soil texture, soil organic content, and soil permeability (Wischmeier and Smith 1965). The soil erodibility factor was calculated using Table 3.2 of different soil textures.

The topographic factor/*LS*-Factor

According to Whichmeier and Smith (1978), the LS-factor indicates the vulnerability of an area to erosion based on the steepness and length of slope in each spatial location compared to a standard site with 9% steepness 22.13 m slope length. The increase of slope length and slope degree increases the water flow rate at the ground surface and causes the acceleration of soil erosion (Haan et al. 1994). The slope of the catchment was calculated in degrees using a DEM of the study area and Eq. 3.9 (Moore and Burch 1986) was used to calculate the LS-factor. The computation was performed in the raster calculator.

$$LS = P(flo\_acc * X''/22.13, 0.4) * (\mathrm{Sin}(''Y'' * 0.01745)/0.0896, 1.4) * 1.4 \tag{3.9}$$

where, P = Power, X = cell resolution, Y = Slope in Degrees.

Crop cover management/*C*-Factor

According to Biesemans et al. (2000), the management factor of vegetation is indicative of the planting effect on agricultural management and the effect of trees, shrub, grass, and ground covers on soil erosion reduction. By increasing the vegetation, soil erosion is reduced. Even though the cover factor is usually determined by empirical equations (Wischmeier and Smith 1978), the most used cover index is the Normalized Difference Vegetation Index (NDVI) which is obtained from remote sensing technology. Table 3.3 was used to classify the land use and land cover images according to their classes.

Table 3.2 Soil erodibility values for common surface textures, source (Stone and Hilborn 2012)

Soil textural class	K-Factor (ton/ha^{-1}/year^{-1})		
	More than 2% organic matter content	Less than 2% Organic matter content	Average organic matter content
Coarse sandy loam	0.16	–	0.16
Fine sand	0.13	0.20	0.18
Fine sandy loam	0.38	0.49	0.40
Loam	0.58	0.76	0.67
Loamy fine sand	0.20	0.34	0.25
Loamy sand	0.09	0.11	0.09
Loamy very fine sand	0.56	0.99	0.87
Sand	0.02	0.07	0.04
Sandy clay loam	0.45	–	0.45
Sandy loam	0.27	0.31	0.29
Silt loam	0.83	0.92	0.85
Silty clay	0.58	0.61	0.58
Silty clay loam	0.67	0.79	0.72
Very fine sand	0.83	1.03	0.96
Very fine sandy loam	0.74	0.92	0.79
Clay	0.47	0.54	0.49
Clay loam	0.63	0.74	0.67
Heavy clay	0.34	0.43	0.38

Table 3.3 Table of cover management factors, adapted from RUSLEFAC (1997)

Land cover type	*C*-Factor
Wetland	0.05
Grassland	0.08
Cultivation	0.5
Bushes/forests	1
Settlements	1

Table 3.4 Table of support practice factors, adapted from RUSLEFAC (1997)

Land cover type	*P*-Factor
Wetland	1
Grassland	1
Cultivation	0.5
Bushes/Forests	1
Settlements	1

Support practice/*P*-Factor

The support practice factor (*P*-Factor) in the Universal Soil Loss Equation is the ratio of soil support practice to the corresponding loss with up and down slope culture. Cultivation on slope lands requires protective measures, which include contour cropping, terracing system, etc., in estimating the practice factor, Table 3.4 adapted from RUSLEFAC (1997) was used.

The above five parameters were then multiplied together in the raster calculator to obtain the potential annual soil loss.

3.2.7 Application of Runoff Plots and Particle Size Analysis

Runoff plots were set up in four sites within the catchment. The area of each runoff plot was

Fig. 3.5 Runoff plots, sediments, and particles sieves used during experiments

100 m^2. They were strategically located in areas of bare land, grassland, and forest/shrubs. These runoff plots were furnished with a rain gauge, and the rainfall events, 260 L containers for collection of runoff and sediment. The sediments collected from these plots were then quantified based on rainfall intensity and converted to annual/seasonal soil loss. Sediments collected from the runoff plots were collected and dried at 105 °C in the laboratory, then weighed and sieved to obtain different classes of sediments generated in order to have a clear picture of the average soil particle sizes to plan for management measures. Basically, the generation of sediments using runoff plots was a method used to prove that soil erosion is occurring in the catchment. Figure 3.5 shows one runoff plot onsite, a tray used for drying samples, and particle sieves.

Particle size analysis is very important in sediment yield studies. It gives an idea of the type of sediments that are likely to be eroded, transported, and deposited in reservoirs, in the process advising on the management actions that are needed for sediment management, such as dredging.

3.2.8 Estimation of Soil Loss Rates

Estimation of soil loss from the model was performed by classifying the classes of soil obtained from the ArcMap into three classes: low, moderate, and high. To do that, the reclassify operation in spatial analyst tools was used in ArcMap. Table 3.5 was developed.

3.2.9 Grab Sampling and Sediment Concentration

500 ml containers were used to take grab samples on three sites of the stream after a rainfall event. One site was upstream, about 1 km before

Table 3.5 Soil loss classification in Mntjuzalala Catchment

Category	Range (t/ha/year)
Low	0–0.5
Moderate	0.5–2
High	> 2

the proposed reservoir site; the other was in the middle (0.6 km) from the proposed reservoir site, while the other was on the reservoir site. There are three methods for determining suspended sediment concentration; evaporation, filtration, and wet sieving method. For this study, the evaporation method was used since the other two methods are applicable in low sediments concentrations. The method is also simple in terms of the equipment needed (Edward and Glysson 1999). This method involved allowing the sediments to settle to the bottom of the sample bottle, decanting the supernatant liquid, washing the sediments into an evaporation dish, and drying it in an oven for 24 h at temperature of 105 °C. The reason for determining the sediment concentration was to collect primary evidence about the potential of siltation of the reservoir. The concentration of the sediments in the sample was estimated against the flow at the time of sampling using the equation as follows:

$$Q = \text{Discharge}(\text{m}^3/\text{s}), A = \text{Cross sectional area}(\text{m}^2), V = \text{Stream Velocity}(\text{m/s}) \tag{3.10}$$

3.2.10 Sediment Delivery Ratio (SDR)

According to de Vente et al. (2011), since not all sediments generated within a catchment end up deposited in water bodies such as reservoirs or rivers, there was, therefore, a need to use the Sediment Delivery Ratio that would account for the deposition of sediments along the way as it gets trapped by vegetation, or even the river channel itself. The Sediment Delivery Ratio (SDR) was calculated using Eq. 3.11 (USDA 1983).

$$\text{SDR} = 0.5656\ \text{CA}^{-0.11} \tag{3.11}$$

where, SDR = Sediment Delivery Ratio (dimensionless), CA = Catchment area (km^2). Once the Sediment Delivery Ratio was determined, the average sediment yield was calculated using Eq. 3.12 by Wischmeier and Smith (1978).

$$\text{SR} = \text{SDR} \times A \tag{3.12}$$

where, SR = Sediment Yield (ton/year), SDR = Sediment Delivery Ratio, A = Average soil loss (ton/year), and Sediment Delivery Ratios can be compared to Roel (1962) graph as shown in Fig. 3.6.

3.2.11 Trap Efficiency and Release Efficiency of Reservoirs

The trap efficiency of a reservoir is defined as the ratio of deposited sediments to the total incoming sediment in the reservoir during its lifespan, and this affects the economic and useful life of the

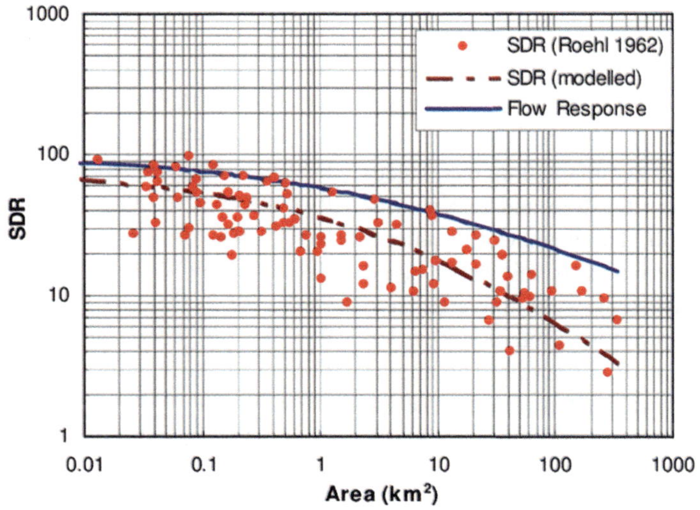

Fig. 3.6 Comparison of SDR % using Roel (1962) estimates

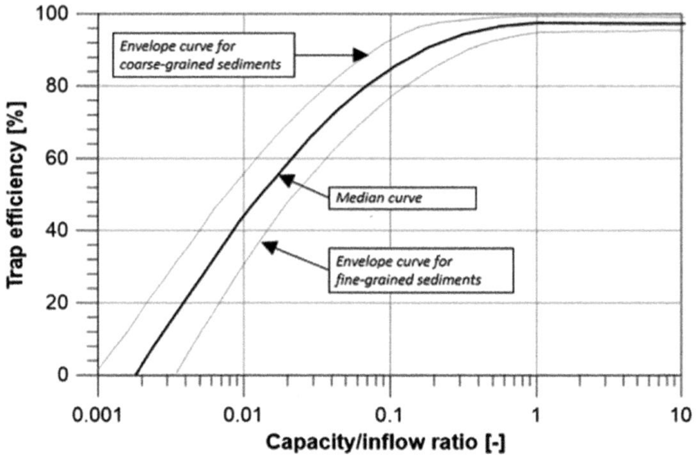

Fig. 3.7 The Brune's Curve for estimating trap efficiency (Brune 1953)

reservoir. The trapping efficiency was estimated using Brune's Curve shown in Fig. 3.7 (Brune 1953). The annual inflow was estimated using Table 3.6, a report by Mkhonta (2017), and Eq. 3.13, developed by Brown (1943), was used to compare the result of Brune's Curve method.

$$\mathrm{TE} = 100\left(1 - \frac{1}{1 + 0.0021D * \frac{C}{W}}\right) \quad (3.13)$$

where, TE = Trap Efficiency in percent, W = Catchment area in Km^2, D = range from 0.046 to 1, depending on reservoir characteristics, Brown (1943) says these values are close to 1.

The release efficiency is basically the sediment that is able to be flushed by the reservoir. Churchill (1948) established Eq. 3.14 to estimate release efficiency:

$$\text{Release Effeciency} = 100 - \text{Trap Efficiency} \quad (3.14)$$

The annual reservoir sedimentation was calculated after considering the trap and release efficiency of the reservoir. The weight of sediments deposited in the reservoir was calculated based on 1.35 t/m^3, and the tons were converted into cubic meters, which were then interpreted as the annual storage loss of the reservoir. The annual storage loss was then calculated as a ratio of the total reservoir capacity and multiplied by 100% to get annual storage loss in percent.

3.2.12 Determining the Influence of Some of the RUSLE Factors on the Soil Loss Rate

A relationship was established between soil loss and some of the RUSLE parameters. These parameters included: rainfall erosivity and cover management factors. Another attribute that was considered was slope/altitude. This was done to see how soil loss varies with these factors/attributes to inform some of the possible management strategies.

3.2.13 Stakeholder Engagement

A focus group discussion was held between community stakeholders, consisting of traditional leaders, chiefdom development committees, and irrigation scheme members. This was aimed at getting stakeholder opinions on the prioritization of sustainable land and water management strategies. There were general issues that were discussed in relation to community natural resources management. And after those discussions, five strategies were discussed in detail, and participants formed five groups in support of each of the five strategies and came up with actions that need to be taken to implement that strategy. Based on numbers per each group of the groups, as chosen by the

Table 3.6 Annual runoff from preliminary reservoir design report

GS	Oct	Nov	Dec	Jan	Feb	Mar	Apr	May	Jun	Jul	Aug	Sep
Station: GS 8 (1305 Km^2)												
Qavg (m^3/s)	1.85	2.91	4.74	4.54	6.32	4.32	3.30	2.08	1.67	1.51	1.24	1.19
Qmed (m^3/s)	1.38	2.58	3.97	2.94	3.80	2.74	2.58	1.79	1.35	1.25	1.07	1.16
Qmax (m^3/s)	6.89	7.30	14.49	12.26	27.53	13.06	15.39	5.48	5.44	7.32	3.24	2.63
Qmin (m^3/s)	0.36	0.79	0.81	0.73	0.57	0.66	0.55	0.39	0.33	0.34	0.26	0.23
Cnt (No. data)	31	34	34	34	30	30	33	32	33	33	33	31
Q75% (m^3/s)	0.93	1.54	2.03	2.13	1.97	1.99	1.57	1.19	1.04	0.88	0.73	0.68
Q50% (m^3/s)	1.38	2.58	3.97	2.94	3.80	2.74	2.58	1.79	1.35	1.25	1.07	1.16
qavg (lps/Km^2)	1.41	2.23	3.63	3.48	4.85	3.31	2.53	1.60	1.28	1.15	0.95	0.91
q75% (lps/Km^2)	0.71	1.18	1.55	1.63	1.51	1.52	1.21	0.91	0.79	0.68	0.56	0.52
Catchment area: 6.11 Km^2												
Q	Oct	Nov	Dec	Jan	Feb	Mar	Apr	May	Jun	Jul	Aug	Sep
q75% (lps/Km^2)	0.71	1.18	1.55	1.63	1.51	1.52	1.21	0.91	0.79	0.68	0.56	0.52
Q75% (lps)	4.3	7.2	9.5	10	9.2	9.3	7.4	5.6	4.9	4.1	3.4	3.2
V75% (m^3)	11,612	13,628	25,416	26,708	22,305	24,943	19,089	14,917	12,573	11,060	9092	8246
ΣV75% (m^3)	11,612	30,241	55,657	82,364	104,669	129,612	148,701	163,618	176,190	187,251	196,343	204,589
Qavg (lps)	8.6	13.6	22.2	21.3	29.6	20.2	15.4	9.8	7.8	7.1	5.8	5.6
Vavg (m^3)	23,146	35,353	59,378	56,957	71,631	54,209	40,018	26,125	20,262	18,884	15,528	14,461
ΣVavg (m^3)	23,146	58,499	117,877	174,834	246,464	300,674	340,691	366,816	387,078	405,962	421,490	435,951

participants, weighting and ranking of the strategies were then done.

3.2.14 Multi-Criteria Analysis in the Development of Management Strategies

Since there are many possible solutions to the problem of soil erosion and possible reservoir siltation, there was the need to use multi-criteria analysis in coming up with the best possible solution/strategy. This was done by first proposing the possible strategies in Table 3.7:

After having developed the attribute table, crisp numbers were allocated based on Table 3.8. The attributes with linguistic term *Low* were given crisp number 0.17, *Medium* crisp number 0.5, and *High* crisp number 0.83, and this is illustrated in Fig. 3.8.

Five attributes were chosen as possible strategies/interventions. These were named as A1, A2, A3, A4, and A5. Each attribute was given a linguistic term as low, medium, and high. After this, the attributes were given crisp number from Fig. 3.8. The parameters which were of concern in evaluating the alternatives were: cost, social acceptance, and political will. This is shown in Table 3.8.

After having allocated crisp numbers to the attribute table, in order to be able to compare the alternatives, the attribute table was normalized through the process of normalization. In this process, all attributes were converted to an equal scale of between 0 and 1. The maximization and minimization methods below were applied using Eq. 3.15 and 3.16, respectively:

$$Y(ij) = \frac{Y(ij)}{Y\max(j)} \tag{3.15}$$

$$Y(ij) = 1 - \frac{Y(ij)}{Y\max(j)} \tag{3.16}$$

where, $Y(ij)$ is the normalized value for *i*th alternative and *j*th attribute, $y_{(ij)}$ is the original value for *i*th alternative and *j*th attribute, $y^{\max}(j)$ maximum value for *j*th attribute. To rank strategies, Eq. 3.17 was applied.

$$Wj = \frac{(m - rj + 1)^p}{\sum(m - rj + 1)^p} \tag{3.17}$$

where Wj is the weight of *j*th attribute, m is the number of attributes, r_j is the rank of *j*th attribute, j is the attribute, p is parameters for weight distribution, which is equal to 1.

Table 3.7 Table showing proposed management strategies

• Soil conservation and runoff control structures	A1
• Implementation and monitoring of chiefdom development plan	A2
• Enforcement of existing laws and policies in relation to environment, land, and water management	A3
• Community awareness raising on sustainable land and water management	A4
• Capacity building to ensure sustainability of the water infrastructure	A5

Table 3.8 Attribute table showing proposed management strategies

Attribute		Cost	Social acceptance	Political will
A1	Infrastructure installation	High	High	High
A2	Implementation of CDP	Low	High	Medium
A3	Enforcement of laws and policies	Low	Medium	Medium
A4	Community awareness	Low	High	High
A5	Capacity building	High	High	High
		Minimize	Maximize	Maximize

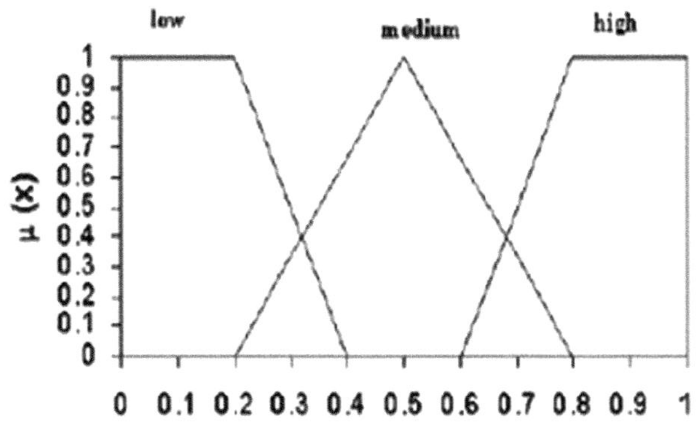

	Crisp number
Low	0.17
High	0.83
Medium	0.5

Fig. 3.8 Crisp numbers allocated to attribute table

3.3 Results and Discussion

3.3.1 Assessing Land Use and Land Cover Changes

In this catchment, it has been observed that since 1988 there has been a huge change in land use/land cover in the catchment, especially in grassland and bushes/forest cover. In 1988 the forest/bushes covered 13.3% of the catchment while grassland was 39.9%. This was probably because there were still few settlements, only 11.5%, thus less cultivation being practiced in the area. The wetland/riverine vegetation from which the stream originates covered 32.2% of the catchment. In 1998, a decline of 7% in grassland was observed, probably due to the increase of about 12% in cultivation and approximately 2% increase in settlements. There was a huge decline in the size of the riverine vegetation, from 32 to 20%; this was attributed to the fact that people started cultivating on the wetland and grazing their livestock. In 2008, the land use/land cover showed to change completely as bushes/forests started to dominate the catchment. And it was observed that the kind of trees/plants dominating the area are classified as invasive alien plant species, which is a problem the country has been facing for over the past two decades, and the ministry of Tourism and Environmental Affairs, Department of Forestry has a program in place aimed at eradicating the problem. In 2008, bush coverage was about 24% from 17.6% in 1998. Grassland had a decline of approximately 10% from 33% in 1998 to 24.4% in 2008, while cultivation increased by approximately 6% from 1998 to about 21% in 2008. Finally, in 2018 there were also noticeable changes in land cover, especially on coverage by bushes, which showed coverage of more than 37% of the catchment compared to 24% in the past 10 years. When interviewed, the local authorities mentioned that they are losing grazing land/grassland over the years due to the encroachment of these invasive plants, and it has been observed that plants/bushes now are covering more area than grassland. Cultivation declined by approximately 5%, from 21.3% to 17.7%, between 2008 and 2018 (Fig. 3.9 and Table 3.9).

3.3.2 Accuracy Assessment of Land Use and Land Cover Classification

The Error matrix and Kappa coefficient were used to validate the LULC results. The results of the error matrix are shown in Table 3.10. The results showed a strong agreement with what is really on the ground. A total of 244 control points were sampled. There were 214 correctly classified cells, and it was observed that the classification accuracy was 87.9%.

The Kappa coefficient was estimated, and the result is shown in Table 3.11. The Kappa value showed a strong agreement of 0.85, which is classified as very good (Landis and Koch 1977).

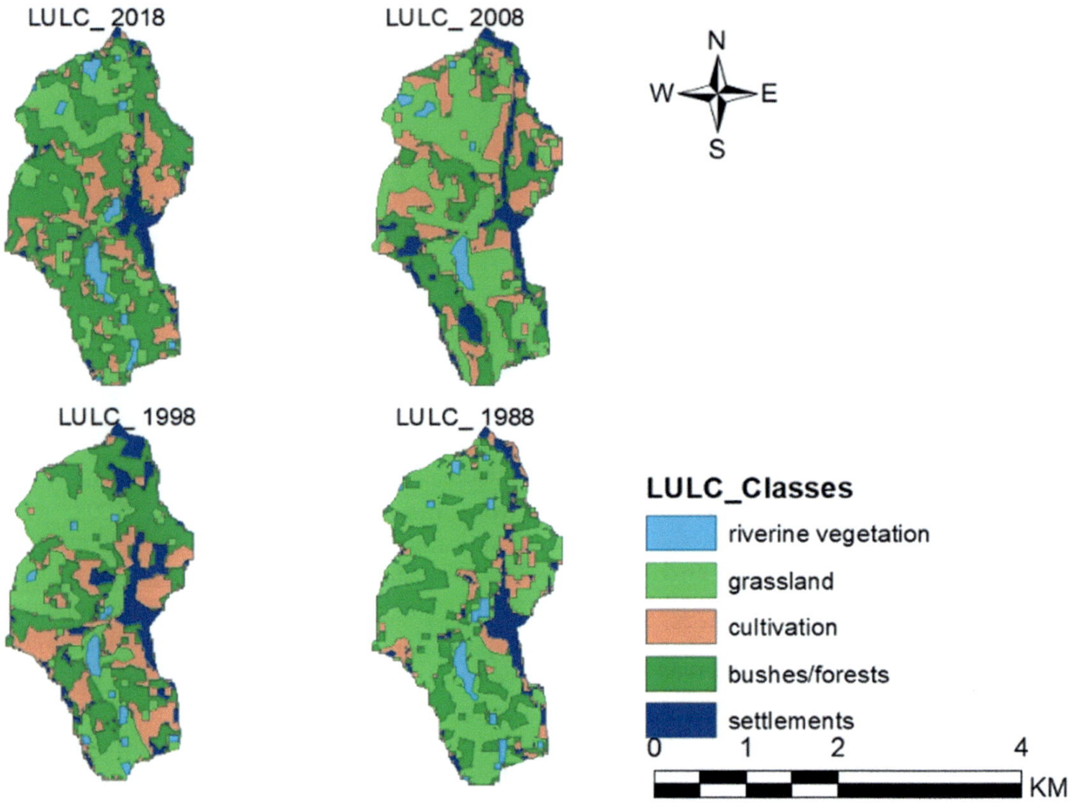

Fig. 3.9 LULC changes from 1988 to 2018 in the catchment

Table 3.9 Table showing area coverage of each class from 1988 to 2018

LULC Class	Area coverage in hectares			
	1988	1998	2008	2018
Riverine vegetation/wetland	179	112	67	25
Grassland	222	184	136	103
Cultivation	18	88	119	91
Bushes/forests	74	98	134	207
Settlements	64	76	102	110

Table 3.10 Error matrix for LULC classification accuracy assessment

LULC	RV	G	C	B/F	S	SUM	Producer's accuracy (%)
Riverine vegetation (RV)	38	6	0	0	0	44	86.4
Grasslands (G)	1	48	0	3	0	52	92.3
Cultivation (C)	1	1	45	1	1	49	91.8
Bushes/forests (B/F)	5	2	0	48	0	55	87.3
Settlements (S)	1	0	6	2	35	44	79.5
SUM of ground control points	46	57	51	54	36	244	
User's accuracy	82.6	84.2	88.2	88.8	97.2		87.5

Table 3.11 Accuracy assessment of LULC classes

LULC class	Class accuracy (%)	Overall accuracy (%)
Riverine vegetation	84.5	
Grasslands	90.3	
Cultivation	90.0	
Bushes/forests	88.1	
Settlements	88.4	
Average	88.3	87.9

3.3.3 Estimation of Soil Loss in the Catchment Using RUSLE Model

Under this section, the results of the RUSLE model inputs are discussed. These inputs are the: rainfall erosivity factor (R) (MJ.mm/ha/h), soil erodibility factor (K) (ton/ha/MJ/mm), topographic factor (LS), cover and management factor (C), support practice factor (P). The final output or result of this model is the annual soil loss (tons/ha/year).

***R*-Factor**

The spatial and temporal variation, when calculating for all the years (1988, 1998, 2008, and 2018), the rainfall/runoff erosivity ranged from as low as 86 to 568 MJ.mm/ha/year. It has been observed that within the individual observed years; there was a variation in the R-factor, where the highest erosivity factor variation was in the year 2008, where the difference between the minimum and maximum *R*-factor values was 482 MJ.mm/ha/year, while the lowest *R*-factor value was in 1988, where the variation between minimum and maximum *R*-factor values is 23 MJ.mm/ha/year. 1998 and 2018 received almost similar variations of erosivity between the maximum and minimum, being 140 and 143 MJ. mm/ha/year. In 2008 there were very few rains received, thus the huge variation between low and high values of erosivity, while 1988 showed higher amounts of rainfall thus variation between low and high erosivity factors was low, while average rainfall was received in 1998 and 2018; thus the average erosivity was experienced.

As observed in Fig. 3.10, the highest erosivity values are experienced in high altitude and northern areas of the catchment where higher rainfall amounts are received, while the lowest values are experienced in low altitudes of the catchment is go toward the dry lowveld. It has been observed that in this period, most of the areas that had high erosivity were in the steeper slopes, which has a possibility of high volumes of runoff and potential of erosion.

In summary, spatio-temporal erosivity variation within the catchment was observed to vary widely between minimum and maximum in 2008, while in 1988, the variation was almost insignificant, while in 1998 and 2018, variation was moderate.

***K*-factor**

The soil type on the North-West part of the catchment was rocky. It showed low soil erodibility of 0.003 t/ha/MJ/mm, and this class was only 5.2% of the catchment. In comparison, marshy/clay soils with erodibility factor of 0.026 t/ha/MJ/mm covered about 10.8% of the catchment. The rest of the soils ranged from loamy to sandy loamy soils observed, especially on the eastern part of the catchment, which is also highly inhabited, and on the parts surrounding the proposed reservoir site. With an erodibility factor ranging from 0.028 to 0.038 t/ha/MJ/mm, these soils are highly erodible. With the combined effect of human activity (settlements and cultivation), medium to high soil erosion rates is experienced. These areas, which consist of highly erodible soils, are also recognized by gullies which even carry sediments to the stream and the proposed

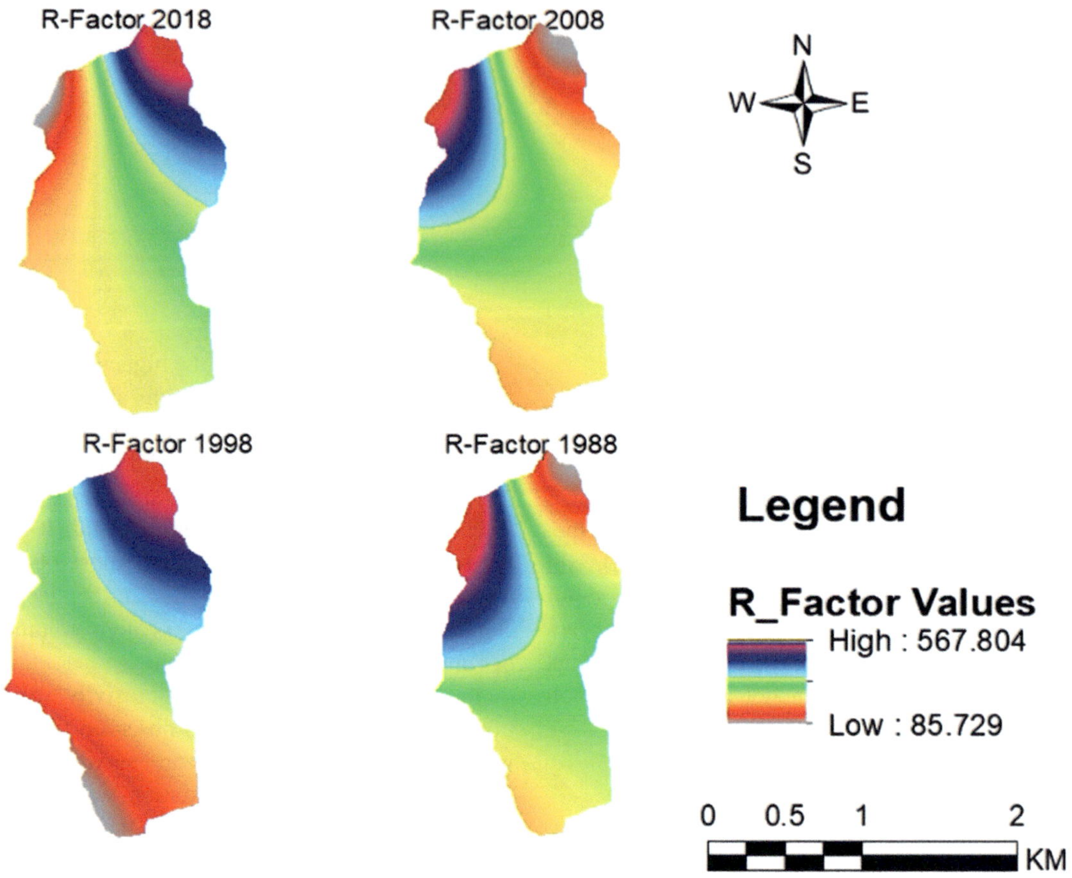

Fig. 3.10 Rainfall erosivity variation

reservoir site, posing a risk of siltation once the reservoir has been constructed (Fig. 3.11).

LS-factor

A digital elevation model was used to calculate the slope length and steepness factor, using slope in degrees and flow accumulation as an input in the raster calculator. The *LS*-factor in the analysis was observed to be ranging from 0 to 5.02. It was also observed that it was high with an increased slope steepness and flow accumulation. The topographic factor/LS-factor indicates the influence of slope length and slope steepness in the occurrence of soil erosion; this is proven by the fact that where the LS-factor is high as a result of the increase in slope, erosion is also predominant, as shown in the soil erosion maps. Most of the watershed has an *LS* value of less than four, which shows a moderate vulnerability to erosion (Beskow 2009) (Fig. 3.12).

C-Factor

The crop cover management factor in the Mntjuzalala Catchment has been increasing over time. It has been observed to be increasing from 1988 to 2008, being constant between 1998 to 2008. In the first 20 years of the observed period, the average C-factor increased from 0.12 to 0.19. This continuous increase could be attributed to an increase in cultivation and settlements, which render the soil vulnerable to erosion. Within this period, cultivation and settlements increased by 84% and 38%, respectively. This low level of C-factor in 2018 could be attributed to an increase in forest cover and a small decrease in cultivation. Forest coverage from 2008 to 2018

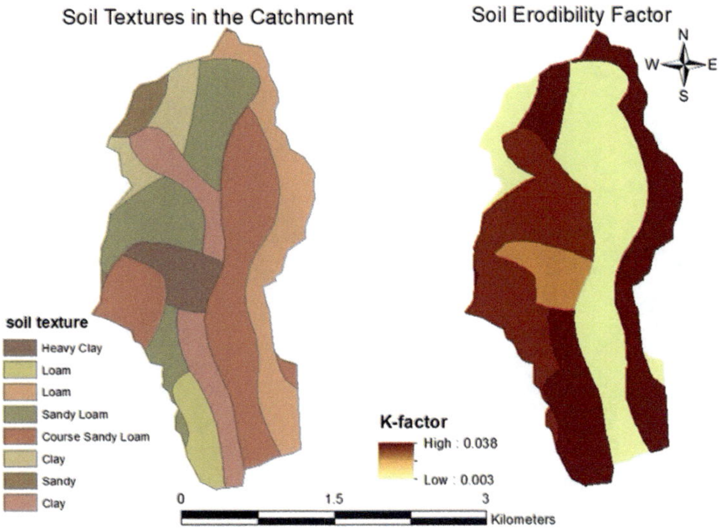

Fig. 3.11 Soil map and *K*-factor variation

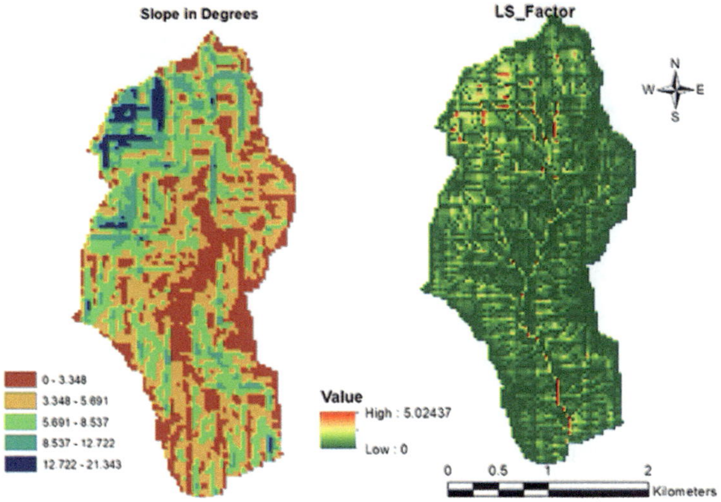

Fig. 3.12 Slope in degrees and *LS*-factor

increased by 13%, while cultivation decreased by 5% in the same period. A low cover factor is observed in forest areas, while a higher cover management factor is observed in the cultivated areas. This means that the soil is more vulnerable to erosion in cropping areas than in forest land. This is shown in Fig. 3.13.

P-Factor

The support practice factor (P-factor) accounts for practices that minimize soil erosion potential by the runoff. The supporting mechanical practices include the effects of contouring, strip cropping, or terracing, especially in cultivated areas. Over the 30-year period, areas with a high P-factor value have been decreasing, while in 1988, such areas covered most of the catchment. This can be attributed to the fact that there were no soil erosion control practices in places such as plowing across the slope and contouring. The spatial and temporal P-factor is shown in Fig. 3.14.

The average *P*-factor value in 1988 was high, possibly because there were no support practices such as contour and strip plowing. This was due to low coverage of cultivation since there were

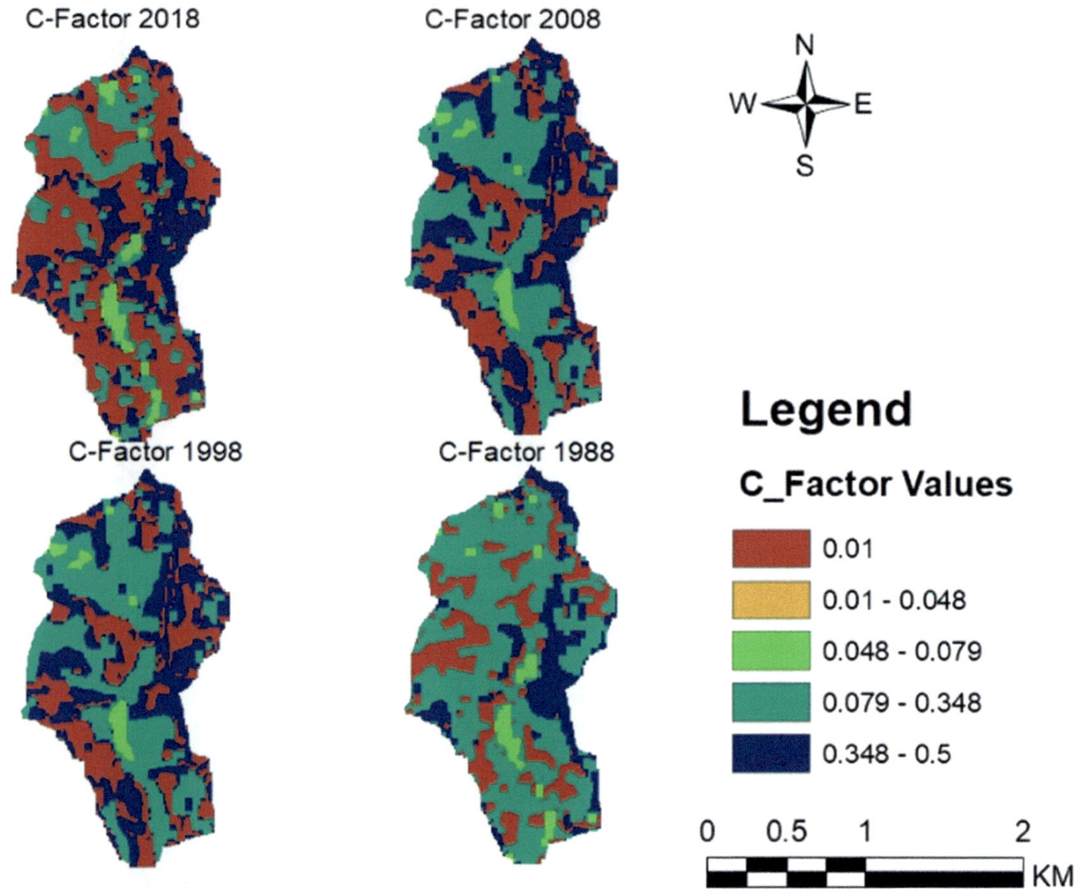

Fig. 3.13 Spatial and temporal variation of *C*-factor

few settlements by that time. In 1998 and 2008, the support practice factor decreased to 0.88, with a standard deviation of 0.21. This decrease was attributed to an increase in cultivation, which resulted in the application of control practices, including plowing across the slope. However, again in 2018, an increase of the support factor, from 0.88 in 2008 to 0.91 in 2018, was observed. This could be attributed to a decline in cultivation, which led to a decrease in the application of support practices.

Spatial and temporal soil loss is shown in Fig. 3.15. The soil loss was reclassified into three classes (low, medium, and high). The spatial and temporal variation of soil loss in the catchment was observed to be highest in 1998, maximum of 6.4 t/ha/year. This is due to high rainfall received in this year compared to the other three individual years under study. This can also result from the increase in cultivation, which can lead to increased soil loss if good farming practice measures are not implemented.

The year 2008 experienced the lowest level of soil loss, with the maximum being 3.3 t/ha/year. This year experienced the lowest rainfall amounts as compared to the others. On the other hand, in 1988, soil loss maximum was 4.6 t/ha/year, while in 2018, it was 5.4 t/ha/year increasing by 38.8% from 2008. A high soil loss rate was observed on steep slopes, and this is confirmed by the fact that these areas are already eroded, while moderate soil rate was within settlements and cultivated areas, and low soil loss was in grasslands and forest areas. The average soil loss in the catchment ranges from 0.089 to 0.192 t/ha/year when considering all the

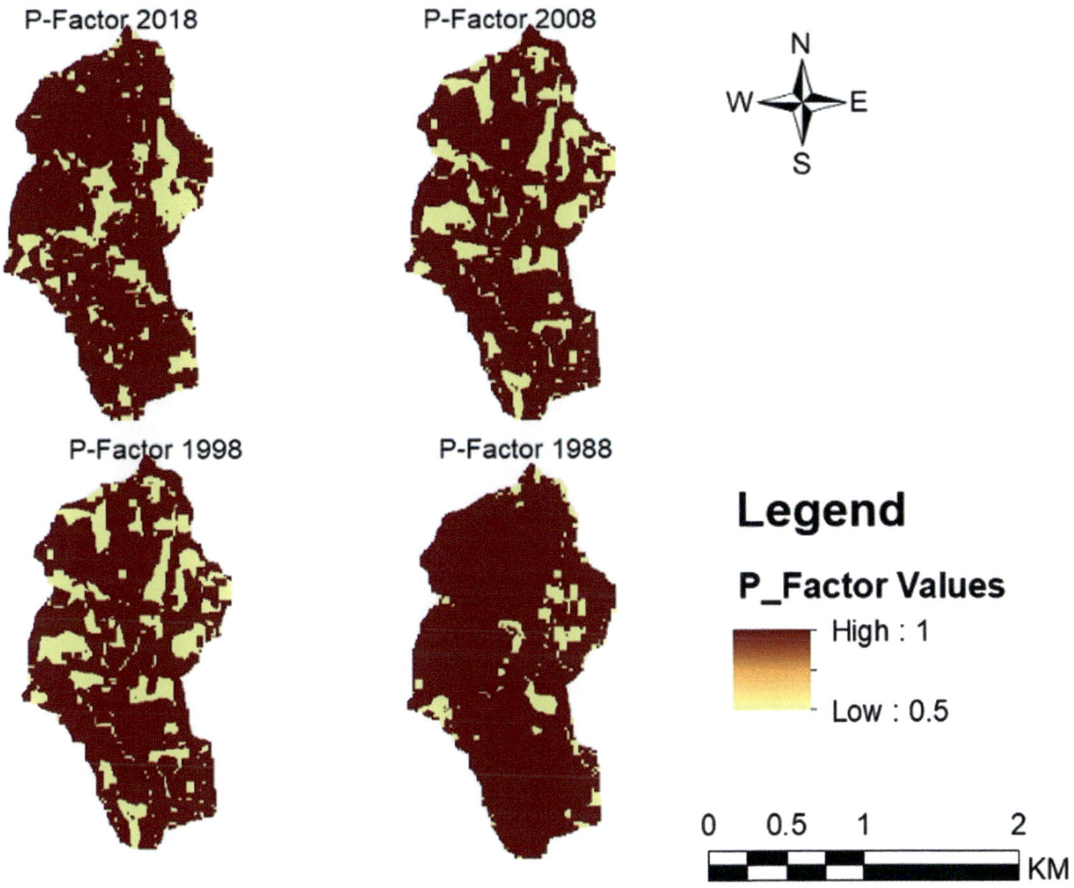

Fig. 3.14 Spatial and temporal variation of *P*-factor

individual years' averages, while the long-term average is 0.13 t/ha/year. A study conducted in the same chiefdom showed an average soil loss of 0.17 t/ha/year, and this is shown in Table 3.12.

Figure 3.16 shows annual soil loss over the observation period. The annual soil loss rates in the catchment for the period ranged from 54.38 t/year to 117.31 t/year. Average soil loss for 2018 was the lowest owing to improved land cover, reduced cultivation, and moderate rainfalls, while 1998 experienced high average soil loss due to increased cultivation and rainfall. Increase in cultivation and rainfall increases soil erosion potential; thus, 1998, compared to the other years, shows high soil loss. The long-term average specific soil loss was estimated based on the averages of the observed period, and it was found to be 0.1340 t/ha/year, which translates to an average annual soil loss of 82.04 t/year. The RUSLE factors' results were summarized in Table 3.13.

3.3.4 Application of Runoff Plots Soil Loss Analysis

Sediment yield from the runoff plots was quantified against rainfall received and land use/land cover classes. It was observed that more sediments were collected where there was high rainfall intensity, and the sediments varied according to land use/land cover classes. The bare land cover class experienced a high volume of sediment yield over the experimentation period, totaling up to 76% of total sediments collected, while thick grassland experienced low sediment

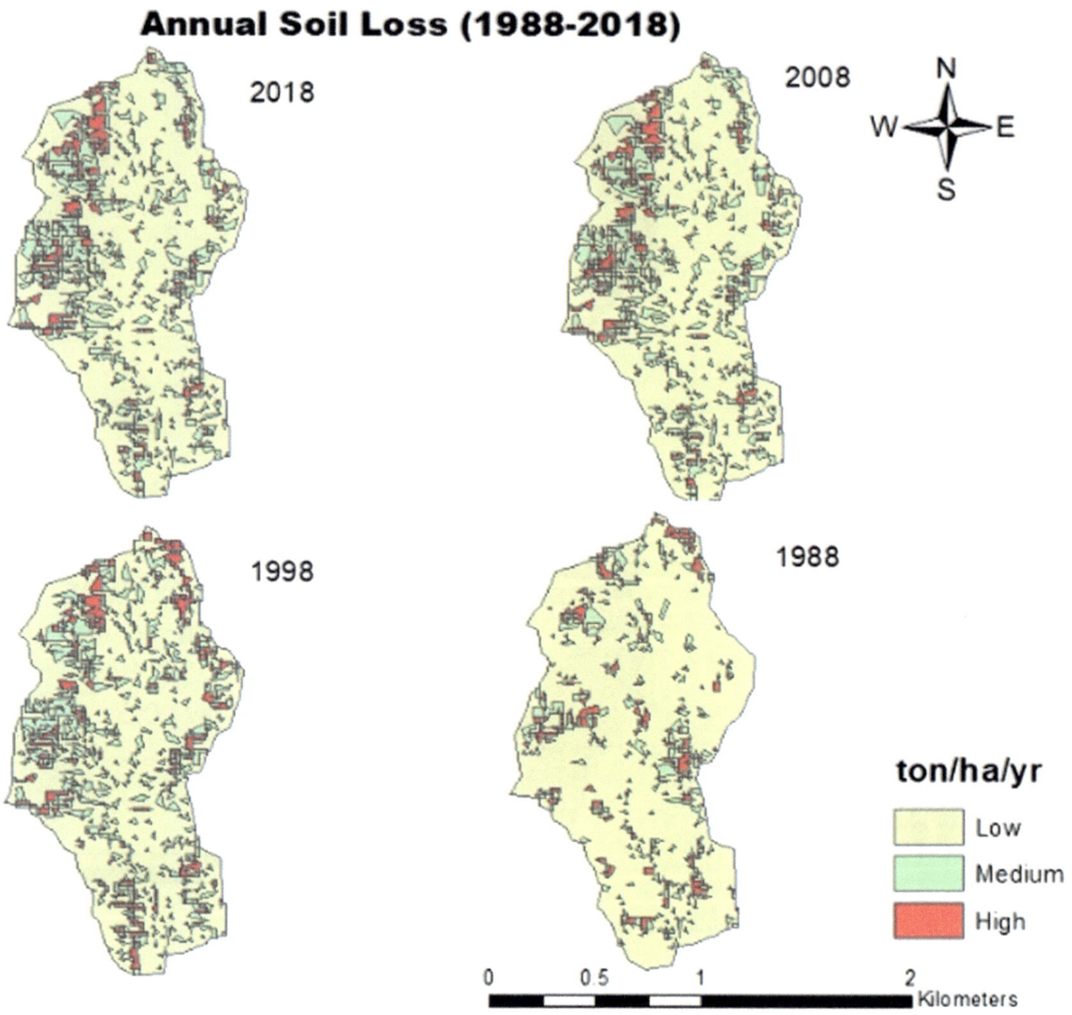

Fig. 3.15 Spatial and temporal variation of soil loss

yield, totaling 10% of the total sediments, the fallow cropland yielded 1.3% and open bushland experienced about 12% of the total sediment yield. This shows that there is a positive relationship between rainfall, land cover class, and sediment yield. The higher the land cover density, the lower the sediment yield, while low the land cover density, the higher the sediment yield. The total sediment yield received during the experimenting period was 3235.3 g, as shown in Table 3.14. This is illustrated in Fig. 3.17.

The sediments which were collected during experimentation were used to validate the RUSLE results. It was observed that the RUSLE model "overestimated" soil loss. This is due to the fact that the experimentation results were observed over a very short duration, two months of rainfall events, while the RUSLE modeled soil loss over a one-year period. To obtain a strong relationship between observed and modeled results, there is a need to collect soil loss data over a full hydrological year period to compare it with RUSLE results. The observed soil loss was 45.8 tons/year, about 15% lower than the modeled soil loss of 2018, which was 54.4 tons/year.

Table 3.12 Soil loss statistics from land and water inventory (MTK sustainable technologies)

Chiefdom	Area (m^2)	Area (ha)	Min (tons/ha/year)	Max (tons/ha/year)	Range (tons/ha/year)	Mean (tons/ha/year)
Kwendzeni	24,620,400	2462	0.000009	3.13	3.13	0.35
Nhletjeni	24,323,400	2432	0.000011	2.76	2.76	0.23
Ngololweni	27,585,000	2759	0.000011	1.97	1.97	0.17
Mbhoke	93,183,300	9318	0.000004	3.50	3.50	0.25
Ndushulweni	12,959,100	1296	0.000006	2.36	2.36	0.26
Zishineni 1	76,950,900	7695	0.000011	3.88	3.88	0.31
Nhlalabantfu	39,025,800	3903	0.000005	2.46	2.46	0.22
KaPhunga	159,104,700	15,910	0.000005	3.97	3.97	0.19
Magele	79,326,900	7933	0.000007	5.23	5.23	0.38
Lulakeni	73,480,500	7348	0.000003	1.96	1.96	0.08
Qomintaba	63,657,000	6366	0.000004	1.67	1.67	0.05
Bhanganoma	12,903,300	1290	0.000012	1.50	1.50	0.23
Zishineni 2 (Manyeveni)	63,842,400	6384	0.000004	2.95	2.95	0.18
Ngobolweni	20,805,300	2081	0.000009	2.47	2.47	0.28
Ondiyaneni	15,474,600	1547	0.000015	2.37	2.37	0.30
Ndlinilembi	106,305,300	10,631	0.000003	5.11	5.11	0.19
Luhlanyeni	76,695,300	7610	0.000010	2.55	2.55	0.10
Mamisa	207,771,300	20,777	0.000015	2.78	2.78	0.02
Nzameye	11,736,900	1174	0.000004	1.92	1.92	0.21
Ndunayithini	120,301,200	12,030	0.000010	2.84	2.84	0.03
Ndinda	60,543,000	6054	0.000004	3.62	3.62	0.27
Ntondozi	59,757,300	5976	0.000005	2.98	2.98	0.26

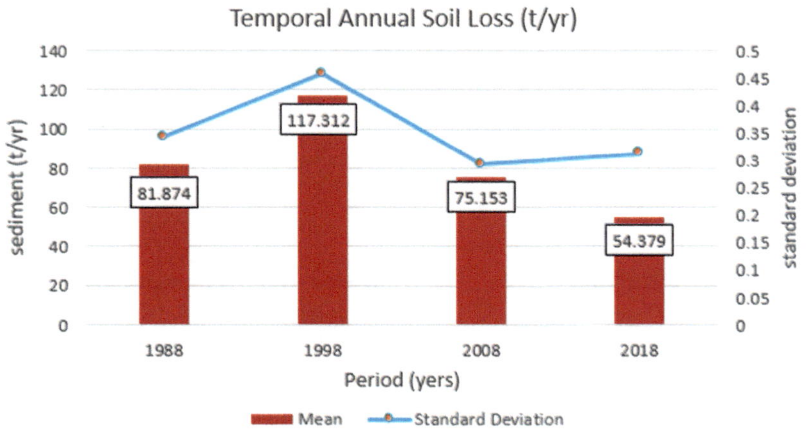

Fig. 3.16 Annual soil loss variation

Table 3.13 Statistics for RUSLE factors

Description	*R*-Factor	*K*-Factor	*LS*-Factor	*P*-Factor	*C*-Factor
Minimum	85.7	0.003	0.00	0.5	0.01
Maximum	567.8	0.038	5.02	1.00	0.5
Mean	374.2	0.02	0.22	0.92	0.2
Standard deviation	38.4	0.01	0.57	0.18	0.13

Table 3.14 Sediment collection from runoff plots

Name	Rainfall (mm)	Duration (h)	Sediment collected (g)	Sediment collected/mm of Rainfall (g/mm)	Erosion rates (t/km^2)
SITE_1	12	4	0	0.00	0.00
SITE_1	7	1.5	0	0.00	0.00
SITE_1	16	6	43.5	2.72	0.03
SITE_2	12	4	0	0.00	0.00
SITE_2	7	1.5	42.4	6.06	0.08
SITE_2	16	6	137	8.56	0.11
SITE_2	33	6	207.7	6.29	0.08
SITE_3	13	4	0	0.00	0.00
SITE_3	8	1.5	0	0.00	0.00
SITE_3	15	6	67.3	4.49	0.06
SITE_3	32	6	272.4	8.51	0.11
SITE_4	12	4	27.1	2.26	0.03
SITE_4	8	1.5	429.3	53.66	0.67
SITE_4	17	6	255.5	15.03	0.19
SITE_4	35	6	1753.1	50.09	0.63

SITE_ID	Rainfall	Hours	Sediments	Sediment/mm rain	Soil-Loss (t/km^2)	Soil-Loss (t/ha)
SITE-1	35	11.5	43.5	2.71875	0.033984	0.0003
SITE_2	68	17.5	387.1	20.91358	0.26142	0.0026
SITE_3	68	17.5	339.7	12.99917	0.16249	0.0016
SITE_4	72	17.5	2465	121.0388	1.512985	0.0051
		Total	3235.3	157.6703	1.970879	0.0096

3.3.5 Particle Size Analysis

Particle size analysis is very important in sediment yield studies. It gives an idea of the type of sediments that are likely to be eroded, transported, and deposited in reservoirs, in the process advising on the management actions that are needed for sediment management, such as dredging. Figure 3.18 shows that the median particle size from the sediments in the catchment ranges from 0.50 to 1.00 mm. This was obtained through locating the 50th percentile on the graph below, and a horizontal line is drawn and the intersection marked.

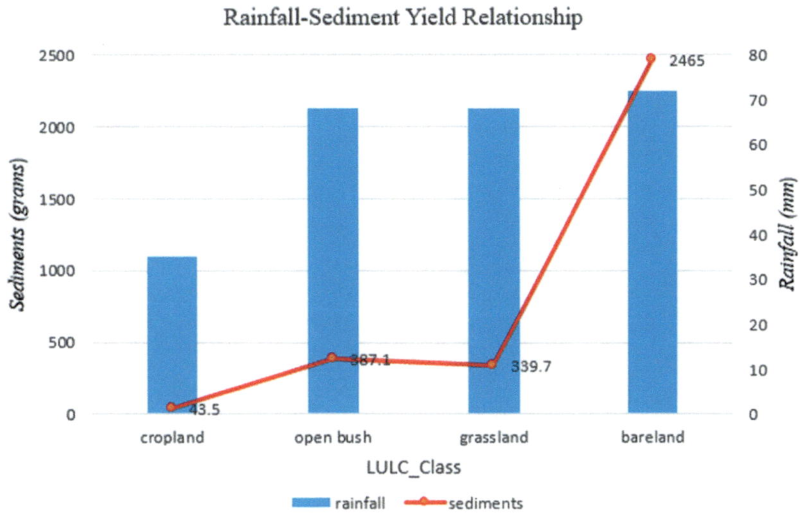

Fig. 3.17 Relationship between sediment yield, rainfall, and land use/land cover class

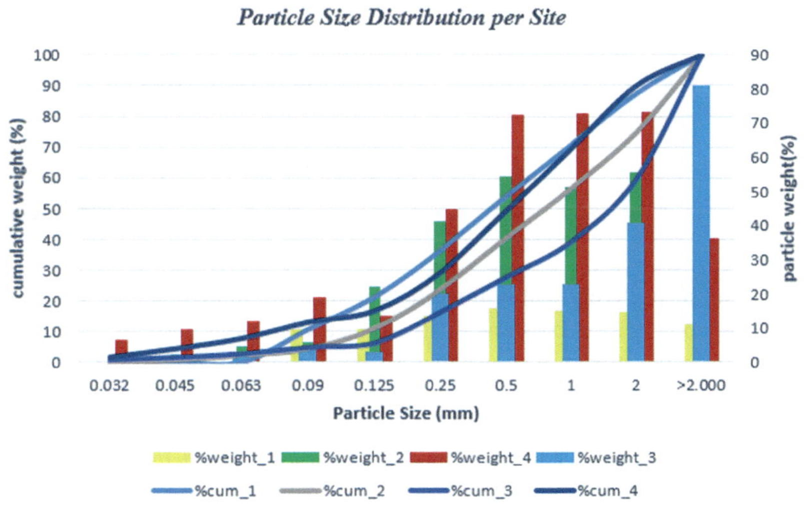

Fig. 3.18 Particle size distribution measured from runoff plots

3.3.6 Estimation of Reservoir Sedimentation Rate and Economic Life

Measuring sediment concentration

Sediment concentration was estimated in three sites: upstream, middle, and downstream. The concentration was high upstream, possibly due to high sediments from the nearby gravel road, which had its stormwater diversion channel taking the runoff directly to the stream. The concentration was 26.24 g/L. On the other hand, the concentration in the middle stream was lower than upstream, possibly due to the thick vegetation cover surrounding that stretch of the stream. It was observed to be 3.14 g/L. Finally, sediment concentration downstream increased, from 3.14 g/L to 6.52 g/L. The sediment concentration increase was attributed to the gullies which were carrying and emptying runoff to this part of the stream. It is worth noting that this part of the stream is within the proposed reservoir site, which shows that there is a potential for siltation of the reservoir over time due to sediment transport from parts of the catchment. Figure 3.19 shows sediment concentration quantities and

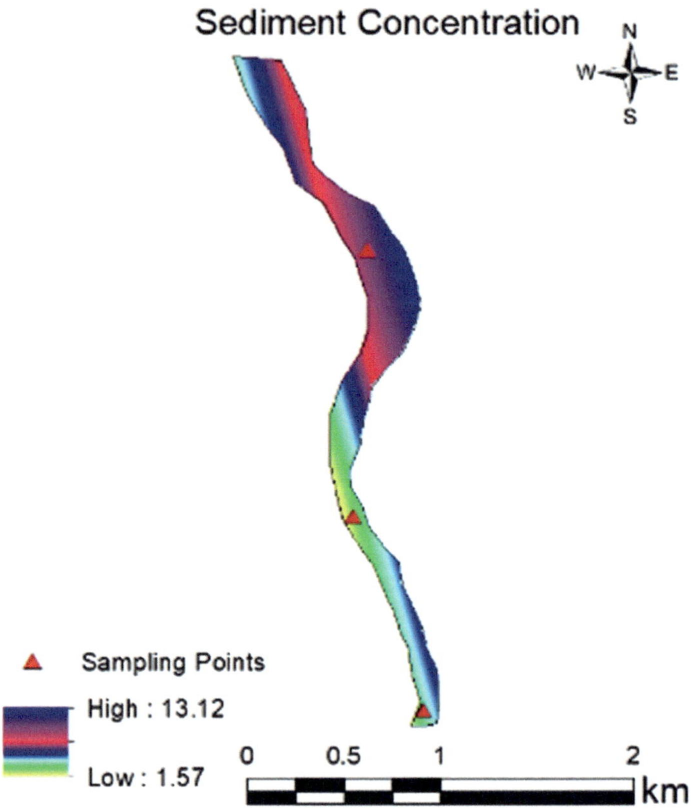

Fig. 3.19 Spatial sediment concentration variation in Mntjuzalala River

location, respectively within the different sites in the stream. Figure 3.20 shows particle size distribution in the stream. It shows that 32.7% of the particles are 0.053 mm while 50% of the deposited sediments are in the 0.063 mm class. This class of soil particles falls under the silt category described in Table 3.5.

Sedimentation rate and Sediment Delivery Ratio (SDR)

The Sediment Delivery Ratio (SDR) was estimated using Eq. 3.11, which is considered as a general equation for estimating Sediment Delivery Ratio, and it was found to be 0.46. This measurement was compared to Roel (1962) measurements, and there was a strong relationship since the SDR fell in between 0.4 and 0.5 of Fig. 3.6. The SDR is inversely proportional to the catchment area; the smaller the catchment, the larger the SDR, while larger catchments have smaller SDR. This is a result of the possibility of sediments being trapped within the catchment and along the stream if they are being transported by moving water. A study conducted by Mkhonta (2016) in a catchment in Swaziland bigger than this one found that the SDR was 0.3.

The annual net soil loss was computed as 37.74 t/year, while the sedimentation rate was 1.55%. The annual sediment deposition was calculated as 30.19 tons per year before considering the proposed reservoir trap efficiency and release efficiency. Using the Brune Curve (Brune 1953) shown in Fig. 3.7, the reservoir trap efficiency was estimated to be 0.80 (80.0%). The capacity-inflow ratio (C/I) was estimated to be 0.04. This was based on reservoir capacity and mean annual runoff calculated from Table 3.12. According to Khan and Khan (2007), on average, small reservoirs have a trap efficiency of 80%, ranging from 77 to 98%, depending on the capacity of the reservoir and catchment area. The release efficiency which was calculated from the trap efficiency using Eq. 3.14 was found to be 0.20.

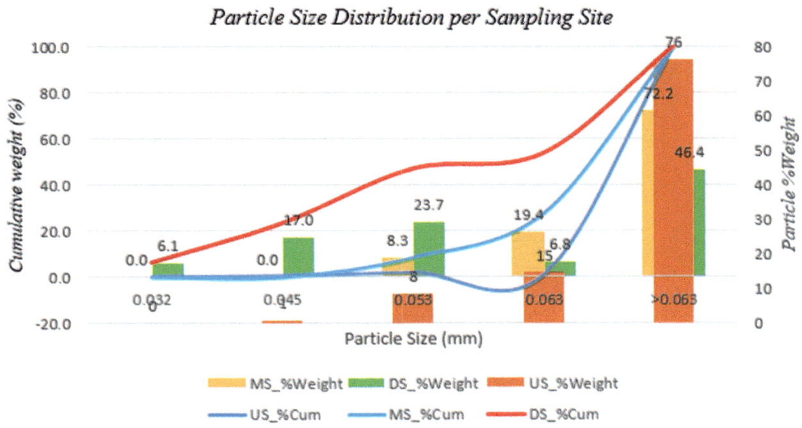

Fig. 3.20 Particle size variation from grab sampling sediments

Loss of storage and economic lifespan of the reservoir

The annual sediment loading was based on 1.35 ton/m^3, and it was found to be 1946.75 m^3. This annual sediment deposition contributes to approximately 1.55% reservoir storage loss per annum. Based on this sediment deposition rate estimate, the proposed reservoir will have lost 50% of its design storage after 32 years. In as much as this conclusion has been reached, it should be taken into consideration that it was based on the current physical conditions of the catchment. Any changes that may occur are likely to increase the sedimentation rate and reduce the lifespan of the reservoir. According to FAO (2010), manual for small reservoir lifespan designs, small reservoirs are designed for a minimum period of 20 years. Table 3.15 shows the lifespan estimate of some reservoirs based on the annual siltation rate, where the average is 38 years while the minimum overall average is as low as 14 years.

The results of this study are supported by studies conducted by Wallingford (2003) in Zimbabwe and Northern Tanzania. The studies were estimating siltation rates of small reservoirs in these two countries, and they concluded that the average lifespan of small reservoirs in these two countries and the region was 38 years, even though it was emphasized that it varies based on catchment and reservoir size and other factors. Again, another study conducted by Wallingford (2004) in small dams in India found that the storage loss of such reservoirs had an average of 2.5% per year, with reservoirs having an average lifespan of 38 years and with capacity ranging from 8 × 103 m^3 to 120 × 103 m^3. The annual storage loss was calculated as 1.55% based on annual sediment deposition. It is noted that the sedimentation rate is less than the average of most reservoirs worldwide, and this can be attributed to the fact that this catchment was found to be mostly low risk in terms of soil loss. According to White (2001), the loss of storage of small reservoirs is higher as compared to larger reservoirs. It has been estimated that the loss of storage of larger reservoirs ranges from 0.5 to 1% per annum. The storage loss of this reservoir is slightly higher than this.

Table 3.15 shows the estimated reservoir storage at different time intervals of 5 years. The last raw is a time interval of 2 years, accumulated annual storage loss and accumulated storage loss rate, and the predicted economic lifespan under each condition. Since the economic life of the reservoir has been estimated at 32 years, the time steps range from 0 to 32 years. It should be noted that after 32 years, the reservoir has not completely been silted, but its capacity has been reduced by 50% from the initial designs, which is said to have lost its economic or useful life. This is illustrated in Fig. 3.21.

The reservoir's full supply capacity (FSC) is 126,000 cubic meters, and to reach its economic life of 50% storage loss, it will take 32 years. However, it should be noted that the reservoir operations might have interfered very early; thus, soon after, the dead storage might be filled.

Table 3.15 The predicted storage changes of the proposed reservoir

Time (Years)	Capacity (m^3) at time (t)	% Storage change	Storage loss
32	126,000	100	1947
27	116,235	92.25	9765
22	106,470	84.5	19,530
17	96,000	76.75	29,295
12	86,940	69	39,060
7	77,175	61.25	48,825
2	67,410	53.5	58,590
0	62,496	49.6	62,496

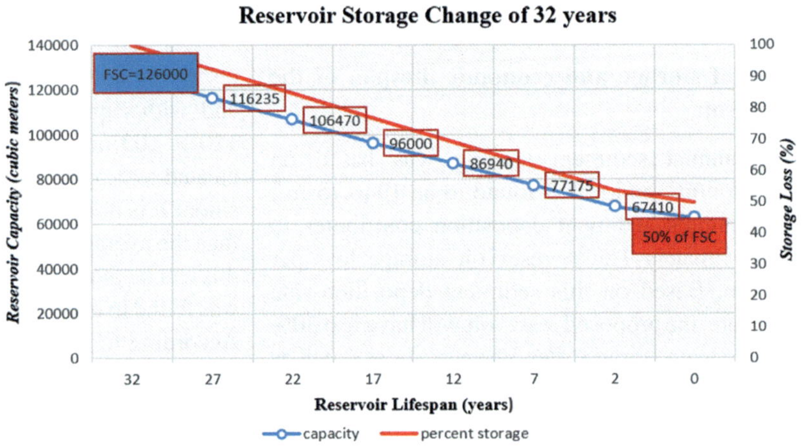

Fig. 3.21 Reservoir storage change due to sedimentation over 32 years

3.3.7 Proposed Soil Loss and Sedimentation Mitigation and Management Strategies

Environmental scan (SWOT analysis)

The development of the management interventions in the Mntjuzalala Catchment has taken into account the existing and the changing environment, circumstances, and the emerging challenges that exist in the main Ngwavuma River Basin where the catchment is located and the country at large. Hence, an environmental scan has been conducted. While doing this, attention has been taken of the fact that the environment that the Ngwavuma River Basin operates in is dynamic and complex and touches all sub-catchment in the basin as a whole. Firstly, to group the internal circumstances within which the basin/catchment operates, an internal environmental scan was conducted. Secondly, it was also considered that it operates in a country and global environment; hence, an external environmental scan was done accordingly. The SWOT analysis shown in Table 3.16 was developed based on Ngwavuma strategic plan, focus group discussions with stakeholders, and field observations.

Selection of the best strategy based on multi-criteria analysis

A focal group discussion was conducted between the traditional local authorities and chiefdom development committees where issues related to natural resources management were discussed with more emphasis on sustainable land and water management. Forty-six (46) participants were interviewed. 37% of the participants considered community awareness on issues of natural resources as key in safeguarding the

Table 3.16 SWOT analysis matrix for the Mntjuzalala Catchment

Internal environment	
Strengths	Weaknesses
– Existence of chiefdom development plan – Strong support from traditional authorities – Existence of chiefdom development committees	– Non-enforcement of relevant and localized policies – Non-existence of local/chiefdom bylaws on environmental management – Poor coordination among stakeholders – Lack of community participation – Lack of knowledge on sustainable land management – Poor and inadequate catchment institutions
External environment	
Opportunities	Threats
– Willingness of government to support sustainable land and water management interventions – Willingness of international donors for financial support on land degradation – Implementation of ESWADE climate-smart and resilient livelihood project – Availability of scientific research funding	– Potential of disputes with neighboring chiefdom and farm owner(s) – Lack of cooperation from relevant stakeholders

environment. They blamed poor environmental practices for lack of knowledge. On the other hand, 22% opined that implementation of the Chiefdom Development Plan (CDP) is the best tool to ensure sustainable management of natural resources; they mentioned that the Chiefdom Development Plan is viewed as an integrated development approach because it accommodates all their aspirations as a community, including social, economic, political and environmental. Moreover, 17% of the participants thought the installation of soil conservation infrastructure should be a priority among the strategies. They justified this by saying already there are eroded areas that need rehabilitation, not just awareness of CDP implementation. 13 and 11% thought enforcement of laws and policies and community capacity building, respectively, were of high priority to ensure sustainable natural resources and ensure long-term benefits of the proposed reservoir. The attribute table was formulated with five attributes, being A1, A2, A3, A4, and A5, as explained in Table 3.6. The assigning of crisp numbers to Table 3.7 resulted in Table 3.17.

In normalization of the decision matrix, Eq. 3.15 and 3.16 were used, resulting in the following normalized Table 3.18. As earlier mentioned, the normalized values range from 0 to 1.

After having ranked the attributes with the proposed order of importance, weights were assigned. Community awareness was assigned rank number one in the attribute table. This was because, for whatever intervention, without engaging and creating awareness of affected and interested parties, which in this case is the community, that intervention is deemed to fail. This is also supported by the fact that even the community members during the focal group discussion they mentioned that most of the time lack awareness or knowledge more than anything (Tables 3.19 and 3.20).

The best strategy was selected based on the outcome of assigning weights to the attribute. Raising awareness in the community/catchment about sustainable land and water management is critical to ensure improved catchment management with minimal negative impacts on the proposed reservoir. It is therefore advised that the community, government, and other development agencies implement management interventions following the ranking in Fig. 3.22 in order to conserve soil and prolong the lifespan of the proposed reservoir.

Table 3.17 Allocation of CRISP numbers

Attribute		Cost	Social acceptance	Political will
A1	Infrastructure installation	0.83	0.83	0.83
A2	Implementation of CDP	0.17	0.83	0.5
A3	Enforcement of laws and policies	0.17	0.5	0.5
A4	Community awareness	0.17	0.83	0.83
A5	Capacity building	0.83	0.83	0.83
		Minimize	Maximize	Maximize

Table 3.18 Normalization of decision matrix

Attribute		Cost	Social acceptance	Political will
A1	Infrastructure installation	0	1	1
A2	Implementation of CDP	0.8	1	0.6
A3	Enforcement of laws and policies	0.8	0.6	0.6
A4	Community awareness	0.8	1	1
A5	Capacity building	0	1	1
		Minimize	Maximize	Maximize

Table 3.19 Assignment of attribute weights to the attribute table

Attribute		Rank	Weight
A3	Community awareness	1	0.33
A2	Implementation of CDP	2	0.27
A1	Infrastructure installation	3	0.20
A4	Enforcement of laws and policies	4	0.13
A5	Capacity building	5	0.07

Table 3.20 Selection of the best alternative or strategy from the attribute table

Attribute		Cost	Social acceptance	Political will	Weight
A1	Infrastructure installation	0	0.2	0.2	0.40
A2	Implementation of CDP	0.23	0.27	0.16	0.66
A3	Enforcement of laws	0.1	0.08	0.08	0.26
A4	Community awareness	0.26	0.33	0.33	**0.92**
A5	Capacity building	0	0.07	0.07	0.14
		Minimize	Maximize	Maximize	High

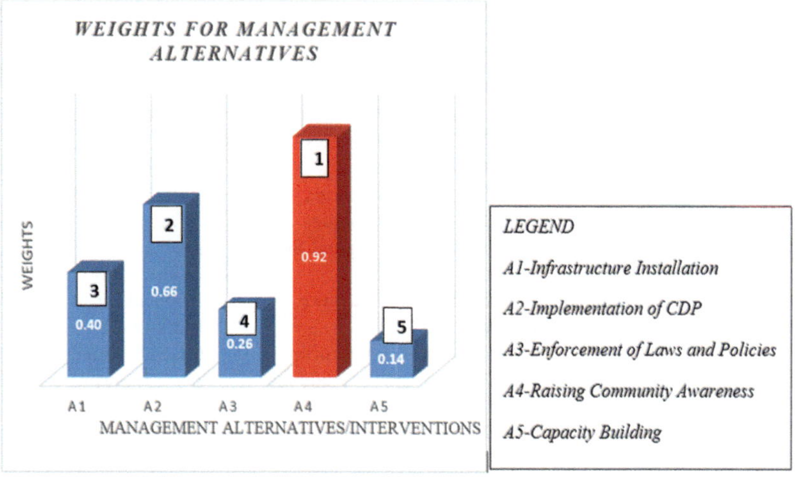

Fig. 3.22 Weighting of strategies, ranking, and selection of the best strategy

3.4 Conclusions

Over the past 30-year period, it has been observed that there have been major changes in land use/land cover. These changes mainly replaced grassland with trees, normally classified as invasive alien plant species. In as much as this does not pose any risk as far as soil erosion rate is concerned, but it is not good for the environment and the survival of other ecosystems. This can be attributed to the fact that the country has been experiencing a decline in rainfall; as a result, people have begun to reduce their efforts in cultivating their fields. There was a strong correlation between land use/land cover (LULC) and sediment yield. This was observed during sediment yield concentrations from runoff plots of varying LULC. Bare areas experienced higher sediment yield under similar rainfall intensities than covered areas like grassland and fallow cropland. The application of the Revised Soil Loss Equation (RUSLE) model in the estimation of soil loss in the Mntjuzalala Catchment showed that average soil loss ranged from 54.4 t/year to 117.3 t/year. From the model results, it can be concluded that the catchment experiences low soil erosion. It has also been observed that soil loss rate is proportional to rainfall received. Those years where high rainfall was received experienced high soil loss (t/ha/year) compared to those which received low rainfall. The storage loss and lifespan of the proposed reservoir are not very different compared to other similar studies conducted for such dams in the region. Similar studies showed an average lifespan of 38 years, while this study showed an average of 32 years when applying a sediment density of 1.35 t/m^3. And this lifespan is greater than the 20-year lifespan recommended by FAO (2010) for such reservoirs, of course, assuming all factors remain equal.

References

Abate S (2011) Evaluating the land use and land cover dynamics in Borena Woreda of South Wollo highlands. Ethiopia J Sustain Dev Afr 13(1):87–105

Beskow S (2009) Soil erosion prediction in the Grande River Basin, Brazil using distributed modelling. CATENA 79:49–59

Biesemans J, Meirvenne MV, Gabriels D (2000) Extending the RUSLE with the Monte Carlo error propagation technique to predict long-term average on-site sediment accumulation. J Soil Water Conserv 35–42

Breetzke GD, Koomen E, Critchley WRS (2013) GIS-Assisted modelling of soil erosion in a South African catchment: evaluating the USLE and SLEMSA approach. InTech J 53–71

Brown CB (1943) Discussion of sedimentation in reservoirs. Proc Am Soc Civ Eng 69:1493–1500

Brune GM (1953) Trap efficiency of reservoirs. Trans Am Geophys Union 34(3):407–418

Churchill MA (1948) Discussion of analysis and use of reservoir sedimentation data. In: Proceedings of the federal interagency sedimentation conference. bureau of reclamation, US Department of the Interior, Washington, DC, pp 139–140

de Vente J, Verduyn R, Verstraeten G, Vanmaercke M, Poesen J (2011) Factors controlling sediment yield at the catchment scale in NW Mediterranean geoecosystems. J Soils Sediments 11:690–707
Ebrahim E (2017) Land use/cover dynamics and its drivers in Gelda catchment, Lake Tana watershed. Ethiopia Environ Syst Res 6(4):1–13
Edwards TE, Glysson GD (1999) Field methods for measurement of fluvial sediment. Techniques of Water-Resources Investigations, U.S. Geological Survey, Reston, Va. (Oct. 5, 2005)
FAO (2010) Manual on small earth dams, a guide to siting, design and construction. 33–34
Haan CT, Barfield BJ, Hayes JC (1994) Design hydrology and sedimentology for small catchments. Academic Press, San Diego, California, USA, p 588
IFAD (2014) Lower usuthu sustainable land management project (LUSIP-GEF). In: Department, international fund for agricultural development
Jeje LK, Ogunkoya OO, Adediji A (1997) Adapting the universal soil loss equation. USLE to South-eastern Nigeria. Nigerian J Sci 31:139–149
Khan MJ, Khan GD (2007) Sediment load assessment of small embankment dams in southern regions of NWFP. Technical report submitted to LPRP/UNDP
Lal R (1990) Soil erosion in the tropics. McGraw-Hill, New York
Landis JR, Koch GG (1977) The measurement of observer agreement for categorical data. Biometrics 33:159–174
Mavhima G, Soropa G, Makurira H, Dzvairo W (2011) Sedimention impacts on reservoir as a result of landuse on a selected catchment in Zimbabwe. Int J Eng Sci Technol 3:6599–6608
Mkhonta B (2017) Ngololweni water and irrigation project preliminary design report
Mkhonta S (2016) Assessing the rate of sedimentation of the Lubovane reservoir and the implications on the lifespan of the LUSIP project in Siphofaneni, Swaziland
Moore LD, Burch GJ (1986) Physical basis of the length-slope factor in the universal soil loss equation. Soil Sci Soc Am 50:1294–1298
Morris GL, Fan J (1998) Reservoir sedimentation handbook: design and management of dams, reservoirs and watersheds for sustainable use. McGraw Hill
Parysow P, Wang GX, Gertner G, Anderson A (2003) Spatial uncertainly analysis for mapping soil erodibility based on joint sequential simulation. CATENA 53:65–78
Patil RJ, Sharma SK, Tignath S (2014) Remote sensing and GIS based soil erosion assessment from an agricultural watershed. 3:6967
Roel JE (1962) Sediment source areas, delivery ratios and influencing morphological factors. In: Symposium of Bari, 202–213. IAHS Publ. no 59
Ruslefac (1997) Revised universal soil equation for application in Canada, Manual
Saha P, Mitra R, Chakraborty K, Roy M (2022) Application of multi layer perceptron neural network Markov Chain model for LULC change detection in the Sub-Himalayan North Bengal. Remote Sens Appl Soc Environ 26:100730. https://doi.org/10.1016/j.rsase.2022.100730
Senzanje A, Boelee E, Rusere S (2008) Multiple use of water and water productivity of communal small dams in the Limpopo Basin, Zimbabwe. Irrig Drainage Syst 22:22537. https://doi.org/10.1007/s10795-008-9053-7
Stone RP, Hilborn D (2012) Universal soil loss equation (USLE). 1–8
Temesgen G (2014) Evaluations of land use/land cover changes and land degradation in Dera District, Ethiopia: GIS and remote sensing-based analysis. Int J Sci Res Environ Sci 2(6):199–208
USDA (1983) Sediment sources, yields, and delivery ratios. National Engineering Handbook, Sec 3: Sedimentation. Washington, D.C.: U.S. Department of Agriculture, Natural Resources Conservation Service
Wallingford HR (2003) Sedimentation in small dams-hydrology and drawdown computation. Rep. ODTN 152. UK. 125
Wallingford HR (2004) Guidelines for predicting and minimizing sedimentation in small dams. Rep. ODTN 152. UK. 125
White R (2001) Evacuation of sediment from reservoirs. Thomas Telford. London, UK
Wischmeier WH, Smith DD (1978) Predicting rainfall erosion losses. Agriculture Handbook No. 537. USDA Science and Education Administration, Washington, DC
Wischmeier WH, Smith DD (1965) Predicting rainfall-erosion losses from cropland east of the Rocky Mountains. Agriculture Handbook No. 282, Washington, DC

Flood Susceptibility Mapping Using GIS and Multi-criteria Decision Analysis in Dibrugarh District of Assam, North-East India

4

Shyam Lochan Bora, Jayanta Das, Kalyan Bhuyan, and Partha Jyoti Hazarika

Abstract

Due to its geographical location, Northeast India (NEI) is prone to water-related disasters. The high precipitation in Assam, an NEI state, results in annual severe floods and delays in the region's overall development. The purpose of this work is to create a flood susceptibility map for the flood-prone district of Dibrugarh in Assam using the analytical hierarchy process (AHP), a multi-criteria decision-making approach. The eight indicators, i.e., elevation, slope, drainage density (DD), distance to river (DR), topographic wetness index (TWI), rainfall intensity (RI), normalized difference vegetation index (NDVI) and stream power index (SPI) were considered as the essential flood conditioning parameters. The multi-collinearity statistics were employed to erase the issues regarding highly correlated parameters. The model's efficiency was judged by applying the ROC-AUC to analyze the better-suited model for mapping the flood susceptibility. The sensitivity results showed that the most effective flood causing factor is the elevation (46.27%) and about 36% of the study area belongs to high and very-high flood-prone areas. The validation of the flood susceptibility model is found good enough with an AUC value of 74%. The findings and conclusions of this study may assist policy-makers in estimating flood hazards and mitigating them in the study area.

Keywords

Flood susceptibility · Geographic information systems (GIS) · Analytical hierarchy process (AHP) · Multi-criteria decision making (MCDM)

4.1 Introduction

Natural disasters are widely regarded as the most serious issue in developed and developing countries, resulting in socio-economic losses (Mitra et al. 2022; Souissi et al. 2020). Floods are one of the most frequent and devasting natural disasters in NEI. The NEI is one of India's wettest regions, and its geographical location makes flooding a possibility (Jain et al. 2013). Climate change

S. L. Bora (✉)
Department of Physics, D.D.R. College, Chabua, Assam 786184, India
e-mail: shyamlochan04@gmail.com

J. Das
Department of Geography, Rampurhat College, Rampurhat, Birbhum 731224, India

S. L. Bora · K. Bhuyan
Department of Physics, Dibrugarh University, Dibrugarh, Assam 786004, India
e-mail: kalyanbhuyan@dibru.ac.in

P. J. Hazarika
Department of Statistics, Dibrugarh University, Dibrugarh, Assam 786004, India

J. Das and S. K. Bhattacharya (eds.), *Monitoring and Managing Multi-hazards*, GIScience and Geo-environmental Modelling, https://doi.org/10.1007/978-3-031-15377-8_4

affects the rate and intensity of precipitation, which is a major contributor to the region's catastrophic floods, impeding agricultural growth, economic development, and industrial development (Praveen et al. 2020). Numerous researchers, including hydrologists, hydrogeologists, and planners, are now interested in assessing and managing floods. The scientists are attempting to develop a reliable and cost-effective method for reducing the risk of flooding, which is critical for ensuring long-term socio-economic development, and mapping flood-prone areas is a crucial component of this effort (Eguaroje et al. 2015; Fenta et al. 2015; Shahabi et al. 2021). Numerous methods have been developed for mapping the susceptibility of natural hazards, including the analytical hierarchy process (AHP) (Choudhury et al. 2022; Ali et al. 2020; Msabi and Makonyo 2021), frequency ratio (FR) models (Shirzadi et al. 2017; Youssef et al. 2016), logistic regression (Mousavi et al. 2011), and machine learning models such as ANN (Pham et al. 2018). Recent years have seen an increase in interest in exploration that incorporates remote sensing (RS) and geographical information systems (GIS) (Mukherjee and Singh 2020). Researchers worldwide employ the AHP method, which is widely regarded as an effective, dependable, and convenient technique (Ali et al. 2020; Mukherjee and Singh 2020). The AHP method weighted and classified the thematic layers in accordance with expert knowledge and site-specific contexts (Basak et al. 2021). The occurrence of flood is associated with numerous factors from physical to anthropogenic. Therefore, it is required to consider the sufficient number of factors including topographic, hydrologic, land cover as well as geological, etc., in order to develop an effective flood susceptibility mapping. The current study focuses on mapping flood-prone areas in Dibrugarh district of Assam integrated with AHP approach and GIS so that it will be helpful to take precautions and minimize flood damage in this region. For this purpose, eight parameters were selected which are associated with the occurrence of floods based on the literature review and similar work conducted earlier in different parts of the world.

The primary aims of the current study are to (a) construct a model based on GIS and AHP techniques for mapping the district's flood-prone zones; (b) assessment of the flood susceptibility model; and (c) identification of the parameters that significantly affect the study area's flood susceptibility. This research will aid planners, regulators, policymakers, and local governments mitigate and managing flood risk in this area through prudent land use and management techniques.

4.2 Methodology and Materials

4.2.1 Study Area

The study area is situated between 27° 5′ 38″ N to 27° 42′ 30″ N latitude and 94° 33′ 46″ E to 95° 29′8″ E longitude covering an area of 3381 km^2 (Fig. 4.1). The Dibrugarh district is located in the northeastern portion of the upper Brahmaputra valley. Dibrugarh district is bounded on the north by Dhemaji and Lakhimpur districts, on the south by Arunachal Pradesh, on the east by Tinsukia, and on the west by Sivasagar district. Dibrugarh's physiography is diverse, encompassing flood plains, swamps and beels, as well as occasional hills. The Brahmaputra River runs along the northern boundary of the Dibrugarh district and is extremely large (10 km) and braided in Dibrugarh compared to other sections of the river. The Dihing River is significant in Dibrugarh because it creates an extraordinary scenery of bamboo orchards, tea gardens, and wet paddy fields. Dibrugarh has a humid subtropical climate, with severely wet summers and comparatively dry winters.

4.2.2 Parameters Used in the Analysis of Flood Susceptibility

Eight parameters (elevation, rainfall intensity, slope, drainage density, distance to river, normalized difference vegetation index, topographic wetness index, and stream power index) were

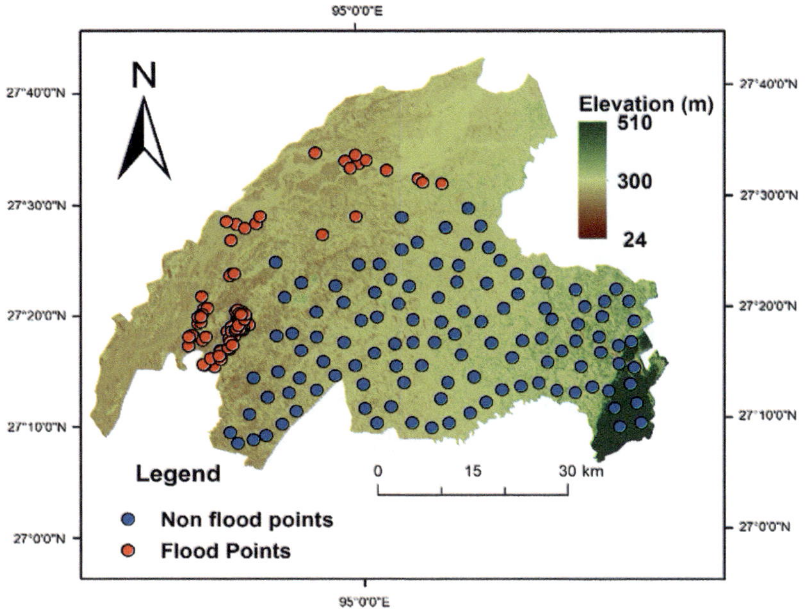

Fig. 4.1 Map of the study area with flood and non-flood points

selected for creating favorable thematic maps using ArcGIS 10.4 based on a review of similar work conducted in other parts of the world (Fig. 4.2).

4.2.2.1 Elevation

Elevations, in general, are a critical factor in determining which areas are at risk of flooding. Flooding is proportional to elevation in an inverse manner (Fernández and Lutz 2010; Shahabi et al. 2021). The higher the elevation, the less likely flooding will occur; conversely, the lower the elevation, the greater the likelihood of flooding. ArcGIS 10.4 was used to create an elevation map of the study area using ASTER DEM. The study area's elevation map is divided into five categories: very low (24–105 m), low (105.01–145 m), medium (145.01–211 m), high (211.01–300 m), and very high (300.01–510 m) (Fig. 4.2a and Table 4.1). As a result, areas with extremely low elevations are extremely vulnerable to flooding.

4.2.2.2 Slope

In flood risk assessment, the slope is very important because the speed of flowing water and the amount of water that can get into the ground is controlled by the area's slope. (Rahmati et al. 2016). The slope map for the study area was created in ArcGIS 10.4 using the ASTER DEM and classified into five categories: very low (0–2°), low (2.01–8°), medium (8.01–15°), high (15.01–30°), and very high (30.01–53.98°) (Fig. 4.2b and Table 4.1). As a result, areas with the lowest slope are disproportionately affected by flooding.

4.2.2.3 Drainage Density (DD)

Drainage density is a significant factor in the occurrence of floods. When drainage density is high, runoff occurs at a significant rate. (Saha et al. 2022). As a result, the likelihood of flooding increases. The DD map of the study area is classified into five categories: very low (0.00–0.28 km/km^2), low (0.29–0.55 km/km^2), medium (0.56–0.83 km/km^2), high (0.84–1.10 km/km^2), and very high (1.11–1.38 km/km^2) (Fig. 4.2c and Table 4.1). As a result, areas with the highest DD are disproportionately affected by flooding.

4.2.2.4 Distance to River (DR)

The distance from the rivers is critical in determining the flood risk zone (Chapi et al. 2017). Flood hazard has a greater impact in areas adjacent to the river network, while its effect diminishes gradually as one moves away from the

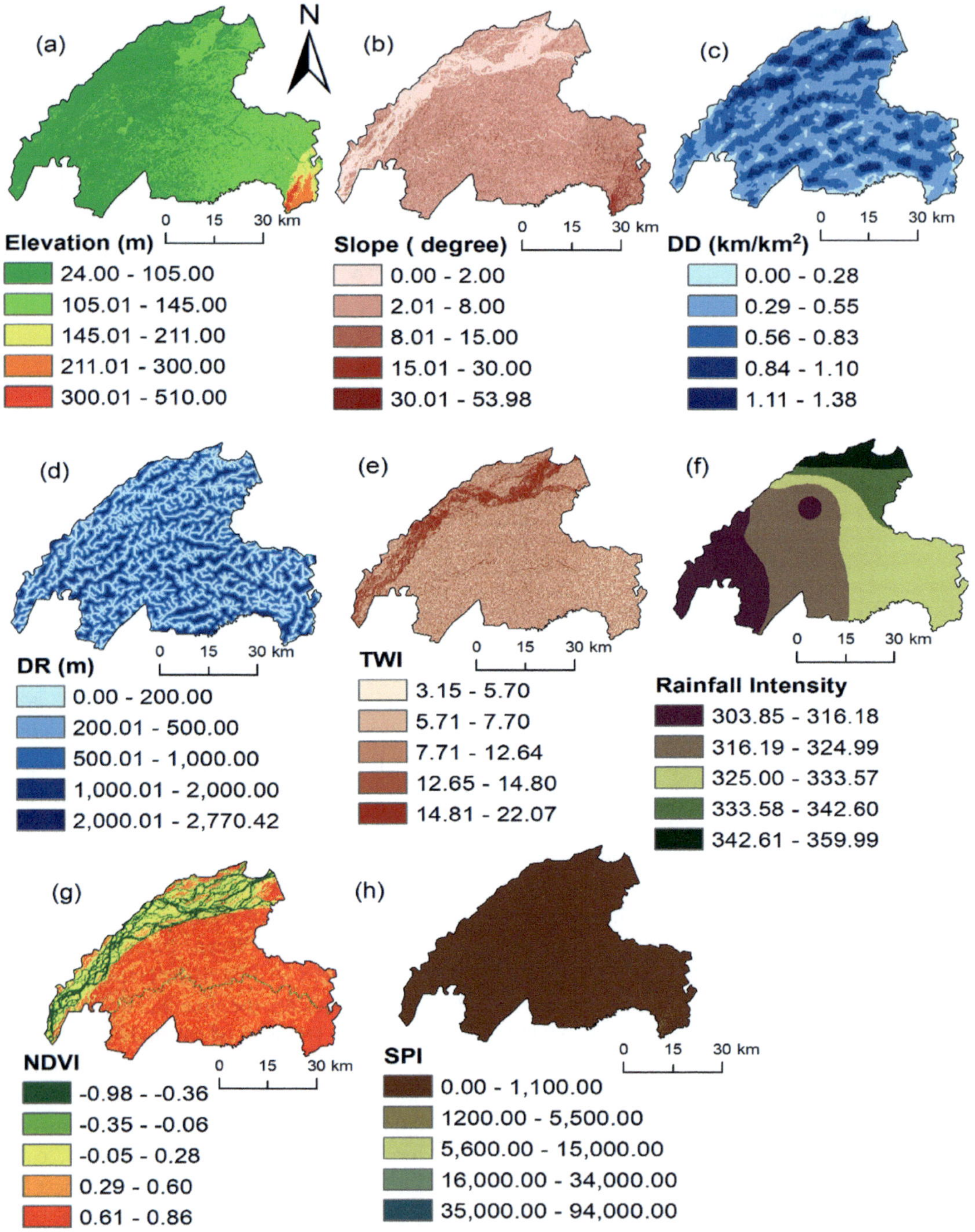

Fig. 4.2 Parameters used for flood susceptibility analysis **a** elevation (m), **b** slope (°), **c** drainage density (DD) (km/km^2), **d** distance to river (DR) (m), **e** topographic wetness index (TWI), **f** rainfall intensity (RI), **g** normalized difference vegetation index (NDVI), and **h** stream power index (SPI)

Table 4.1 Selected parameters for flood susceptibility with their AHP weights and rating of sub-classes of each parameter

Parameters	AHP weight	Class	Class range	Flood level	Area in km^2	Area in %	Rating
Elevation	0.35	1	24–105	Very high	1949.52	57.47	0.44
		2	105.01–145	High	1321.34	38.95	0.26
		3	145.01–211	Medium	63.47	1.87	0.16
		4	211.01–300	Low	48.74	1.44	0.09
		5	300.01–510	Very low	9.44	0.28	0.06
Slope	0.23	1	0–2	Very High	909.40	27.00	0.44
		2	2.01–8	High	1997.26	59.00	0.26
		3	8.01–15	Medium	425.84	13.00	0.15
		4	15.01–30	Low	56.93	2.00	0.09
		5	30.01–53.98	Very low	3.08	0.00	0.05
Drainage density	0.13	1	0.00–0.28	Very low	224.97	6.63	0.06
		2	0.29–0.55	Low	1497.70	44.14	0.10
		3	0.56–0.83	Medium	1284.13	37.85	0.16
		4	0.84–1.10	High	349.81	10.31	0.26
		5	1.11–1.38	Very high	36.08	1.06	0.42
Distance to river	0.12	1	0.00–200.00	Very high	798.40	23.53	0.44
		2	200.01–500.00	High	1019.72	30.06	0.26
		3	500.01–1000.00	Medium	1189.25	35.05	0.15
		4	1000.01–2000.00	Low	382.54	11.28	0.09
		5	2000.01–2770.42	Very low	2.78	0.08	0.05
TWI	0.07	1	3.15–5.70	Very low	460.06	13.56	0.05
		2	5.71–7.70	Low	1031.47	30.40	0.09
		3	7.71–14.80	Medium	1621.64	47.80	0.16
		4	14.81–14.90	High	20.99	0.62	0.27
		5	14.91–22.07	Very high	258.34	7.61	0.43
Rainfall intensity	0.04	1	303.85–316.18	Very low	627.80	18.50	0.05
		2	316.19–324.99	Low	1098.52	32.38	0.09
		3	325.00–333.57	Medium	1203.59	35.48	0.16
		4	333.58–342.60	High	277.32	8.17	0.27
		5	342.61–359.99	Very high	185.42	5.47	0.43
NDVI	0.03	1	−0.98–−0.36	Very high	275.07	8.00	0.42
		2	−0.35–−0.06	High	70.29	2.00	0.26
		3	−0.05–0.28	Medium	472.59	14.00	0.16
		4	0.29–0.60	Low	1093.07	32.00	0.10
		5	0.61–0.86	Very low	1481.67	44.00	0.06
SPI	0.02	1	0.0–1100	Very high	3380.97	99.66	0.54
		2	1200–5500	High	9.88	0.29	0.22
		3	5600–15,000	Medium	1.31	0.04	0.12
		4	16,000–34,000	Low	0.29	0.01	0.08
		5	35,000–94,000	Very low	0.06	0.00	0.05

river. We have used the Euclidean distance method in ArcGIS 10.4 to produce the thematic map of distance to the river. The thematic layer is categorized into five categories: very low (0.00–200.00 m), low (200.01–500.00 m), medium (500.01–1000.00 m), high (1000.01–2000.00 m), and very high (2000.01–2770.42 m) (Fig. 4.2d and Table 4.1). As a result, areas near the main river are more vulnerable to flooding.

4.2.2.5 Topographic Wetness Index (TWI)

The topographic wetness index (TWI), which is calculated by combining the specific area of a basin and the slope angle, is widely utilized to measure the topographic influence on hydrological processes (Shahabi et al. 2021). Channels have the highest values, while ridges have the lowest. The TWI is given by

$$\mathrm{TWI} = \ln\left(\frac{a}{\tan\beta}\right) \tag{4.1}$$

where a denotes the local upslope area draining through a given point per unit contour length and $\tan\beta$ denotes the local slope.

The thematic map is classified into five classes as follows: very low (3.15–5.70), low (5.71–7.70), medium (7.71–14.80), high (14.81–14.90), and very high (14.91–22.07) (Fig. 4.2e and Table 4.1). As a result, areas with a higher TWI are more susceptible to flooding.

4.2.2.6 Rainfall Intensity (RI)

RI is a critical factor because it significantly affects the spatiotemporal flood magnitude. The modified Fourier index (MFI) methodology was used to calculate RI, and the thematic map was created in ArcGIS 10.4 using the inverse distance weighted (IDW) method (Mitra and Das 2022).

$$\mathrm{MFI} = \sum_{i=1}^{12} \frac{\mathrm{R}_i^2}{\mathrm{R}} \tag{4.2}$$

where R_i is the average monthly rainfall for month i (mm) and R is the annual average rainfall (mm). The thematic map is classified into five classes as: very low (303.85–316.18 mm/year), low (316.19–324.99 mm/year), medium (325.00–333.57 mm/year), high (333.58–342.60 mm/year), and very high (342.61–359.99 mm/year) (Fig. 4.2f and Table 4.1). Accordingly, higher rainfall intensity means high susceptibility to flood.

4.2.2.7 Normalized Difference Vegetation Index (NDVI)

The NDVI is a statistical indicator used to determine the amount of green on the land surface and the presence or absence of bodies of water. Changes in vegetation and surface water cover are indicated by the changes in NDVI over time and can provide insight into the relationship between flooding and vegetation in a region (Shahabi et al. 2021; Tehrany et al. 2013). An increment of the density of vegetation cover is believed to decrease the likelihood of flooding in the region. The thematic layer (map) was created using Landsat 8 satellite data in ArcGIS 10.4 (Fig. 4.2g and Table 4.1).

4.2.2.8 Stream Power Index (SPI)

The SPI quantifies the erosive power of a discharge in relation to a watershed's specific area (Cao et al. 2016). It measures the water flow's force at a particular location in a watershed. The greater the SPI value, the more powerful the flow. SPI can be calculated in the following manner:

$$\mathrm{SPI} = A_S \tan\alpha \tag{4.3}$$

where α and A_s represent slope angle in degree as well as specific catchment area (m/m^2) respectively. This study classified SPI into five categories using ArcGIS 10.4's natural breaks classification method. (Fig. 4.2h and Table 4.1).

4.2.3 Multicollinearity Analysis

Multicollinearity occurs due to the correlation among the independent variables in a regression model. This correlation results in inaccuracies in

Table 4.2 Saaty's fundamental 9-point intensity scale for measuring relative importance between two parameters

Relative importance scale		
Intensity of relative importance	Definition	Explanation
1	Equal importance	Two activities contribute equally to 1
3	Moderate importance	Experience and judge slightly favor one activity over another activity
5	Strong or essential importance	Experience and judge strongly favor one activity over another activity
7	Very strong importance	An activity is strongly favored, and its dominance is exhibited in practice
9	Extreme importance	The evidence of favoring one activity over another is of highest possible order of affirmation
2, 4, 6, 8	Intermediate	Applied when comparison is required
Reciprocals	Inverse comparison	Applied in inverse comparison

the model's output. As a result, prior to constructing a regression model, it is required to check for and eliminate multicollinearity among the input parameters. (Mukherjee and Singh 2020).

Using the 'Create Random Points' tool in ArcGIS 10.4, we have selected 1000 random points (N = 1000) from the research region, and from each thematic layer, we have collected data for those random points using the 'Extract Data' tool in ArcGIS 10.4. Using SPSS software, the multicollinearity test was performed, considering elevation as the dependent parameter and the remaining factors as independent parameters.

4.2.4 Assignment of Weight and Rank to Each Thematic Layer

The AHP is a MCDM (multi-criteria decision-making technique) for hazard susceptibility analysis and is employed in this study (Saaty 1987). Because the produced thematic layers are thought to be key elements impacting the occurrence of floods in the research region, this method was utilized to integrate them. Each layer, however, had a different effect. As a result, each thematic layer was assigned a relative weight of 1–9 using the scale given by Saaty (Table 4.2). The evaluation of the importance of one thematic layer relative to another is based on expert knowledge that forms the basis of AHP (Das et al. 2017). Therefore, as a consequence of earlier research and field experience in various geographical regions, proportionate weights were assigned (Mu and Pereyra-Rojas 2018).

To evaluate the weight of each layer, a pairwise comparison matrix is constructed, as shown in Table 4.3. Each sub-class of each thematic layer was given a relative rank from 0–9 (Table 4.2) on the basis of their relative impact on flooding, and the pairwise comparison matrices for the sub-classes within each parameter are shown in Table 4.5. Table 4.4 shows the normalized vector for mapping flood susceptibility.

4.2.5 Consistency Analysis

The consistency ratio (CR) is used to validate the judgment which can be obtained from the pairwise comparison matrix between the thematic layers (Saaty 1990). The CR is calculated using

$$\text{Consistency Ratio (CR)} = \frac{\text{Consistency Index (CI)}}{\text{Random Consistency Index (RCI)}} \quad (4.4)$$

Table 4.3 Pairwise comparison matrix of the parameters for flood susceptibility mapping

Parameters	E	S	DD	DR	TWI	RI	NDVI	SPI
E	1	3	4	4	5	7	8	9
S	1/3	1	3	4	4	5	7	8
DD	¼	1/3	1	1	3	4	5	6
DR	¼	1/4	1	1	3	4	5	6
TWI	1/5	1/4	1/3	1/3	1	2	3	4
RI	1/7	1/5	1/4	1/4	1/2	1	2	3
NDVI	1/8	1/7	1/5	1/5	1/3	1/2	1	2
SPI	1/9	1/8	1/6	1/6	1/4	1/3	0.5	1

E elevation, *S* slope, *DD*: drainage density, *DR* distance to river, *TWI*: topographic wetness index, *RI* rainfall intensity, *NDVI*: normalized difference vegetation index, and *SPI*: stream power index

Table 4.4 Normalized vector for flood susceptibility

Parameters	E	S	DD	DR	TWI	RI	NDVI	SPI
E	0.41	0.57	0.40	0.37	0.29	0.29	0.25	0.23
S	0.14	0.19	0.30	0.37	0.23	0.21	0.22	0.21
DD	0.10	0.06	0.10	0.09	0.18	0.17	0.16	0.15
DR	0.10	0.05	0.10	0.09	0.18	0.17	0.16	0.15
TWI	0.08	0.05	0.03	0.03	0.06	0.08	0.10	0.10
RI	0.06	0.04	0.03	0.02	0.03	0.04	0.06	0.08
NDVI	0.05	0.03	0.02	0.02	0.02	0.02	0.03	0.05
SPI	0.05	0.02	0.02	0.02	0.01	0.01	0.02	0.03
Sum	1.00	1.00	1.00	1.00	1.00	1.00	1.00	1.00

As illustrated in Table 4.6, we have used Saaty's standard table for the calculation of RCI, and the values of consistency index (CI) were calculated using the following equation:

$$\mathrm{CI} = \frac{(\lambda_{\max} - N)}{N - 1} \quad (4.5)$$

where the principal eigenvalue of the pairwise comparison matrix is denoted by $\lambda_{\max}$. A CR of ≤ 0.1 implies that the model of AHP is acceptable. If, however, a CR greater than 0.1 is obtained, it is necessary to perform the model again so that a CR less than 0.1 is obtained Saaty (1990). The CR is in this study is found to be 0.05, which shows good consistency (Table 4.7).

4.2.6 Weighted Overlay Analysis

After weighting and rating each thematic layer and each sub-parameters of thematic layers (Tables 4.1 and 4.5), the flood-prone zones can be determined by overlapping all thematic layers in using the equation

$$\mathrm{FSZ} = \sum_{i=1}^{n} (W_i \times R_i) \quad (4.6)$$

where FSZ denotes the flood susceptible zones, W_i denotes the AHP weight assigned to each thematic layer, and R_i denotes the rating assigned to each thematic layer's sub-classes (Ali et al.

Table 4.5 AHP rating of sub-criteria for flood susceptibility analysis

Elevation									
Class	VH	H	M	L	VL	Rating	Lamda max	CI	CR
24–105	1	2	3	5	6	0.44	5.05	0.01	0.01
105.01–145		1	2	3	4	0.26			
145.01–211			1	2	3	0.16			
211.01–300				1	2	0.09			
300.01–510					1	0.06			
Slope									
Class	VH	H	M	L	VL	Rating	Lamda max	CI	CR
0–2	1	2	3	5	7	0.44	5.03	0.01	0.01
2.01–8		1	2	3	5	0.26			
8.01–15			1	2	3	0.15			
15.01–30				1	2	0.09			
30.01–53.98					1	0.05			
Drainage density (DD)									
Class	VH	H	M	L	VL	Rating	Lamda max	CI	CR
0.00–0.28	1	1/2	1/3	1/4	1/5	0.06	5.07	0.02	0.02
0.29–0.55		1	1/2	1/3	1/4	0.10			
0.56–0.83			1	1/2	1/3	0.16			
0.84–1.10				1	1/2	0.26			
1.11–1.38					1	0.42			
Distance to river (DR)									
Class	VH	H	M	L	VL	Rating	Lamda max	CI	CR
0.00–200.00	1	2	3	5	7	0.44	5.03	0.01	0.01
200.01–500.00		1	2	3	5	0.26			
500.01–1000.00			1	2	3	0.15			
1000.01–2000.00				1	2	0.09			
2000.01–2770.42					1	0.05			
TWI									
Class	VH	H	M	L	VL	Rating	Lamda max	CI	CR
3.15–5.70	1	1/2	1/3	1/5	1/7	0.05	5.03	0.01	0.01
5.71–7.70		1	1/2	1/3	1/4	0.09			
7.71–14.80			1	1/2	1/3	0.16			
14.81–14.90				1	1/2	0.27			
14.91–22.07					1	0.43			
Rainfall intensity									
Class	VH	H	M	L	VL	Rating	Lamda max	CI	CR
303.85–316.18	1	1/2	1/3	1/5	1/7	0.05	5.03	0.01	0.01
316.19–324.99		1	1/2	1/3	1/4	0.09			
325.00–333.57			1	1/2	1/3	0.16			
333.58–342.60				1	1/2	0.27			
342.61–359.99					1	0.43			

(continued)

Table 4.5 (continued)

Elevation									
Class	VH	H	M	L	VL	Rating	Lamda max	CI	CR
NDVI									
Class	VH	H	M	L	VL	Rating	Lamda max	CI	CR
−0.98–−0.36	1	2	3	4	5	0.42	5.07	0.02	0.02
−0.35–−0.06		1	2	3	4	0.26			
−0.05–0.28			1	2	3	0.16			
0.29–0.60				1	2	0.10			
0.61–0.86					1	0.06			
SPI									
Class	VH	H	M	L	VL	Rating	Lamda max	CI	CR
0.0–1100	1	3	5	7	9	0.54	5.05	0.01	0.01
1200–5500		1	2	3	5	0.22			
5600–15,000			1	2	3	0.12			
16,000–34,000				1	2	0.08			
35,000–94,000					1	0.05			

VH very high, *H* high, *M* medium, *L* low, and *VL* very low

Table 4.6 Random consistency index (RCI) to check consistency ratio for different matrices

1	2	3	4	5	6	7	8	9	10	11	12
0.00	0.00	0.58	0.90	1.12	1.24	1.32	1.41	1.45	1.49	1.51	1.48

Table 4.7 Consistency check of aggregated used for flood hazard zonation

Lambda max	N	CI	CR
8.45	8	0.06	0.05

2020). According to their flood susceptibility, the zones were categorized as very low, low, moderate, high, and very high. This is performed using the weighted overlay method in ArcGIS 10.4.

4.2.7 Sensitivity Analysis

There were two methods used in this study: SPSA (single parameter sensitivity analysis) and map removal to deduce the most influential thematic layers in the obtained FSZs and the influence of each class's rank and weights. We conducted these analyses in ArcGIS 10.4.

4.2.7.1 Analysis of Single Parameter Sensitivity

In this technique, the weighting factor assigned to every thematic layers is compared to the weighting (empirical) given to the identical layer to produce the FSZ map to ascertain the effect of each of the layers (thematic) on the flood susceptibility zonation map (Fenta et al. 2015). The effective weight (W) for all the thematic layer was calculated in this study using

$$W = \frac{P_r P_w}{\mathrm{FSZ}} \times 100 \qquad (4.7)$$

where FSZ is the flood susceptible zones calculated using all the thematic layers. P_r is the rating and P_w is the weight of each thematic layer.

4.2.7.2 Map Removal Sensitivity Analysis (MRSA)

The MRSA examines the thematic layers which are used in the FSZ map computation are removed. A new FSZ map is produced if we remove one layer (thematic) at a time. The sensitivity index (SI) when a layer is excluded is evaluated using Eq. (4.8) (Fenta et al. 2015; Mukherjee and Singh 2020).

$$\mathrm{SI} = \frac{\left|\left(\frac{\mathrm{FSZ}}{N}\right) - \left(\frac{\mathrm{FSZ'}}{n}\right)\right|}{\mathrm{FSZ}} \times 100 \qquad (4.8)$$

where SI: sensitivity index, N: number of layers (thematic) which produce the map (FSZ), FSZ: flood susceptible zone, FSZ′: flood susceptible zone obtained by removing a thematic layer, n: number of layers (thematic) which produce the FSZ′ map.

4.3 Results and Discussion

4.3.1 Analysis of Multicollinearity

Table 4.8 summarizes the results of the multicollinearity analysis. As a result of the findings, there is no evidence of multicollinearity among the eight input thematic layers (VIF $\leq$ 10 and tolerance $\geq$ 0.1). As a result, all parameters are used in the study.

4.3.1.1 Flood Susceptibility Zonation and Validation

The flood susceptibility map is created using the weighted overlay method in ArcGIS 10.4 by integrating all of the input thematic layers (Fig. 4.3). Using the natural break method in ArcGIS 10.4, the flood-prone zones are classified as: very low, low, medium, high, and very high. The results indicate that Dibrugarh district's 149.81 km^2 (4.43%) is extremely vulnerable to flooding, while 434.56 km^2 (12.86%) is extremely vulnerable to flooding. Similarly, 907.42 km^2 (26.86%), 1072.48 km^2 (31.74%), and 814.50 km^2 (24.11%) of the Dibrugarh district are susceptible to flooding on a low, medium, or high level. The study's findings are validated by comparing them to a flood inventory map created using ISRO's Bhuvan portal data. Receiver operating characteristic curve (ROC) is used for validation purposes (Fig. 4.4). The area under the curve (AUC) is a metric for evaluating a parameter's ability to discriminate between two diagnostic groups (here, flood points and non-flood points). The AUC values imply the prediction accuracy as low: 0.50–0.60, average: 0.60–0.70, good: 0.70–0.80, very good: 0.80–0.90, and excellent: 0.90–1.00 (Andualem and Demeke 2019; Mukherjee and Singh 2020). The result obtained from validation show that the AUC of the FSZ map is 0.740, which indicates that the model has a high degree of accuracy.

4.3.2 Sensitivity Analysis of FSZ Map

Tables 4.9 and 4.10 contain descriptive statistics for single parameter sensitivity analysis and map removal sensitivity analysis, respectively. The

Table 4.8 Collinearity statistics of flood susceptibility parameters

SP	S	DD	DR	TWI	RI	NDVI	SPI
Tolerance	0.75	0.85	0.90	0.71	0.99	0.77	0.81
VIF	1.33	1.17	1.11	1.40	1.01	1.30	1.23

SP susceptibility parameters, *S* slope, *DD* drainage density, *DR* distance to river, *TWI*: topographic wetness index, *RI* rainfall intensity, *NDVI*: normalized difference vegetation index, and *SPI* stream power index

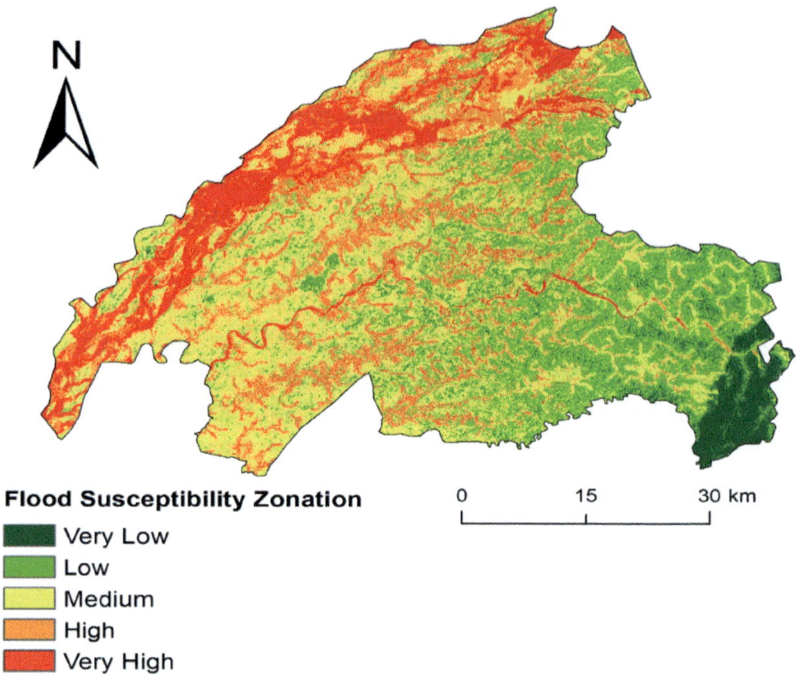

Fig. 4.3 Flood susceptibility zonation (FSZ) map of Dibrugarh

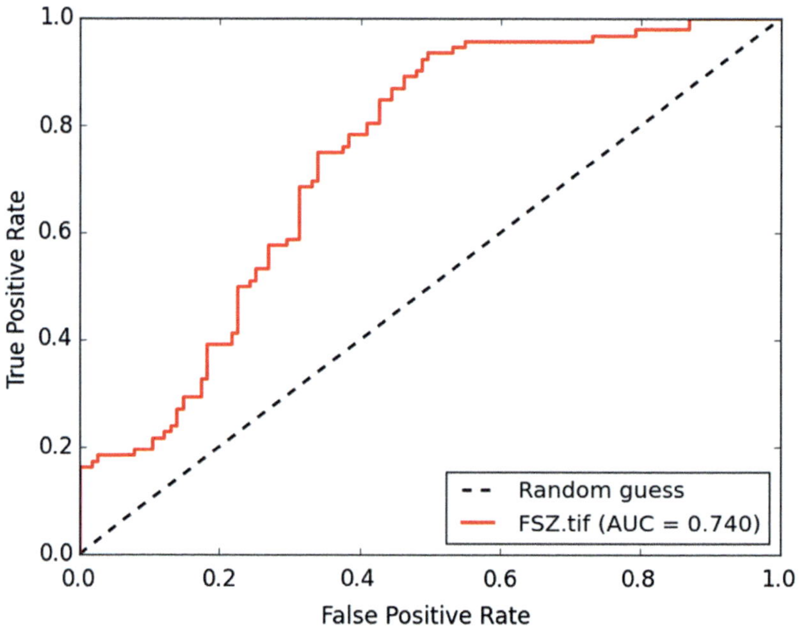

Fig. 4.4 Model validation by ROC curve

results of the single parameter sensitivity analysis indicate differences in the effective weights of the parameters used for flood susceptibility mapping compared to the empirical weights. Elevation and slope are the most influential factors in mapping flood susceptibility in the study region, followed by distance to the river. In comparison with their empirical weights, drainage density, TWI, rainfall intensity, and NDVI are typically less effective influential factors.

Table 4.9 Descriptive statistics of sensitivity analysis of single parameter

Thematic layers	Empirical weight (%)	Effective weight (%)			
		Min	Max	Mean	SD
Elevation	35	10.38	72.85	46.27	8.42
Slope	23	3.99	60.93	24.54	6.04
Drainage density	13	2.02	25.07	6.86	2.67
Distance to river	12	1.88	41.87	10.90	5.10
TWI	7	0.88	30.22	3.74	1.99
Rainfall intensity	4	0.50	10.00	2.21	1.46
NDVI	3	0.44	8.17	1.29	0.88
SPI	2	0.28	14.67	4.19	1.03

Table 4.10 Descriptive statistics of map removal sensitivity analysis

Thematic layer	SI variation in %			
	Minimum	Maximum	Mean	SD
Elevation	0.00	8.01	1.56	1.15
Slope	0.00	5.91	1.71	0.99
Drainage density	0.00	3.98	2.92	0.41
Distance from river	0.00	3.75	2.23	0.80
TWI	0.00	3.11	2.52	0.32
RI	1.02	2.77	2.34	0.26
NDVI	0.99	2.59	2.24	0.18
SPI	0.01	2.54	1.83	0.20

Table 4.11 Percentage of changes of flood susceptibility mapping with the removal of each thematic layer

Thematic layers excluded	FSZ (%)				
	Very low	Low	Medium	High	Very high
Elevation	197.69	55.37	−27.88	−43.96	−32.53
Slope	−40.23	5.28	−24.59	5.59	53.08
Drainage density	10.85	−3.57	−5.91	2.84	12.56
Distance from river	−36.74	−40.87	−36.49	70.82	55.35
TWI	−14.31	−3.12	−14.89	12.00	25.70
RI	2.55	0.58	−6.28	3.42	9.08
NDVI	−1.34	−1.73	−8.08	4.90	14.94
SPI	1.32	−2.07	−5.73	4.74	9.14

The results of the map removal analysis indicate that the FSZ map is sensitive to all of the parameters used in this study, with the greatest variation in the sensitivity index occurring when drainage density is removed and the least variation occurring when elevation is removed. Additionally, as shown in Table 4.11, removing the elevation layer reduces the very highly susceptible areas by −32.53%, while removing all other parameters increases them significantly. Similarly, removing each layer results in significant variation in very low, low, medium, and high susceptible areas, demonstrating that the FSZ is map is dependent on all the factors used in this study.

4.4 Conclusions

The AHP method is used in this study to create a flood susceptibility zonation map for the Dibrugarh district, which is based on expert opinion and prior knowledge of parameters that influence flooding. The ROC curve, which shows good accuracy of 0.74, validates the study's findings. Furthermore, the sensitivity analysis revealed that the flood susceptibility map is sensitive to all of the parameters used in this study, indicating that all of the factors taken into account influence the flooding situation in the study region. Because this type of analysis is uncommon in this region and is a flood-prone area, this study will benefit policymakers, planners, and administrative bodies responsible for planning flood response and emergency services.

References

Ali SA, Parvin F, Pham QB, Vojtek M, Vojteková J, Costache R, Linh NTT, Nguyen HQ, Ahmad A, Ghorbani MA (2020) GIS-based comparative assessment of flood susceptibility mapping using hybrid multi-criteria decision-making approach, Naïve Bayes tree, bivariate statistics and logistic regression: a case of Topľa basin, Slovakia. Ecol Ind 117(June):106620. https://doi.org/10.1016/j.ecolind.2020.106620

Andualem TG, Demeke GG (2019) Groundwater potential assessment using GIS and remote sensing: a case study of Guna tana landscape, upper blue Nile Basin, Ethiopia. J Hydrol Reg Stud 24. https://doi.org/10.1016/j.ejrh.2019.100610

Basak A, Das J, Rahman ATM, Pham QB (2021) An integrated approach for delineating and characterizing groundwater depletion hotspots in a coastal state of India. J Geol Soc India 97(11):1429–1440. https://doi.org/10.1007/s12594-021-1883-z

Cao C, Xu P, Wang Y, Chen J, Zheng L, Niu C (2016) Flash flood hazard susceptibility mapping using frequency ratio and statistical index methods in coalmine subsidence areas. Sustainability (Switzerland) 8(9). https://doi.org/10.3390/su8090948

Chapi K, Singh VP, Shirzadi A, Shahabi H, Bui DT, Pham BT, Khosravi K (2017) A novel hybrid artificial intelligence approach for flood susceptibility assessment. Environ Model Softw 95. https://doi.org/10.1016/j.envsoft.2017.06.012

Choudhury S, Basak A, Biswas S, Das J (2022) Flash flood susceptibility mapping using GIS-based AHP method. In: Pradhan B, Shit PK, Bhunia GS, Adhikary PP, Pourghasemi HR (eds) Spatial modelling of flood risk and flood hazards. GIScience and geo-environmental modelling. Springer, Cham. https://doi.org/10.1007/978-3-030-94544-2_8

Das J, Gayen A, Saha S, Bhattacharya SK (2017) Modelling of alternative crops suitability to tobacco based on analytical hierarchy process in Dinhata subdivision of Koch Bihar district, West Bengal. Model Earth Syst Environ 3(4):1571–1587. https://doi.org/10.1007/s40808-017-0392-y

Eguaroje O, Alaga T, Ogbole J, Omolere S, Alwadood J, Kolawole I, Muibi KH, Nnaemeka D, Popoola DS, Samson SA, Adewoyin JE (2015) Flood vulnerability assessment of Ibadan city, Oyo state, Nigeria. World Environ 5(4)

Fenta AA, Kifle A, Gebreyohannes T, Hailu G (2015) Spatial analysis of groundwater potential using remote sensing and GIS-based multi-criteria evaluation in Raya Valley, northern Ethiopia. Hydrogeol J 23 (1):195–206. https://doi.org/10.1007/s10040-014-1198-x

Fernández DS, Lutz MA (2010) Urban flood hazard zoning in Tucumán Province, Argentina, using GIS and multicriteria decision analysis. Eng Geol 111(1–4). https://doi.org/10.1016/j.enggeo.2009.12.006

Jain SK, Kumar V, Saharia M (2013) Analysis of rainfall and temperature trends in northeast India. Int J Climatol 33(4). https://doi.org/10.1002/joc.3483

Mitra R, Das J (2022) A comparative assessment of flood susceptibility modelling of GIS-based TOPSIS VIKOR and EDAS techniques in the Sub-Himalayan foothills region of Eastern India. Environ Sci Pollut Res. https://doi.org/10.1007/s11356-022-23168-5

Mitra R, Saha P, Das J (2022) Assessment of the performance of GIS-based analytical hierarchical process (AHP) approach for flood modelling in Uttar Dinajpur district of West Bengal, India. GeomatS, Nat Hazards Risk 13(1):2183–2226

Mousavi SZ, Kavian A, Soleimani K, Mousavi SR, Shirzadi A (2011) GIS-based spatial prediction of landslide susceptibility using logistic regression model. Geomatics Nat Hazards Risk 2(1). https://doi.org/10.1080/19475705.2010.532975

Msabi MM, Makonyo M (2021) Flood susceptibility mapping using GIS and multi-criteria decision analysis: a case of Dodoma region, central Tanzania. Remote Sens Appl Soc Environ 21:100445. https://doi.org/10.1016/j.rsase.2020.100445

Mu E, Pereyra-Rojas M (2018) Practical decision making using super decisions. Springer

Mukherjee I, Singh UK (2020) Delineation of groundwater potential zones in a drought-prone semi-arid region of east India using GIS and analytical hierarchical process techniques. CATENA 194

(May):104681. https://doi.org/10.1016/j.catena.2020.104681

Pham BT, Shirzadi A, Tien Bui D, Prakash I, Dholakia MB (2018) A hybrid machine learning ensemble approach based on a radial basis function neural network and rotation forest for landslide susceptibility modeling: a case study in the Himalayan area, India. Int J Sedim Res 33(2). https://doi.org/10.1016/j.ijsrc.2017.09.008

Praveen B, Talukdar S, Shahfahad, Mahato S, Mondal J, Sharma P, Islam ARMT, Rahman A (2020) Analyzing trend and forecasting of rainfall changes in India using non-parametrical and machine learning approaches. Sci Rep 10(1). https://doi.org/10.1038/s41598-020-67228-7

Rahmati O, Zeinivand H, Besharat M (2016) Flood hazard zoning in Yasooj region, Iran, using GIS and multi-criteria decision analysis. Geomatics Nat Hazards Risk 7(3). https://doi.org/10.1080/19475705.2015.1045043

Saaty RW (1987). The analytic hierarchy process-what it is and how it is used. Math Model 9(3–5). https://doi.org/10.1016/0270-0255(87)90473-8

Saaty TL (1990) How to make a decision: the analytic hierarchy process. European J Oper Res 48(1). https://doi.org/10.1016/0377-2217(90)90057-I

Saha S, Das J, Mandal T (2022) Investigation of the watershed hydro-morphologic characteristics through the morphometric analysis: a study on Rayeng basin in Darjeeling Himalaya. Environ Challenges 100463. https://doi.org/10.1016/j.envc.2022.100463

Shahabi H, Shirzadi A, Ronoud S, Asadi S, Pham BT, Mansouripour F, Geertsema M, Clague JJ, Bui DT (2021) Flash flood susceptibility mapping using a novel deep learning model based on deep belief network, back propagation and genetic algorithm. Geosci Front 12(3):101100. https://doi.org/10.1016/j.gsf.2020.10.007

Shirzadi A, Chapi K, Shahabi H, Solaimani K, Kavian A, Ahmad BB (2017). Rock fall susceptibility assessment along a mountainous road: an evaluation of bivariate statistic, analytical hierarchy process and frequency ratio. Environ Earth Sci 76(4). https://doi.org/10.1007/s12665-017-6471-6

Souissi D, Zouhri L, Hammami S, Msaddek MH, Zghibi A, Dlala M (2020) GIS-based MCDM–AHP modeling for flood susceptibility mapping of arid areas, southeastern Tunisia. Geocarto Int 35(9):991–1017. https://doi.org/10.1080/10106049.2019.1566405

Tehrany MS, Pradhan B, Jebur MN (2013) Spatial prediction of flood susceptible areas using rule based decision tree (DT) and a novel ensemble bivariate and multivariate statistical models in GIS. J Hydrol 504. https://doi.org/10.1016/j.jhydrol.2013.09.034

Youssef AM, Pradhan B, Sefry SA (2016) Flash flood susceptibility assessment in Jeddah city (Kingdom of Saudi Arabia) using bivariate and multivariate statistical models. Environ Earth Sci 75(1). https://doi.org/10.1007/s12665-015-4830-8

Effects of Climatic Hazards on Agriculture in the Teesta Basin of Bangladesh

5

Md. Abdullah Al Mamun,
A. T. M. Sakiur Rahman,
Most. Shayda Shamsea Aziz Shabee,
Jayanta Das, G. M. Monirul Alam,
M. Mizanur Rahman, and Md. Kamruzzaman

Abstract

This study presents the damages to the agricultural sector due to various climatic hazards at the Teesta basin of Bangladesh. Survey data were collected from 276 farmers from Lalmonirhat, Rangpur and Nilphamari districts. Remote sensing maps have also been used to analyze changes in land use and river channel. The results reveal that flood is the most catastrophic natural disaster in the study area. Inundation of *Aman* rice at the sowing stage is the main problem which causes huge losses of the farmers. Hailstorms affect jute cultivation in Kursamari and Azam khan villages. Cold waves caused serious damage to *Rabi* crops. Diseases associated with cold and fog also lessen farmer's winter crops. The finding of this study will support the government to design future strategies to reduce the damages as well as prepare the people to adapt to the adverse climatic conditions in the Teesta basin.

Keywords

Climatic hazards · Agriculture · Mainland · *Char* land · Teesta basin · Bangladesh

5.1 Introduction

Climate change has been a steady process on earth, but it has ended up the foremost characterizing concern in the present world as it is changing the earth's environment rapidly. Extreme climatic events (e.g., floods, storms, droughts and heat waves) have accounted for the increase in worldwide natural catastrophes (Basak et al. 2022; Thomas et al. 2014). Moreover, people are persistently living and building infrastructure within the zones that are vulnerable to floods, storms and severe heat or cold

Md. A. Al Mamun · Md. Kamruzzaman (✉)
Institute of Bangladesh Studies, University of Rajshahi, Rajshahi, Bangladesh
e-mail: mkzaman@ru.ac.bd

A. T. M. S. Rahman
Hydrology Lab, Department of Earth and Environmental Science, Kumamoto University, Kumamoto, Japan

Most. S. S. A. Shabee
3-1-56-510, Minatojima Nakamachi, Kobe City, Hyogo 650-0046, Japan

J. Das
Department of Geography, Rampurhat College, Birbhum, West Bengal 731224, India

G. M. M. Alam
Department of Agribusiness, Bangabandhu Sheikh Mujibur Rahman Agricultural University, Gazipur, Bangladesh

M. M. Rahman
Department of Geography and Environmental Studies, University of Rajshahi, Rajshahi, Bangladesh

J. Das and S. K. Bhattacharya (eds.), *Monitoring and Managing Multi-hazards*,
GIScience and Geo-environmental Modelling, https://doi.org/10.1007/978-3-031-15377-8_5

(Easterling et al. 2000a, b). Climate change is mostly associated with anthropogenic emissions of greenhouse gases (GHGs), i.e., carbon dioxide (CO_2), methane, nitrous oxide and chlorofluorocarbons, which is rushing and is a cause for international concern (Young et al. 2010). The concentration of CO_2 in the atmosphere has increased by 30% since pre-industrial time. CO_2 is still mounting at a rate of 0.4% per year which is very frightening (Lal 2003). Atmospheric GHGs have persistent to rise, with an escalating trend, throughout the twentieth century (Sanderson et al. 2011), and it will direct to changes in climate over the twenty-first century (Arnell and Lloyd–Hughes 2014). Deforestation has lessened the worldwide storage of carbon in the soil. It has also declined the capacity to bind CO_2 that has released more CO_2 into the atmosphere. Thus, the concentrated GHGs increase the global temperature by absorbing radiant energy (Aydinalp and Cresser 2008). As a result, the worldwide average surface temperature has risen by about 0.6 °C since the late nineteenth century (Easterling et al. 2000a, b). Temperature is estimated to increase by 1.4–5.8 °C from 1990 to 2100 globally (Carter and Rovere 2011). As a result, most glaciers of the world retreated from 1961 to 1990. Though precipitation will increase by 10% per degree of global warming, it will not prevent rapid ice-melting (Oerlemans et al. 1998). Global ocean heat content has already risen considerably since the late 1950s. As a result, the ice sheets in the Arctic Ocean have declined by 10% in the area during the twentieth century (Lal 2003). The mean sea level is likely to rise by 16 cm in 2050, especially in Asia and Africa (Tol and Wagner 2010; Arnell et al. 2016).

Anthropogenic worldwide warming will lead to an increase in the intensity and frequency of heavy precipitation (Dankers and Feyen 2009). Precipitation in tropics, mid-latitudes and high latitudes are likely to increase over the twenty-first century. However, precipitation in subtropical latitudes will decrease (Weltzin et al. 2003; Sanderson et al. 2011). The extended rainfall is the main reason for river floods in large river basins (Dankers and Feyen 2009; Kundzewicz et al. 2014). For example, Elbe and Danube river basins suffered from severe floods in 2006 (Dankers and Feyen 2009). The possible increase in river floods is likely to be the most severe impact of future climate change as around 450 million people are likely to be exposed to river flooding in the world (Arnell and Gosling 2016; Arnell et al. 2016). Changes in rainfall patterns in the monsoon have turned into prolonged rainfall in South Asia (Dewan 2014). Generally, intensive and prolonged rainfall creates flash floods (Choudhury et al. 2022). Frequent floods inundate the coastal areas of Asian countries during the monsoon period due to the geographical location (Dutta 2011).

An increase in atmospheric CO_2 levels (Parry et al. 2004) and changes in temperature and precipitation will create an influence on global agriculture (Porter and Semenov 2005), especially in the tropical regions (Aydinalp and Cresser 2008). Within the HadCM2 and HadCM3 climate change scenarios, the expanded concentration of atmospheric CO_2 that results in higher temperatures that are likely to reduce grain production all over the world except in a few parts of Canada and Russia, where the growing season is generally cold (Parry et al. 1999). Generally, according to the Special Report on Emissions Scenarios (SRES) of the Intergovernmental Panel on Climate Change (IPCC), crop yield decreases in developing countries, and it increases in developed countries (Parry et al. 2004). Higher temperatures have the potential to reduce grain yield (Moore et al. 2012; Nuttall et al. 2018). The repeated days with high temperatures refer to heat waves (Vautard et al. 2013). It decreased the number of grains per wheat plant by 60% over the last 30 years in Denmark (Porter and Semenov 2005). It also reduced wheat and maize yields, respectively, by 1.23 and 1.55% from 1980–2008 in China (Tao et al. 2012). In Africa, it is likely to decline in maize and wheat yield too (Lobell et al. 2008). More than 3 mm daily rainfall and around 30 °C temperature increased the cloud cover, and it reduced the solar radiation as well as depleted soil moisture and nutrients in the southeastern zones of the United States. It reduced the yield of cotton too (Liang et al. 2012). Rainfall variability also played a vital role in crop production. Reduced rainfall declined the maize yield considerably in central Uganda (Moore et al. 2012). However, crop yield not only depends on climatic factors but also depends on education, technology and agriculture policies (Thirtle et al. 2003; Brown and Funk 2008).

The antagonistic impact of climate change will affect a large proportion of the world's population, particularly in South Asia (Haque et al. 2019). Generally, temperature and the frequency of extreme weather events increased over the last twentieth century in South Asia. Moreover, the temperature is likely to increase significantly in this area, especially in the Himalayan highlands in the twenty-first century. Temperature showed a rising trend (0.02–0.042 °C per year) in the southern, southeastern, northeastern and northwestern parts of Bangladesh from 1971 to 2010. The temperatures in Cox's Bazar, Sylhet and Dinajpur districts showed maximum positive trends. In addition, using the autoregressive integrated moving average (ARIMA) time series model, the mean temperature is likely to increase by 0.98 °C by 2020 compared to that of 1971 in Bangladesh (Rahman and Lateh 2017). Rainfall has increased in the Ganges basin in Bangladesh. It is favorable to rain-fed agriculture but constrained to a specified level. On the contrary, both increasing and decreasing rainfall trend were found over Bangladesh (Das et al. 2021a). It causes longer inundation in the *Haor* areas of the country (Mirza et al. 1998).

Bangladesh is regularly alluded to as one of the most vulnerable countries to extreme climatic events, i.e., droughts, floods, riverbank erosion, sea-level rise, and salinity entrance (Paudyal 2002; Etzold et al. 2014; Alam 2016). Moreover, the geographical location and low elevation of coastal areas have made Bangladesh exposed to intense tropical cyclones and severe floods (Ali et al. 2013; Lu et al. 2016). Besides, Bangladesh gets relatively lower output from the agriculture sector due to its huge population, scarcity of land and inadequate research facilities (Thirtle et al. 2003). Kamruzzaman et al. (2016b) developed an index of drought proneness (DI) to identify the highest probability of drought in an area by computing three consecutive dry weeks. It showed that Rajshahi meteorological station got the highest score (0.851) during the Pre-*Kharif* season in Bangladesh. The annual rainfall also decreases at -8.017 and -6.571 mm/year by Sen's slope estimator and linear regression analysis, respectively, in Rajshahi (Kamruzzaman et al. 2015, 2021). It indicates the severe drought condition in the west-central part of Bangladesh. On the contrary, the rainfall during the *Kharif* season at Dinajpur, Rangpur and Khepupara stations shows a considerable escalation. It increases the potential flood in these areas results in damages to *Aman* cultivation (Kamruzzaman et al. 2016a). To address the reason of severe floods in Bangladesh, Mirza (2002) has spotted the river system that drains out vast monsoon (June–September) runoff that surpasses its capacity resulting in severe riverine floods. In this case, floods may submerge about 70% areas of the country. Apart from riverine floods, flash floods also occur in the eastern and northern river basins, along the borders of Bangladesh. Intense precipitation over nearby hills and mountains in India creates ash floods in Bangladesh. Glacier melts in the Himalayan regions also occur in flash floods at the foothills of Nepal that extend to the northern part of Bangladesh (Dewan 2014). These floods often affect the Teesta basin as it locates in the northwestern part of the country. It is very devastating due to its sudden and rapid characteristics (Karim and Thiel 2017). Ruane et al. (2013) have focused on deforestation in the case of *char* (sand bar in the river channel) formation in the Teesta River. Deforestation leads to extensive soil erosion in the catchment area. It accelerates rapid siltation on the river bed that forms *chars* (Alam et al. 2018). *Chars* make obstacles to river flow, which is responsible for lengthy floods and riverbank erosion. *Chars* become very fragile to riverbank erosion due to their unstable soil structure. However, sedimentation in crop fields is another hazard in *chars*. Deep sand deposits make the land unsuitable for cultivation. Amir and Ahmed (2013) referred to the changes in agricultural yield under the threats of climatic variability at

Kalapara Upazila in coastal Bangladesh. They marked that the production of crops, fisheries and livestock was hampered due to natural hazards, i.e., floods, salinity intrusion and diseases. Veron et al. (2015) studied the impact of precipitation and temperature on wheat, maize and soya yield in the Pampas area of Argentina from 1971 to 2012. Their study revealed a significant negative impact of climate change on crop yields. The production of maize, wheat and soya declined by 5.4%, 5.1% and 2.6%, respectively, in their study area. Those drops in production could make loss of USD 1.1B, which reduced 5% grain export in 2013. Olesen et al. (2011) noticed damages due to climate changes in Europe. Unusual frost and extreme snow cover damaged grain maize, grapevine and spring barley. Besides, floods created continuing problems to grassland, winter wheat, spring barley and cereal production. Drought and heat stress also damaged cereals, seed crops, wheat, barley and maize in deferent parts of Europe.

Increasing temperature and rainfall-induced climate change result in natural hazards, which are threats to agricultural activities in different parts of the world, including Bangladesh. However, the agricultural damages in a flash flood-prone river basin in plain land do not get much consideration. In this case, *chars* in a braided river need more attention. Thus, this study has explored the effects of climatic threats to the farming of the Teesta basin in Bangladesh.

5.2 Data and Methods

5.2.1 Study Area

Lalmonirhat, Rangpur, Kurigram, Nilphamari and Gaibandha districts (district comprises two or more upazilas and the second-order administrative unit in Bangladesh) are considered to be Teesta basin area of Bangladesh. Lalmonirhat, Rangpur and Nilphamari districts have been selected by lottery method to ensure the randomness. Lalmonirhat Sadar, Dimla and Kaunia upazilas (upazila comprises two or more unions and the third-order administrative unit in Bangladesh) from Lalmonirhat, Nilphamari and Rangpur districts, respectively, have been selected in similar way. Khaga Kharibari, Gokunda, and Tepa Modhupur unions (union comprises villages and the smallest unit of local government in rural Bangladesh) from Dimla, Lalmonirhat Sadar, Kaunia upazilas, respectively, have also been selected in lottery process. Finally, Kismat Chatnai, Kursamari and Azam Khan villages have been selected, respectively, from Khaga Kharibari, Gokunda and Tepa Modhupur unions through lottery method (without replacement). It is well known that the three forms of SRS, such as simple random sampling without replacement, simple random sampling with replacement and fixed cost simple random sampling constitute the sample mean both acceptable and unbiased (Mitra and Pathak 1984). Thus, this study has used the SRS without replacement technique. The SRS technique of sample selection was also followed by Manandhar et al. (2011) in Nepal. To pinpoint the location of the study area, selected districts, upazilas and unions have been shown in Fig. 5.1.

5.2.2 Sample Size and Sampling

A questionnaire survey has been conducted from September 2017 to November 2017 (3 months) to select 276 sample households from three sample villages. 130, 76 and 70 households from Azam Khan, Kursamari and Kismat Chatnai villages, respectively, have been considered as sample respondents for this study. Sample households have been selected from sample villages with the help of a random number table to ensure the representativeness of the sample. The most experienced person involved in agricultural activities has been selected from every household.

5.2.3 Data Analysis and Interpretation

Both qualitative and quantitative data have been used in this study. Quantitative data have been

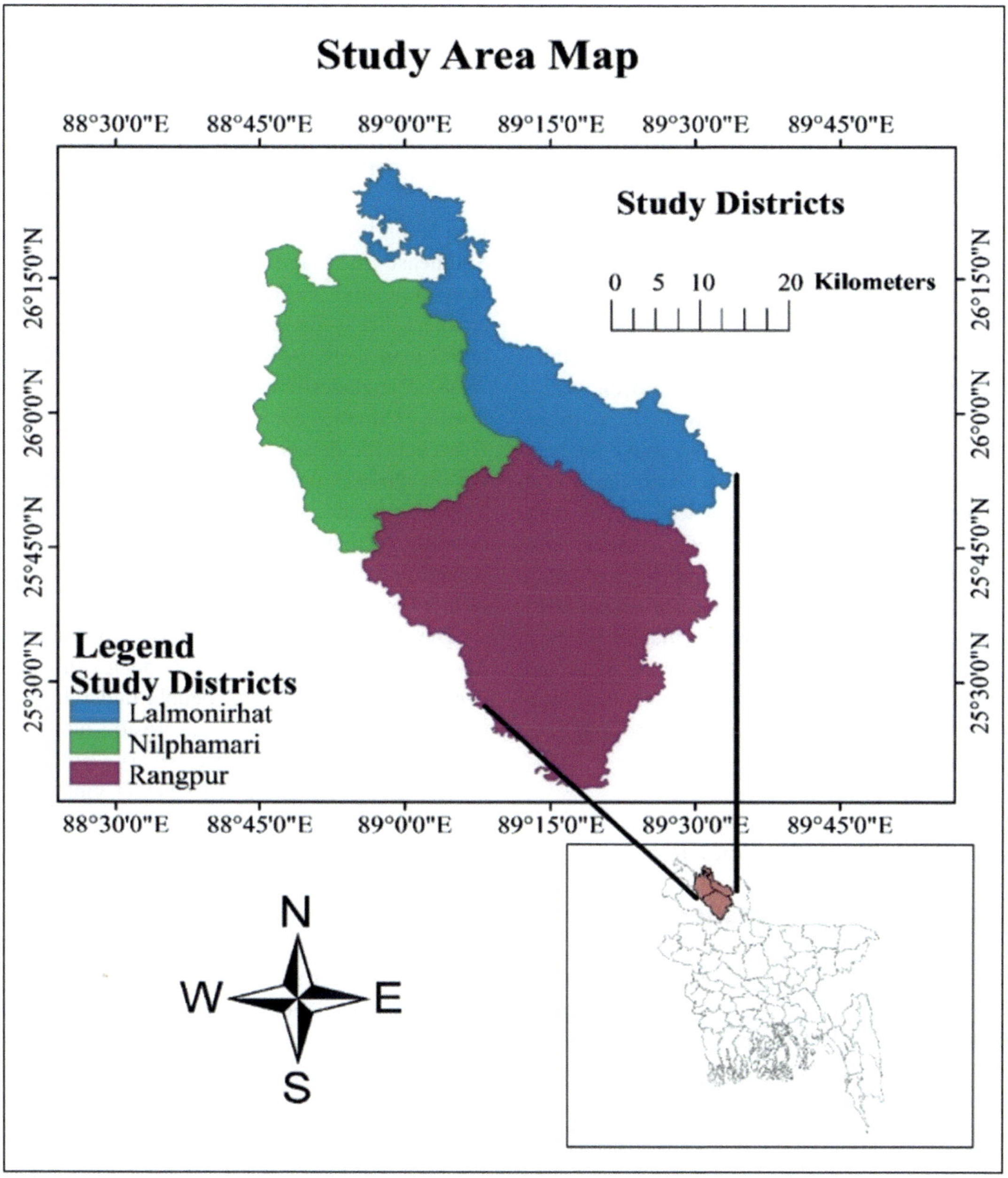

Fig. 5.1 Study area map with location of districts, upazilas and unions

collected through face to face interviews, and qualitative data have been collected through in-depth interview. Qualitative data have been an addition to quantitative data and for cross-checking the data and information. Primary data have been collected from the respondents (experienced farmers) through questionnaire survey and in-depth interviews. A semi-structured questionnaire has been developed to collect the primary data through face to face interviews. This technique of questionnaire development has been used in different studies worldwide (Mubaya and Mafongoya 2017; Partey et al. 2020). The duration of interviews ranged from 30 to 40 min. The farmers who have struggled with natural hazards have been prioritized to be interviewed in-depth. Secondary data have been collected from relevant books, journal articles, research papers and research reports. Data analyses have been carried out by using IBM SPSS (Statistical Package for Social Sciences) version 22. For primary data, inferential statistics have been used to explore the existing situation. Frequency distributions in LS (Likert Scale) have been used for understanding the degree of damages to agriculture by natural hazards. Likelihood ratio tests have been used to find out the associations between the damages and the climatic hazards. All the above analyses have been carried out to understand the overall scenario in the study area as well as the comparative situation in three study villages. Data have been presented in tables and charts by using Microsoft Word 2013 and Microsoft Excel 2013. In some cases, various types of maps have been developed by using Arc GIS software of version 10.2.2. Remote sensing maps have also been used to analyze changes in land use and river channel.

5.3 Results and Discussion

5.3.1 Vulnerable Location to Flood

Teesta floodplain is a piedmont alluvial plain in Bangladesh. This floodplain is located relatively in higher altitude. The land elevation increases by about 90 m above the mean sea level toward the north–west region of this floodplain. This area is a gently sloping land that is composed of mainly sand deposits. Kismat Chatnai village locates in the northern part of the Teesta basin at Dimla Upazila in Nilphamari district. This village is a kind of island *char* (Fig. 5.2). Kursamari village locates at Gokunda union in Lalmonirhat Sadar upazila. This village is north to the downstream of Teesta River (Fig. 5.3). This mainland village stands on the bank of Suti river (a tributary of Teesta River). Azam Khan village stands on the south bank of Teesta River at Kaunia upazila in Rangpur district (Fig. 5.3). This village is the part of attached char of Teesta River.

5.3.2 Effects of Climatic Hazards on Agriculture

People of low law latitude countries in South, Southeast Asia and Africa are vulnerable to starvation and extreme hardship. Indeed the smallest decay in crop yields could be very harmful in these regions (Aydinalp and Cresser 2008). Bangladesh is a low latitude country and hostile climatic events affect its agriculture. However, the degree of these hazards is dissimilar in diverse cropping seasons and different villages in the study area. Flood, excessive rainfall, riverbank erosion, storm, hailstorm, cold waves, and drought are the main climatic hazards in this country (Figs. 5.4, 5.5 and 5.6).

The flood is the most devastating natural disaster in the study area. It inundates this area every year. Almost every people undergo damages due to flood. Moreover, erosion along the riverside is a common scenario every year during floods, especially in Kismat Chatnai village. As an island char, this village is very vulnerable to riverbank erosion. Life and wealth of every people in this village overwhelms every year due to this disaster. Besides, majority of respondents have considered excessive rainfall as a disaster. It is responsible for severe flood too. Furthermore, the test of association between flood and rainfall results in significant relationship between flood and increased rainfall (p-value = 0.022) that indicates the increase of rainfall is responsible for the devastating floods during the rainy season. Flood is harmful to every village during *Kharif* season. In the case of Kismat Chatnai village, 38.6% of respondents have shown *Aman* as the affected crop (Table 5.1). Almost one-fourth of respondents have mentioned that flood inundates *Aman* plants severely at the sowing stage during the *Pre-Kharif* season and (Table 5.2). Floods also affect *Aman* at the germination stage during this season. Besides, flood and riverbank erosion also affect *Aman* during at the vegetation and flowering stages during the *Kharif* season

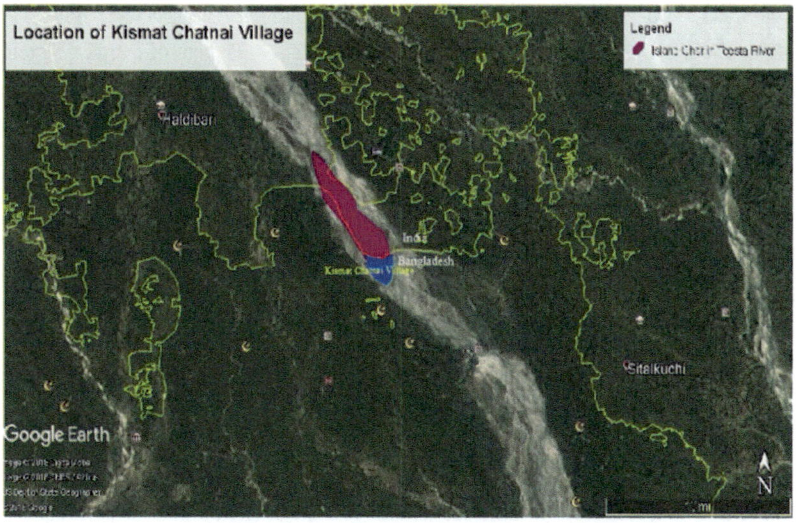

Fig. 5.2 Location of Kismat Chatnai village at an island *Char* in Teesta river. *Source* Extracted from Google Earth on August 29, 2018

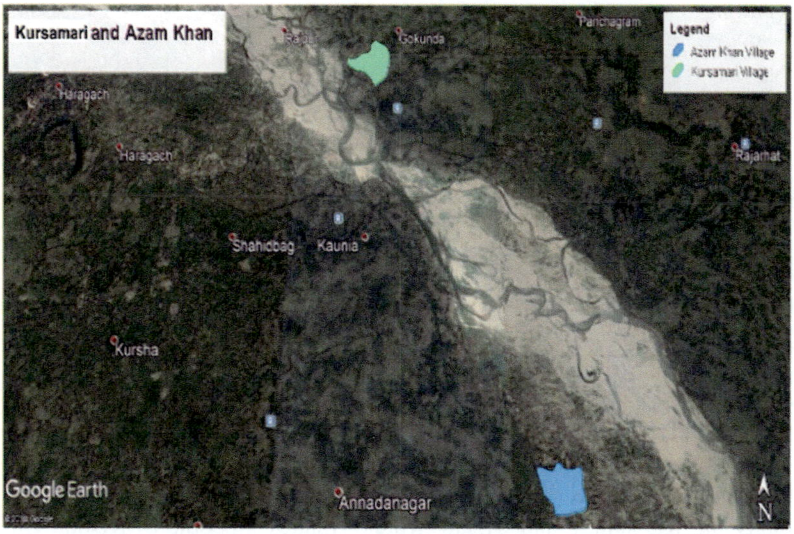

Fig. 5.3 Location of Kursamari and Azam Khan villages adjacent to Teesta river. *Source* Extracted from Google Earth on August 29, 2018

(Fig. 5.7). The flood has hampered *Aman* cultivation in three villages. This hazard has suffered more than thirty percent of farmers (Table 5.1).

Sedimentation is also a big problem in Kismat Chatnai village. Deep sandy sediments destroy the *Aman* crop fields at the sowing stage during *Pre-Kharif* season. Then, farmers have two options to choose from. Firstly, they have to leave the land unused for that season. Or, they have to use excessive fertilizer and irrigation to make the land fertile for cultivation. Both options are not profitable for the farmers in this village. On the other hand, sustainable groundwater management is challenging due to excessive groundwater irrigation (Das et al. 2020), and also shallow pump systems will not be worked in near future (Das et al. 2021b).

The bank of the Teesta River is very erosion-prone (Fig. 5.5). Azam Khan and Kursamari villages are protected from riverbank erosion by embankments (Fig. 5.2). On the other hand, Kismat Chatnai village is vulnerable to riverbank erosion (Figs. 5.2 and 5.6).

Table 5.1 Crops which are more affected by climatic hazards by peoples' opinion

Agricultural products	Number of total respondents	Total percent	Percent of peoples' opinion in three villages		
			Kursamari	Azam Khan	Kismat Chatnai
Aus	90	32.6	53.9	33.1	8.6
Aman	92	33.3	34.2	30.0	38.6
Jute	44	15.9	10.5	20.0	14.3
Green pepper	5	1.8	0.0	3.8	0.0
Onion	2	0.7	0.0	1.5	0.0
Potato	30	10.9	1.3	11.5	20.0
Maize	13	4.7	0.0	0.0	18.6
Total	276	100.0	100.0	100.0	100.0

Source Field survey from September to November 2017

Table 5.2 Affected crops in different seasons and growing stages in three villages

	Affected crops in different seasons and growing stages		
Growing seasons	Pre-*Kharif* season	*Kharif* season	*Rabi* season
Natural hazards	Flood, sedimentation, storms and hailstorms	Flood and riverbank erosion	Cold waves, pests and diseases
Villages			
Kursamari	*Aman* (sowing stage)	*Aman* (vegetation and flowering stage)	*Aus* (vegetation, flowering and ripening stage)
Azam Khan	*Aman* (sowing stage) Jute (vegetation stage)	*Aman* (vegetation and flowering stage)	*Aus* (vegetation, flowering and ripening stage)
Kismat Chatnai	*Aman* (sowing stage)	*Aman* (vegetation and flowering stage)	Potato (vegetation stage) Maize (vegetation, flowering and ripening stage)

Source Field survey from September to November 2017

During summer, storm and hailstorm cause most damages to crops in the study area, especially in Kursamari and Azam Khan villages during the *Pre-Kharif* season. "I have suffered from hailstorms by getting 100–120 kg less *Aus* production in every 30 decimal lands". A farmer has mentioned it in Kursamari village. Hailstorms have severely affected jute cultivation at the vegetation stage in both Azam Khan and Kursamari villages during the Pre-*Kharif* season. The situation has become worse in Azam Khan village. Storms and hailstorms have damaged majority of farmer's jute production at the vegetation stage in this village. Nevertheless, hailstorms do not create severe problems in Kismat Chatnai village.

However, people do not consider drought and heat waves as major catastrophes except the people of Kismat Chatnai village. People of this village are deeply affected by both drought and heat waves due to its landuse pattern. Landuse of this villages changes regularly due to its unstable formation. So, this village does not have a permanent vegetation. Besides, sandy soils of this island char cause the scarcity of moisture in the soil.

Though winter temperature has increased in northern Bangladesh, yet cold waves have appeared to be a disaster to the Teesta basin. Kismat Chatnai village is more affected by this disaster than those of the other two villages because of its location. This village stands north

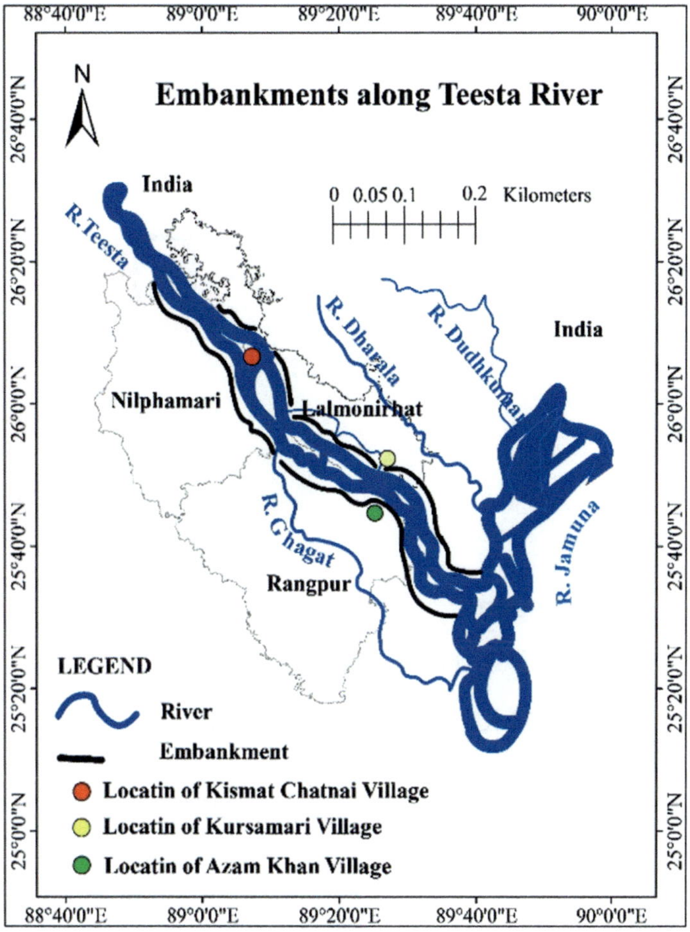

Fig. 5.4 Location of flood protection embankments in the study area (Developed by the Researcher)

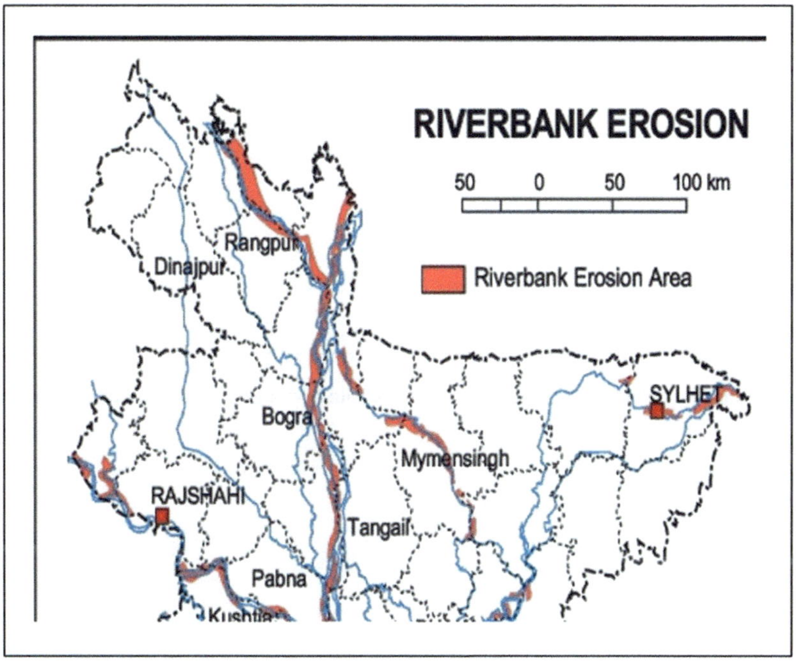

Fig. 5.5 Erosion-prone area in the study area. *Source* BWDB

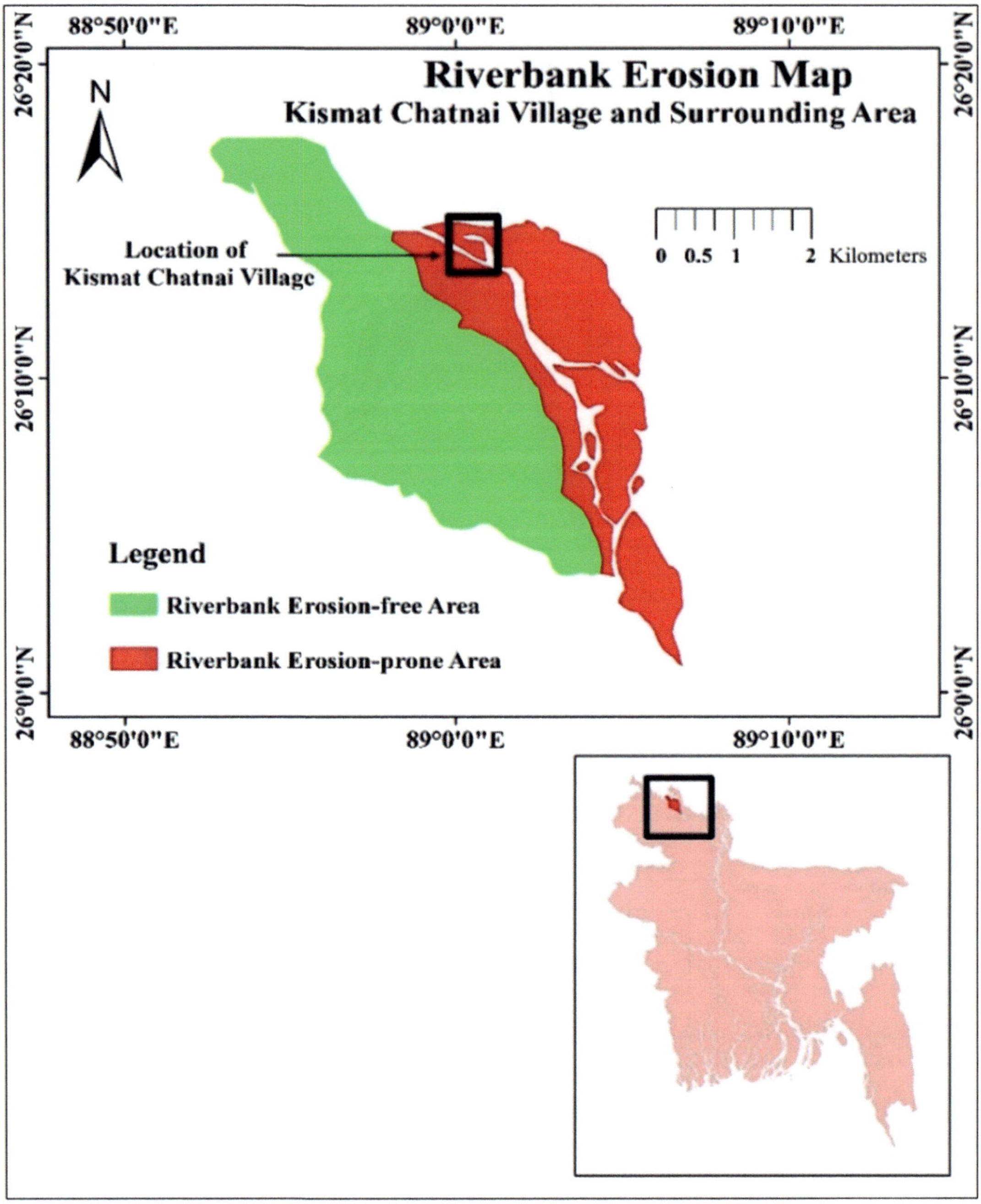

Fig. 5.6 Riverbank erosion map of Kismat Chatnai village and surrounding areas. *Source* BWDB

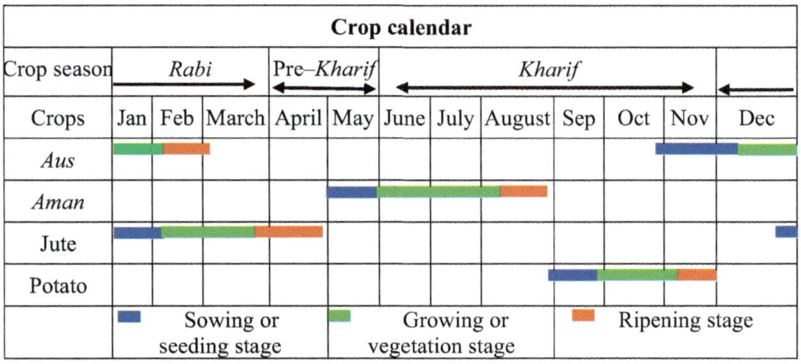

Fig. 5.7 Crop calendar of main crops according to peoples' opinion in the study area. *Source* Field survey from September to November 2017

to the other two villages. Thus, this village is very vulnerable to cold waves during the *Rabi* season. More than 85% of respondents in this village have pointed cold waves as the hazards. Cold waves have become a disaster to Azam Khan and Kursamari villages too. The shifts in precipitation and temperature patterns will spread out the pests and diseases (Aydinalp and Cresser 2008). Thus, various pests and diseases have also hampered some crops and that will increase in the future. More than 75% of respondents in Azam Khan village have mentioned that severe cold and fog have caused the break out of some diseases of plants. Similar results were found by Kamruzzaman et al. (2022).

Similarly, *Aus* and Transplanted *Aman* (T. *Aman*) have been affected by different diseases, such as tungro, head blight, blast, bacterial blight, brown spot, bakanae, ufra, leaf scald and stem rot in Bangladesh (Bazlur Rashid et al. 2011). People have mentioned blast disease, which is responsible for damaging *aus* cultivation at the vegetation, flowering and ripening stages during the *Rabi* season (Table 5.2). In Kursamari and Azam Khan villages, 53.9% and 33.1% of respondents, respectively, have told that different diseases have hampered *Aus* cultivation at the vegetation, flowering and ripening stages. However, only 8.6% of respondents in Kismat Chatnai village have stated about diseases (Table 5.1).

Potato is a vital crop in the study area. Different pests and diseases are the main constraints to potato cultivation. Late blight caused by *Phytophthora infestans* is an acute disease of potato in Bangladesh (Rahman et al. 2008). People have mentioned that the winter rainfall is responsible for late blight disease during the *Rabi* season. Thus, late blight disease has hampered potato cultivation at the vegetation stage. It has also caused damage to maize severely at the vegetation, flowering and ripening stages in Kismat Chatnai village (Table 5.2). The association between pests attack on crops and fall of temperature during the winter season is significant (p-value = 0.011) at 5% level of significance. Thus, the colder winter is very suitable to the pests for reducing the production of potato and maize.

Climatic hazards (e.g., floods, riverbank erosion, sedimentation, storms, hailstorms, cold waves, pests and diseases) hamper farming in all three study villages. They damage *Aman*, *Aus*, jute, potato and maize at different stages of cultivation (Figs. 5.8, 5.9, 5.10 and 5.11).

5.4 Conclusions

Climatic hazards affect the farming of Bangladesh due to its vulnerable geographical location. This study investigates the damages to the agricultural sector due to various climatic hazards at the Teesta basin of Bangladesh. The climatic effects differ from village to village due to location and different socio-economic conditions. In the study villages, *Aman rice* is the most affected crop during the *Kharif* season.

Fig. 5.8 Settlement at unstable

Fig. 5.9 Sandy sedimentation at cropfield

Many farmers cultivate *Aman* rice second time after inundating the first planted sapling by the flash flood. *Aman* rice fields often erode during this season. Deep and rapid sedimentation by sands has also damaged *Aman* rice in Kismat Chatnai village. Riverbank erosion affects this village due to its mid-river location. Storm and hailstorms have damaged jute in Kursamari and Azam Khan villages during the Pre-*Kharif* season. Pests have damaged crops too in Kursamari village. Severe cold, fog and diseases caused suffered potato, maize and *Aus* rice cultivation during the *Rabi* season in Kismat Chatnai village. Winter rainfall is responsible for late blight disease during the *Rabi* season which hampers potato cultivation at the vegetation stage. Thus, this study has successfully identified major natural disasters to various crops in particular cropping seasons. It has spotted on the critical farming stages of different crops. It also highlights *chars,* which are very vulnerable to flood and riverbank erosion. The finding of this study will support the government to design future strategies to reduce the damages as well as

Fig. 5.10 Flood water in the Teesta river

Fig. 5.11 Riverbank erosion

prepare the people to adapt to the adverse climatic conditions in the Teesta basin.

References

Alam GMM (2016) An assessment of the livelihood vulnerability of the riverbank erosion hazard and its impact on food security for rural households in Bangladesh. Ph.D. thesis, University of Southern Queensland, Queensland, Australia

Alam GMM, Alam K, Mushtaq S, Khatun MN, Filho WL (2018) Strategies and barriers to adaptation of hazard-prone rural households in Bangladesh. In: Filho WL, Nalau J (eds) Limits to climate change adaptation. Springer International Publishing AG

Ali MS, Mahjabin T, Hosoda T (2013) Impact of climate change on floods of Bangladesh and introducing flood intensity index to characterize the flooding scenario. J Eng Sci 4(1):23–34

Amir KI, Ahmed T (2013) Climate change and its impact on food security in Bangladesh: a case study on Kalapara, Patuakhali, Bangladesh. Earth Sci Clim Change 4(5):155. https://doi.org/10.4172/2157-7617.1000155

Arnell NW, Lloyd-Hughes B (2014) The global-scale impacts of climate change on water resources and flooding under new climate and socio-economic scenarios. Clim Change 122:127–140. https://doi.org/10.1007/s10584-013-0948-4

Arnell NW, Gosling SN (2016) The impacts of climate change on river flood risk at the global scale. Clim Change 134:387–401. https://doi.org/10.1007/s10584-014-1084-5
Arnell NW, Brown S, Gosling SN, Gottschalk P, Hinkel J, Huntingford C, Lloyd-Hughes B, Lowe JA, Nicholls RJ, Osborn TJ, Osborne TM, Rose GA, Smith P, Wheeler TR, Zelazowski P (2016) The impacts of climate change across the globe: a multi-sectoral assessment. Clim Change 134:457–474. https://doi.org/10.1007/s10584-014-1281-2
Aydinalp C, Cresser MS (2008) The effects of global climate change on agriculture. Am Eurasian J Agric Environ Sci 3(5):672–676
Basak A, Rahman ATMS, Das J, Hosonod T, Kisi O (2022) Drought forecasting using the prophet model in semi-arid climate region of western India. Hydrol Sci J. https://doi.org/10.1080/02626667.2022.2082876
Bazlur Rashid AQM, Rahman MME, Islam MZ, Hossain MA (2011) Management of cereal diseases in Bangladesh: a review. Int J Sustain Crop Prod 6 (2):36–41
Brown ME, Funk CC (2008) Food security under climate change. Science 319:580–581
Carter TR, Rovere EL (2011) Developing and applying scenarios. In: Climate change 2001: impacts, adaptation, and vulnerability, contribution of working group II to the third assessment report of the intergovernmental panel on climate change, pp 145–190. Retrieved from http://www.ipcc.ch/ipccreports/tar/wg2/pdf/wg2TARannexB.pdf. On 22 Dec 2016
Choudhury S, Basak A, Biswas S, Das J (2022) Flash flood susceptibility mapping using GIS-based AHP method. In: Pradhan B, Shit PK, Bhunia GS, Adhikary PP, Pourghasemi HR (eds) Spatial modelling of flood risk and flood hazards. GIScience and geo-environmental modelling. Springer, Cham. https://doi.org/10.1007/978-3-030-94544-2_8
Dankers R, Feyen L (2009) Flood hazard in Europe in an ensemble of regional climate scenarios. J Geophys Res 114:D16108. https://doi.org/10.1029/2008JD011523
Das J, Rahman AS, Mandal T, Saha P (2020) Challenges of sustainable groundwater management for large scale irrigation under changing climate in Lower Ganga river basin in India. Groundw Sustain Dev 11:100449. https://doi.org/10.1016/j.gsd.2020.100449
Das J, Mandal T, Rahman AS, Saha P (2021a) Spatio-temporal characterization of rainfall in Bangladesh: an innovative trend and discrete wavelet transformation approaches. Theoret Appl Climatol 143(3):1557–1579. https://doi.org/10.1007/s00704-020-03508-6
Das J, Rahman AS, Mandal T, Saha P (2021b) Exploring driving forces of large-scale unsustainable groundwater development for irrigation in lower Ganga River basin in India. Environ Dev Sustain 23(5):7289–7309. https://doi.org/10.1007/s10668-020-00917-5
Dewan TH (2014) Societal impacts and vulnerability to floods in Bangladesh and Nepal. Weather Clim Extremes 7:36–42. Retrieved from https://doi.org/10.1016/j.wace.2014.11.001. On 20 May 2019
Dutta D (2011) An integrated tool for assessment of flood vulnerability of coastal cities to sea-level rise and potential socio-economic impacts: a case study in Bangkok, Thailand. Hydrol Sci J 56(5):805–823. https://doi.org/10.1080/02626667.2011.585611
Easterling DR, Evans JL, Groisman PY, Karl TR, Kunkel KE, Ambenje P (2000a) Observed variability and trends in extreme climate events: a brief review. Bull Am Meteor Soc 81(3):417–425. https://doi.org/10.1175/1520-0477(2000)081%3c0417:OVATIE%3e2.3.CO;2
Easterling DR, Meehl GA, Parmesan C, Changnon SA, Karl TR, Mearns LO (2000b) Climate extremes: observations, modeling, and impacts. Science 289:2068–2074. https://doi.org/10.1126/Science.289.5487.206
Etzold B, Ahmed AU, Hassan SR, Neelormi S (2014) Clouds gather in the sky, but no rain falls. Vulnerability to rainfall variability and food insecurity in Northern Bangladesh and its effects on migration. Clim Dev 6(1):18–27. https://doi.org/10.1080/17565529.2013.833078
Haque AKE, Lohano HD, Mukhopadhyay P, Nepal M, Shafeeqa F, Vidanage SP (2019) NDC pledges of South Asia: are the stakeholders onboard? Clim Change 155:237–244. https://doi.org/10.1007/s10584-019-02417-6
Kamruzzaman M, Rahman ATMS, Jahan CS (2015) Adapting cropping systems under changing climate in NW Bangladesh. Lambert Academic Publishing
Kamruzzaman M, Rahman ATMS, Kabir ME, Jahan CS, Mazumder QH, Rahman MS (2016a) Spatio-temporal analysis of climatic variables in the western part of Bangladesh. Environ Dev Sustain 18(6). https://doi.org/10.1007/s10668-016-9872-x
Kamruzzaman M, Kabir ME, Rahman ATMS, Mazumder QH, Rahman MS, Jahan CS (2016b) Modeling of agricultural drought risk pattern using Markov chain and GIS in the western part of Bangladesh. Environ Dev Sustain 18(6). https://doi.org/10.1007/s10668-016-9898-0
Kamruzzaman M, Mandal T, Rahman ATM, Khalek A, Alam GMM, Rahman MS (2021) Climate modeling, drought risk assessment and adaptation strategies in the western part of Bangladesh. In: Alam et al. (eds) Climate vulnerability and resilience in the global south, pp 103–129
Kamruzzaman M, Rahman ATMS, Basak A, Alam J, Das J (2022) Assessment of climate change impacts and adaptation strategies through the prism of farmers' perception: a case study. Int J Environ Sci Technol 1–20. https://doi.org/10.1007/s13762-022-04254-0
Karim MR, Thiel A (2017) Role of community based local institution for climate change adaptation in the Teesta riverine area of Bangladesh. Clim Risk Manag 17:92–103. Retrieved from https://doi.org/10.1016/j.crm.2017.06.002. On 22 July 2020

Kundzewicz ZW, Kanae S, Seneviratne SI, Handmer J, Nicholls N, Peduzzi P, Mechler R, Bouwer LM, Arnell N, Mach K, Muir-Wood R, Brakenridge GR, Kron W, Benito G, Honda Y, Takahashi K, Sherstyukov B (2014) Flood risk and climate change: global and regional perspectives. Hydrol Sci J 59 (1):1–28. https://doi.org/10.1080/02626667.2013.857411

Lal M (2003) Global climate change: India's monsoon and its variability. J Environ Stud Policy 6(1):1–34. Retrieved from https://www.researchgate.net/publication/281402625. On 18 May 2020

Liang X, Xu M, Gao W, Reddy KR, Kunkel K, Schmoldt DL, Samel AN (2012) Physical modeling of U.S. cotton yields and climate stresses during 1979 to 2005. Agron J 104(3):675–683. https://doi.org/10.2134/agronj2011.0251

Lobell DB, Burke MB, Tebaldi C, Mastrandrea MD, Falcon WP, Naylor RL (2008) Prioritizing climate change adaptation needs for food security in 2030. Science 319(1):607–610

Lu X, Wrathall DJ, Sundsøy PR, Nadiruzzaman M, Wetter E, Iqbal A, Qureshi T, Tatem AJ, Canright GS, Engø-Monsen K, Bengtsson L (2016) Detecting climate adaptation with mobile network data in Bangladesh: anomalies in communication, mobility and consumption patterns during cyclone Mahasen. Clim Change 138:505–519. https://doi.org/10.1007/s10584-016-1753-7

Manandhar S, Vogt DS, Perret SR, Kazama F (2011) Adapting cropping systems to climate change in Nepal: a cross-regional study of farmers' perception and practices. Reg Environ Change 11:335–348. https://doi.org/10.1007/s10113-010-0137-1

Mirza MMQ (2002) Global warming and changes in the probability of occurrence of floods in Bangladesh and implications. Glob Environ Chang 12:127–138

Mirza MQ, Warrick RA, Ericksen NJ, Kenny GJ (1998) Trends and persistence in precipitation in the Ganges, Brahmaputra and Meghna river basins. Hydrol Sci J 43(6):845–858. https://doi.org/10.1080/02626669809492182

Mitra SK, Pathak PK (1984) The nature of simple random sampling. Ann Stat 12(4):1536. https://doi.org/10.1214/aos/1176346810

Moore N, Alagarswamy G, Pijanowski B, Thornton P, Lofgren B, Olson J, Andresen J, Yanda P, Qi J (2012) East African food security as influenced by future climate change and land use change at local to regional scales. Clim Change 110:823–844. https://doi.org/10.1007/s10584-011-0116-7

Mubaya CP, Mafongoya P (2017) The role of institutions in managing local level climate change adaptation in semi-arid Zimbabwe. Clim Risk Manag 16:93–105. https://doi.org/10.1016/j.crm.2017.03.003

Nuttall JG, Barlow KM, Delahunty AJ, Christy BP, O'Leary GJ (2018) Acute high temperature response in wheat. Agron J 110(4):1296–1308. https://doi.org/10.2134/agronj2017.07.0392

Oerlemans J, Anderson B, Hubbard A, Huybrechts P, Johannesson T, Knap WH, Schmeits M, Stroeven AP, van de Wal RSW, Wallinga J, Zuo Z (1998) Modelling the response of glaciers to climate warming climate dynamics. Clim Dyn 14:267–274. Retrieved from https://epic.awi.de/id/eprint/3664/1/Oer1998b.pdf. On 30 Apr 2020

Olesen JE, Trnka M, Kersebaum KC, Skjelvåg AO, Seguin B, Peltonen-Sainio P, Rossi F, Kozyra J, Micale F (2011) Impacts and adaptation of European crop production systems to climate change. Eur J Agron 34:96–112. https://doi.org/10.1016/j.eja.2010.11.003

Parry M, Rosenzweig C, Iglesias A, Fischer G, Livermore M (1999) Climate change and world food security: a new assessment. Glob Environ Chang 9: S51–S67

Parry ML, Rosenzweig C, Iglesias A, Livermore M, Fischer G (2004) Effects of climate change on global food production under SRES emissions and socio-economic scenarios. Glob Environ Change 14:53–67. https://doi.org/10.1016/j.gloenvcha.2003.10.008

Partey ST, Dakorah AD, Zougmoré RB, Ouédraogo M, Nyasimi M, Nikoi GK, Huyer S (2020) Gender and climate risk management: evidence of climate information use in Ghana. Clim Change 158:61–75. https://doi.org/10.1007/s10584-018-2239-6

Paudyal GN (2002) Forecasting and warning of water-related disasters in a complex hydraulic setting—the case of Bangladesh. Hydrol Sci J 47(S1):S5–S18. https://doi.org/10.1080/02626660209493018

Porter JR, Semenov MA (2005) Crop responses to climatic variation. Philos Trans R Soc B 360:2021–2035. https://doi.org/10.1098/rstb.2005.1752

Rahman MR, Lateh H (2017) Climate change in Bangladesh: a spatio-temporal analysis and simulation of recent temperature and rainfall data using GIS and time series analysis model. Theoret Appl Climatol 128:32–38. https://doi.org/10.1007/s00704-015-1688-3

Rahman MM, Day TK, Ali MA, Khalequzzaman KM, Hussain MA (2008) Control of late blight disease of potato by using new fungicides. Int J Sustain Crop Prod 3(2):10–15

Ruane AC, Major DC, Yu WH, Alam M, Hussain SG, Khan AS, Hassan A, Rosenzweig C (2013) Multi-factor impact analysis of agricultural production in Bangladesh with climate change. J Glob Environ Change 23:338–350. Retrieved from https://ntrs.nasa.gov/archive/nasa/casi.ntrs.nasa.gov/20150002677.pdf. On 26 Mar 2017

Sanderson MG, Hemming DL, Betts RA (2011) Regional temperature and precipitation changes under high-end ($\geq$ 4 °C) global warming. Phil Trans R Soc A 369:85–98. https://doi.org/10.1098/rsta.2010.0283

Tao F, Zhang Z, Zhang S, Zhu Z, Shi W (2012) Response of crop yields to climate trends since 1980 in China. Clim Res 54:233–247. https://doi.org/10.3354/cr01131

Thirtle C, Lin L, Piesse J (2003) The impact of research-led agricultural productivity growth on poverty reduction in Africa, Asia and Latin America. World Dev 31 (12):1959–1975. https://doi.org/10.1016/j.worlddev.2003.07.001

Thomas V, Albert JRG, Hepburn C (2014) Contributors to the frequency of intense climate disasters in Asia-Pacific countries. Clim Change 126:381–398. https://doi.org/10.1007/s10584-014-1232-y

Tol RSJ, Wagner S (2010) Climate change and violent conflict in Europe over the last millennium. Clim Change 99:65–79. https://doi.org/10.1007/s10584-009-9659-2

Vautard R, Gobiet A, Jacob D, Belda M, Colette A, Déqué M, Fernández J, García-Díez M, Goergen K, Güttler I, Halenka T, Karacostas T, Katragkou E, Keuler K, Kotlarski S, Mayer S, Meijgaard EV, Nikulin G, Patarčić M, Scinocca J, Sobolowski S, Suklitsch M, Teichmann C, Warrach-Sagi K, Wulfmeyer V, Yiou P (2013) The simulation of European heat waves from an ensemble of regional climate models within the EURO-CORDEX project. Clim Dyn 41:2555–2575. https://doi.org/10.1007/s00382-013-1714-z

Veron SR, Abelleyra DD, Lobell DB (2015) Impact of precipitation and temperature on crop yields in the Pampas. Clim Change 130(2):235–245. Retrieved from http://www.altemetric.com/detailes/3726019. On 30 July 2016

Weltzin JF, Loik ME, Schwinning S, Williams DG, Fay PA, Haddad BM, Harte J, Huxman TE, Knapp AK, Lin G, Pockman WT, Shaw MR, Small EE, Smith MD, Smith SD, Tissue DT, Zak JC (2003) Assessing the response of terrestrial ecosystems to potential changes in precipitation. BioScience 53(10):941–952. https://doi.org/10.1641/0006-3568(2003)053[0941:ATROTE]2.0.CO;2

Young G, Zavala H, Wandel J, Smit B, Salas S, Jimenez E, Fiebig M, Espinoza R, Diaz H, Cepeda J (2010) Vulnerability and adaptation in a dryland community of the Elqui Valley, Chile. Clim Change 98:245–276. https://doi.org/10.1007/s10584-009-9665-4

6 Mizoram, the Capital of Landslide: A Review of Articles Published on Landslides in Mizoram, India

Jonmenjoy Barman, Brototi Biswas, and Jayanta Das

Abstract

Mizoram, a northeastern hilly state of India, suffers from heavy landslides every year during monsoon season. Geologically, Mizoram has been formed in the tertiary era predominantly made of unconsolidated sandstone and shale, which are the dominant rock types in these areas. Unplanned and non-scientific urban sprawls are the triggering factors of landslides in Mizoram. Various researches have been conducted on landslide susceptibility analysis and slope instability. The present work presents a review of various such research works conducted in Mizoram to understand the overall nature of landslide's geological, geomorphological, and anthropological aspects. The review revealed that most of the landslides in Mizoram happen along the roadside and are rainfall induced. So, government policy and PWD must consider various factors like slope holding capacity, geological strength, and rainfall distribution at the time of road construction and expansion. We hope the present work will help the researchers, and planners take proper steps toward slope stability in Mizoram.

Keywords

Mizoram · Landslide · Hunthar Veng · Lunglei · Hazard

6.1 Introduction

Natural hazards can be termed as either the likelihood of a relatively stable landmass suddenly altering its state (Scheidegger 1994) or the possibility of a destructive event in a given area and time (Varnes et al. 1984). Among the various natural hazards, landslides are the most destructive, having wide socioeconomic and environmental ramifications (Mengistu et al. 2019). It results from two main groups of factors—the triggering and conditioning factors, which are a combination of various factors like topography, geomorphology, anthropogenic, seismic factors, and geology (Pourghasemi et al. 2018). In landslide susceptibility mapping, the area of study is divided or categorized into various zones based on the likelihood or probability of landslide events occurring depending on the specific

J. Barman · B. Biswas
Department of Geography and Resource Management, Mizoram University, Aizawl 796004, India
e-mail: jonmenjoybarman07@gmail.com

B. Biswas
e-mail: brototibiswas@gmail.com

J. Das (✉)
Department of Geography, Rampurhat College, Rampurhat, Birbhum 731224, India
e-mail: jayanta.daas@gmail.com

J. Das and S. K. Bhattacharya (eds.), *Monitoring and Managing Multi-hazards*, GIScience and Geo-environmental Modelling, https://doi.org/10.1007/978-3-031-15377-8_6

area's triggering and conditioning factors. This type of prediction method, along with landslide inventory based on past landslides, is an integral part of the landslide hazard mitigation technique since they act as warning systems for necessary steps to minimize the after-effect. It is estimated that worldwide, "3.7 million Km^2 of the surface of the land" and about 5% of the world's population are in constant threat of landslides (Pourghasemi et al. 2018).

Various techniques have been applied over the past 20–25 years to derive the landslide susceptibility evaluation, which can be broadly classified into "inventory-based, expert evaluation (Raghuvanshi et al. 2014; Guzzetti et al. 1999; Turrini and Visintainer 1998; Sarkar et al. 1995, Anbalagan 1992; etc.), statistical (Girma et al. 2015, Biswas et al. 2021; etc.), deterministic (Fall et al. 2006), and distribution-free approaches (Kanungo et al. 2006)." Each of the techniques has produced its own set of landslide susceptibility maps, each having its own set of strengths and lacunae. Very few review studies have been conducted to compare the various techniques and find out the limitations of each. The majority of research projects focus on determining the landslide hazard or susceptibility zonation of a specific study area.

The main objective of the present research work is to provide a detailed review of the various researches that have been conducted on landslide susceptibility or zonation in Mizoram, which happens to fall under seismic zone 5. No such work has been conducted till date for the study area, and the present work will thus be a pioneering work toward the same.

6.2 Background

Mizoram, the land of Mizos, is a hilly state situated in northeast India. Mizoram is geographically located between 21° 58′ N and 24° 35′ N latitude and 92° 15′ E to 93° 29′ E longitude. Undulating and rugged topography is the common physiography here. Including Tlawng, Mizoram has 23 major watersheds, of which some are north flowing, while the rest are south flowing rivers (Verma 2018). Mat, Tlawng, Tuivai, Khawthlang -tuipui, Tuirai, Tuirini, Tuirial, Tiau, and Tuivawl are the major drainage systems in Mizoram. Barman (2021) identified drainage systems with trellis, parallel, and dendritic patterns, indicating a strong geological relationship. Mizoram belongs to the Tertiary Barail, Surma, and Tipam groups, with a thickness of 800 m (Barman and Rao 2021). The Mizo hills are characterized in a north–south direction by having a continuous and discontinuous syncline and anti-syncline formation with a strike direction of N15 E to S15° W. Work on the Sangiang fault area, which has an active tectonic collision footprint between the Indo-Burmese plate (Barman and Rao 2021), has been conducted. The Geological Survey of India demarked Mizoram as Zone V which indicates a very high damage risk zone. Mizoram comes under the direct influence of the southeast monsoon. The average annual rainfall is 2500 mm. It received heavy rainfall in July (439.4 mm) and the least in the month of March (3.2 mm), as recorded in 2020 (Directorate of Economics & Statistics, Government of Mizoram).

Mizoram faces numerous natural hazards as well as landslides due to diversity in geology, geomorphology, rapid urban growth, and heavy rainfall during the monsoon season, as shown in Table 6.1. For disaster mitigation and RH, the Mizoram government classified Mizoram into five landslide hazard zones, as shown in Table 6.2.

The majority of landslides in Mizoram occur along the roadways, as observed in Fig. 6.1 and are rainfall induced. Most landslides were caused by rapid urbanization and road construction along oblique to dip slopes. Due to heavy rainfall in Mizoram and many fracture zones in the rocks, the rain enters the cracks and crevices and acts as a lubricating and driving force for landslides.

6.3 Discussion

According to the reviews and inventory data provided by the Geological Society of India, it has been found that the whole of Mizoram is a

Table 6.1 Hazard in Mizoram

Nature of calamity	Human causality		House damaged		
	Deaths	Injuries	Fully	Severely	Partially
Landslide	68	37	600	313	659
Fire	20	21	775	161	386
Flood	14	–	1016	84	1327
Hail storm	–	–	50	115	1432
Storm	–	–	142	102	1496
Cyclone	5	2	687	586	2813
Cloud burst	-	1	3	1	11
Lightening	4	–	–	2	2
Earthquake	–	–	1	–	–
Total	111	61	3237	1364	8126

Source Pachuau et al. (2020)

Table 6.2 Landslide hazard class and their corresponding area

Hazard classes	Area	Percentage (%)
Very high	1822.48	8.65
High	4263.79	20.22
Moderate	8903.47	42.24
Low	5011.57	23.77
Very low	968.72	4.60
Water body	111.97	0.53

Source Pachuau et al. (2020)

landslide-prone area. In the state, the capital district—Aizawl has been ranked first in landslide frequency, followed by the Lunglei district (Fig. 6.2). After Aizawl, Lunglei is the second largest and fastest growing district in terms of urbanization. In both the districts, rainfall and unscientific slope uses have led to increased cases of landslides. The majority of landslides were sliding, followed by fall, flow, and subsidence.

In his case study, Chenkual (2015) studied the incidences of landslides in Laipuitlang, Aizawl. He ascertained that human activities in the hill slopes are the primary reason behind the landslide.

Geologically, the Laipuitlang area falls under the middle Bhuban formation of the Surma group, having a steep slope (43–47) with a strike direction of N40° E and N45° E. Heavy rainfall with 2064 mm was recorded in Laipuitlang in 2012, resulting in a long crack in the rock bed. Further, heavy construction was also observed, and this has been learned to be the main cause behind the landslides in this region. This tragedy resulted in the total destruction of 15 buildings and the loss of the lives of 8 people.

Verma (2014) also did landslide hazard zonation on the Laipuitlang location. He found that excessive seepage between joints and heavy construction on a hill slope triggered the landslide in the Laipuitlang area. Rao and Verma (2017) reported nine prominent landslide sites along with Aizawl city to Lengpui airport road. They further added that lack of proper drainage and rainfall monitoring systems was the primary reasons for landslides along this section. Hunthar Veng, a subsidence and landslip site region, present along the above-mentioned road section, has been classified as a highly susceptible zone (Plate 6.1). Another study was done by Pandey et al. (2017) in the Aizawl district. Due to slope cutting, major landslides also happened along the road side within the upper Bhuban formation. It has been found that landslide densities are higher

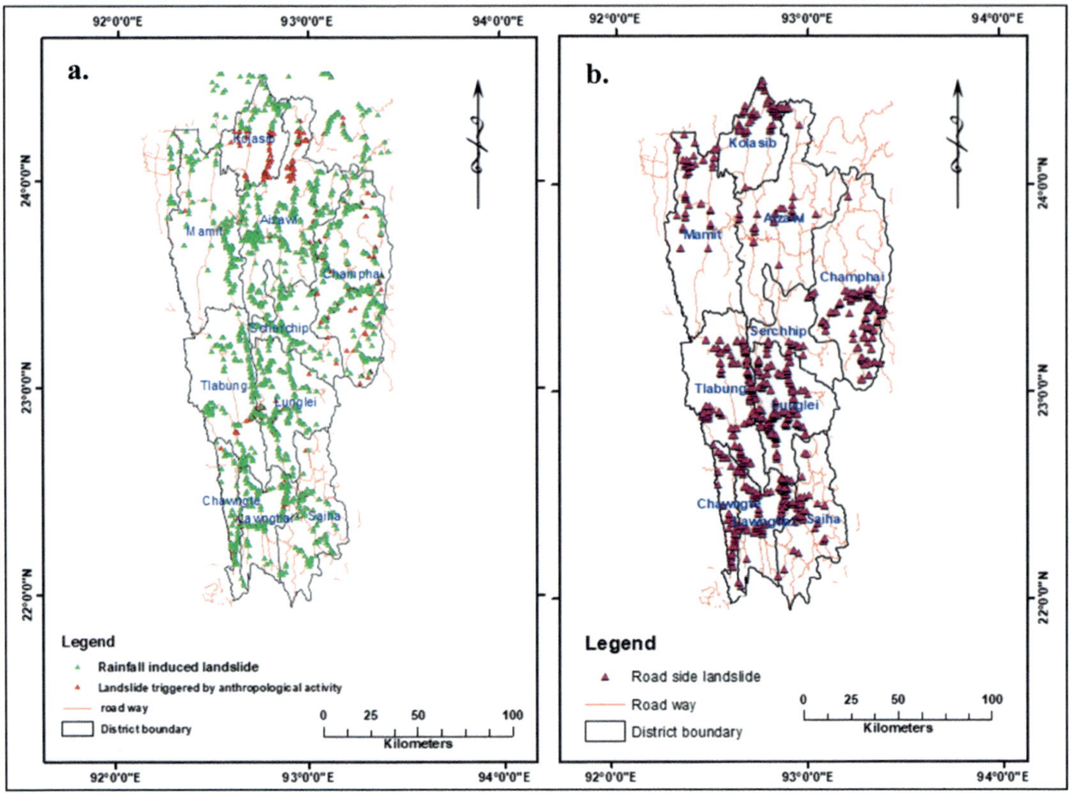

Fig. 6.1 Landslide distribution in Mizoram, **a** distribution of landslide that induced by rainfall and anthropological factors, **b** distribution of landslide that happen along roadside

Fig. 6.2 District wise landslide distribution in Mizoram

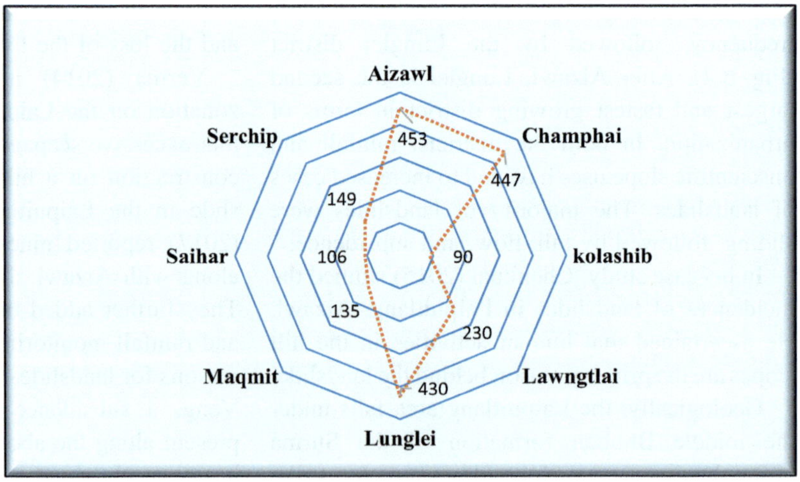

in areas in between high and low drainage density areas. Srivastava et al. (2002) considered geomorphology, lithology, slope, transport network, and settlement locations while finding out the landslide hazard zones of Aizawl district. They emphasized plantation and a ban on any type of construction to increase slope stability. Rakshit and Bharali (2018) studied the rock fall

Plate 6.3 Hunthar Veng subsidence area

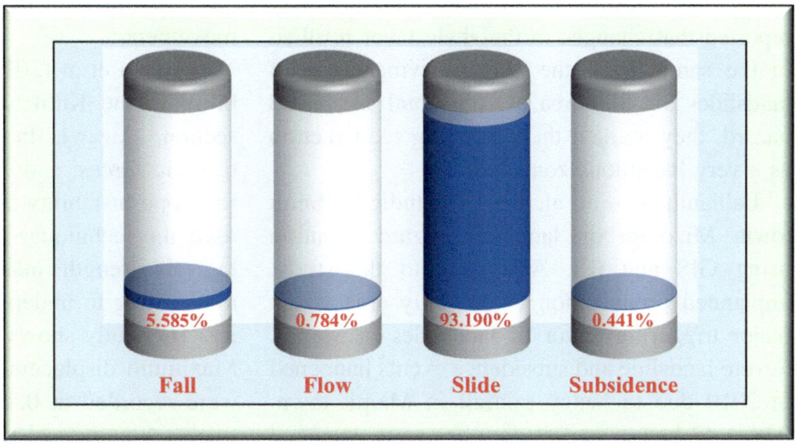

Fig. 6.3 Types of Landslide in Mizoram

hazard and risk assessment of the Aizawl-Durtlang road section. It is a sandstone area, with the bedrock being 10 m thick. They considered topography, slope, stream pattern, slope direction, and rock strength for risk assessment. The in situ geotechnical technique was used to measure the rock strength by them. Their study found that wedge failure was the most dominant type of failure in this road section (Fig. 6.3).

Kumar and Singh (2008) studied landslide hazard and development planning in Lunglei town using total estimated hazard (TEHD) values and discovered that only 0.61% of Lunglei town is in a high hazard zone, followed by 2.44% in a moderate hazard zone, 6.94% in a low hazard zone, 79.17% in a very low hazard zone, and 10.17% in a very low hazard zone. Lallianthanga et al. (2013) have also done a study on Lunglei town using the BIS guidelines of the Geological Society of India. Their study revealed that Rashi Veng, Hauruang, College Veng, the eastern side of Bazar Veng, and Theirat areas are the most prone to landslides.

Khullar and Chakraborty (2002) did a study in the Lawngtlai and Saiha town areas, Mizoram. According to their study, slope gradient was the most important causative factor of landslides. They used gradient changing points as the basis of landslide hazard zonation. An equal slide contour value of more than 25 indicates a high hazard zone; an equal slide contour value of 10–25 indicates a medium landslide zone, and a contour value of less than 10 indicates a low landslide hazard zone.

Verma and Blick (2018) did landslide hazard zonation along the Thakthing Veng road section using the landslide hazard evolution factor rating scheme. Geologically, the Thakthing Veng area consists of soft shale, siltstone, shale, and silty shale. Due to insufficient drainage planning, they reported that changes in the shale layer resulted in the same being the major driving force of landslides in the area. Using total estimated hazard, they marked the Thakthing road section as a very hazardous zone.

Lallianthanga et al. (2013) studied Mamit town, Mizoram on landslide hazard zonation using GIS and RS. According to the study, unplanned urbanization in the hilly area is the major triggering factor of landslides in Mamit. Severe landslide and subsidence events happened in 2010 due to heavy rainfall in Mamit town. About 43 houses and 50 families were damaged due to landslides that occurred in 2010. Unconsolidated slope material with a high slope angle was found to be the triggering factor of landslides in Mamit town. They divided the whole town into five susceptibility classes, and found that only, 1.78% of the area was in the very low susceptible zone. They suggested that the town could be pushed toward a severe landslide-prone area without proper monitoring and planning.

Pachuau (2019) worked on susceptibility and risk assessment of landslides in Serchip, Mizoram. He used the technique of GIS weight approach, considering geological structure, slope, house type, lithology, etc. He divided the Serchip town area into five susceptible classes.

Kumar et al. (1996) discussed the instability problem in important sandstone quarries in south Hlimen, Mizoram. On August 9th, 1992, the region faced severe rockfall. Their study emphasized the presence of joints, and they suggested excavation of sandstone initiated from top to bottom instead of piedmont.

Sarkar and Samanta (2017) did a stability study at the Keifang landslip area. A dominant landslide site, Keifang, is situated 66 km south-east of Aizawl city. The landslide was recorded in February 2014. Although no rainfall occurred in February 2014, excavation works led to landslides due to slope instability. Another slide occurred in the month of April 2014. The main cause of the slide was identified as cracks at 2–3 m depth. A RCC study revealed that the Keifang landslide area has an unstable slope. Retaining wall or gabion wall was suggested at the foothill of the slope to resist future movement.

Sardana et al (2019) discussed the slope stability in the Kulikawn to Saikhamakawn road section, Aizawl. In Mizoram, heavy rainfall, tectonic forces, and earthquakes are responsible for slope instability as well as a landslide. They used the techniques of slope mass rating, geological strength index, and continuous slope mass rating to understand the nature of instability. The study shows that the area is not stable. Maximum displacement and maximum velocity were recorded at 0.79 m and 0.76 m/s, respectively. A numerical two-dimensional model was applied in 11 locations to determine the failure types. Wedge failure was the dominant sliding type observed along the abovementioned road section.

6.4 Conclusions

This review paper was done on landslides in the geological, geomorphological, and anthropological aspects of Mizoram. The review reveals that Mizoram is situated in a moderate-to-high susceptible landslide hazard zone. Proper monitoring, management, and awareness can reduce this risk. From the various reviews and inventory of landslides from Bhukosh, it is found that most of the landslides occur along the road section and are rainfall induced. In Mizoram, Aizawl has the highest incidence of landslides, followed by Lunglei. The reason can be attributed to unscientific urban growth without considering slope stability measures. Vegetation cover can protect soil erosion and restrict landslides. The use of Vetiver grass and bamboo, both of which grow abundantly in Mizoram, can help work as a binding agent for the landslide-prone slopes of Mizoram. Improvement and awareness of indigenous knowledge (Changkham) and community-based work are suggested to reduce landslide risk for Mizoram.

Acknowledgements The authors would like express their heartfelt gratitude to Md. Jiyarul Hoque, Dr. Binoy kumar Barman, and Subhom Narjinary for sharing their constructive knowledge.

References

Anbalagan R (1992) Landslide hazard evaluation and zonation mapping in mountainous terrain. Eng Geol 32(4):269–277

Barman BK, Rao KS (2021) Geology of sedimentary formation in the state of Mizoram, NE India: a review. In: Sati VP (ed), Today and Tomorrow`s Printers and Publishers, New Delhi-110002. ISBN: 9788170197010

Barman BK (2021) Drainage Morphometry and Morphotectonics of Upper Tuirial River Basin, Mizoram. Unpublished PhD thesis, Department of Geology, Mizoram university

Biswas B, KS V, Ranjan R (2021) Landslide susceptibility mapping using intregrade approach of multi-criteria and geospatial techniques at Nilgiris district of India. Arab J Geosci (14)

Chenkual L (2015) Some issues and concern from 2013 rockslide disaster at Laipuitlang, Aizawl, Mizoram India. Sci Vis 15(1):34–38

Fall M, Azzam R, Noubactep C (2006) A multi-method approach to study the stability of natural slopes and landslide susceptibility mapping. Eng Geol 82:241–263

Girma F, Raghuvanshi TK, Ayenew T, Hailemariam T (2015) Landslide hazard zonation in Ada Berga district, Central Ethiopia—a GIS based statistical approach. J Geom 9(i):25–38

Guzzetti F, Carrara A, Cardinali M, Reichenbach P (1999) Landslide hazard evaluation: a review of current techniques and their application in a multi-scale study, Central Italy. Geomorphology 31:181–216

Kanungo DP, Arora MK, Sarkar S, Gupta RP (2006) A comparative study of conventional, ANN black box, fuzzy and combined neural and fuzzy weighting procedures for landslide susceptibility zonation in Darjeeling Himalayas. Eng Geol 85:347–366

Khullar VK, Chakraborty U (2002) Landslide Hazard Zonation in Lunglei and Chhimuipui districts, Mizoram. In: Dubey CS, Saklani PS (ed), Pilgrims Publishing, Varanasi-221010. ISBN-81-7769-187-2

Kumar P, Singh CD (2008) Landslide hazard zonation and evaluation of Lunglei Town, Mizoram-its utility in planning and development. J Engg Geol 35(1–4):313–319

Kumar S, Singh TN, Sawmliana C (1996) On the instability problems in South Hlimen landslide, Mizoram. Indian J Eng Mat Sci 4:78–80

Lallianthanga RK, Lalbiakmawia F, Lalramchuana F (2013) Landslide hazard zonation of Mamit Town, Mizoram, India using remote sensing and GIS techniques. Int J Geol Earth Environ Sci 3(1):184–194

Mengistu F, Suryabhagavan KV, Raghuvanshi TK, Lewi E (2019) Landslide Hazard zonation and slope instability assessment using optical and InSAR data: a case study from Gidole town and its surrounding areas, southern Ethiopia. Remote Sens Land 3(1):1–14

Pachuau L (2019) Zonation of landslide susceptibility and risk assessment in Serchhip town, Mizoram. JIndian Soc Remote Sens 47(9):1587–1597

Pachuau Z, Hamar Z, Chenkual L, Zoenmawii (2020) Mizoram State disaster management plan, 2020. Disaster management and rehabilitation department. Govt. of Mizoram, Chawupui

Pandey AC, Gulab S (2017) Evaluating geo-aspects implications on land sliding in himalayan terrain in a part of aizwal district, mizoram state using remote sensing and gis. International Archives of the Photogrammetry, Remote Sens Spat Inf Sci 34(Part XXX)

Pourghasemi, H.R., Teimoori Yansari, Z., Panagos, P. (2018) Analysis and evaluation of landslide susceptibility: a review on articles published during 2005–2016 (periods of 2005–2012 and 2013–2016). Arab J Geosci 11, 193.https://doi.org/10.1007/s12517-018-3531-5

Raghuvanshi TK, Ibrahim J, Ayalew D (2014) Slope stability susceptibility evaluation parameter (SSEP) rating scheme—an approach for landslide hazard zonation. J Afr Earth Sci 99:595–612

Rakshit R, Bharali B (2018) Rockfall Hazard and risk assessment along Aizawl-Durtlang road section, Mizoram, India. Senhri J Multi Stud III(1):19–31. ISSN 2456-3757

Rao CUB, Verma R (2017) Micro-zonation of landslide hazards between Aizawl City and Lengpui Airport, Mizoram, India, using Geoinformatics. Int J Basic Appl Sci IJBAS-IJENS 17(05):7–17

Sardana S, Verma AK, Verma R, Singh TN (2019) Rock slope stability along road cut of Kulikawn to Saikhamakawn of Aizawl, Mizoram India. Nat Hazards 99(2):753–767

Sarkar S, Kanungo DP, Mehrotra GS (1995) Landslide hazard zonation: a case study in Garhwal Himalaya India. Moun Res Dev 15(4):301–309

Sarkar S, Samanta M (2017) Stability analysis and remedial measures of a landslip at Keifang, Mizoram–a case study. J Geol Soc India 89(6):697–704

Scheidegger AE (1994) Hazards: singularities in geomorphic systems. Geomorphology 10"19–25

Srivastava HS, Ravindran KV, Chhetri P, Singh OP, Manchanda ML (2002) Land slide hazard zone mapping of Aizawl district using remote sensing and GIS techniques. J Ind Geophys Union 6(4):183–185

Turrini CT, Visintainer P (1998) Proposal of a method to define areas of landslide hazard and application to an area of the Dolomites, Italy. Eng Geol 50:255–265

Varnes DJ (1984) IAEG commission on landslides and other mass-movements landslide hazard zonation: a review of principles and practice. UNESCO Press, Paris, p 63

Verma R (2018) The major drainage systems in the northeastern region of India. Indian Rivers 429–464

Verma R, Blick J (2018) Landslide hazard estimation in Thakthing Veng Road Section, Aizawl, Mizoram. Int J Civil Eng Technol 9(4):298–307

Verma R (2014) Landslide hazard in Mizoram: case study of Laipuitlang landslide, Aizawl. Int J Sci Res (IJSR) 2319–7064. ISSN (Online)

Application of Geostatistical and Geospatial Techniques for Groundwater Quality Vulnerability Assessment Using Hydrogeochemical Parameters: A Case Study of NCT Delhi

7

Mohd Sayeed Ul Hasan, Sufia Rehman, Nadeem Akhtar, Abhishek Kumar Rai, Samina Wasi, Md Nashim Akhtar, and Shams Tabrez

Abstract

We use an integrated approach for evaluating groundwater vulnerability using the multi-criteria decision making (MCDM), statistical, and geographic information system (GIS). Hydrogeochemical data from 50 sample points were converted to a surface raster using the inverse distance weightage method. The literature survey indicates absence of systematic study of the water quality vulnerability in the study area. Ten thematic layers namely pH, total dissolved solid (*TDS*), bicarbonate (HCO_3^-), chloride (Cl^-) sulfate (SO_4^{2-}) nitrate (NO_3^-), fluoride (F^-), calcium (Ca^{2+}), magnesium (Mg^{2+}), and total hardness (TH) as calcium carbonate were used for the weighted overlay analysis. Our analysis indicates that the NCT Delhi region can be classified into five classes based on groundwater vulnerability i.e. very low, low, moderate, high, and very high. The result shows that percentage of the area of groundwater vulnerability is very low (17.9%), low (22.0%), moderate (20.7%), high (19.9%), and very high (19.5%) respectively. The land use pattern of the study area was found to be 1.53% of the land covered with water, 5.46% trees, 0.36% grass, 0.01% of flooded vegetation, 28.69% of crops, 3.64% of shrubs, 59.97% as the built-up area, and 0.35% as the bare land, for the year 2020.

Keywords

MCDM · Geostatistics · GIS · Groundwater · Hydrogeochemical · Vulnerability

M. S. U. Hasan · N. Akhtar · M. N. Akhtar
Department of Civil Engineering, Aliah University, New Town, Kolkata 7000160, India

S. Rehman
Department of Civil Engineering, National Institute of Technology Karnataka, Surathkal 575025, India

M. S. U. Hasan · A. K. Rai (✉)
Centre for Oceans, Rivers, Atmosphere and Land Sciences, Indian Institute of Technology Kharagpur, Kharagpur, West Bengal 721302, India
e-mail: abhishek@coral.iitkgp.ac.in

S. Wasi
Department of Biochemistry, College of Medicine, Imam Abdulrahman Bin Faisal University, Dammam 31952, Saudi Arabia

S. Tabrez (✉)
King Fahd Medical Research Center, King Abdulaziz University, Jeddah 21589, Saudi Arabia
e-mail: shamstabrez1@gmail.com

Department of Medical Laboratory Sciences, Faculty of Applied Medical Sciences, King Abdulaziz University, Jeddah 21589, Saudi Arabia

J. Das and S. K. Bhattacharya (eds.), *Monitoring and Managing Multi-hazards*,
GIScience and Geo-environmental Modelling, https://doi.org/10.1007/978-3-031-15377-8_7

7.1 Introduction

Groundwater is an essential natural resource for the domestic, industrial, and agricultural purposes. It accounts for approximately 43% of global irrigation use (Das et al. 2021). Groundwater plays a crucial role in maintaining the river flows and water level in the wetlands and lakes. Estimating groundwater volume is difficult in a country or region as it depends on the permeability of rocks and characteristics of the aquifer (Margat and Van der Gun 2013). Rapid industrialization, changing climate and growing population have created tremendous pressure on groundwater resources, resulting in the crisis of this resource (Basak et al. 2021). In developing nations, pollutants from inadequate water sources, trash, unmanaged waste, and insufficient fecal coliform drainage clog the waterways. The other contaminants originating from pharmaceuticals and plastics are also ejecting dangerous chemicals and choking the waterways (Damania et al. 2019). The scarcity of clean water globally is a significant problem the population faces. This water strain is likely to increase by 34% in 2050 and could be further aggravated by the increasing population and diminishing freshwater resources. Currently, nearly 47% population are residing in areas with severe water scarcity for at least a month in a year. It is also projected that by 2050, more than half of the world population (57%) will live in water-scarce regions at least for a month every year (Boretti and Rosa 2019). The main factors responsible for water stress are rapid and unplanned urbanization, climate change, and hydrological variability. Many areas in Asia and Africa are experiencing severe water stress mainly due to extreme hydrologic variability (AID 2008). The bioremediation process is an effective way to clean and detoxify groundwater.

Earlier, surface water was considered vulnerable to groundwater, but increasing waste pollution has depleted the groundwater resources. Moreover, the rural population in India is dependent on groundwater for irrigating their fields (Madhav et al. 2018). Thus, monitoring and assessing groundwater quality are essential for the critical management of groundwater resources. Remote sensing or drone-based water quality assessment is comparatively a recent technique becoming prominent due to their growing accuracy. Moderate and high-resolution satellite datasets, including Envisat MERIS, Sentinel, etc., and models have allowed precise prediction of water quality parameters. It further helps establish a relationship between water quality and satellite data by comparing the water quality parameters with the in-situ measurements (Damania et al. 2019). The development of remote sensing-based water quality analysis started in 1970s. Remote sensing establishes an empirical relationship by estimating the spectral and thermal differences of the reflected energy from water bodies. Remote sensing datasets are inexpensive and reveal the spatial variation in the water quality. New algorithms and high-resolution datasets also effectively make water quality assessment possible in rivers and lakes (Khan and Jhariya 2018). Variation in groundwater quality spatially depends on the geological formation and the human-induced activities in the region of the groundwater basin. Several studies have been carried out to use the geographical information system (GIS) to demarcate the groundwater potential zone and assess groundwater quality (Kawo and Karuppannan 2018; Singh et al. 2018; Azimi et al. 2018; Ahmad et al. 2020). Various scholars have used different methods to assess groundwater quality. Remote sensing with GIS techniques, hydrogeochemical methods, geostatistical, multivariate statistical analysis, and multi-criteria decision-making models are used for groundwater quality assessments (Goswami and Rai 2022).

Kawo and Karuppannan (2018) utilized a GIS-based water quality index (WQI) to examine the groundwater quality in the Modjo River Basin of Central Ethiopia. Duraisamy et al. (2019) carried out GIS and in-situ measurements to assess the groundwater quality in Kangayam taluk, Tirupur district of Tamil Nadu. El-Rawy et al. (2019) examined the groundwater quality using hydrogeochemistry, factor analysis, and

GIS techniques in Qena governorate, Egypt. They emphasized that GIS effectively analyzes the spatial distribution of ions and groundwater quality. Islam et al. (2018) suggested that the groundwater quality index (GWQI) is essential in identifying groundwater quality in an area.

Further, the use of integrated hydro-chemical techniques, multivariate statistics, and GIS has increased exponentially during the past few decades (Nayak et al. 2017; Islam et al. 2018). Hasan and Rai (2020) effectively used the entropy information theory and the GIS to assess the groundwater quality in the lower Ganga basin. The analytical hierarchy process (AHP) is a widely known multi-criteria technique and is used prominently in GIS to analyze suitability and zonation mapping. It helps in an effective decision-making process by analyzing the multiple criteria in a pairwise-comparison matrix. It assigned the weightage to the factors as per their relative significance and aggregated them based on weightage similar to the weighted linear combination. Each criterion was ranked from 1 to 9 on their relative significance. The criteria were then used to prepare the pairwise-comparison matrix to compare the various factors (Bera et al. 2020). Researchers have used fuzzy AHP, frequency ratio, geostatistical methods, and frequency ratio to assess the groundwater quality at spatial scales (Karan et al. 2018; Lad et al. 2019; Mahammad and Islam 2021). Bioremediation refers to the microbial species used for cleaning the chemically contaminated groundwater and soil (Wasi et al. 2013a, b). The bioremediation process encourages microbes to thrive that disintegrate the chemical and food contaminants. In this process, the growth of pollutant-eating organisms, including protists, fungi, bacteria, etc., is enhanced for decomposing the contaminants (Speight 2018). Several experimental studies have been carried out using bioremediation to improve groundwater quality (Pous et al. 2013; Majone et al. 2015). Depletion of groundwater quality is a significant problem in metropolitan areas. Studies have stressed the depleting groundwater quality; however, evidence-based assessment is limited, especially in densely populated areas. In this study, a concerted effort has been made to examine the groundwater quality in Delhi using a GIS-based analytical hierarchy process.

7.2 Study Area

Delhi is one of the most densely populated metropolitan cities globally and is also known as the National Capital Territory (NCT) and the capital of India. The angular extent of NCT Delhi lies as 28^0 40′ to 28^0 87′ north latitude, and from 76^0 84′ to 77^0 34′ lies along the river Ganga. Figure 7.1 shows the study region's spatial location and the sample point location of hydrogeochemical parameters. The climatic condition of NCT Delhi shows a wide range of hot and winter seasons with humid and semi-arid nature. The region is highly vulnerable to groundwater scarcity as well as quality perspective. The groundwater availability shows a large spatial variability ranging from shallow depth of 5 m to 20–64 m deep below the ground level. The objective of this work was to delineate the vulnerable groundwater zones which could lead to deteriorating human health and the socio-economic status of Delhi.

7.3 Data and Methodology

The hydrogeochemical parameters were collected from Central Groundwater Board (CGWB 2021) under the aegis of the Ministry of Jal Shakti, Govt. of India. The point data of ten hydrogeochemical parameters such as pH, total dissolved solid (*TDS*), bicarbonate (HCO_3^-), chloride (Cl^-) sulfate (SO_4^{2-}), nitrate (NO_3^-), fluoride (F^-), calcium (Ca^{2+}), magnesium (Mg^{2+}), and total hardness (TH) as calcium carbonate. The point data are converted into a spatial raster database using a geostatistical technique of inverse distance weightage method (IDW). All the 50 samples point data are now converted to a thematic layer for the overlay analysis achieved by the analytical hierarchy process based on the subjective nature of multi-criteria decision-making. The spatial distribution of hydrogeochemical parameters is

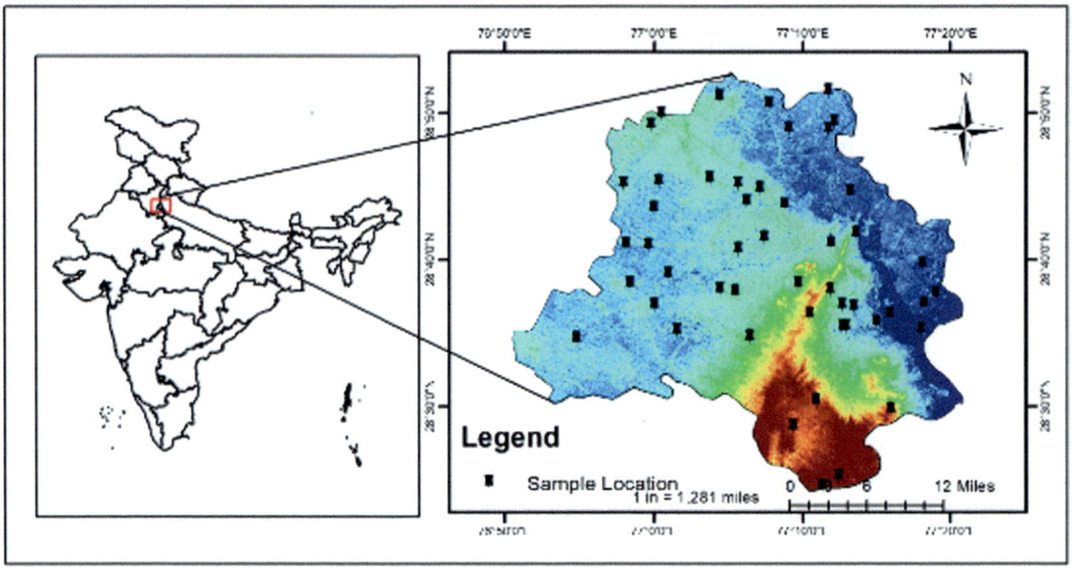

Fig. 7.1 Spatial location of study area (NCT Delhi)

shown in Fig. 7.2a, b, and the descriptive statistics are provided in Table 7.1.

7.3.1 Groundwater Vulnerability Mapping

We use the analytical hierarchical process (AHP), a widely used and effective multi-criteria decision-making method (Rehman et al. 2021). This technique was proposed by Thomas L. Saaty (Saaty and Vargas 1980) for determining the criteria weightage for the multi-criteria decision-making model (MCDM). AHP examines the multiple indicators in the decision-making process by quantifying and creating a hierarchical structure as per the mathematical and psychological logic for integrating quantitative and qualitative decisions (Meng and Pian 2016). AHP considers structural characteristics and alternatives for calculating attributes within a hierarchical structure (Mitra and Das 2022). Weightage and ranking preferences were assigned to attributes at a scale ranging from 1 to 9 using the pairwise-comparison matrix. Consistency ratio (CR) is estimated by dividing the set preferences by a random index matrix to the consistency index (Das et al. 2017). The linear algebraic procedure assigns the relative ranking to a series of hierarchical matrices (Dyson 2017). AHP accepts the inconsistency and effectively identifies suitable solutions in the decision-making process (Qazi et al. 2018).

AHP is a widely accepted method for assessing groundwater vulnerability Gangadharan and Vinoth 2016; Sahoo et al. 2016; Das and Pal 2020). Thus, AHP is used to examine the influence of water quality parameters for analyzing the groundwater vulnerability in NCT Delhi. Identifying the parameters influencing the quality of groundwater is essential for sustainable development of any region. In this study, groundwater quality vulnerability assessment was carried out in five steps: construction of hierarchies, formulation of the pairwise-comparison matrix, estimation of weightage, groundwater vulnerability estimation and determination of consistency matrix. Integration of all the parameters using their respective weights was carried out to prepare the groundwater vulnerability map for the study region.

A pairwise-comparison matrix (designated by X) was prepared to examine the relative importance of the parameters (Choudhury et al. 2022)

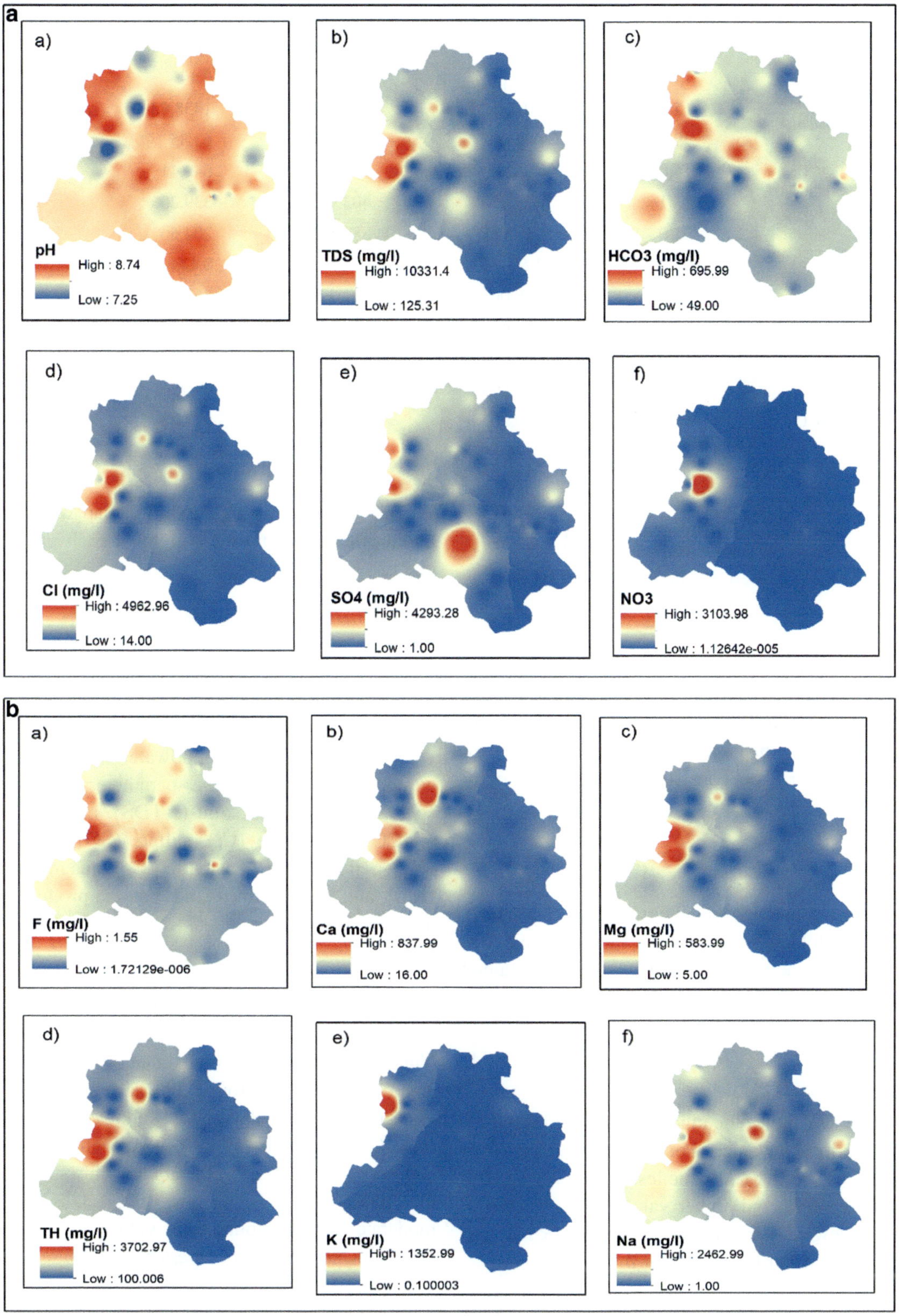

Fig. 7.2 a Spatial variations of hydro-geochemical parameters (a) pH, (b) TDS (mg/l), (c) HCO_3^- (mg/l), (d) Cl^-(mg/l), (e) SO_4^{2-} (mg/l), and (f)NO_3^- (mg/l). **b** Spatial variations of hydro-geochemical parameter (a) F^- (mg/l), (b) Ca^{2+} (mg/l), (c) Mg^{2+} (mg/l), (d) TH (mg/l), (e) K^+ (mg/l), and (f) Na^+ (mg/l)

Table 7.1 Descriptive statistics of hydrogeochemical parameters

Parameter	Min	Max	Mean	Standard deviation
pH	7.25	8.74	8.16	0.14
TDS	125.32	10,331.37	2138.91	1458.83
HCO_3	49.00	696.00	235.09	63.35
Cl	14.01	4962.96	725.53	642.41
SO_4	1.01	4293.28	403.90	339.20
NO_3	0.00	3103.98	88.45	204.50
F	0.00	1.55	0.52	0.17
Ca	16.00	838.00	121.37	86.60
Mg	5.00	584.00	97.39	77.29
Na	1.00	2462.99	438.99	283.58
K	0.10	1352.99	35.90	99.22
TH as $CaCO_3$	100.01	3702.97	697.01	512.23

$$X = \begin{pmatrix} x_1 & x_2 \cdots & x_3 \\ x_{11} & x_{12} \cdots & x_{13} \\ x_{n1} & x_{n2} \cdots & x_{nn} \end{pmatrix} \tag{7.1}$$

$$X_{ij} = \frac{w_i}{w_j} \tag{7.2}$$

where X is the relative importance and n denotes the total number of groundwater quality parameters. Wj is the weight of each parameter of alternative j. The consistency index was computed as

$$CI = \frac{\lambda \max - n}{n-1} \tag{7.3}$$

where λ = maximum eigenvalue of the matrix and n = total number of water quality parameters

Consistency of the matrix was ascertained through consistency ratio (CR) computed as

$$CR = \frac{CI}{RI} \tag{7.4}$$

where RI denotes the random index which varies for various n values. For ten parameters, RI is taken as 1.49. A CR value lesser than 0.1 or 10% is acceptable, and higher values are subjected to be reviewed. The computed value of CR is found as 0.08, which is lesser than 0.10.

The groundwater vulnerability map was prepared after assessing the weight of each parameter as

$$\begin{aligned} GWV = {} & [(\mathrm{pH}\,0.23) + (\mathrm{TDS}\,0.15) + (\mathrm{HCO_3}\,0.25) \\ & + (\mathrm{Cl}\,0.10) + (\mathrm{SO_4}\,0.10) + (\mathrm{NO_3}\,0.06) + (\mathrm{F}\,0.06) \\ & + (\mathrm{Ca}\ 0.03) + (\mathrm{Mg}\,0.02) + (\mathrm{TH}\,0.01)] \end{aligned} \tag{7.5}$$

where GWV is the groundwater vulnerability computed by adding the parameters, namely (pH) total dissolved solid (s(TDS), bicarbonate ((HCO_3)), chloride (Cl), sulfate (SO_4), nitrate (NO_3), fluoride (F), calcium (Ca), magnesium (Mg), and total hardness (TH) as the product of their weights.

7.4 Results and Discussion

Vulnerability analysis revealed that out of the total area of NCT Delhi, around 22% area came under low groundwater vulnerability, followed by moderate (20.7%), high (19.9%), very high (19.5%), and very low (17.9%) vulnerability. This suggests that more than half the area (60%) falls under moderate to high groundwater vulnerability zone. Wards located in the northwestern and central part

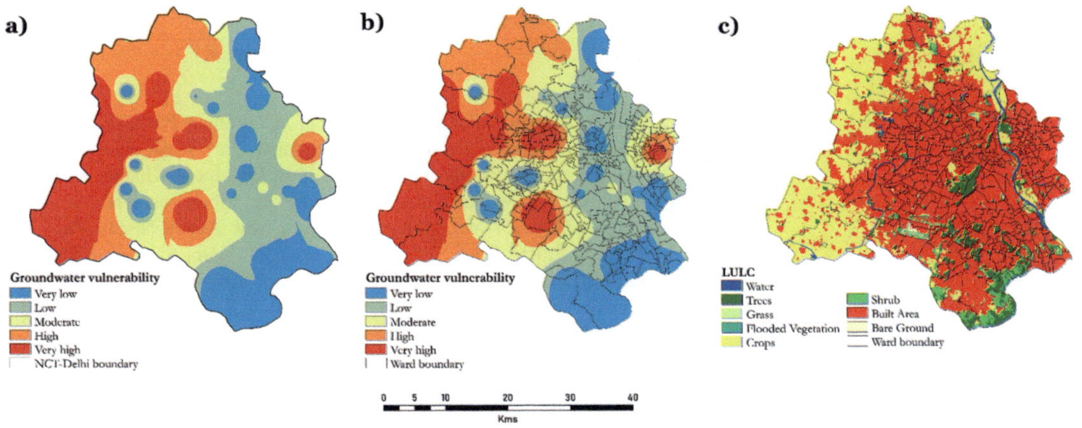

Fig. 7.3 Final groundwater vulnerability map: (**a**) spatial distribution of groundwater vulnerability, (**b**) various wards under groundwater vulnerability, and (**c**) land use/land cover map

Table 7.2 Computed pairwise matrix for the weight calculations of the respective parameters

Parameter	pH	TDS	HCO_3	Cl	SO_4	NO_3	F	Ca	Mg	TH as $CaCO_3$	Weight
pH	1										0.23
TDS	0.33	1									0.15
HCO_3	1	1	1								0.25
Cl	0.33	0.5	0.2	1							0.1
SO_4	0.25	1	0.2	0.5	1						0.1
NO_3	0.33	0.33	0.14	0.5	0.5	1					0.06
F	0.17	0.2	0.25	0.5	0.5	0.5	1				0.06
Ca	0.25	0.14	0.14	0.2	0.14	0.5	0.25	1			0.03
Mg	0.17	0.14	0.14	0.17	0.14	0.2	0.17	0.33	1		0.02
TH as $CaCO_3$	0.11	0.13	0.11	0.25	0.17	0.17	0.17	0.25	0.33	1	0.01

of the study area are likely to experience high vulnerability (Table 7.3 and Fig. 7.3). The land use pattern map was obtained from the ESRI, having a spatial resolution of 10 m for 2020. The concentration of TDS, chloride, fluoride, calcium, magnesium, and total hardness were identified as the potent influencing parameters in the high and very high vulnerability areas of Alipur, Anand Vihar, Anarkali, Bakhtawarpur, Banker, Bawana, Begumpur, Delhi Cantt, Dilshad colony, Dilshad Garden, Guru Harikishan Nagar, Kapashera, Karala, Krishna Nagar, Manglapuri, Nangli Sakravati, Nangloi East, Narela, Nilothi, Palam, Paschim Vihar South, Pooth Kalan, Punjabi Bagh, Quammruddin Nagar, Raja Garden, Rampura, Shahdara, Shakurpur, etc (Table 7.3).

Central and northeastern part was identified as moderate groundwater vulnerability zones which includes areas of Alipur, Ashok Vihar, Babarpur, Bakhtawarpur, Bhalaswa, Jahagirpur, Bijwasan, Chauhan Banger, Dharampura, Gandhi Nagar, Geeta Colony, Gokalpur, Hari Nagar, Harsh Vihar, Hastsal, I.P Extention, Karam Pura, Kardam Puri, Khyala, Krishna Nagar, Kunwar Singh Nagar, Madhu Vihar, Mahavir enclave, Mahipalpur, Major Bhupinder Singh Nagar, Mansarovar Garden, Matiala, Maujpur, Mohan Garden, Nanak Pura, Naraina,

Table 7.3 Area and different wards under vulnerability classes

Vulnerability classes	Area (ha)	Area (%)	Wards
Very low	26,516.5	17.9	Aya Nagar, Badarpur, Bakhtawarpur, Bhati, Chitranjan Park, G.T.B. Nagar, Hastsal, Jaitpur, Janak Puri North, Janak Puri West, Jharoda, Kakraula, Kamla Nagar, Khanpur, Khichripur, Madanpur Khadar, Mahavir Nagar, Malkaganj, Mayur Vihar Phase-I, Mayur Vihar Phase II, Meetheypur, Model Town, Nangli Sakravati, New Ashok Nagar, Pul Pehlad, Rana Pratap Bagh, Roshanpura, Said Ul Ajaib, Sangam Vihar, Sangam Vihar Central, Sangam Vihar East, Sangam Vihar West, Sarita Vihar, Tekhand, Tigri, Trilok Puri, Tughlakabad, Tughlakabad Extn, and Vikas Nagar
Low	32,595.9	22.0	Amar colony, Andrewsganj, Bakhtawarpur, Ballimaran, Bazar Sitaram, Bhajanpura, Bhalaswa Jahagirpur, Bhalswa, Bhogal, Brahampuri, Burari, Chandni Chowk, Chhatarpur, Chirag Delhi, Daryaganj, Deoli, Deputyganj, Dhirpur, East of Kailash, Greater Kailash-I, Hauz Khas, Idgah Road, Kalkaji, Karawal Nagar East, Karawal Nagar West, Karol Bagh, Kashmere Gate, Khajoori Khas, Kishan Ganj, Kotla Mubarakpur, Kucha Pandit, Ladosarai, Lajpat Nagar, Madanpur Khadar, Malviya Nagar, Mehrauli, Minto Road, Model Basti, Mukherjee Nagar, Mukundpur, NDMC Charge 1, NDMC Charge 2, NDMC Charge 3, NDMC Charge 4, NDMC Charge 5, NDMC Charge 6, NDMC Charge 7, NDMC Charge 9, New Usmanpur, Nizamuddin, Okhla, Paharganj, Pusa, Qasabpura, Rajender Nagar, Ram Nagar, Rohini, Rohini North, Safdarjang Enclave, Samaypur Badli, Sarai Pipal Thala, Sarita Vihar, Shalimar Bagh North, Shapurjat, Shiv Vihar, Sri Niwaspuri, Timarpur, Tukhmirpur, and Zakir Nagar
Moderate	30,660.7	20.7	Alipur, Ashok Vihar, Babarpur, Bakhtawarpur, Bhalaswa, Jahagirpur, Bijwasan, Chauhan Banger, Delhi Cantt Charge 1, Delhi Cantt Charge 2, Dharampura, Gandhi Nagar, Geeta Colony, Gokalpur, Hari Nagar Harsh Vihar, Hastsal, I.P Extention, Karam Pura, Kardam Puri, Khyala, Krishna Nagar, Kunwar Singh Nagar, Madhu Vihar, Mahavir enclave, Mahipalpur, Major Bhupinder Singh Nagar, Mansarovar Garden, Matiala, Maujpur, Mohan Garden, Nanak Pura, Naraina, Nawada, New Seema Puri, Pitampura North, Preet Vihar, Pusa, Quammruddin Nagar, R. K. Puram, Raghubar Pura, Rajouri Garden, Rohini, Rohini central, Sahibabad, Daulat Pur, Sawan Park, Shalimar Bagh South, Sunder Nagari, Trinagar, Vasantkunj, Welcome Colony, Yamuna Vihar, and Zaffrabad
High	29,542.4	19.9	Alipur, Anand Vihar, Anarkali, Bakhtawarpur, Banker, Bawana, Begumpur, Delhi Cantt Charge 3, 4, 5, and 8, Dilshad colony, Dilshad Garden, Guru Harikishan Nagar, Kapashera, Karala, Krishna Nagar, Manglapuri, Nangli Sakravati, Nangloi East, Narela, Nilothi, Palam, Paschim Vihar South, Pooth Kalan, Punjabi Bagh, Quammruddin Nagar, Raja Garden, Rampura, Shahdara, Shakurpur, Vijay Vihar, and Vivek Vihar
Very high	28,984.5	19.5	Anand Vihar, Begumpur, Chhawla, Delhi Cantt Charge 6, Dichaon Kalan, Jhilmil, Karala, Khera, Kohat Enclave, Mangolpuri West, Mundaka, Paschim Vihar North, Peragharhi, Rani Bagh, Rohini East, Rohini Central, Rohini South, Shahdara, Saraswati Vihar, and Vishwas Nagar
Total	148,300	100	

Nawada, New Seema Puri, Pitampura North, Preet Vihar, Pusa, Quammruddin Nagar, R. K. Puram, Raghubar Pura, Rajouri Garden, Rohini, Rohini central, Sahibabad, Daulat Pur, Sawan Park, Shalimar Bagh South, Sunder Nagari, Trinagar, Vasantkunj, Welcome Colony, Yamuna Vihar, and Zaffrabad wards (Table 7.3). Low to very low vulnerability is found in the

Table 7.4 Regression analysis between groundwater vulnerability and the influencing parameters

Parameters	Unstandardized coefficients		Standardized coefficients	t	Sig	95.0% confidence interval for B	
	B	Std. error	Beta			Lower bound	Upper bound
pH	− 1.912	1.190	− 0.001	− 1.608	0.116	− 4.318	0.494
TDS	0.157	0.001	0.696	117.171	0.000	0.154	0.160
HCO_3^-	0.244	0.003	0.052	92.270	0.000	0.239	0.249
Cl^-	0.087	0.003	0.170	32.562	0.000	0.082	0.093
SO_4^{2-}	0.091	0.001	0.117	65.583	0.000	0.088	0.094
NO_3^-	0.062	0.001	0.049	82.615	0.000	0.061	0.064
F^-	− 0.540	0.823	0.000	− 0.657	0.515	− 2.205	1.124
Ca^{2+}	0.027	0.033	0.008	0.829	0.412	− 0.039	0.094
Mg^{2+}	0.017	0.065	0.004	0.268	0.790	− 0.114	0.149
TH as CaCO	0.015	0.015	0.024	0.986	0.330	− 0.016	0.046

Dependent Variable: Groundwater vulnerability

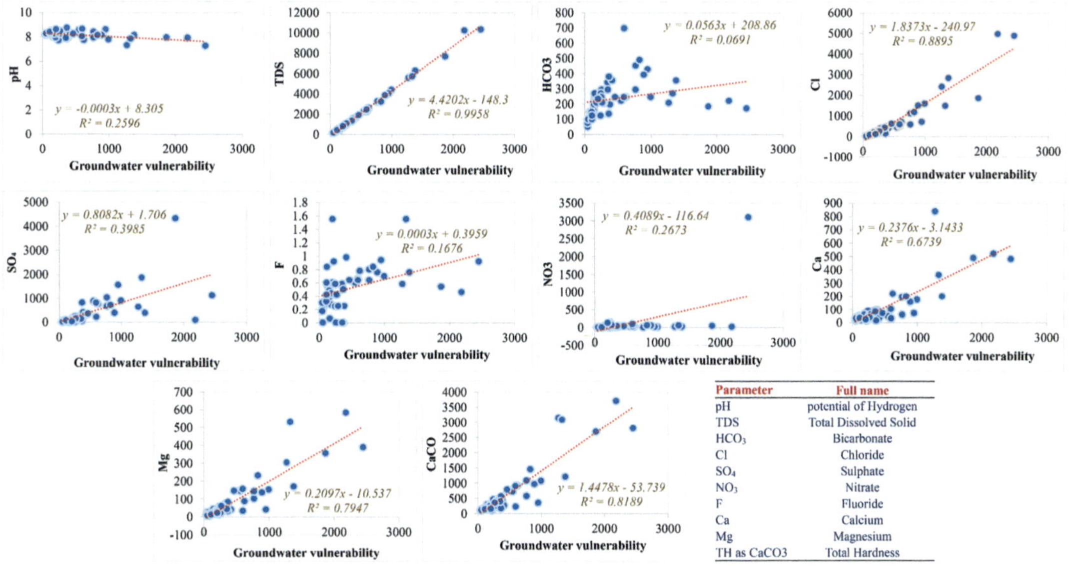

Fig. 7.4 Relationship between groundwater vulnerability and the influencing parameters

wards located in the vicinity of Yamuna River, covering a long stretch in the eastern part of the study area. Low concentrations of TDS, chloride, fluoride, calcium, magnesium, total hardness, and moderate pH resulted in low to very low groundwater vulnerability.

Most of the central and northern part of NCT Delhi has densely built-up areas, leading to low infiltration due to impervious surface. Increased built-up areas are also responsible for the increasing groundwater vulnerability in the study area (Fig. 7.3c). Multilinear regression analysis was also carried out between groundwater vulnerability and influencing parameters (Table 7.4). All the factors except pH and fluoride were positively associated with groundwater vulnerability. However, TDS, chloride, calcium, total hardness, and magnesium showed a strong relationship,

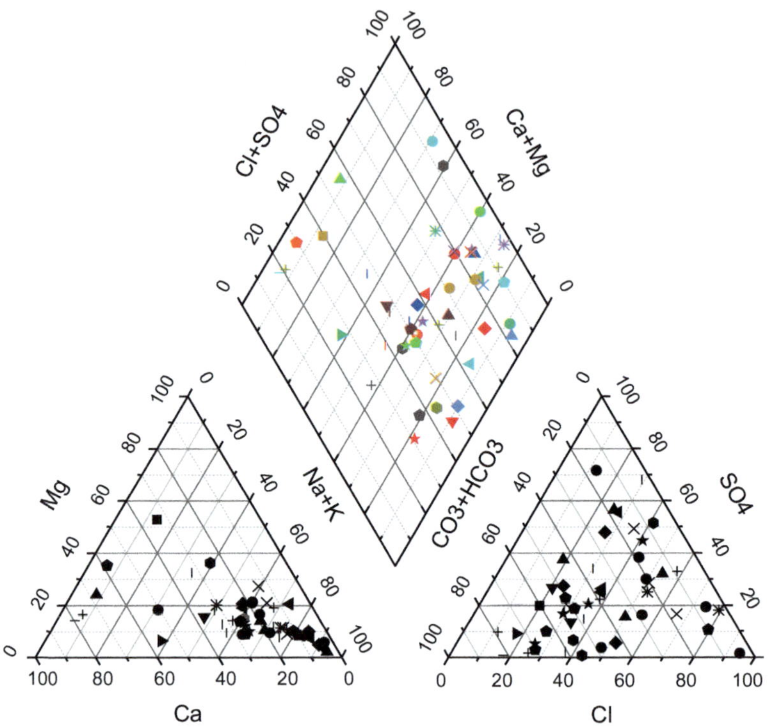

Fig. 7.5 Piper plot showing the cationic and anionic composition of groundwater

while bicarbonate and nitrate revealed a weak association with GWV (Table 7.4 and Fig. 7.4).

7.4.1 Hydrogeochemical Nature of Groundwater Quality Parameters

The piper trilinear diagrams shows the strong statistical visual relationship between the hydrogeochemical parameters of cation and anion in groundwater (Piper 1994). The trilinear chart is the composition of three ternary diagrams based on the ionic nature of the hydrogeochemical parameters. The lower left ternary diagram shows the cation (Ca^{2+}, Mg^{2+}, and $Na^+ + K^+$). The bottom right ternary diagram shows the anions (Cl^-, CO_3^- + HCO_3^-, and SO_4^{2-}) and projected transformed diamond plots in the middle of the cation, and the anion shows the water types. The ionic compositions projected on the diamond plot helps in grouping of the water quality based on the position of the samples on the plot. In our study, most of the samples lie in the right quadrant of diamond plot showing mostly sodium chloride water influenced by water-rock interactions and ion-exchange processes in the subsurface (Goswami et al. 2022). Figure 7.5 shows the trilinear plot of the NCT Delhi describing the nature of groundwater.

7.5 Conclusions

This work attempted to integrate groundwater vulnerability mapping in the NCT Delhi using geostatistical, MCDM, and geographic information systems. The vulnerability of groundwater quality is classified into five classes: very low, low, moderate, high, and very high. The percentage of the area contributing to very high vulnerability is 19.05%, and very low vulnerability covers the area of 17.9% of the total area. The groundwater vulnerability shows a strong linear relationship with TDS, Cl^-, and $CaCO_3^-$ like 0.99, 0.88, and 0.81, respectively. The resulting plot from Piper's trilinear plots represents the sodium chloride water type in the right quadrant. The results are helpful for the policymakers and the departments working for the sustainable management of groundwater for further contamination. Therefore, it is essential to take preventive measures to monitor the groundwater under changing climatic conditions by analyzing the current scenario of groundwater.

Acknowledgements The authors express their sincere thanks to the Central Groundwater Board (CGWB) for providing the groundwater data through public platforms and ESRI for providing the land use map for 2020.

References

Ahmad I, Dar MA, Andualem TG, Teka AH, Tolosa AT (2020) GIS-Based multi-criteria evaluation for deciphering of groundwater potential. J Indian Soc Remote Sens 48(2):305–313

AID U (2008) Addressing water challenges in the developing world: a framework for action. US agency for industrial development. An annex of the 2008 Report to congress for the senator Paul Simon water for the poor act

Azimi S, Moghaddam MA, Monfared SAH (2018) Spatial assessment of the potential of groundwater quality using fuzzy AHP in GIS. Arab J Geosci 11(7):1–22

Basak A, Das J, Rahman ATM, Pham QB (2021) An integrated approach for delineating and characterizing groundwater depletion hotspots in a coastal State of India. J Geol Soc India 97(11):1429–1440. https://doi.org/10.1007/s12594-021-1883-z

Bera A, Mukhopadhyay BP, Barua S (2020) Delineation of groundwater potential zones in Karha river basin, Maharashtra, India, using AHP and geospatial techniques. Arab J Geosci 13(15):1–21

Boretti A, Rosa L (2019) Reassessing the projections of the world water development report. NPJ Clean Water 2(1):1–6

Choudhury S, Basak A, Biswas S, Das J (2022) Flash flood susceptibility mapping using GIS-Based AHP method. In: Pradhan B, Shit PK, Bhunia GS, Adhikary PP, Pourghasemi HR (eds) Spatial modelling of flood risk and flood hazards. GIScience and geo-environmental modelling. Springer, Cham. https://doi.org/10.1007/978-3-030-94544-2_8

Damania R, Sébastien D, Rodella A-S, Russ J, Zaveri E (2019) Quality unknown: the invisible water crisis. Washington, DC: World Bank. https://doi.org/10.1596/978-1-4648-1459-4. License: Creative Commons Attribution CC BY 3.0 IGO

Das B, Pal SC (2020) Assessment of groundwater vulnerability to over-exploitation using MCDA, AHP, fuzzy logic and novel ensemble models: a case study of Goghat-I and II blocks of West Bengal India. Environ Earth Sci 79(5):1–16

Das J, Gayen A, Saha S, Bhattacharya SK (2017) Modelling of alternative crops suitability to tobacco based on analytical hierarchy process in Dinhata subdivision of Koch Bihar district, West Bengal. Model Earth Syst Environ 3(4):1571–1587. https://doi.org/10.1007/s40808-017-0392-y

Das J, Rahman AS, Mandal T, Saha P (2021) Exploring driving forces of large-scale unsustainable groundwater development for irrigation in lower Ganga River basin in India. Environ Dev Sustain 23(5):7289–7309. https://doi.org/10.1007/s10668-020-00917-5

Duraisamy S, Govindhaswamy V, Duraisamy K, Krishinaraj S, Balasubramanian A, Thirumalaisamy S (2019) Hydrogeochemical characterization and evaluation of groundwater quality in Kangayam taluk, Tirupur district, Tamil Nadu, India, using GIS techniques. Environ Geochem Health 41(2):851–873

Dyson B (2017) Integration of life cycle assessment into decision-analytic approaches for sustainable technologies. In: Martin A (ed) Abraham in "Encyclopedia of sustainable technologies". Elsevier, pp 81–89. ISBN 9780128047927, https://doi.org/10.1016/B978-0-12-409548-9.10037-5

El-Rawy M, Ismail E, Abdalla O (2019) Assessment of groundwater quality using GIS, hydrogeochemistry, and factor statistical analysis in Qena Governorate Egypt. Desalin Water Treat 162:14–29

Gangadharan R, Vinoth S (2016) Assessment of groundwater vulnerability mapping using AHP method in coastal watershed of shrimp farming area. Arab J Geosci 9(2):107

Goswami S, Rai AK (2022) Estimating suitability of groundwater for drinking and irrigation, in Odisha (India) by statistical and WQI methods. Environ Monit Assess 194(7):1–18

Goswami S, Rai AK, Tripathy S (2022) Re-visiting geothermal fluid circulation reservoir depth and temperature of geothermal springs of India. J Hydrol 612128131-S0022169422007065 128131. https://doi.org/10.1016/j.jhydrol.2022.128131

Groundwater Year Book, National Capital Territory Delhi, 2021. Central Ground Water Board Ministry of Water Resources, Government of India, Report

Hasan MSU, Rai AK (2020) Groundwater quality assessment in the Lower Ganga Basin using entropy information theory and GIS. J Clean Prod 274:123077

Islam AT, Shen S, Haque MA, Bodrud-Doza M, Maw KW, Habib MA (2018) Assessing groundwater quality and its sustainability in Joypurhat district of Bangladesh using GIS and multivariate statistical approaches. Environ Dev Sustain 20(5):1935–1959

Karan SK, Samadder SR, Singh V (2018) Groundwater vulnerability assessment in degraded coal mining areas using the AHP–Modified DRASTIC model. Land Degrad Dev 29(8):2351–2365

Kawo NS, Karuppannan S (2018) Groundwater quality assessment using water quality index and GIS technique in Modjo River Basin, central Ethiopia. J Afr Earth Sc 147:300–311

Khan R, Jhariya DC (2018) Assessment of land-use and land-cover change and its impact on groundwater quality using remote sensing and GIS techniques in Raipur City, Chhattisgarh, India. J Geol Soc India 92 (1):59–66

Lad S, Ayachit R, Kadam A, Umrikar B (2019) Groundwater vulnerability assessment using DRASTIC model: a comparative analysis of conventional, AHP, Fuzzy logic and Frequency ratio method. Model Earth Syst Environ 5(2):543–553

Madhav S, Ahamad A, Kumar A, Kushawaha J, Singh P, Mishra PK (2018) Geochemical assessment of groundwater quality for its suitability for drinking and irrigation purpose in rural areas of Sant Ravidas Nagar (Bhadohi), Uttar Pradesh. Geology Ecology Landscapes 2(2):127–136

Mahammad S, Islam A (2021) Evaluating the groundwater quality of Damodar fan delta (India) using fuzzy-AHP MCDM technique. Appl Water Sci 11 (7):1–17

Majone M, Verdini R, Aulenta F, Rossetti S, Tandoi V, Kalogerakis N, Agathos S, Puig S, Fava F (2015) In situ groundwater and sediment bioremediation: barriers and perspectives at European contaminated sites. New Biotechnol 32(1):133–146

Margat J, Van der Gun J (2013) Groundwater around the world: a geographic synopsis. Crc Press

Meng X, Pian Z (2016). Low-Voltage risk assessment of the distribution network based on vulnerability information coordination among buses. In: Meng, X, Pian Z (eds) Intelligent coordinated control of complex uncertain systems for power distribution network reliability. Elsevier, pp 127–141, ISBN 9780128498965. https://doi.org/10.1016/B978-0-12-849896-5.00006-4

Mitra R, Das J (2022) A comparative assessment of flood susceptibility modelling of GIS-based TOPSIS VIKOR and EDAS techniques in the Sub-Himalayan foothills region of Eastern India. Environ Sci Pollut Res. https://doi.org/10.1007/s11356-022-23168-5

Nayak P, Rai AK, Subhasish T (2017) Evaluating groundwater prospects using GIS techniques. Sustain Water Resour Manage 3(2):129–139. https://doi.org/10.1007/s40899-017-0082-y

Piper AM (1994) A graphical procedure in the geochemical interpretation of water analysis. Am Geophys Union Trans 25:914–923

Pous N, Puig S, Coma M, Balaguer MD, Colprim J (2013) Bioremediation of nitrate-polluted groundwater in a microbial fuel cell. J Chem Technol Biotechnol 88 (9):1690–1696

Qazi WA, Abushammala MF, Azam MH (2018) Multi-criteria decision analysis of waste-to-energy technologies for municipal solid waste management in Sultanate of Oman. Waste Manage Res 36(7):594–605

Rehman S, Hasan MSU, Rai AK, Avtar R, Sajjad H (2021) Assessing flood-induced ecological vulnerability and risk using GIS-based in situ measurements in Bhagirathi sub-basin India. Arab J Geosci 14(15):1–17

Saaty TL, Vargas LG (1980) Hierarchical analysis of behavior in competition: prediction in chess. Behav Sci 25(3):180–191

Sahoo S, Dhar A, Kar A, Chakraborty D (2016) Index-based groundwater vulnerability mapping using quantitative parameters. Environ Earth Sci 75(6):522

Singh LK, Jha MK, Chowdary VM (2018) Assessing the accuracy of GIS-based multi-criteria decision analysis approaches for mapping groundwater potential. Ecol Ind 91:24–37

Speight JG (2018) Mechanisms of transformation. In: Butterworth-Heinemann (ed) Reaction mechanisms in environmental engineering, Butterworth-Heinemann. https://doi.org/10.1016/B978-0-12-804422-3.00010-9. ISBN 9780128044223

Wasi S, Tabrez S, Ahmad M. (2013a) Use of pseudomonas spp. for the bioremediation of various environmental pollutants: a review. Environ Monit Assess 185(10):8147–55

Wasi S, Tabrez S, Ahmad M (2013b) Toxicological effects of major environmental pollutants: an overview. Environ Monit Assess 185(3):2585–2593

A Literature Review of the Impact of COVID-19 Pandemic on Land Surface Temperature and Air Quality of India

8

Vanlalchhuanga, Brototi Biswas, and Jonmenjoy Barman

Abstract

The unprecedented upsurge of the COVID-19 has disrupted human life in many ways. Although the pandemic has severely crashed the global economy, the envisaged positive results were experienced in many countries, especially in the case of the natural environment. India, the country which experienced one of the worst environmental pollutions, has witnessed the growth of positive impact on land surface temperature and air quality. India's since March 25, 2020, the federal and state governments have imposed a countrywide total and partial lockdown at various times and locations. This paper highlights the various research literatures in land surface temperature (LST) and air quality monitoring during India's pandemic lockdown. According to the study, the primary air pollutants' concentration levels had reduced significantly across the country: PM10 (33%), PM2.5 (34%), CO (21%), NO_2 (47%), and SO_2 (21%). The lockdown time is related with better air quality in most cities, the Central Pollution Control Board NAQI result reveals that the prominent pollutants are found to be particulate matter, during the lockdown, Kolkata (62%) saw the greatest reduction in PM2.5 levels, other major cities also experienced high reduction, New Delhi had witness 26% reduction, likewise, Chennai (34%) and Mumbai (49%). In addition to air quality, the LST also decreased during the lockdown with major cities experiencing decreases of mean surface temperature during the lockdown in India.

Keywords

Air quality · LST · COVID-19 · Lockdown · Pollutants

8.1 Introduction

The novel pandemic SARS-COVID-19 is a transmittable infectious disease associated with severe respiratory illness (Islam et al. 2020). COVID-19 has profound global consequences, resulting in an increase in death and widespread economic losses. (Bukhari and Jameel 2020). India enacted a mandatory community lockdown on March 22, 2020. Following this several parts of the nation were put under lockdown or partial lockdown over a year, with distinct rates of regulations. The high-density areas and megacities were highly affected in terms of social and economic activity. Nonetheless, COVID-19 had

Vanlalchhuanga · B. Biswas · J. Barman (✉)
Department of Geography and RM, Mizoram (Central) University, Aizawl, Mizoram, India
e-mail: jonmenjoybarman07@gmail.com

J. Das and S. K. Bhattacharya (eds.), *Monitoring and Managing Multi-hazards*,
GIScience and Geo-environmental Modelling, https://doi.org/10.1007/978-3-031-15377-8_8

a substantial impact on major cities environmental aspects which including air quality and land surface temperature. A study between 1901 and 2001 reveals that the India's overall urban population rose by more than tenfold, from 26 to 285 million (Mohan 1996). Environmental degradation has been intensifying in India as a result of unregulated urbanization, resulting in a slew of issues (Uttara et al. 2012). The impact may be seen in the rise of harmful greenhouse gases and urban heat island, with India having one of the worst pollution issues in the world. This paper does a detailed study of the works of several scholars who studied the impact and implications of corona virus induced confinement on the air quality and urban heat island through LST in India using various techniques. The studies revolve across India's different cities. The findings of the studies appear to suggest that lockdown had enhanced the air quality index and led to a decrease in mean urban heat island (Nanda et al. 2021).

8.2 Land Surface Temperature

Land surface temperature was previously known as conventional surface-air temperature and is recorded by a protected thermometer. However, with the advancement in satellite technology, new forms of LST, termed as skin temperature has become available worldwide (Dickinson 1994). The level of LST is a significant factor in the surface energy budget because it regulates the amount of sensible heat transfer and, as a direct consequence, the air temperature (Frey and Kuenzer 2014). LST can become a useful indicator of monthly average air temperature over a long period of time as well as, relative humidity, and total precipitation (Ozelkan et al. 2014). It depends on LULC types with vegetation playing an important role (Mallick et al. 2008). LST has such a negative relationship with the abundance of green space (Li et al. 2013). Negative relationship between vegetation area and LST is observed with a 1.750 °C lower when compared to the built-up area (John et al. 2020). LST has a negative relationship with vegetation index and built up (positive) (Guha et al. 2018). The highest increase in LST has been about 20 °C over the Delhi metropolitan built-up regions (Chakraborty et al. 2017). NDVI trends can be utilized to establish an urban heat island (Grover and Singh 2015).

The heterogeneity of land cover and other atmospheric conditions causes LST to vary spatially (Talukdar 2020). The influence of change in elevation at various sites must also be addressed when analyzing the spatial distribution of LST across a large area (Khandelwal et al. 2017). In high altitude places, the LST-elevation scatterplot reveals an adverse correlation (Mathew et al. 2017). Shaded relief plays a significant role and is positively related to LST. (Peng et al. 2020). Despite the fact that there are still numerous research gaps in atmospheric composition, air quality modeling, and forecasting studies (Baklanov and Zhang 2020), Sen's slope estimator and basic linear regression are the two approaches and models that are dependable for studying. In terms of estimating the long-term variation trend of the mean monthly LST, the results show that there was no significant difference between the two models. (Firoozi et al. 2020).

In order to determine distinct land surface classifications, LST is generated from satellite-thermal IR data using a single-channel (SC) method utilizing atmospheric correction parameters, as well as the land surface emissivity (LSE) derived by a hybrid-ground-based model (Chatterjee et al. 2017). The LST can also be predicted using the multivariate adaptive regression spline (MARS), adaptive-network-based fuzzy inference system (ANFIS), and dynamic evolving neural-fuzzy inference systems (DENFIS) models. (Mustafa et al. 2020). The machine learning linear regression model for prediction of particulate matter shows the model was performed better as compared to the traditional model (Doreswamy et al. 2020). A novel hybrid model for predicting the air quality is also reliable. This combines two recurrent neural network (RNN) models, and the result has shown overall improvement in model performance (Hossain et al. 2020).

8.3 Air Quality

Air is the elixir of life, and hence, the increasing level of air pollution is a cause of great concern. (Reddy et al. 2020). According to the World Health Organization, air pollution is a serious cause of premature death in the general public, culminating in millions of deaths each year. Air pollution is a complex mixture of gases, particles, aerosols, and water vapor caused by human activity as well as other anthropogenic and natural processes (Nigam et al. 2016). The air quality index (AQI) is a statistical index that government agencies use to notify about how polluted the air is or will become. (Tiwari 2015). Many organizations and nations across the world have created and developed AQI standards. In India, the Central Pollution Control Board (CPCB) is in charge of setting the standard for ambient air quality.

8.4 Impact of COVID-19 Lockdown on LST and Air Quality in India

The COVID-19 shutdown has resulted in a massive drop in global economic activity and related air pollution levels, owing mostly to a decrease in inland transport mobility (Venter et al. 2021). The Indian government imposed a countrywide shutdown on March 25, 2020, as a result of COVID-19 fast and devastating spread throughout India. This led to a sharp reduction in the air pollution levels of India, with a countrywide decrease of 33% in average "PM10, PM2.5, carbon monoxide (CO), nitrogen dioxide (NO_2), and sulfur dioxide (SO_2)" of 21% and 21%, of 21%, respectively, (Verma and Kamyotra 2020). Particulate matter and various human-induced greenhouse gases are significant contributors to air pollution in India (Nasir et al. 2016).

Prior to the shutdown, India's air quality showed alarmingly high levels of air pollution in various places throughout the country. In Agra city, foundries using coal as fuel, coupled with vehicular emissions have yielded to the city's high pollution rates. The levels of nitrogen dioxide (NO_2), respirable suspended particulate matter (RSPM), and suspended particulate matter (SPM) levels were discovered to be constantly above average National Ambient Air Quality Standards (NAAQS). Pollution levels were found to be higher in the winter, but lower in the summer and monsoon seasons. (Nasir et al. 2016). If the concentration of pollutants of eight sample cities is compared with data from the prior year during the lockdown period, as well as the NAAQS, lockdown time is linked to better air quality (Dasgupta and Srikanth 2020) PM2.5 and PM10 levels were reduced by 52% and 53.5%, respectively, during the initial lockdown in the foremost cities of Punjab and Chandigarh paralleled to pre-lockdown levels. With relaxation on lockdown measures, a slow rise in the pollutant levels was noted again (Sahoo et al. 2021). During the lockdown, Kolkata experienced the greatest reduction in PM2.5 (62%), followed by Mumbai (49%), Chennai (34%), and New Delhi (26%). The level of PM2.5 in Hyderabad, in comparison, fell slightest by 10%. (Ravindra et al. 2021). This is true for the majority of pollutants and cities in the study, with pollution levels being significantly lower than in 2019. CO, NO_2, and SO_2 levels have all dropped drastically in Kolkata, during the lockdown in 2020. Meanwhile, due to the complete halt of vehicular movement, biomass burning, and dust particles from construction projects, during lockdown, the percentage decrease in PM10 and PM2.5 was around 17.5% as compared to previous year (Bera et al. 2020). Automobile, power stations and home emissions are the major sources of emission in Kolkata, suggesting that local emissions are dominated by the concentration of SO_2, NO_2, NH_3, and PM10 (Gupta et al. 2008). In Kolkata, the mean LST was 28.76 °C before lockdown (March 13, 2020); however, after lockdown (30 April 2020) it fell to 26.56 °C (Sahani et al. 2020). Substantial improvement in the nature of air pollution and AQI was the consequence of the shutdown, according to research. The AQI has decreased during the first and second phase of lockdown and rose during the third phase of lockdown owing to government relaxations during this phase; however, the AQI declined again during the fourth phase of lockdown in 2020 (Tripathi 2020). During Delhi's post-lockdown period, the percentages of "PM10, PM2:5, NO_2, and SO_2" dropped by 55%, 49%, 60%, and 19%, respectively, whereas in Mumbai, they decreased by 44%, 37%, 78%, and 39%, respectively, (Kumari and Toshiwal 2020).

In 2020, the air quality in 39 Indian cities improved (including ten cities in India that are among the global highest 20 worst polluted). During the lockdown period in 2020, only 15% of cities had an AQI in the "unhealthy" category (151–200), whereas 56% of cities had an AQI in the "unhealthy" category in 2019 (Kumari et al. 2020). The pollution levels in the northern part of India were poor during shutdown (including Delhi which is one of the most polluted) improved up to 30–46.67% (Gautam et al. 2020). The "Central Pollution Control Board" data about PM 2.5 emissions in the Delhi area show a significant decrease during the shutdown period (Fig. 8.1), in Delhi, PM10 and PM2.5 dwindled by 50% during the lockdown period from the pre-lockdown period along with NO_2 (52.68%) and CO (30.35%). National Air Quality Index (NAQI) reported a reduction in the Central, Eastern, South, Western, and Northern portions of Delhi by 54%, 49%, 43%, and 37%, respectively, (Mohato et al. 2020). Pan India an improvement in air quality of 40–50% has been observed by 40–50% just four days into lockdown.

The "European Space Agency Sentinel" 5p mission image shows the lockdown impact has clearly visible as the western and northern portion of the sub-continent has witness a swept clean the emission concentration spot which shows tremendous decreases in nitrogen dioxide emission (Fig. 8.2). The mean tropospheric NO_2 concentration in India, according to the data, was 214.4×1013 cm^2 from March 1 to March 21, 2020, and then decreased by 12.1% over the next four weeks. As a result, the decline in tropospheric NO_2 levels is attributable to less anthropogenic activities. In 2019, tropospheric NO_2 concentration had increased by 0.8%. Several research works done during the lockdown were studied, and biomass emissions were discovered to be a significant source of tropospheric NO_2. (Biswal et al. 2020). Significant shift of 7% to 67% in AQI has been recorded in Chandigarh and Ballygunge (Kolkata). There has been a huge change from 25.76% to 68.55% in AQI of Ahmedabad, Worli, Talkatora, and Lalbagh (Alok et al. 2020).

8.5 Conclusions

The COVID-19 pandemic-induced shutdown has already had a favorable influence on environmental concerns. The present work is a review of articles published during lockdown of 2020 revolving around various environmental issues in the backdrop of diminished anthropogenic

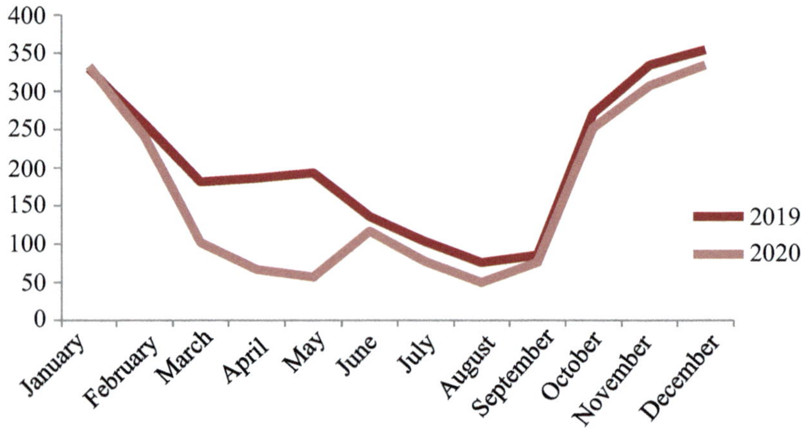

Fig. 8.1 Delhi PM 2.5 emission. *Source* Central Pollution Control Board

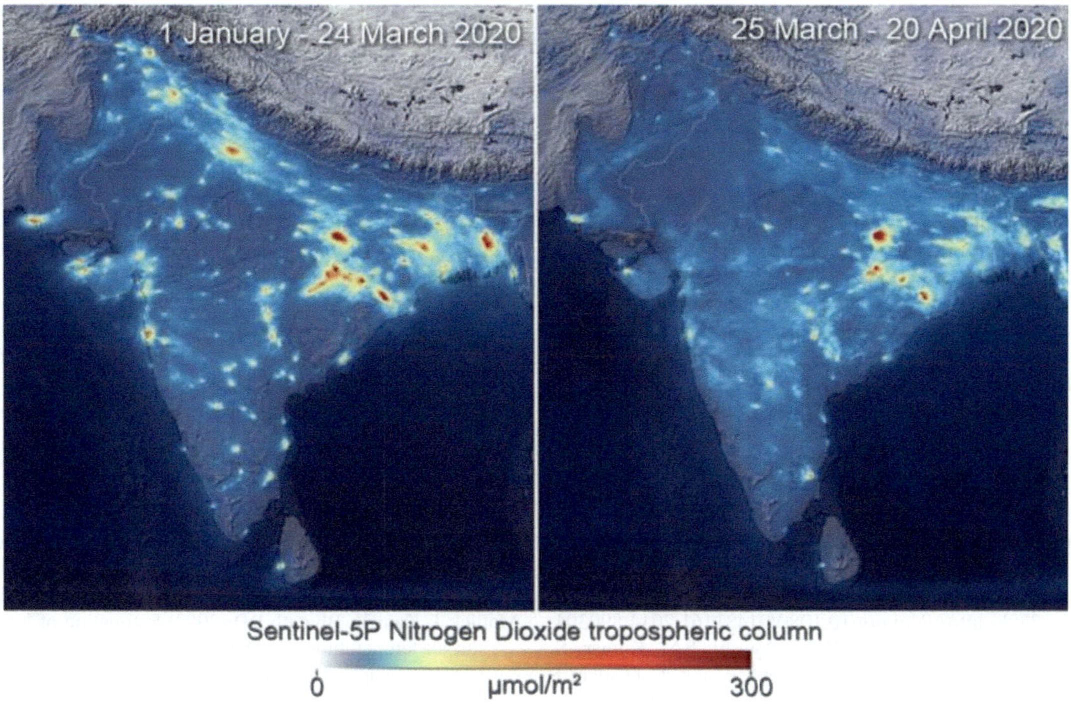

Fig. 8.2 Nitrogen dioxide drops due to lockdown in India. *Source* European Space Agency (2020)

activities owing to COVID-19 induced lockdown. During the COVID-19 lockdown, many socio-economic activities came to a complete halt, resulting to have positive results in the air quality and also decrease in land surface temperature pan India. This gives us hope for environmental sustainability, not only for India, but for the world's megacities that are notoriously polluted. All the polluted Indian cities, within just four days into implementing lockdown, saw air quality has improved by 40–50%. There has been a reduction by 60–39% in the PM10 and PM2.5 level, respectively. The lockdown phase significantly lowered NO_2 (52.68%) and CO (30.35%) among other pollutants.

References

Alok SG, Kumar S, Gautam S, Aryan AA, Kumar R, Abhishek JA, Bauddh K, Singh K (2020) Pandemic induced lockdown as a boon to the environment: trends in air pollution concentration across India. Asia-Pacific J Atmos Sci. https://doi.org/10.1007/s13143-021-00232-7

Baklanov A, Zhang Y (2020) Advances in air quality modelling and forecasting. Global Transitions 2:261–270.https://doi.org/10.1016/j.glt.2020.10.002

Bera B, Bhattacharjee S, Pravat KS, Sengupta N, Saha S (2020) Significant impacts of COVID-19 lockdown on urban air pollution in Kolkata (India) and amelioration of environmental health. Environment, development and sustainability. Springer Nature B.V. https://doi.org/10.1007/s10668-020-00898-5

Biswal A, Singh T, Singh V, Ravindra K, Mor S (2020) COVID-19 lockdown and its impact on tropospheric NO_2 concentrations over India using satellite-based data. Heliyon 6.https://doi.org/10.1016/j.heliyon.2020.e04764

Bukhari Q, Jameel Y (2020) Will coronavirus pandemic diminish by summer? SSRN Electron J. https://doi.org/10.2139/ssrn.3556998

Chakraborty SD, Kant Y, Bharath BD (2014) Study of land surface temperature in Delhi city to managing the thermal effect on urban developments. Int J Adv Sci Techn Res 4(1). www.rspublication.com/ijst/index.html

Chatterjee RS, Singha N, Thapa S, Sharma D, Kumar D (2017) Retrieval of land surface temperature (LST) from landsat TM6 and TIRS data by single channel radiative transfer algorithm using satellite and ground-based inputs. Int J Appl Earth Obs Geoinf 58:264–277. https://doi.org/10.1016/j.jag.2017.02.017

Dasgupta P, Srikanth K (2020) Reduced air pollution during COVID-19: learnings for sustainability from Indian Cities. Global Transitions 2. https://doi.org/10.1016/j.glt.2020.10.002

Dickinson RE (1994) Satellite systems and models for future climate change. Future climates of the world: a modelling perspective. In: Henderson-Sellers A (Eds) World survey of climatology, vol 27. Elsevier

Harishkumar KS, Yogesh KM, Gad I (2020) Forecasting air pollution particulate matter (PM2. 5) using machine learning regression models. Procedia Comput Sci 171:2057–2066

Firoozi F, Mahmoudi P, Amir SM, Shahi J, Tavousi T, Liu Y, Liang Z (2020) Modeling changes trend of time series of land surface temperature (LST) using satellite remote sensing productions (case study: Sistan plain in east of Iran). Arab J Geosci 13:367. https://doi.org/10.1007/s12517-020-05314-w

Frey CM, Kuenzer C (2014) Land-surface temperature dynamics in the upper Mekong basin derived from MODIS time series. Int J Remote Sens 35(8):2780–2798. https://doi.org/10.1080/01431161.2014.890304

Gautam AS, Dilwaliya NK, Ayushi SA, Kumar S, Bauddh K, Siingh D, Shah MA, Singh K, Gautam S (2020) Temporary reduction in air pollution due to anthropogenic activity switch–off during COVID-19 parts of India, lockdown in northern parts of India. Environ Dev Sustain. https://doi.org/10.1007/s10668-020-00994-6

Grover A, Singh RB (2015) Analysis of Urban Heat Island (UHI) in relation to normalized difference vegetation index (NDVI): a comparative study of Delhi and Mumbai. Environments 2:125–138. https://doi.org/10.3390/environments2020125

Guha S, Govil H, Dey A, Gill N (2018) Analytical study of land surface temperature with NDVI and NDBI using Landsat 8 OLI and TIRS data in Florence and Naples city, Italy. Eur J Remote Sens 51(1):667–678. https://doi.org/10.1080/22797254.2018.1474494

Gupta AK, Karar K, Ayoob S, John K (2008) Spatio-temporal characteristics of gaseous and particulate pollutants in an urban region of Kolkata, India. Atmos Res 87:103–115. https://doi.org/10.1016/j.atmosres.2007.07.008

Hossain E, Uddin SMA, Hossain MS, Andeson K (2020) A novel deep learning approach to predict air quality index. In: Advances in intelligent systems and computing. Springer, pp 367–381.https://doi.org/10.1007/978-981-33-4673-4_29

Islam MS, Tusher TR, Roy S, Rahman M (2020) Impacts of nationwide lockdown due to COVID-19 outbreak on air quality in Bangladesh: a spatiotemporal analysis. Air Qual Atmos Health 14:351–363. https://doi.org/10.1007/s11869-020-00940-5

John J, Bindu G, Srimuruganandam B, Wadhwa A, Rajan P (2020) Land use/land cover and land surface temperature analysis in Wayanad district, India, using satellite imagery. Ann GIS.https://doi.org/10.1080/19475683.2020.1733662

Khandelwal S, Goyal R, Kaul N, Mathew A (2017) Assessment of land surface temperature variation due to change in elevation of area surrounding Jaipur, India. Egypt J Remote Sens Space Sci 21:87–94. https://doi.org/10.1016/j.ejrs.2017.01.005

Kumari P, Toshniwal D (2020) Impact of lockdown measures during COVID-19 on air quality—a case study of India. Int J Environ Health Res. https://doi.org/10.1080/09603123.2020.1778646

Kumari S, Lakhani A, Kumari KM (2020) COVID-19 and air pollution in Indian cities: world's most polluted cities. Aerosol Air Qual Res 20:2592–2603. https://doi.org/10.4209/aaqr.2020.05.0262

Li X, Zhou W, Ouyang Z (2013) Relationship between land surface temperature and spatial pattern of greenspace: what are the effects of spatial resolution?. Landsc Urban Plan 114:1–8

Mahato S, Pal S, Ghosh KG (2020) Effect of lockdown amid COVID-19 pandemic on air quality of the megacity Delhi. Science of the Total Environment, India. https://doi.org/10.1016/j.scitotenv.2020.139086

Mallick J, Kant Y, Bharath BD (2008) Estimation of land surface temperature over Delhi using Landsat-7 ETM+. J Indian Geophys Union 12(3):131–140

Mathew A, Khandelwal S, Kaul N (2017) Investigating spatial and seasonal variations of urban heat island effect over Jaipur cityand its relationship with vegetation, urbanization and elevation parameters. Sustain Cities Soc. https://doi.org/10.1016/j.scs.2017.07.013

Mohan R (1996) Urbanisation in India: patterns and emerging policy issues in the Urban transformation of the developing world. In: Gugler J (ed) Oxford University Press, Oxford

Mustafa EK, Co Y, Liu G, Kaloop MR, Beshr AA, Zarzoura F, Sadek M (2020) Study for predicting land surface temperature (LST) using landsat data: a comparison of four algorithms. Adv Civil Eng. https://doi.org/10.1155/2020/7363546

Nanda D, Deepk RM, Debadatta S (2021) COVID-19 lockdowns induced land surface temperature variability in mega urban agglomerations in India. Environ Sci Process Impacts 23:144–159. https://doi.org/10.1039/D0EM00358A

Nasir H, Goyal K, Prabhakar D (2016) Review of air quality monitoring: case study of India. Indian J Sci Technol 9(44)

Nigam S, Rao BPS, Kumar N, Mhaisalkar VA (2016) Air quality index—a comparative study for assessing the status of air quality. Res J Eng Tech 6(2)

Ozelkan E, Bagis S, Cem E, Burak O, Berk UB, Ormeci C (2014) Land surface temperature retrieval for climate analysis and association with climate data. Eur J Remote Sens 47(1):655–669. https://doi.org/10.5721/EuJRS20144737

Peng X, Wu W, Zheng Y, Jingyi SJ, Tangao HT, Wang P (2020) Correlation analysis of land surface temperature and topographic elements in Hangzhou, China.

Sci Rep 10:10451. https://doi.org/10.1038/s41598-020-67423-6
Ravindra K, Singh T, Biswal A, Singh V, Mor S (2021) Impact of COVID-19 lockdown on ambient air quality in megacities of India and implication for air pollution control strategies. Environ Sci Pollut Res 28:21621–21632. https://doi.org/10.1007/s11356-020-11808-7
Reddy S, Verma P, Waghulade M (2020) Air quality index—a study to assess the air quality. Int Res J Eng Technol 7(2)
Sahani N, Goswami SK, Saha A (2020) The impact of COVID-19 induced lockdown on the changes of air quality and land surface temperature in Kolkata city. Spatial Information Research, India. https://doi.org/10.1007/s41324-020-00372-4
Sahoo PK, Chauhan AK, Mangla S, Pathak AK, Garg VK (2021) COVID-19 pandemic: an outlook on its impact on air quality and its association with environmental variables in major cities of Punjab and Chandigarh, India. Environ Forensics 22(2).143–154.https://doi.org/10.1080/15275922.2020.1836082
Talukdar KK (2020) Land surface temperature retrieval of Guwahati city and suburbs, Assam, India using Landsat Data. Int J Eng Res Technol 9(5)
Tiwari P (2015) Computational methods of air quality indices: a literature review. J Environ Sci Toxicol Food Technol 1(5):46–49
Tripathi A (2020) Air pollution in four Indian cities during the Covid-19 pandemic. Int J Environ Stud. https://doi.org/10.1080/00207233.2020.1858578
Uttara S, Bhuvandas N, Aggarwal V (2012) Impacts of urbanization on environment. Int J Res Eng Appl Sci 2:2
Venter ZS, Kristin AK, Chowdhury S, Jos LJ (2021) Air pollution declines during COVID-19 lockdowns mitigate the global health burden. Environ Res 192. https://doi.org/10.1016/j.envres.2020.110403
Verma RL, Kamyotra JS (2020) Impacts of COVID-19 on air quality in India. Taiwan Assoc Aerosol Res 21(4). https://doi.org/10.4209/aaqr.200482

Seasonal and Inter-Annual Variation of Chlorophyll and Sea Surface Temperature in Northern and Southern Arabian Sea, India

9

S. Abhilash, R. Nirmala, and Aravind Sahay

Abstract

Arabian Sea is a dynamic and productive sea due to its long borders and open ocean processes, which evident as upwelling during summer and cooled in winter, especially in the northern regions. The present study aims to discuss the seasonal, inter-annual variations and distinct differences of chlorophyll-a (Chl-a) between Northern Arabian Sea (NAS) and Southern Arabian Sea (SAS) using the MODIS data from 2003 to 2019. The annual mean Chl-a values in NAS are 1.89 mg/m^3 which is large difference compared to 0.44 mg/m^3 in SAS. Seasonal variations in February, March and September, the average Chl-a in the NAS were 3.07 mg/m^3, 2.79 mg/m^3 and 2.51 mg/m^3 and the SAS found average Chl-a value in July, August and September were 1.01 mg/m^3, 1.20 mg/m^3 and 1.20 mg/m^3, respectively. From December to March, which are the cooler months with mean SST of 24.2–25.5 °C in the NSA and 26.7–27.8 °C in the SAS, coincided with the peak months of Chl-a concentration. The SAS recorded the highest monthly SST in May at 29.9 °C and lowest at 26.2 °C in August and 29.3 °C in June and least 23.6 °C in February, NAS showing more complexity than the SAS. The results show that trend of chlorophyll concentration and SST are not consistent for the year and vary greatly with the seasons and that Chl-a concentration shows an inverse relationship with SST. Convection and surface cooling in northern region by increased evaporation and reduced solar radiation makes NAS more productive compared to SAS in the winter and also productive in the summer due to coastal fluctuations.

Keywords

Arabian Sea · Chlorophyll · Sea surface temperature

9.1 Introduction

Plankton is an indicator of climate change (Hays et al. 2005; Taylor et al. 2002; Adrian et al. 2009; Richardson 2008) because their growth and development directly impacts the temperature change of the plankton population during seasonal during seasonal periods of growth (Adrian et al. 2006; Berger et al. 2010; Schulter et al. 2010 and Theckeray et al. 2012). Recent work has shown that in the case of phytoplankton, it

S. Abhilash · R. Nirmala (✉)
Department of Marine Geology, Mangalore University, Karnataka, India 574199, India
e-mail: nirmala.agara@gmail.com

A. Sahay
Space Application Centre (SAC), ISRO, Ahmedabad, Gujarat 380015, India

J. Das and S. K. Bhattacharya (eds.), *Monitoring and Managing Multi-hazards*, GIScience and Geo-environmental Modelling, https://doi.org/10.1007/978-3-031-15377-8_9

changes not only in temperature but also in the seasonal cycles that have been shown to contribute to nutrient load (Theckeray et al. 2008; and Theckery 2012). The inherent chlorophyll depth derived from the satellite is associated with the amount of particles present in the water and the coefficient of extinction of the water (Smith 1981; Sergey et al. 2012; Gordon and Mccluney 1975).

The presence of chlorophyll at appropriate concentrations indicates a healthy ecosystem. However, under certain environmental conditions, it tends to overgrow in the short term (Sathyenadranath et al. 1996). High concentration of chlorophyll is visible to naked eye, and water can looks blue, blue-green, green, and red depending on the type of pigments in the particular plankton. Each plankton species has different favourable conditions that promote its growth and reproduction. Also the ocean colour studies are using remote sensing techniques to understand phytoplankton concentration and their impact on climate change (Morel 1980; Sturm 1993; Gordon and Wang 1994). Chlorophyll growth depends on SST and nutrient levels in the sea and is more abundant in cold water. This impact study may help predict the occurrence of oceanographic conditions in the study area for future outbreaks of these events, and may be helpful in implementing fisheries related management measures in this particular study area.

9.2 Materials and Methods

9.2.1 Study Area

The Arabian Sea has an area of 6.225×10^6 km^2. and extends from 0° to 25° N and 50° to 80° E (Qasim 1977). It is bordered by Oman at the west, Iran at the northwest and India in the east (Fig. 9.1). The Arabian Sea is the one of the world's most biologically productive regions and subject to seasonal reverse monsoon wind system, with blooms in the winter monsoon season (December–March) and the summer monsoon (June–September) under strong seasonality. The summer blooming is driven by upwelling process such as lateral advection, wind mixing, mesoscale eddies, filaments and Ekman pumping (Margulis et al., 1990; Deuser et al., 1990 and Lee et al., 2000) which play a key role in supplying nutrients to the euphotic zone during the summer.

In the spring-inter monsoon (April–May), the water is mostly oligotrophic and has low chlorophyll pigment concentration (Madhupratap et al, 2001), but the fall-inter monsoon season (October–November) represents the tipping point of the summer monsoon, with chlorophyll pigment concentration rapidly declining (Banse, 1987; Smith 1998; Wiggert et al. 2005 and Prasanna Kumar et al. 2000). The inter-annual variation of the chlorophyll in the Arabian Sea is poorly explored and understood. The present study mainly focuses on the northern and southern part of Arabian Sea.

9.2.2 Methods

Moderate resolution imaging spectroradiometer (MODIS) on board aqua satellite data from NASA-OPG from 2003 to 2019 was used to obtain monthly chlorophyll-ocx level-3 data with spatial resolution of 4 km (NASA Ocean colour Web) (Table 9.1). MODIS data is useful for tracking changes in the ocean over time (NASA MODIS Web). The SeaWiFS Data Analysis System (SeaDAS) is a software for extracting ocean colour data for chlorophyll and SST analysis. The monthly climatological images are merged in SeaDAS to form seasonal climatological images and statistical analysis is done by Microsoft Excel to find out the values, and then, seasonal variations are plotted. The same analysis pattern applied in case of SST is followed in this case also. The SST values of pre-monsoon, summer monsoon, winter monsoon and post-monsoon seasons of each year and inter-annual variations are plotted to determine differences by season.

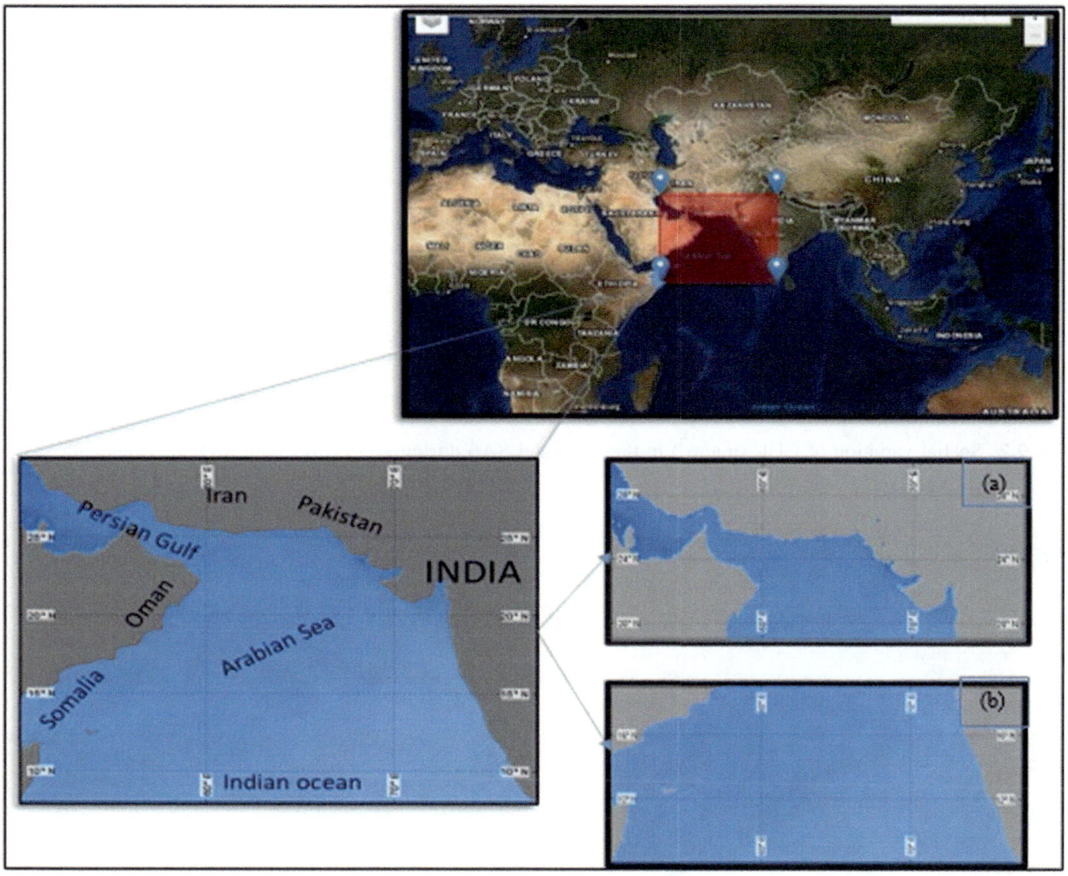

Fig. 9.1 Study area map **a** Northern Arabian Sea and **b** Southern Arabian Sea

Table 9.1 Channel of MODIS Aqua

Uses	Bands	Band width (nm)
Ocean colour/phytoplankton	8–16	405–877
Surface temperature	31	10.780–11.280
	32	11.770–12.270

9.3 Results and Discussion

The present analysis aims at accessing Chl-a and SST variability in the NAS and SAS from 2003 to 2019 by SeaDAS software (Figs. 9.2 and 9.3). Using ocean colour observations reveals that the Chl-a distribution exerts a controlling influence on seasonal evolution of SST (Kamykowsi 1987; Hood et al. 1990; Sathyendranath et al. 1991; Muller-Karger et al. 1991; Chaturvedi et al. 1998a, b).

9.3.1 Analysis of Northern Arabian Sea (NAS)

Monthly variation of the mean concentration of chlorophyll-a and SST (Fig. 9.4) between 2003 to 2019 is inversely correlated, i.e. sea surface temperature increases but chlorophyll concentration decreases and vice-versa. Chlorophyll concentration variability changes by season due to coastal upwelling and wind mixing affect the chlorophyll concentration (Banse 1987; Prasanna Kumar et al. 2000; Smith 1998 and Wiggert et al.

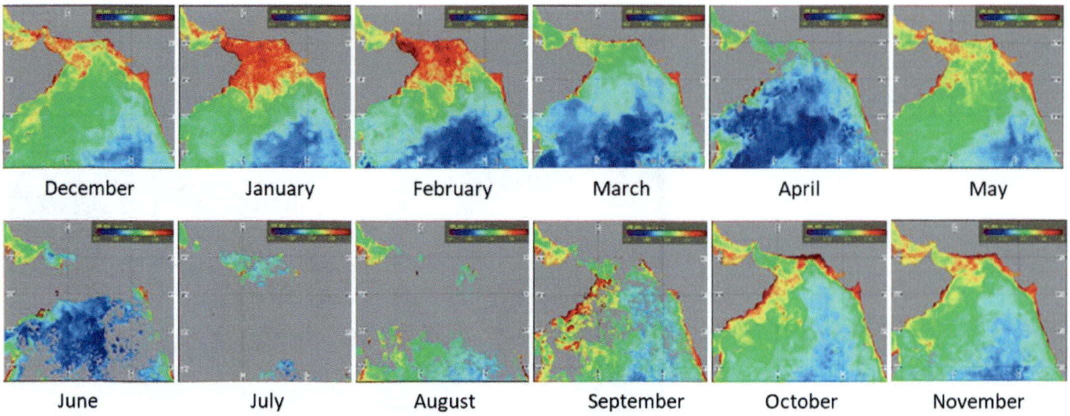

Fig. 9.2 Spatial variation of CHL (mg/m^3) in the NAS and SAS during 2003–2019 with MODIS Aqua data

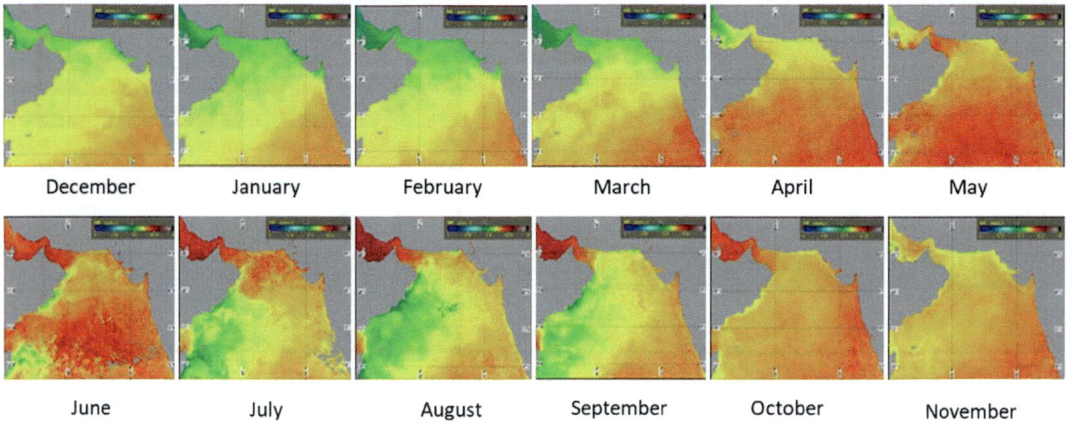

Fig. 9.3 Spatial variation of SST (°C) in the NAS and SAS during 2003–2019 with MODIS Aqua data

2005) and the NAS (the coastal region of Oman and Arabian Gulf) has shown high chlorophyll pigment density.

9.3.2 Seasonal Variation of Chlorophyll and SST (NAS)

Monthly and seasonal mean are calculated for the spring, winter, summer and fall-inter monsoon (Fig. 9.4). Their monthly mean chlorophyll pigment concentration in the NAS from 2003 to 2019 shows second peak with the highest chlorophyll concentration in September after a strong season with a primary peak in February. The summer monsoon season ranges from 0.47 mg/m^3 to 3.4 mg/ m^3 (Sept. 2004), then from 0.59 mg/m^3 to 2.93 mg/m^3 (April, 2019) in the spring-inter monsoon, from 1.04 mg/m^3 to 5.79 mg/m^3 (Mar 2017) in winter season and from 0.97 to 2.87 mg/ m^3 (Nov. 2006) during fall-inter monsoon season.

Analysing all the averages of chlorophyll concentrations in NAS from 2003 to 2019, the highest concentration were found in February (3.07 mg/m^3) and March (2.79 mg/m^3) with SST minimum values and the lowest Chl-a values in May (0.78 mg/m^3) with highest values in the case of SST. The maximum SST value is found in the year 2019 (29.5 °C) to least 2013 (27.8 °C). The months of October and November fall in

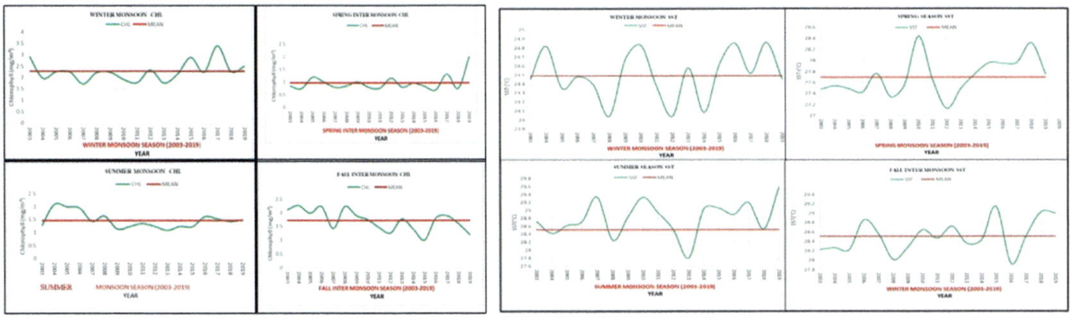

Fig. 9.4 Variation in Chl-a and SST during four different seasons from 2003 to 2019 of NAS

the post-monsoon season period (Fall-inter) and this shows the mean value (28.51 °C), which is higher than the winter monsoon and lower than the summer monsoon and depicts the high and low values for 2015 (29.15 °C) and 2008 (28.02 °C), respectively (Fig. 9.5).

9.3.3 Inter-Annual Variation of Chlorophyll (NAS)

The inter-annual variation of chlorophyll over the year 2003–2019 shows a slightly decrease with a negative slope of − 0.0042 (R^2 = 0.0075), with some notable positive and negative anomalies in some years (Fig. 9.6). The highest peak found in 2008 shifted from the grand mean (mean of 2003–2009) with the chlorophyll value in 2017, of which 2017 has the most positive anomaly (i.e. + 2.37 mg/m^3). The 2007 maximum value is depicted negative anomaly during the study period (i.e. −1.49 mg/m^3) and from 2009 to 2014 show a negative anomaly. The inter-annual variation of SST over the year 2003–2019 shows an increasing trend with positive slope of 0.0248 with some significant positive and negative anomalies over the years. The years 2015, 2016, 2017, 2018 and 2019 showed an increase in SST from the grand mean (mean of 2003–2019), of which 2015 had the highest positive anomaly (i.e. + 0.52 °C). The 2008 depicts a peak negative anomaly of − 0.64 °C.

9.3.4 Analysis of Southern Arabian Sea (SAS)

The Southern Arabian Sea is affected by the south West Monsoon (SWM), with a lower chlorophyll concentration compared to northern part and higher chlorophyll found in winter and summer season (Khole 2005) while the

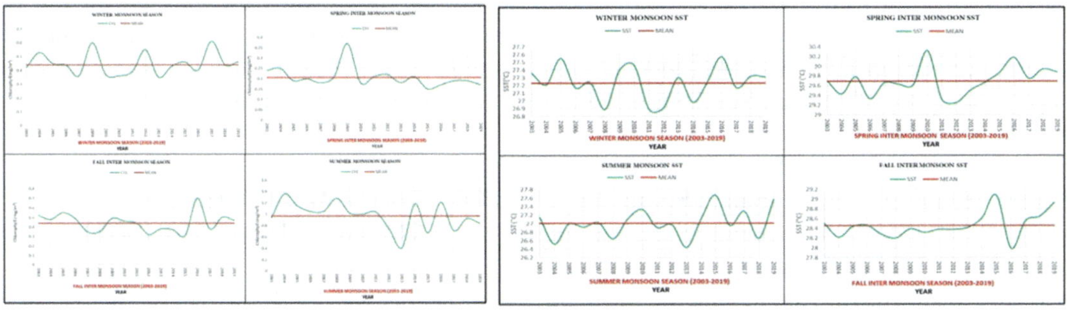

Fig. 9.5 Variation in Chl-a and SST during four different seasons from 2003 to 2019 of SAS

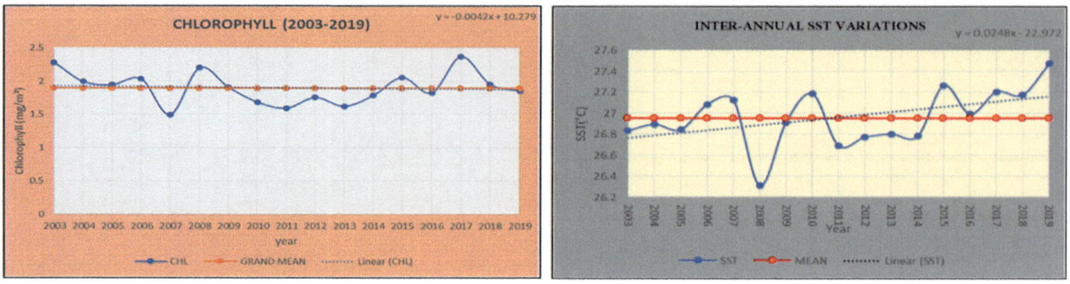

Fig. 9.6 Inter-annual chlorophyll and SST variability for the period 2003 to 2019 of NAS

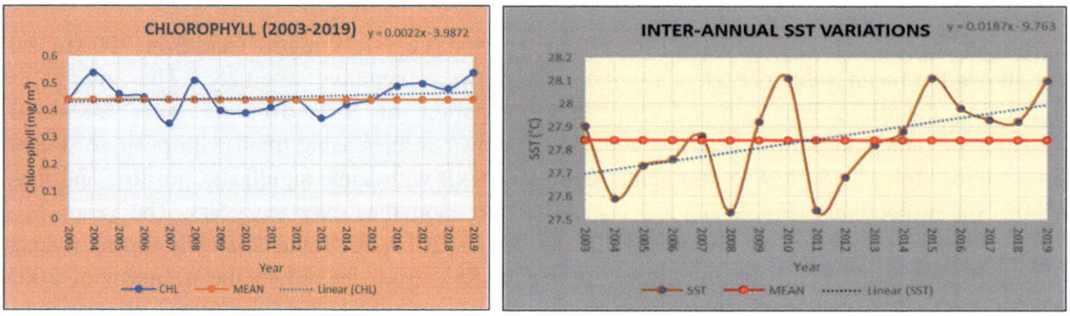

Fig. 9.7 Inter-annual chlorophyll and SST variability for the period 2003 to 2019 of SAS

chlorophyll concentration during the summer monsoon is higher than in winter. The highest value of chlorophyll concentration was recorded in February 2008 (0.91 mg/m^3), showing normal chlorophyll concentration in winter, with no significant differences this season. The same chlorophyll value was 0.41 mg/m^3 Jan and Feb 2007 but there were changes in SST (26.3 °C and 27.2 °C), respectively. The highest SST observed in March 2004 (27.9 °C) with a chlorophyll concentration of 0.31 mg/m^3 was due to the end of winter season, and the lowest SST was 0.48 mg/m^3 in Jan 2011.

Strong winds of the SWM are generally present at the centre of Arabian Sea from mid-June to late-September (Banse et al. 2000), affecting the southwestern Arabian Sea more than southeastern region. The SAS from 2003 to 2019, most of the chlorophyll is found in the year of July 2008 (2.54 mg/m^3) and the remaining years was less than 1 mg/m^3. From June to July, the chlorophyll concentration increased rapidly with an increase of 2.19.

9.3.5 Seasonal Variation of Chlorophyll and SST (SAS)

The winter monsoon season shows the least values of chlorophyll (i.e. values range from 0.35 to 0.61 mg/m^3) compared to the summer season and greater than the other seasons. Year 2009 shows the maximum chlorophyll concentration, which then depicts a negative anomaly. The value of the summer monsoon season ranges from 0.4 mg /m^3 to 1.36 mg /m^3 slightly greater compared to the other season. From 2003 to 2011, the values are slightly higher, later it shows a declining trend when compared to their mean values.

9.3.6 Inter-Annual Variation Chlorophyll and SST (SAS)

The inter-annual variation of chlorophyll from 2003 to 2019 shows a slightly increasing trend with a positive slope of + 0.0022 (R^2 = 0.0396) and depicts positive and negative anomalies in few years and from 2009 to 2015, chlorophyll has been shown to decrease from the grand mean (Fig. 9.7). Both 2004 and 2019 showed a very high peak in chlorophyll values (both 0.54 mg/m^3) with positive anomaly from the grand mean (i.e.0.1 mg/m^3). But in 2007 shows maximum negative anomaly during the study period (i.e. − 0.09 mg/m^3).

The inter-annual variability of SST in SAS shows a large difference in some years with great negative and positive anomalies. From 2003 to 2019, the linear trend curve shows an increasing trend with a positive slope of 0.0187. The 2010, 2015 and 2019 show same maximum values of SST (i.e. 28.1 °C). In 2008 and 2011, it showed maximum negative anomaly from the grand mean (2003–2019), with a negative anomaly of − 0.31 °C in 2008 and a positive anomaly of + 0.27 °C in 2010.

9.3.7 Chlorophyll and SST Variation

The spatial distribution of chlorophyll-a concentrations is completely different in the two regions; the northern region has high chlorophyll-a concentrations in February–March and low in August–September, but is completely opposite in the case of SAS. Additionally, there was a decrease in maximum and minimum monthly values in the two regions offshore. The highest chlorophyll-a concentration was recorded in the Northern Arabian Sea, was far higher than the Southern Arabian Sea.

The 16-year monthly SST from MODIS data showed an annual average of 27.5 °C for the entire region, 29.65° C in the NAS, and 27.84 °C in the SAS (Fig. 9.5). The minimum monthly SST in the NAS was recorded in February and SAS have found in August. The NAS showed relatively lower SST compared to the SAS and April is the transition from a relatively mild winter SST to warm summer SST. The monthly variation of SST in the NAS and SAS behaves differently. Remote sensing-to-ocean communication is the most effective tool for analysing global oceanic physical and biological changes with coastal zone colour scanner (CZCS). This is the only option to study the expected changes in future (Chaturvedi. 2005).

9.4 Conclusions

The behaviour of the Arabian Sea is differs from what is traditionally understood as a tropical basin due to its high salinity, reverse of monsoons, fluctuations of the coastal and open sea, and the effects of winter season cooling with a variety physical processes in summer and winter. The annual mean chlorophyll-a values in NAS showed a better contrast with higher average chlorophyll-a compared to SAS concentrations in February, March and September along the coast. The SAS showed higher average chlorophyll-a concentration in December–February and decreased in April and May. In NAS, the maximum single month chlorophyll-a value was recorded at 5.76 mg/m^3 in March 2017 and at least 0.53 mg/m^3 in July 2003, while the SAS in July 2008 recorded a maximum of 2.54 mg/m^3 and a minimum of 0.15 mg/m^3 in June 2015. The monthly variation of SST in NAS and SAS behaves differently, with the least monthly SST in the NAS recorded in February and SAS in August. The NAS showed relatively low SST compared to SAS, and April transitioned from a relatively mild winter SST to a warm summer SST. In the winter, the NAS becomes a biologically highly productive area due to winter cooling and enhanced evaporation from dry continental air under low incoming solar radiation. Since the ambient salinity of the NAS is high and sufficient cooling to increase the density and initiate convection. In south central waters, high productivity is caused by nutrient rich and high water advection from Somalia.

Acknowledgements The authors would like to thank Mr. Nilesh Desai, Director, Space Applications Centre (SAC) of Ahmedabad, for providing the opportunity to conduct research work and support is deeply acknowledged.

References

Adrian R, O'Reilly CM, Zagarese H, Baines SB, Hessen DO, Keller W, Winder M (2009) Lakes as sentinels of climate change. Limnol Oceanogr 54:2283–2297

Adrian R, Wilhelm S, Gerten D (2006) Life history traits of lake plankton species may govern their phonological response to climate warming. Global Change Biol 12:652–661

Banse K (1987) Seasonality of phytoplankton chlorophyll in the central and northern Arabian Sea. Deep-Sea Res 34:713–723

Banse K, English DC (2000) Geographical differences in seasonality of CZCS derived phytoplankton pigment in the Arabian Sea for 1978–1986. Deep-Sea Res ll 47:1623–1677

Berger SA, Diehl S, Stibor H, Trommer G, Ruhenstroth M (2010) Water temperature and stratification depth independently shift cardinal events during plankton spring succession. Glob Change Biol 16(7):1954–1965

Chaturvedi N (2005) Variability of chlorophyll concentration in the Arabian Sea and Bay of Bengal as observed from SeaWiFS data from 1997–2000 and its interrelationship with Sea surface temperature (SST) derived from NOAA AVHRR. Int J Remote Sens 26(17):3695–3706

Chaturvedi N, Narain A, Pandey PC (1998a) Remote sensing applications in oceanography with special references to bio-physical coupling: a review. Indian J Marine Sci 27:1–9

Chaturvedi N, Narain A, Pandey PC (1998b) Phytoplankton pigment temperature relationship in the Arabian Sea. Indian J Marine 27:286–291

Deuser WG, Muller-Karger FE, Evans RH, Brown OB, Esaias WE, Feldman GC (1990) Surface-ocean color and deep sea carbon fluxes: how close a connection. Deep-Sea Res 37:1331–1343

Gordon HR, Mccluney WR (1975). Estimation of the depth of sunlight penetration in the Sea for remote sensing. Appl Optics 14:413–416.https://doi.org/10.1364/AO.14.000413

Gordon HR, Wang M (1994) Retrieval of water-leaving radiance and aerosol optical thickness over the oceans with SeaWiFS: a preliminary algorithm. Appl Optics 33:443–452

Barber RT, Mooers NK, Bowman MJ, Zeitzschel B (eds) A Review. Lecture Notes on Coastal and Estuarine Studies. Springer, New York, 114 p

Hays CG, Richardson AJ, Robinson C (2005) Climate change and marine plankton. Trends Ecol Evol 20 (6):337–344

Hood RR, Abbott MR, Huyer A, Korso PM (1990) Temperature, Flow, Phytoplankton biomass and species composition in the coastal transition zone off Northern California. J Geophys Res 95:18081–18094

Kamykowski D (1987) A preliminary biophysical model of relationship between temperature and plant nutrients in the upper ocean. Deep-Sea Res 34:1067–1079

Khole M (2005) Inter-annual and decadal variability of Sea surface temperature (SST) Over Indian Ocean. Mausam 56(4):804–810

Kumar SP, Roshin RP, Narvekar J, Kumar PKD, Vivekanandan E (2010) What drives the increased phytoplankton biomass in the Arabian Sea? Curr Sci 99(1):101–106

Lee CM, Jones BH, Brink KH, Fischer AS (2000) The upper-ocean response to monsoonal forcing in the Arabian Sea: seasonal and spatial variability. Deep-Sea Res II 47:1177–1226

Madhupratap M, Nair KNV, Gopalakrishnan TC, Haridas P, Nair KKC, Venugopal P, Mangesh G (2001) Arabian Sea oceanography and fisheries of the west coast of India. Curr Sci 81(4):355–361

Margulis L, Corliss JO, Melkonian M, Chapman DJ (1990) Handbook of Protoctista. Jones and Bartlett publishers, Boston

Morel A (1980) In-water and remote measurement of ocean color Boundary-layer. Meteorol 18:177–201

Muller-Karger FE, Walsh JJ, Evans RH, Meyers MB (1991) On the Seasonal phytoplankton concentrations and Sea surface temperature cycles of the Gulf of Mexico as determined by satellite. J Geophys Res 96:12645–12665

NASA Ocean color web. Ocean Biology Processing Group (OBPG) http://oceancolour.gsfc.nasa.gov

NASA MODIS Web. Moderate Resolution Imagine Spectroradiometer. http://modis.gsfc.nasa.gov

Prasanna Kumar S, Madhupratap M, Kumar MD, Gauns M, Muraleedharan PM, Sarma VVSS, De Souza SN (2000) Physical control of primary productivity on a seasonal scale in central and eastern Arabian Sea. Proc Indian Acad Sci Earth Planet Sci 109:433–441

Qasim SZ (1977) Biological productivity of Indian Ocean. Indian J Mar Sci 6:122137

Richardson AJ (2008) In hot water: Zooplankton and climate change. ICES J Mar Sci 65:279–295

Sathyendranath S, Gouveia AD, Shetye SR, Platt T (1991) Biological controls of Sea surface temperature in the Arabian Sea. Nature 349:54–56

Satyendranath S, Platt T, Stuart V, Irwin BD, Veldhuis MJW, Kraay GW, Harison WG (1996) Some bio-optical characteristics of phytoplankton in NW Indian Ocean. Marine Ecol Prog Series 132(1–3):299–311

Schlueter MH, Merico A, Reginatto M, Boersma M, Wiltshire KH, Greve W (2010) Phenological shifts of three interacting zooplankton groups in relation to climate change. Glob Change Biol 16(11):3144–3153

Sergey F, Ryan JP, Chavez FP (2012) Predicting euphotic-depth-integrated chlorophyll-a from discrete-depth and satellite-observable chlorophyll-a off central California. J Geophys Res 117:C05042. https://doi.org/10.1029/2011JC007322

Smith RC (1981) Remote Sensing and depth distribution of Ocean chlorophyll. Marine Ecological Progress Series 5:359–361

Smith SL (ed) (1998) The 1994–1995 Arabian Sea expedition: oceanic response to monsoonal forcing, Part 1. Deep-Sea Res II(45):1905–2501

Sturm B (1993) CZCS data processing algorithms. In: Barale V, Schlittenhardt PM (eds) ocean colour: theory and applications in a decade of CZCS experience. Kluwer Academic Publishers, Dordrecht, pp 95–116

Taylor AH, Allen JI, Clark PA (2002) Extraction of a weak climatic signal by an ecosystem. Nature 416 (6881):629–632

Theckeray SJ (2012) Mismatch revisited: what is trophic mismatching from the perspective of the phytoplankton? J Plankton Res 34:1001–1010

Theckeray SJ, Jones ID, Maberly SC (2008) Long term change in the phenology of spring phytoplankton: spices specific responses to nutrient enrichment and climate change. J Ecol 96:523–535

Barale V, Schlittenhardt PM (eds), Kluwer Academic Publishers, Dordrecht, pp 33–77

Wiggert JD, Hood RR, Banse K, Kindle JC (2005) Monsoon-driven biogeochemical processes in the Arabian Sea. Prog Oceanogr 65(2–4):176–213

10 Application of a Geospatial-Based Subjective MCDM Method for Flood Susceptibility Modeling in Teesta River Basin, West Bengal, India

Indrajit Poddar, Jiarul Alam, Amiya Basak, Rajib Mitra, and Jayanta Das

Abstract

Flood is one of the most dangerous natural hazards in India due to their devastating effects in every sector. The present study was designed to develop flood-susceptible map using geospatial-based AHP method in Teesta River basin of West Bengal, India. At first, we selected ten flood factors such elevation, slope, TPI, SPI, NDVI, drainage density, mean annual rainfall, MNDWI, lithology, and distance from the river for developing flood susceptibility map. The most important flood influencing factors were elevation (27%), slope (19%), distance from the river (14%), and drainage density (14%). The obtained suitability map identified 11.99%, 13.90%, 21.48%, 23.93%, and 28.70% of study areas as very high, high, moderate low, and very low susceptible areas, respectively. The accuracy of flood susceptibility map was tested using ROC and AUC analysis. The obtained AUC value was 0.739, confirming the reliability of applied method in flood modeling. The finding of this study will be useful in preparing the effective flood management strategies.

Keywords

Flood susceptibility · Teesta river · Multicriteria · AHP · Validation

10.1 Introduction

Natural hazards like floods, earthquakes, cyclones, tsunamis, storms, volcanic eruptions, and landslides have become the most well-recognized calamities around the world (Kafle 2017) in recent decades. It increases due to global climate change and human activity (Arya and Singh 2021). Among the several natural hazards, floods are considered as severe natural catastrophes, occurring every year (Ahmadlou et al. 2021; Arya and Singh 2021), resulting in severe injuries and fatalities, extensive infrastructure

I. Poddar · J. Alam · R. Mitra
Department of Geography and Applied Geography, University of North Bengal, NBU, Darjeeling 734013, India
e-mail: indrajit.poddar.cob@gmail.com

J. Alam
e-mail: geojia94@gmail.com

R. Mitra
e-mail: rajibmitrageo@gmail.com

J. Das
Department of Geography, Rampurhat College, Rampurhat, Birbhum 731224, India
e-mail: jayanta.daas@gmail.com

A. Basak (✉)
Department of Geography, Cooch Behar Panchanan Barma University, University in Cooch Behar, Cooch Behar, West Bengal 736101, India
e-mail: amiyabasak4@gmail.com

J. Das and S. K. Bhattacharya (eds.), *Monitoring and Managing Multi-hazards*, GIScience and Geo-environmental Modelling, https://doi.org/10.1007/978-3-031-15377-8_10

damages, significant socioeconomic losses globally (Vignesh et al. 2021; Das and Gupta 2021). Unanticipated changes in morphological (He et al. 2008), hydrological, and geomorphological processes result in increased monsoonal rainfall of high intensity, melting of snow, and dam overflow, all of which lead to flooding (Roy et al. 2021a, b). Flooding is induced due to high-intensifying rainfall for a short period (Simon Wang et al. 2015); moreover, the various physical forces, in combination with different human activity, have accelerated it and threatening the environment, ecology on a worldwide scale (Vignesh et al. 2021; Ali et al. 2020). Furthermore, without a warning mechanism, floods in monsoon-dominated areas (eastern India) are extremely regular and unexpected (Roy et al. 2021a, b). Appropriate measures to deal with circumstances that may arise due to climate change have been established as a result of the increase in hazards globally (Arya and Singh 2021). In the case of India, which is fittingly known as the "Land of Rivers," it is crisscrossed by numerous rivers. Floods in India have drawn more attention because they are quite regular and expensive in respect of losses of property. It negatively affects the economic sectors of the country (Ali et al. 2019; Panwar and Sen 2020). Although high-intensity floods are seen as a risk, they also help to replenish river biodiversity and soil quality, rendering floods both a blessing and a curse for the economy (Das 2019; Vignesh et al. 2021; Gourav et al. 2020). However, in the rainy season, the high-intensity rainfall for a shorter time causes a sudden peak in runoff that exceeds the absorptive capacity of the existing drainage system, causing flooding primarily in urban and sub-urban areas (Dimitrova and Bora 2020). It has been reported that in India, around 40 million hectares of areas are extremely vulnerable to floods, and 32 million populations are suffered every year from floods (Das and Gupta 2021; Ray et al. 2019). Srivastava (2011) has mentioned in his study that varied climatic and topography factors of India make distinct places vulnerable to a variety of natural dangers. Moreover, according to Andrew et al. (2018), India is classified as a high-risk country regarding hazards, which has a risk management score of 5.7 and a flood hazard rating of roughly 8.5(Dilley 2005). Hence, though catastrophic floods cannot be avoided, using a proper management approach can help to mitigate the damages. That is why identifying the most susceptible areas is a critical strategic aspect for reducing the damages from disasters (Bubeck 2012; Chapi et al. 2017).

Susceptibility assessment is a difficult method for mapping possible susceptible regions and initiating positive measures for long-term management (Merz et al. 2007). Plenty of literatures identifies a "flood susceptibility map" as a critical flood-prevention measure (Choudhury et al. 2022; Mishra and Sinha 2020). Preparing the flood susceptibility maps for the management, planning, and identification of very high flood-prone areas is necessary (Das 2020a, b). The purpose of the susceptibility assessment was to evaluate the probability over decades to aid risk management actions (Tehrany et al. 2013). Various environmental elements, such as topography, land use, geology, climate, and hydrological factors, can influence a flood's outburst and should be taken into account when creating flood susceptibility maps (Das 2020a, b). Though natural disasters cannot be prevented, but yet, their risk and consequences can be lessened with good preparation and strategy.

Among all the methodologies for studying flood hazards, the Analytical Hierarchy Process (AHP) proposed by Saaty in 1980 and developed within the context of MCDM (Multi-criteria decision-making). The AHP is recommended when segregating many substitutes to a set of pairwise comparisons and then incorporating the outcome (Basak et al. 2021). Moreover, it is a method of structured measuring that involves a sophisticated collection of various decision-making criteria (Satty 1990; Allafta et al. 2020; Vignesh et al. 2021). These criteria are organized in a hierarchical order based on the decision makers preferences, which aids in evaluating the relative relevance of the characteristics in a precise and systematic manner (Velasquez and Hester 2013; Seejata et al. 2018; Boulomytis et al. 2019).

The River Teesta originates in the Eastern Himalayas and flows through the Sikkim and

West Bengal of India. It is known as the lifeline of the Sikkim and North Bengal region (Tsering et al. 2019). Though the Teesta River's course and the system of that as a whole changed after 1787, it had effects on the drainage systems and the floodplains (Hanif et al. 1995). The Teesta River, which used to hold water all year, now dries up shortly after the monsoon. On the river bed, there are several chars and shoals. Due to water extraction and the flow of heavy silts from the upper catchments, Teesta's discharge capacity has been substantially diminished. However, flooding is the most frequently occurring natural hazard in a deltaic floodplain like the Teesta River basin. A complicated set of circumstances causes flooding in the Teesta River basin. A large inflow of water from upstream regions, which coincided with substantial monsoon rainfall across the country, a low floodplain gradient, and clogged drainage systems are among them. Different combinations of these components result in distinct types of floods (Al-Hussain et al. 2021).

The primary purpose of this study is to determine flood-susceptible areas in Teesta river basin, West Bengal, by the integration of the multicriteria decision model (AHP model) based and the geospatial techniques where flooding is a typical occurrence due to poor drainage conditions in the lower reaches of the river (Mandal and Chakrabarty 2016; Goyal and Goswami 2018). Many appropriate criteria have been taken to alleviate or control this disastrous circumstance. The creation of a flood susceptibility map is an effective management tool that aids in the identification of sensitive areas as well as the requirement for construction management to minimize flooding in flood-prone areas and conserve the earth's resources. GIS (Geographical information system) implementation aids in the analysis and manipulation of vast amounts of spatial data. Flood models are generally applied to construct flood susceptibility maps which would help the policymakers and can use it to assess future threats spatially. In this case, physical modeling is one of the most trusted and accepted ways for analyzing flood safety.

10.2 Study Area

Teesta is a tributary of the Brahmaputra River on its right bank that drains about 8600 km^2 in Sikkim Darjeeling Himalaya (Mandal and Chakrabarty 2016) which is drained from Chombo Chu, at a height of 17,322.10 ft in the Indian state of Sikkim's northeastern region (Rahaman and Abdullah-Al-Mamun 2020). The top reaches of the Teesta River's catchment region are largely covered in glaciers and snow, while lower reaches are commonly covered in forest. The geography of the Teesta River is highly variable, with an undulating elevated contour. This river is formed when two rivers, the Lachen Chu and the Lachung Chu, converge near Chungthang to form the Teesta River after that it has widens and joins the Rangit river near Sikkim's Teesta Bazar (Goyal and Goswami 2018; Rahaman and Abdullah-Al-Mamun 2020) and enters into the plain near Sevoke, 14 miles northeast of Siliguri, where the Coronation Bridge connects India's northeastern states to the rest of the nation (Al-Hussain et al. 2021; Mondal et al. 2020). Our study area confined within the 78°37′49″E to 78°45′23″E longitudes and 23°23′42″N to 25°56′12″N latitudes (Fig. 10.1), covered 10,049.075 km^2 area.

The Teesta River runs south from Jalpaiguri in three channels: the Punarbhaba on the west, the Karotaya on the east, and the Atrai in the middle. These three rivers possibly origin in the names Trisrota and Teesta. The hydrological regime of this river is complicated moreover rainfall, melting glaciers and snow, as well as groundwater all contribute to the river's nourishment. (Wiejaczka et al. 2014). This river basin area is a risk of flooding on a regular basis (Mandal and Chakrabarty 2016). In addition, due to water extraction and the flow of heavy silts from the upper catchments, Teesta's discharge capacity has been substantially diminished (Karmokar and de 2020). During the monsoon season, the Teesta River basin is characterized by severe rains and floods (Goyal and Goswami 2018). Moreover, flash floods are caused by runoff from nearby upland areas during periods of unusually high

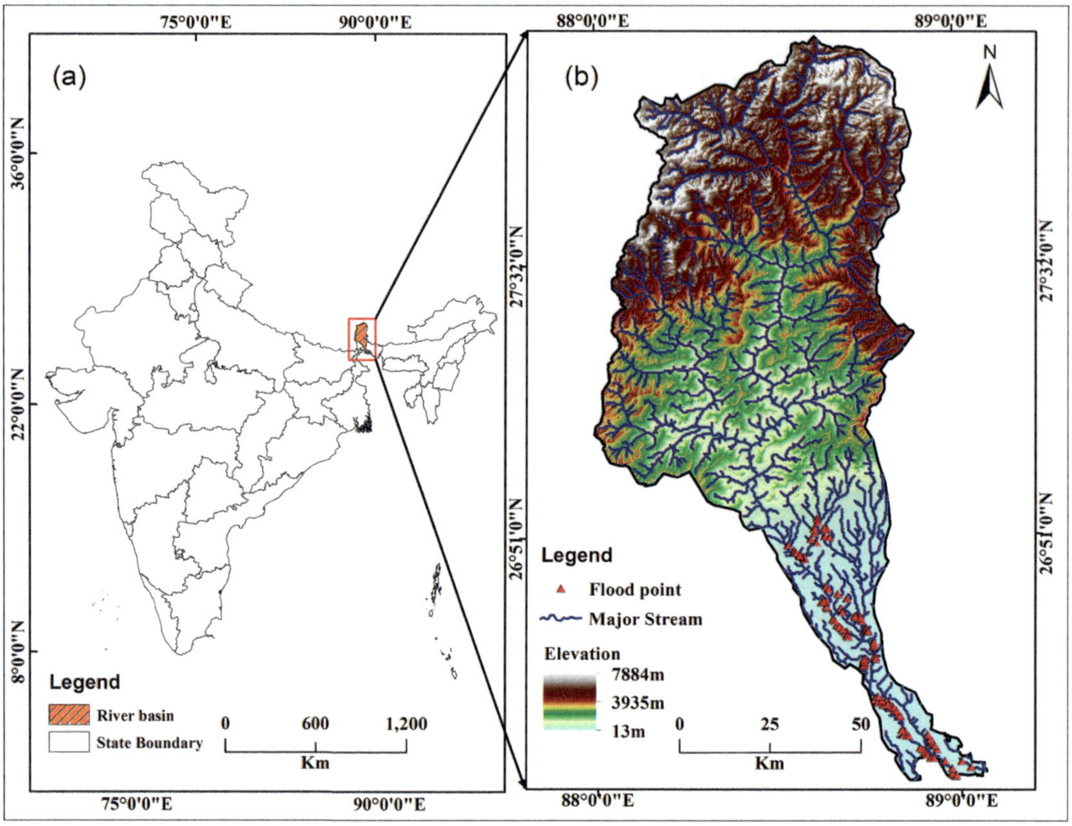

Fig. 10.1 Location map of study area

rainfall. Such floods occur when high-velocity waters from the mountainous upstream surge to the lowlands, devastating standing crops and physical infrastructure (Ahmed and Mirza 2000). Rainwater floods, on the other hand, are generated by high rainfall across the Teesta River flood plain. Meander floodplains, main floodplain basins, and historic piedmont and estuarine floodplains are all prone to rainwater flooding.

10.3 Materials and Methods

10.3.1 Data Sources

The sources of the different factors which are adopted in this study are shown in Table 10.1. Following a comprehensive literature search and expert viewpoints, ten theme layers were chosen to analyze the Teesta River basin's flood susceptibility map. Elevation normalized difference vegetation index (NDVI), distance from rivers, modified normalized difference water index (MNDWI), mean annual rainfall, topographic wetness index (TWI), topographic position index (TPI), slope, drainage density, stream power index (SPI), and lithology are all connected to flood conditioning or susceptibility characteristics. All of the layers were created in the GIS environment after a comprehensive evaluation of field investigation and a review of the literature. These parameters are produced with the help of digital elevation model. The 30-m resolution digital elevation model (DEM) was downloaded from the NASA Earth data login search. From the spatial analyst tool in ArcGIS, the line density interface was used to create the drainage density map, and the Euclidian distance interface was used to create the distance from the river. For the preparation of TWI, DEM was

Table 10.1 Data sources of different flood influencing factors

Sl. no.	Factors	Types of datasets	Spatial resolution/description	Source
1	Elevation	ASTER GDEM	Version 3, 30 m × 30 m	earthdata.nasa.gov
2	Slope	ASTER GDEM	Version 3, 30 m × 30 m	earthdata.nasa.gov
3	Drainage density	ASTER GDEM	Version 3, 30 m × 30 m	earthdata.nasa.gov
4	Distance to Rivers	ASTER GDEM	Version 3, 30 m × 30 m	earthdata.nasa.gov
5	TPI	ASTER GDEM	Version 3, 30 m × 30 m	earthdata.nasa.gov
6	SPI	ASTER GDEM	Version 3, 30 m × 30 m	earthdata.nasa.gov
7	Mean Annual Rainfall	High-resolution gridded data,	PARSIANN-CCS, 0.04° × 0.04°	chrsdata.eng.uci.edu/
8	NDVI	Landsat 8 OLI/TIRS	1:1,091,958	earthexplorer.usgs.gov
9	MNDWI	Landsat 8 OLI/TIRS	USGS, 30 m × 30 m	earthexplorer.usgs.gov
10	Lithology	Vector layer	Geological survey of India (GSI) data	gsi.gov.in

filled by using the fill interface from the hydrology tool after that tan slope, and upslope was done with the help of raster calculator interface using map algebra. For the preparation of lithology map, vector layer was downloaded from the Bhukosh Geological survey of India (GSI); after that the raster layer was created from the conversion tool. Thereafter, rainfall raster data were obtained from a high resolution of 0.4° × 0.4° PARSIAN CCS Web site; after that IDW technique was used for final mean annual rainfall. We have used IDW technique for preparing the final layer of rainfall. Landsat 8 OLI (30-m spatial resolution, June, 2021) imagery was obtained from the Earth Explorer Web site of the US Geological Survey in order to develop some layers such as the normalized difference vegetation index (NDVI) and the modified normalized difference water index (MNDWI).

10.3.2 Flood Influencing Factors

10.3.2.1 Elevation

The elevation is considered the most important parameter to evaluate flooding. The lands with lower elevations are mainly flood-susceptible lands (Waqas et al. 2021). Here, river discharge was observed highest, and the lands nearer to the river are facing waterlogged conditions during the rainy season. This study has prepared an elevation map using the ASTER DEM (30-m resolution), which was collected from the USGS Web site, as shown in Fig. 10.2a. The elevation of the study area ranges from 13–7884 m.

10.3.2.2 Slope

The slope of any region reflects the variation in topography. The topographic slope impacts the infiltration and runoff capacity because it controls the velocity of water flow (Adiat et al. 2012). It observed that the steeper sloping areas are less flooded than the gently sloping areas; hence, the potentiality of flooding events directly depends on the slope (Tang et al. 2020). The slope map has been prepared using ASTER DEM, as shown in Fig. 10.2b, and here, the slope angle ranges from 0 to 80°.

10.3.2.3 TPI

The topographic position index (TPI) in this study is considered as one of the important parameters. This index value always represented the altitudinal variation (Popa et al. 2019). It is also used to classify the slope and watershed metrics of any region (Weiss 2001). If the value of the targeted cells is positive, then it shows the greater altitudinal difference in respect of adjoining cells, whereas negative values indicate that the altitude of the targeted cells is lower than the adjoining cells (De Reu et al. 2013). The TPI was categorized into four classes, ranging from −4.07 to 4.28 (Fig. 10.2c).

10.3.2.4 SPI

The stream power index (SPI) is computing using Eq. 10.1, where Ai depicts the specific area and $\tan\beta$ depicts the gradient (Moore et al. 1991).

$$\mathrm{SPI} = \mathrm{Ai} * \tan\beta \tag{10.1}$$

The region with lower SPI value implies the higher capacity to introduce flood phenomena of the river. In this region, SPI value is found between 0–48,976,092 (Fig. 10.2d).

10.3.2.5 NDVI

Another important component influencing flooding is the normalized difference vegetation index (NDVI) (Lee et al. 2018). It is a useful index for evaluating vegetation cover of any region, which varies between + 1 and −1. The NDVI here computed from the satellite image Landsat 8 OLI using the following equation:

$$\mathrm{NDVI} = \frac{\mathrm{Band5} - \mathrm{Band4}}{\mathrm{Band5} + \mathrm{Band4}} \tag{10.2}$$

The high values of NDVI showed the areas with vegetation covered, while low values indicated the areas with less vegetation. The study area exhibits complex vegetation patterns in respect of height. The produced NDVI map is represented in Fig. 10.2e, and the index ranges from −0.27 to 0.57.

10.3.2.6 Drainage Density

The drainage density map of any region depicts the spatial variation in drainage segment lengths per unit area, which is considered an important factor in determining floods. A region with high river density shows greater flood peaks, and it is frequently flooded during the rainy season (Ogden et al. 2011; Fraser and Schumer 2012; Das and Pardeshi 2018). The extracted rivers from ASTER DEM were used to delineate river density map in the ArcGIS platform. The region was classified into five river density zones (Fig. 10.2f).

10.3.2.7 Annual Rainfall (2003–2020)

The intensity of rainfall directly influences flood hazards (Ghosh and Kar 2018), and it is a crucial factor in assessing flood susceptibility. It contributed to recharging the region and also affect the aquifer level (Knebl et al. 2005). The entire basin is frequently flooding during the rainy season due to the southwest monsoon. Historical records revealed that the study area is experiencing heavy orographic rainfall, and it caused an adverse effect on flood hazards. Here, the rainfall map was prepared from the PARSIANN-CCS data using the IDW interpolation method. The annual rainfall map shows in Fig. 10.2g, where the value ranges from 339 to 1926 mm.

10.3.2.8 Modified Normalized Difference Water Index (MNDWI)

The modified normalized difference water index (NDWI) has played a significant role as a flood triggering factor. Several studies (Jain et al. 2005; Das and Pal 2017; Talukdar and Pal 2018) depicted the correlation between MNDWI and flood hazard. This index gives information on the presence of water bodies, soil characteristics, and vegetation cover. Generally, the positive MNDWI value demonstrated the areas associated with water bodies (Paul et al. 2019). It was calculated using the following formula:

$$\mathrm{MNDWI} = \frac{\mathrm{Green} - \mathrm{MIR}}{\mathrm{Green} + \mathrm{MIR}} \tag{10.3}$$

where MIR represents a middle infra-red band. Here, the MNDWI value varies from −0.41 to + 74 (2 h).

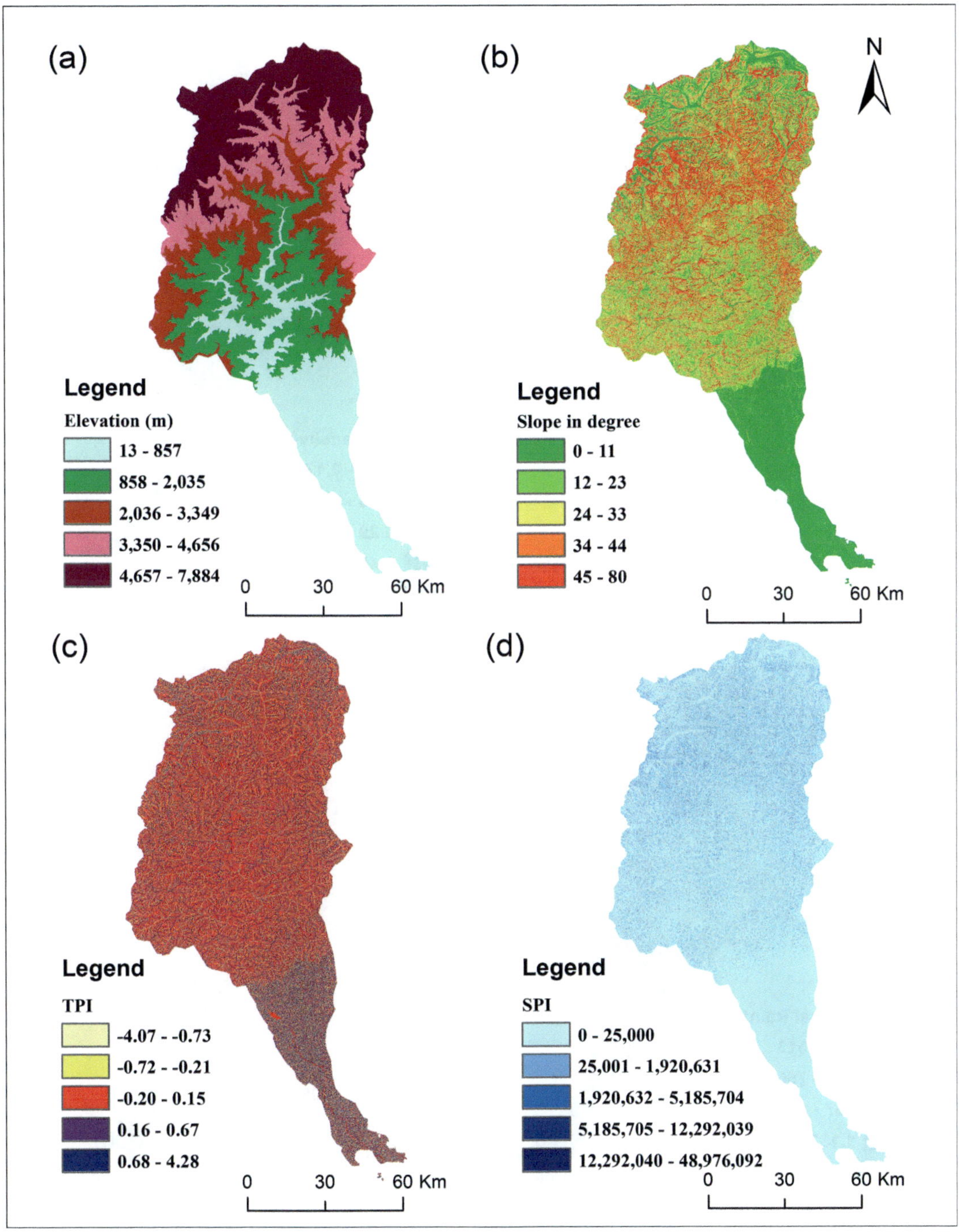

Fig. 10.2 Thematic layers of **a** Elevation, **b** Slope, **c** TPI, and **d** SPI, **e** NDVI, **f** Drainage density, **g** Mean annual rainfall, and **h** MNDWI, **i** Lithology, and **j** Distance from river

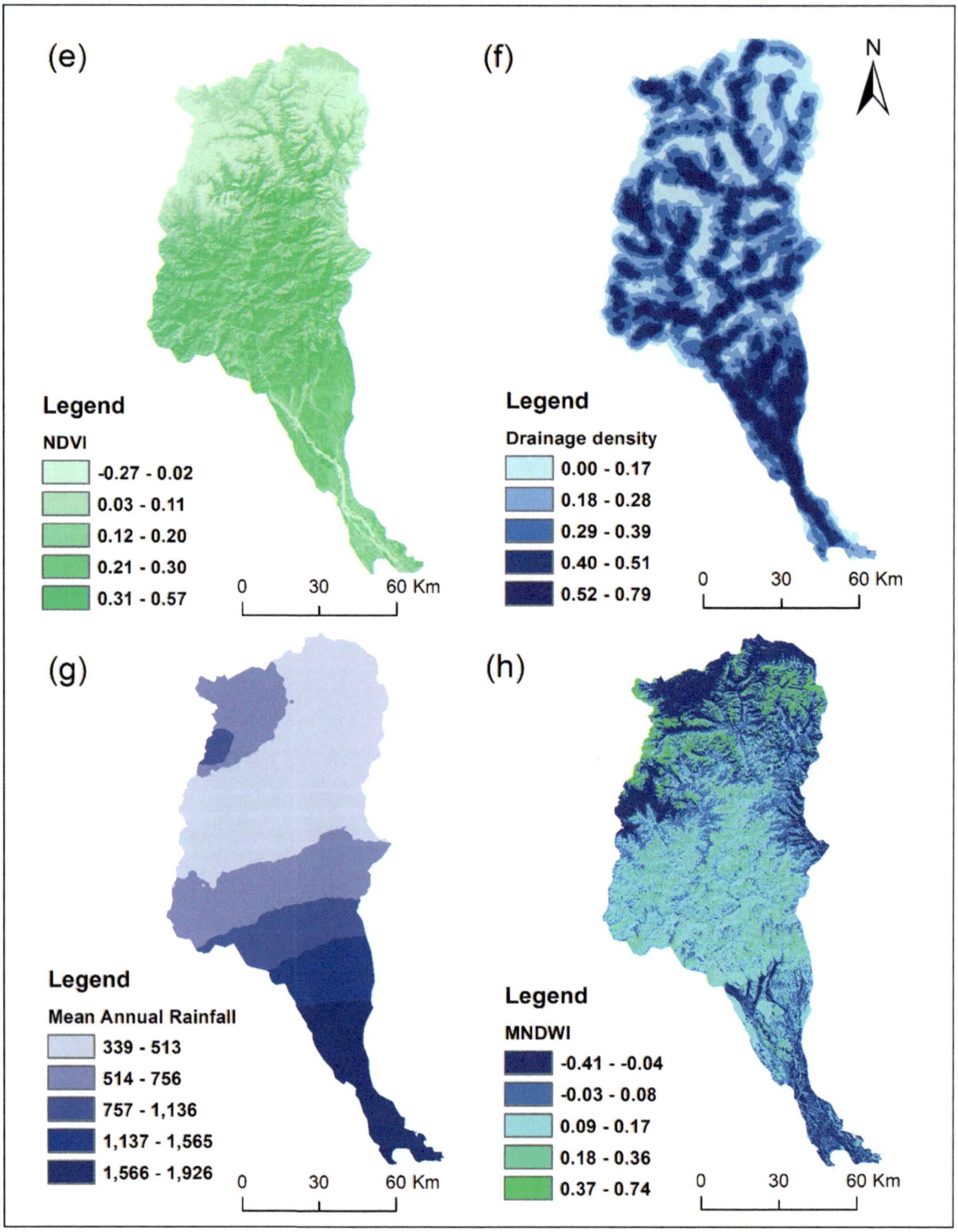

Fig. 10.2 (continued)

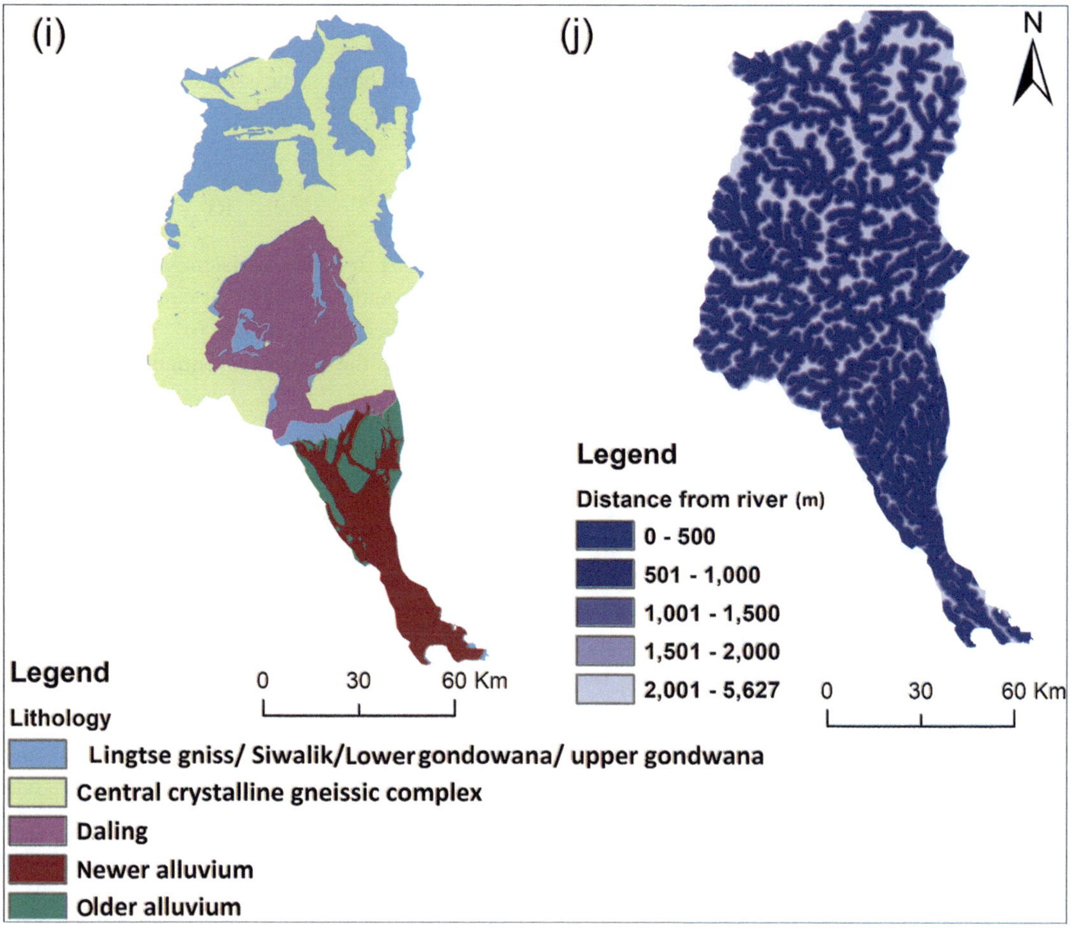

Fig. 10.2 (continued)

10.3.2.9 Lithology

The geological formation has its relation with the flooding. It affected the drainage network and hydrological character of any region. The variation in surface hydrology has been occurred by the differences in rock type (Saha et al. 2022). The areas lie on non-permeable rocks, which have higher river density, and it is frequently flooding (Das 2020a, b). The geological map of the study area was prepared using the data from the Geological Survey of India, and it depicted that the present study area has five geologic units, like lingtse gneiss/siwalik/lower gondowana upper gondowana, central crystalline gnissic coplex, dalig, newer alluvium, and older alluvium (Fig. 10.2i).

10.3.2.10 Distance from River

Distance from the river also affects the intensity of flooding. The regions nearer to the rivers are more frequent in flooding than the regions away from the rivers. Generally, the river adjoining areas are the active flood plain zone; hence, proximity to the river is considered a vital flood conditioning factor (Ghosh and Kar 2018; Pham et al. 2020). In this study, the river network has been extracted from the ASTER DEM. Then, using multi-ring buffer tools in ArcGIS represented the distance to river map by five classes, i.e., 0–500, 501–1000, 1001–1500, 1501–2000, and 2001–5627 as illustrated in Fig. 10.2j.

10.3.3 The Analytical Hierarchy Process (AHP)

We applied AHP method do develop flood susceptibility map in Teesta river basin and popularly used in flood susceptibility mapping. The AHP method was first propounded by Thomas L. Saaty in 1980. It is a framework for resolving various sorts of multicriteria decision challenges based on relative precedence assigned to each criterion in reaching a specific goal (Handfield et al. 2002). This model relies on a big number of subjective marginal inputs and a big number of criteria. These inputs are transformed to scores while evaluating each criterion (Das et al. 2017). Here, the preferences of individual indicators are presented using Saaty's scale of preference, as shown in Table 10.2 (Saaty 2008). Generally, the AHP method consists of four steps, i.e., assignments of weight, pairwise comparison matrix, normalization of weights, and checking of consistency. In this study, all steps were preciously done using the extensive related literature, some professional viewpoints, and practical field knowledges.

In this study, a pairwise comparison matrix has been computed (Table 10.3) to determine the relative relevance of 10 attributes in relation to the model's purpose. In addition to that, we also find out consistency ratio (CR) value to check the consistency of the judgments (Harker and Vargas 1987). The consistency ratio is expressed as follow:

$$\mathrm{CR} = \frac{\mathrm{CI}}{\mathrm{RI}} \quad (10.4)$$

where CI refers consistency index and RI means the random index that can be represented by the number of parameters (n). The value of CI is defined by the following equation:

$$\mathrm{CI} = \frac{\lambda_{\max} - n}{n - 1} \quad (10.5)$$

where $\lambda_{\max}$ is shown by the largest eigenvalue of the comparison matrix, described as:

$$\lambda_{\max} = \sum_{j=1}^{n} aij \frac{\mathrm{wj}}{\mathrm{wi}} = n \quad (10.6)$$

A matrix is consistent if CR = 0, and inconsistency is revealed when CR > 0. The study reveals that the computed consistency ratio (CR) was 0.076; the consistency index was 0.11, and the random index for $n(10)$ was 1.49. The consistent comparison matrix has been used to assign weights to reclassified raster parameters.

10.3.4 ROC and AUC Analysis

Model validation is the procedure by which we evaluate the competence of the applied model in the given work. In the literature, a variety of statistical techniques have been identified for

Table 10.2 Saaty's scale of preference

Relative importance between factors	Identification
1	Equal importance
3	Moderate importance
5	Strong importance
7	Very strong importance
9	Extreme importance
2, 4, 6, 8	Intermediate values
Reciprocals	Inverse comparison

Table 10.3 Pairwise comparison matrix

Parameters	E	S	DR	DD	RI	MNDWI	NDVI	L	TPI	SPI
E	1	2	4	4	6	6	7	7	8	9
S	1/2	1	3	4	5	6	7	7	8	8
DR	1/4	1/3	1	1	4	5	6	6	7	7
DD	1/4	1/4	1	1	4	5	6	6	7	7
RI	1/6	1/5	1/4	1/4	1	3	4	4	5	5
MNDWI	1/6	1/6	1/5	1/5	1/3	1	3	3	4	4
NDVI	1/7	1/7	1/6	1/6	¼	1/3	1	1	3	3
L	1/7	1/7	1/6	1/6	¼	1/3	1	1	3	3
TPI	1/8	1/8	1/7	1/7	1/5	1/4	1/3	1/3	1	1
SPI	1/9	1/8	1/7	1/7	1/5	1/4	1/3	1/3	1	1

validating the applied models. Model validation through ROC and AUC is the most reliable techniques. Given the robustness of this validation technique, the performance of the applied model (AHP) in this study was validated using ROC and AUC analysis. Generally, ROC graph is two dimensional, with the *X*-axis denoting 1-specificity (false positive) and the *Y*-axis representing sensitivity (true positive). AUC, on the other hand, is the area under the ROC curve, which aids in determining how well applied models perform in classification tasks. AUC value ranges from 0.5 to 1, with lower values indicating poor performance (accuracy) and higher values showing good performance (accuracy) of applied methods.

10.4 Results and Discussion

10.4.1 Weightage of Flood Influencing Factors

In developing a susceptibility map, the selection of relevant and sufficient factors is critical. Furthermore, the performance of the applied model is also dependent on the precise weighting to these parameters. In this study, most influencing factors have been identified using the expert-based opinion from the previous literatures (Periyasamy et al. 2018; Talukdar et al. 2020; Idowu and Zhou 2021). In this study, the assigned weights of the ten flood triggering parameters have been illustrated in Table 10.4. The higher weights were assigned to the most relevant factors which are determined for flood susceptibility studies and vice versa. In several studies, like Nachappa et al. (2020), Souissi et al. (2020), Ali et al. (2020), and Roy et al. (2021a, b), elevation is considered as the most triggering factor, whereas Periyasamy et al. (2018) and Idowu and Zhou (2021) assigned rainfall as the highest influencing factor. Therefore, in the present study, elevation has been assigned as the highest weight (27%), and TPI and SPI get the lowest weight (2%). Slope is also considered as an important factor; hence, it gets the second-highest weight (19%). Distance from river and drainage density both parameters assigned as similar weight (14%). Mean annual rainfall and MNDWI are obtained moderate weights, i.e., 9% and 6%, respectively.

10.4.2 Flood Susceptibility Map

The natural break approach is utilized to outline the flood susceptibility map in this study. Despite the fact that numerous criteria influence map accuracy, the accuracy of the map is not completely determined by them. It is dependent on field observations, expert opinions, and correct

Table 10.4 Weightage of different factors used in the study

Parameters	AHP weight (%)	Range	Flood level	Rating
Elevation	27	13–857	Very high	0.54
		858–2035	High	0.35
		2036–3349	Medium	0.22
		3350–4656	Low	0.14
		4657–7884	Very low	0.08
Slope	19	0–11	Very high	0.631
		12–23	High	0.392
		24–33	Medium	0.222
		34–44	Low	0.142
		45–80	Very low	0.077
Distance from river (m)	14	0–500	Very high	0.885
		501–1000	High	0.511
		1001–1500	Medium	0.293
		1501–2000	Low	0.164
		2001–5627	Very low	0.066
Drainage density (km/km^2)	14	0.00–0.17	Very low	0.082
		0.18–0.28	Low	0.158
		0.29–0.39	Medium	0.300
		0.40–0.51	High	0.468
		0.52–0.79	Very high	0.779
Mean annual rainfall	9	339–513	Very low	0.083
		514–756	Low	0.158
		757–1136	Medium	0.300
		1137–1565	High	0.468
		1566–1926	Very high	0.767
MNDWI	6	−0.41–−0.04	Very low	0.083
		−0.03–0.08	Low	0.160
		0.09–0.17	Medium	0.300
		0.18–0.36	High	0.456
		0.37–0.74	Very high	0.655
NDVI	4	−0.10–0.11	Very high	0.540
		0.03–0.11	High	0.351
		0.12–0.20	Medium	0.222
		0.21–0.30	Low	0.142
		0.31–0.57	Very low	0.083

(continued)

Table 10.4 (continued)

Parameters	AHP weight (%)	Range	Flood level	Rating
Lithology	3	Lingtse Gneiss, Siwalik, Lower Gondwana	Very low	0.070
		Central crystalline gneissic complex	Low	0.501
		Daling	Medium	0.159
		Newer alluvium	Very high	0.035
		Older alluvium	High	0.235
TPI	2	−4.07–−0.73	Very low	0.822
		−0.72–−0.21	Low	0.503
		−0.20–0.15	Medium	0.252
		0.16–0.67	High	0.170
		0.68–4.28	Very high	0.067
SPI	2	0–25,000	Very high	0.885
		25,001–1,920,631	High	0.511
		1,920,632–5,185,704	Medium	0.293
		5,185,705–12,292,039	Low	0.164
		12,292,040–48,976,092	Very low	0.066

practical and logical knowledge regarding conventional information, and this is a very crucial for a researcher to complete the outstanding outputs (Ghosh and Kar 2018; Das 2019). The final map has been categorized into five classes: very low (28.70%), low (23.93%), moderate (21.48%), high (13.90%), and very high susceptible areas (11.99%) (Fig. 10.3). If we look at the map closely, we can see that the southern regions of the river basin are particularly susceptible to flooding since they are located in the foothills of the Himalayas, where the break of slope is quite steep and most of the little streams converge. Because the elevation and slope are both quite low, water accumulation is also very high. This is the primary cause of locations with extremely high flood susceptibility. Furthermore, it is clear that highly flood-susceptible places have a characteristics lower elevation, lower slope, higher drainage density, higher TWI, closeness to rivers. Highest flood susceptibility falls under the southern part of the river basin because here the lowest elevation (13 m), negative NDVI (−0.10), positive MNDWI, negative TPI (−4.07), highest drainage density (0.79 km/km^2), lowest distance (500 m) from river, mean annual rainfall (1926 mm), and newer alluvium lithology. All these flood occurring factors are responsible for the highest flood susceptibility. And lowest flood susceptibility zone is found in the upper portion of the river basin because there are highest elevation (7884 m), slope, positive NVDI (+0.79), negative MNDWI (−0.41), positive TPI means ridge (4.28), and lowest rainfall (339 m) and older Gondwana lithology are found. Newer alluvium soil also indicates the high flood-susceptible areas because newer alluvial soils are found in those areas which are closer to river.

10.4.3 Justification of Model Performance

The accuracy of the applied model can be assessed using the validation technique. We identified a total of 240 points as the ground truth points. The location data for these ground truth points were obtained from the http://wbdmd.gov.in/pages/district_dm_plan.aspx website. Among

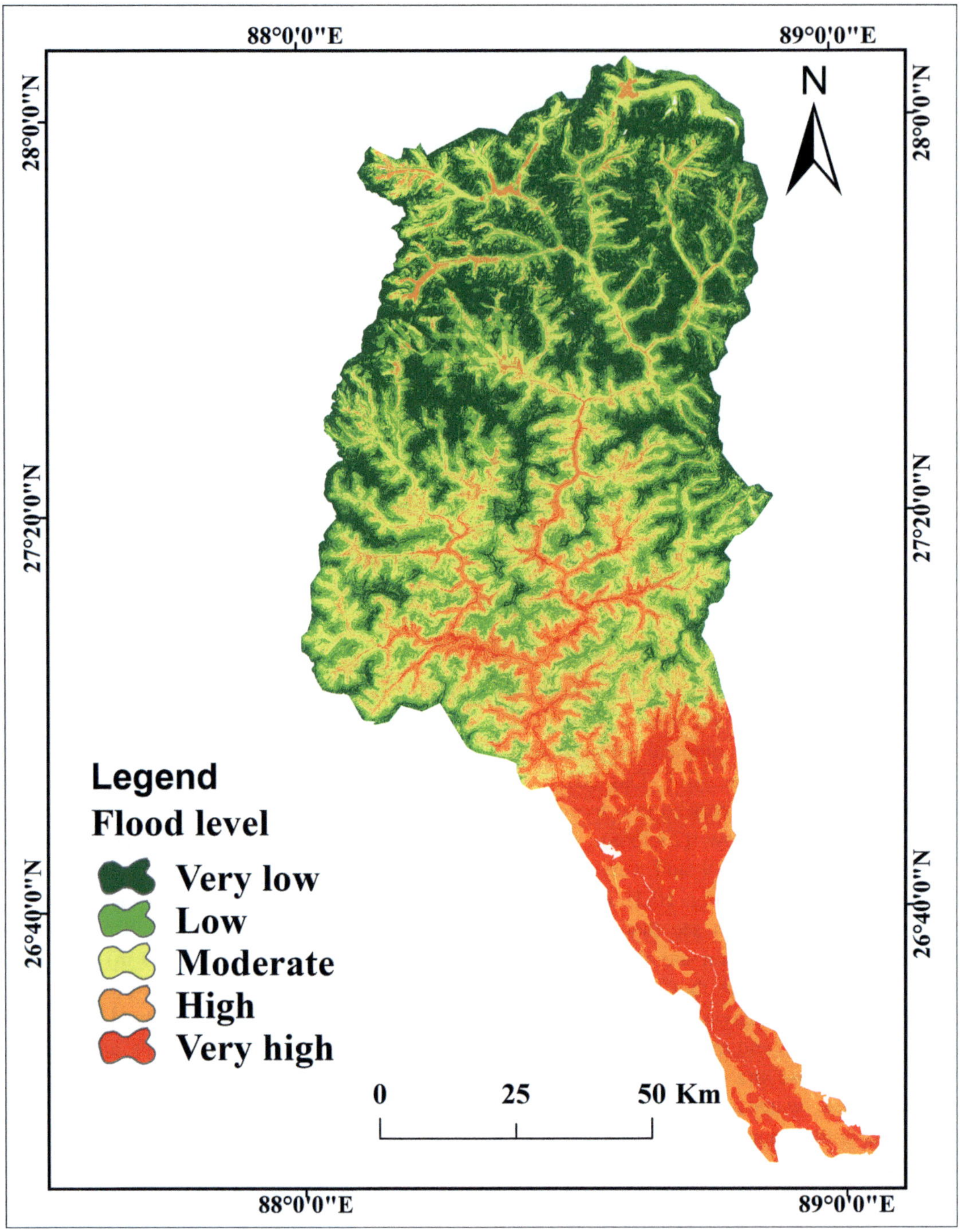

Fig. 10.3 Flood susceptibility map of Teesta River basin of West Bengal

these points, 135 and 105 were recognized as flood and non-flood points, respectively. Following that, we processed these points in the ArcGIS environment using the geographical coordinate system. Thereafter, we used AUC and ROC analysis to assess the accuracy of AHP

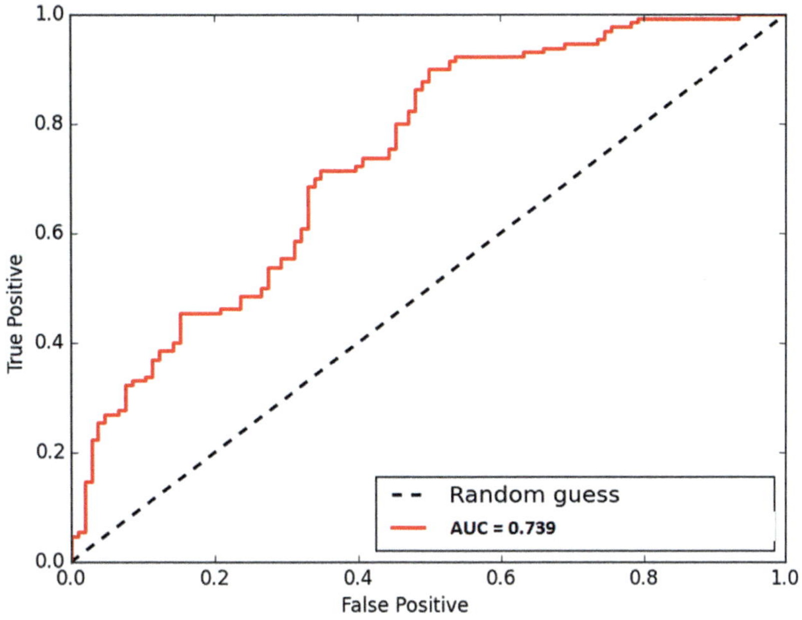

Fig. 10.4 ROC and AUC graph

model. The obtained AUC value showed that the AHP model efficiently developed (AUC =0.739) the flood susceptibility map (Fig. 10.4).

10.5 Conclusions

The goal of this research was to analyze flood-susceptible locations in the Teesta River basin in India. This flood catastrophe susceptible zone mapping is one of the most effective approaches for reducing flood hazard impacts and assisting planners, stakeholders, and decision makers in maintaining correct control over flood-susceptible areas, assuring proper and long-term socioeconomic development. The multi-parametric AHP-GIS technology, which is based on remote sensing and GIS, is a viable alternative for identifying flood susceptibility zones. Elevation, slope, drainage density, Lithology, mean annual rainfall, distance from river, TPI, SPI, MNDWI, NDVI were among the ten variables studied. The geography of the Teesta River is highly variable; during the monsoon season, the Teesta River basin is characterized by severe rains and floods. Flash floods are caused by runoff from nearby upland areas during periods of unusually high rainfall. The importance of flooding phenomena from a spatially and temporally perspective varies greatly in this area. As a result, the methodology of this paper is based on a combination of geospatial techniques (remote sensing and GIS) and MCDM (AHP) approach. The mapping of the area's flood susceptibility was classified into five groups, ranging from "Very Low" to "Very High," based on the superimposition of each parameter. The red, orange, and yellow portion on the map suggested a heightened level of susceptible. 11.99%, 13.90%, 21.48%, 23.93%, and 28.70% of study areas were characterized as very high, high, moderate low, and very low susceptible areas on the suitability map, respectively. To evaluate the outcomes and measure the efficiency of the proposed method, the ROC curve was used. For long-term flood management and mitigation, the findings of this study may be useful to researchers, planners, decision makers, and administrations in the studied region. Due to the generic classification of numerous factors, a critical assessment of flood susceptibility for a specific basin in the study area may communicate more accurate and detailed flood likelihood at the micro-level.

References

Adiat KAN, Nawawi MNM, Abdullah K (2012) Assessing the accuracy of GIS-based elementary multi criteria decision analysis as a spatial prediction tool–a case of predicting potential zones of sustainable groundwater resources. J Hydrol 440:75–89. https://doi.org/10.1016/j.jhydrol.2012.03.028

Ahmadlou M, Al-Fugara AK, Al-Shabeeb AR, Arora A, Al-Adamat R, Pham QB, Al-Ansari N, Linh NT, Sajedi H (2021) Flood susceptibility mapping and assessment using a novel deep learning model combining multilayer perceptron and autoencoder neural networks. J Flood Risk Manag 14(1):e12683. https://doi.org/10.1111/jfr3.12683

Ahmed AU, Mirza MMQ (2000) Review of causes and dimensions of floods with particular reference to flood

Al-Hussain SJ, Paul SK, Hossain MN (2021) Comparative assessment of susceptibility to drought and flood in the lower Teesta River Basin: a SWOT analysis. Asian J Geogr Res 20–33. https://doi.org/10.9734/ajgr/2021/v4i130124

Ali SA, Parvin F, Pham QB, Vojtek M, Vojteková J, Costache R, Linh NT, Nguyen HQ, Ahmad A, Ghorbani MA (2020) GIS-based comparative assessment of flood susceptibility mapping using hybrid multi-criteria decision-making approach, Naïve Bayes tree, bivariate statistics and logistic regression: a case of Topľa basin, Slovakia. Ecol Ind 117:106620. https://doi.org/10.1016/j.ecolind.2020.106620

Allafta H, Opp C, Patra S (2020) Identification of groundwater potential zones using remote sensing and GIS techniques: a case study of the Shatt Al-Arab Basin. Remote Sens 13(1):112. https://doi.org/10.3390/rs13010112

Andrew T, Luca V, Montserrat MF, Brian D (2018) Inform global risk index. https://doi.org/10.2760/754353

Arya AK, Singh AP (2021) Multi criteria analysis for flood hazard mapping using GIS techniques: a case study of Ghaghara River basin in Uttar Pradesh, India. Arab J Geosci 14(8):1–12. https://doi.org/10.1007/s12517-021-06971-1

Basak A, Das J, Rahman ATM, Pham QB (2021) An integrated approach for delineating and characterizing groundwater depletion hotspots in a coastal state of India. J Geol Soc India 97(11):1429–1440. https://doi.org/10.1007/s12594-021-1883-z

Boulomytis VTG, Zuffo AC, Imteaz MA (2019) Detection of flood influence criteria in ungauged basins on a combined Delphi-AHP approach. Oper Res Perspect 6:100116. https://doi.org/10.1016/j.orp.2019.100116

Bubeck P (2012) A review of risk perceptions and other factors that influence flood mitigation behavior. Risk Anal: Official Publ Soc Risk Anal 32:1481–1495. https://doi.org/10.1111/j.1539-6924.2011.01783.x

Chapi K, Singh VP, Shirzadi A, Shahabi H, Bui DT, Pham BT, Khosravi K (2017) A novel hybrid artificial intelligence approach for flood susceptibility assessment. Environ Model Softw 95:229–245. https://doi.org/10.1016/j.envsoft.2017.06.012

Choudhury S, Basak A, Biswas S, Das J (2022) Flash flood susceptibility mapping using GIS-based AHP method. In: Pradhan B, Shit PK, Bhunia GS, Adhikary PP, Pourghasemi HR (eds) Spatial modelling of flood risk and flood hazards. GIScience and geo-environmental modelling. Springer, Cham. https://doi.org/10.1007/978-3-030-94544-2_8

Das S (2019) Geospatial mapping of flood susceptibility and hydro-geomorphic response to the floods in Ulhas basin, India. Remote Sens Appl Soc Environ 14:60–74. https://doi.org/10.1016/j.rsase.2019.02.006

Das S (2020a) Flood susceptibility mapping of the Western Ghat coastal belt using multi-source geospatial data and analytical hierarchy process (AHP). Remote Sens Appl Soc Environ 20:100379. https://doi.org/10.1016/j.rsase.2020.100379

Das S (2020b) Flood susceptibility mapping of the Western Ghat coastal belt using multi-source geospatial data and analytical hierarchy process (AHP). Remote Sens Appl Soc Environ 20:100379. https://doi.org/10.1016/j.rsase.2020.100379

Das RT, Pal S (2017) Exploring geospatial changes of wetland in different hydrological paradigms using water presence frequency approach in Barind Tract of West Bengal. Spat Inf Res 25(3):467–479. https://doi.org/10.1007/s41324-017-0114-6

Das S, Pardeshi SD (2018) Integration of different influencing factors in GIS to delineate groundwater potential areas using IF and FR techniques: a study of Pravara basin, Maharashtra, India. Appl Water Sci 8(7):1–16. https://doi.org/10.1007/s13201-018-0848-x

Das S, Gupta A (2021) Multi-criteria decision based geospatial mapping of flood susceptibility and temporal hydro-geomorphic changes in the Subarnarekha basin, India. Geosci Front 12(5):101206. https://doi.org/10.1016/j.gsf.2021.101206

Das J, Gayen A, Saha S, Bhattacharya SK (2017) Modelling of alternative crops suitability to tobacco based on analytical hierarchy process in Dinhata subdivision of Koch Bihar district, West Bengal. Model Earth Syst Environ 3(4):1571–1587. https://doi.org/10.1007/s40808-017-0392-y

De Reu J, Bourgeois J, Bats M, Zwertvaegher A, Gelorini V, De Smedt P, Chu W, Antrop M, De Maeyer P, Finke P, Van Meirvenne M, Crombé P (2013) Application of the topographic position index to heterogeneous landscapes. Geomorphology 186:39–49

Dilley M (2005) Natural disaster hotspots: a global risk analysis, vol 5. World Bank Publications

Dimitrova A, Bora JK (2020) Monsoon weather and early childhood health in India. PLoS ONE 15(4):e0231479

Fraser N, Schumer R (2012) Low stream density watersheds produce flashier floods than high stream density watersheds in ephemeral streams across the southwestern United States. In: AGU fall meeting abstracts, vol 2012, pp H41F-1240
Ghosh A, Kar SK (2018) Application of analytical hierarchy process (AHP) for flood risk assessment: a case study in Malda district of West Bengal, India. Nat Hazards 94(1):349–368. https://doi.org/10.1007/s11069-018-3392-y
Gourav P, Kumar R, Gupta A, Arif M (2020) Flood hazard zonation of Bhagirathi river basin using multicriteria decision-analysis in Uttarakhand, India. Int J Emerg Technol 11(1):62–71
Goyal MK, Goswami UP (2018) Teesta river and its ecosystem. In: The Indian rivers. Springer, Singapore, pp 537–551. https://doi.org/10.1007/978-981-10-2984-4_37
Handfield R, Walton SV, Sroufe R, Melnyk SA (2002) Applying environmental criteria to supplier assessment: a study in the application of the analytical hierarchy process. Eur J Oper Res 141(1):70–87. https://doi.org/10.1016/S0377-2217(01)00261-2
Hanif M, Mobarak MR, Ronan A, Rahman D, Donovan JJ, Bennish ML (1995) Fatal renal failure caused by diethylene glycol in paracetamol elixir: the Bangladesh epidemic. BMJ 311(6997):88–91. https://doi.org/10.1136/bmj.311.6997.88
Harker PT, Vargas LG (1987) The theory of ratio scale estimation: Saaty's analytic hierarchy process. J Manag Sci 33(11):1383–1403
He H, Bai Y, Garcia EA, Li S (2008) ADASYN: adaptive synthetic sampling approach for imbalanced learning. In: 2008 IEEE international joint conference on neural networks, IEEE world congress on computational intelligence, pp 1322–1328
Idowu D, Zhou W (2021) Land use and land cover change assessment in the context of flood hazard in Lagos State, Nigeria. Water 13(8):1105. https://doi.org/10.3390/w13081105
Jain SK, Singh RD, Jain MK, Lohani AK (2005) Delineation of flood-prone areas using remote sensing techniques. Water Resour Manage 19(4):333–347. https://doi.org/10.1007/s11269-005-3281-5
Kafle SK (2017) Disaster risk management systems in South Asia: natural hazards, susceptibility, disaster risk and legislative and institutional frameworks. J Geogr Nat Disasters 7(3):2167–0587
Karmokar S, De M (2020) Flash flood risk assessment for drainage basins in the Himalayan foreland of Jalpaiguri and Darjeeling Districts, West Bengal. Model Earth Syst Environ 6(4):2263–2289. https://doi.org/10.1007/s40808-020-00807-9
Knebl MR, Yang ZL, Hutchison K, Maidment DR (2005) Regional scale flood modeling using NEXRAD rainfall, GIS, and HEC-HMS/RAS: a case study for the San Antonio River Basin summer 2002 storm event. J Environ Manage 75(4):325–336. https://doi.org/10.1016/j.jenvman.2004.11.024
Lee S, Lee S, Lee MJ, Jung HS (2018) Spatial assessment of urban flood susceptibility using data mining and geographic information system (GIS) tools. Sustainability 10(3):648. https://doi.org/10.1016/j.scitotenv.2018.04.282
Mandal SP, Chakrabarty A (2016) Flash flood risk assessment for upper Teesta river basin: using the hydrological modeling system (HEC-HMS) software. Model Earth Syst Environ 2(2):59
Merz B, Thieken AH, Gocht M (2007) Flood risk mapping at the local scale: concepts and challenges. In: Flood risk management in Europe. Springer, Dordrecht, pp 231–251
Mishra K, Sinha R (2020) Flood risk assessment in the Kosi megafan using multi-criteria decision analysis: a hydro-geomorphic approach. Geomorphology 1(350):106861. https://doi.org/10.1016/j.geomorph.2019.106861
Mondal MSH, Murayama T, Nishikizawa S (2020) Assessing the flood risk of riverine households: a case study from the right bank of the Teesta River, Bangladesh. Int J Disaster Risk Reduction 51:101758. https://doi.org/10.1016/j.ijdrr.2020.101758
Moore ID, Grayson RB, Ladson AR (1991) Digital terrain modelling: a review of hydrological, geomorphological, and biological applications. Hydrol Process 5(1):3–30. https://doi.org/10.1002/hyp.3360050103
Nachappa TG, Piralilou ST, Gholamnia K, Ghorbanzadeh O, Rahmati O, Blaschke T (2020) Flood susceptibility mapping with machine learning, multicriteria decision analysis and ensemble using Dempster Shafer theory. J Hydrol 125275. https://doi.org/10.1016/j.jhydrol.2020.125275
Ogden FL, Raj Pradhan N, Downer CW, Zahner JA (2011) Relative importance of impervious area, drainage density, width function, and subsurface storm drainage on flood runoff from an urbanized catchment. J Water Resour Res 47(12)
Paul GC, Saha S, Hembram TK (2019) Application of the GIS-based probabilistic models for mapping the flood susceptibility in Bansloi sub-basin of Ganga-Bhagirathi river and their comparison. Remote Sens Earth Syst Sci 2(2):120–146. https://doi.org/10.1007/s41976-019-00018-6
Periyasamy P, Yagoub MM, Sudalaimuthu M (2018) Flood vulnerable zones in the rural blocks of Thiruvallur district, South India. Geoenviron Disasters 5(1):1–16. https://doi.org/10.1186/s40677-018-0113-5
Pham BT, Avand M, Janizadeh S, Phong TV, Al-Ansari N, Ho LS, Das S, Le HV, Amini A, Bozchaloei SK, Jafari F, Prakash I (2020) GIS based hybrid computational approaches for flash flood susceptibility assessment. Water 12(3):683. https://doi.org/10.3390/w12030683
Popa MC, Peptenatu D, Drăghici CC, Diaconu DC (2019) Flood hazard mapping using the flood and flash-flood potential index in the Buzău River catchment, Romania. Water 11(10):2116. https://doi.org/10.3390/w11102116

Rahaman MM, Abdullah-Al-Mamun AAM (2020) Hydropower development along Teesta river basin: opportunities for cooperation. Water Policy 22 (4):641–657. https://doi.org/10.2166/wp.2020.136

Ray K, Pandey P, Pandey C, Dimri AP, Kishore K (2019) On the recent floods in India. Curr Sci 117(2):204–218

Roy P, Pal SC, Arabameri A, Rezaie F, Chakrabortty R, Chowdhuri I, Saha A, Malik S, Das, B (2021a) Climate and land use change induced future flood susceptibility assessment in a sub-tropical region of India. Soft Comput 25(8):5925–5949. https://doi.org/10.1007/s00500-021-05584-w

Roy S, Bose A, Chowdhury IR (2021b) Flood risk assessment using geospatial data and multi-criteria decision approach: a study from historically active flood-prone region of Himalayan foothill, India. Arab J Geosci 14(11):1–25

Saaty TL (1990) Decision making for leaders: the analytic hierarchy process for decisions in a complex world. RWS publications

Saaty TL (2008) Decision making with the analytic hierarchy process. Int J Serv Sci 1(1):83–98

Saha S, Das J, Mandal T (2022) Investigation of the watershed hydro-morphologic characteristics through the morphometric analysis: a study on Rayeng basin in Darjeeling Himalaya. Environ Challenges 100463. https://doi.org/10.1016/j.envc.2022.100463

Seejata K, Yodying A, Wongthadam T, Mahavik N, Tantanee S (2018) Assessment of flood hazard areas using analytical hierarchy process over the Lower Yom Basin, Sukhothai Province. Procedia Eng 212:340–347. https://doi.org/10.1016/j.proeng.2018.01.044

Simon Wang SY, Huang WR, Hsu HH, Gillies RR (2015) Role of the strengthened El Niño teleconnection in the May 2015 floods over the southern Great Plains. Geophys Res Lett 42(19):8140–8146

Souissi D, Zouhri L, Hammami S, Msaddek MH, Zghibi A, Dlala M (2020) GIS-based MCDM–AHP modeling for flood susceptibility mapping of arid areas, southeastern Tunisia. Geocarto Int 35(9):991–1017. https://doi.org/10.1080/10106049.2019.1566405

Srivastava RK (2011) Disaster management of India. UNDP, MHA (GOI), Delhi

Talukdar S, Ghose B, Salam R, Mahato S, Pham QB, Linh NTT, Costache R, Avand M (2020) Flood susceptibility modeling in Teesta River basin, Bangladesh using novel ensembles of bagging algorithms. Stoch Env Res Risk Assess 34(12):2277–2300

Talukdar S, Pal S (2018) Impact of dam on flow regime and flood plain modification in Punarbhaba River Basin of Indo-Bangladesh Barind tract. Water Conserv Sci Eng 3(2):59–77. https://doi.org/10.1007/s41101-017-0025-3

Tang X, Li J, Liu M, Liu W, Hong H (2020) Flood susceptibility assessment based on a novel random Naïve Bayes method: a comparison between different factor discretization methods. CATENA 190:104536. https://doi.org/10.1016/j.catena.2020.104536

Tehrany MS, Pradhan B, Jebur MN (2013) Spatial prediction of flood susceptible areas using rule based decision tree (DT) and a novel ensemble bivariate and multivariate statistical models in GIS. J Hydrol 504:69–79. https://doi.org/10.3390/su10030648

Tsering T, Wahed MSA, Iftekhar S, Sillanpää M (2019) Major ion chemistry of the Teesta River in Sikkim Himalaya, India: chemical weathering and assessment of water quality. J Hydrol Reg Stud 24:100612. https://doi.org/10.1016/j.ejrh.2019.100612

Velasquez M, Hester PT (2013) An analysis of multi-criteria decision making methods. Int J Oper Res 10 (2):56–66

Vignesh KS, Anandakumar I, Ranjan R, Borah D (2021) Flood vulnerability assessment using an integrated approach of multi-criteria decision-making model and geospatial techniques. Model Earth Syst Environ 7 (2):767–781. https://doi.org/10.1007/s40808-020-00997-2

Waqas H, Lu L, Tariq A, Li Q, Baqa MF, Xing J, Sajjad A (2021) Flash flood susceptibility assessment and zonation using an integrating analytic hierarchy process and frequency ratio model for the Chitral District, Khyber Pakhtunkhwa, Pakistan. Water 13 (12):1650. https://doi.org/10.3390/w13121650

Weiss A (2001) Topographic position and landforms analysis. In: Poster presentation, ESRI user conference, vol 200. San Diego, CA

Wiejaczka Ł, Bucała A, Sarkar S (2014) Human role in shaping the hydromorphology of Himalayan rivers: study of the Tista River in Darjeeling Himalaya. Curr Sci 717–724

Flood Frequency Analysis of Baitarani River Using Three Probability Distributions

11

Rebati Sinam

Abstract

Baitarani River is one of the important rivers of India falling in the state of Odisha and Jharkhand. It has been affected by the problem of floods and inundation annually. At site Flood Frequency Analysis using three probability distribution models—Log Pearson III, Generalized Extreme Value Distribution and Gumbel Distribution- have been analyzed to find the most suitable and reliable model for the river based on two station data (Anandapur and Champua) using an annual maximum series of 43 years (1976–2018) and 28 years (1991–2018) respectively. Three Goodness of fit (GOF) tests—Anderson–Darling, Kolmogorov–Smirnov and Chi-Square tests are executed to find the best fit model at 95% significance level. From the study, Generalized Extreme Value distribution is the best distribution model for the sub-basin. The significance of the study lies in the fact that finding a suitable distribution model is a pre-requisite for developing an accurate hydraulic structure and efficient flood management strategy. Nevertheless, the study proposed further studies using various other multiparametric models and new advanced techniques for finding the most accurate distribution model for the river sub-basin.

Keywords

Flood frequency analysis · Baitarani River · Gumbel distribution · Generalized extreme value · Log Pearson III

11.1 Introduction

In flood hydraulic engineering, the hydrograph of an extreme event provides a vast amount of valuable data, which can be very vital for hydrologic design. These data are then used to develop a series of models that can be used for developing hydraulic structures like spillways, dams, barrages, bridges and culvert waterways, etc. Flood frequency analysis is an important field of hydrology that works to develop a design flood of a specific basin or river channel. Design flood is the flood considered for designing any hydraulic structures of any desired recurrence interval depending on the degree of protection and cost–benefit analysis of the structure to the desired flood stage (Raghunath 2014). Flood frequency analysis studies flood events to determine how frequently it occurs. It involves finding a suitable probability model to fit a set of annual flood series for a watershed-based on possible future scenarios inferred. In a nutshell, flood

R. Sinam (✉)
Centre for the Study of Regional Development, School of Social Sciences, Jawaharlal Nehru University, New Delhi, India
e-mail: rebati31_ssf@jnu.ac.in

J. Das and S. K. Bhattacharya (eds.), *Monitoring and Managing Multi-hazards*, GIScience and Geo-environmental Modelling, https://doi.org/10.1007/978-3-031-15377-8_11

frequency analysis can determine 50 years or 100 years return period flood. Flood frequency models entail that there is a unique Discharge-Time relationship for every catchment. For estimating this association, good quality and a consistent hydrometric record of data are required and to do the same, "a suitable statistical or stochastic model of the continuous hydrograph relevant to the discharge-time relationship is a pre-requisite" (Cunnane 1989a, b). There are three models, namely, the annual maximum (AM) series model, partial duration (PD) series, and time series (TS) model, depending on the type of data used. According to Gumbel (1941), the "annual maximum series is the largest value of annual flood flows". Several probability distributions models are fitted on the AM series to find the best suitable model. A goodness of fit (GOF) test is usually administered to find the most accurate model. After the final selection of the suitable probability distributions, the model can be used to estimate expected floods for desired recurrence intervals like 5 years, 40 years or 100 years. These results can be used for the development of hydraulic structures and decision-making. In short, flood frequency modeling involves the selection of the type of model, selection of distributions, selection of methods of parameter estimation or quantile estimation and selection of scheme of use of at site data/regional data/both (Cunnane 1989a, b).

The selection of distributions for a given watershed is under a lot of debate and received widespread attention. Given the innumerable types of distributions developed, most distributions for an AM series are selected on the basis of its popularity, conveniences, flexibility, and robustness, theoretically well-based and well documented. However, Cunnane (1989a, b) opined that before selecting any distribution and its method of parameter or quantile estimation, "it is essential to confirm that it yields quantile estimates with small bias and root mean square errors over a reasonable range of flood-like distributions". Most commonly used distributions include Extreme value series (EV1, EV2, EV3) of Gumbel (1941) and Jenkinson (1969), Generalized extreme value (Jenkinson 1955, 1969), Pearson type 3 (Foster 1924), Weibull (Wu and Goodridge 1976), Log-Normal (Hazen 1914), Log Pearson type 3 by US Water Resources Council (1967), Log-logistic (Cunnane 1989a, b) among others.

There are typically two broad methods of parameter estimation: Bayesian and non-Bayesian methods. "Non-Bayesian methods include moments (MOM), maximum likelihood (ML), least squares (LS), probability-weighted moments (PWM) and sextiles" (Greenwood et al. 1979). L-moments and TL-moments are some of the well-known methods of moments used in parameter estimation. The suitability of the different distributions can be evaluated based on their descriptive and predictive ability (Cunnane 1989a, b). The GOF test is one such method to determine the best fit model for the data series. Some common GOF tests includes normal regression coefficient, root mean square estimates, chi-squared test, Kolmogorov–Smirnov (KS) test and Anderson–Darling (AD) test. The latter three have been the most cited tests used in many of the flood frequency analysis studies. With the arrival of modern and advanced technology, the regional study of flood frequency analysis has improved over the years. New studies using soft computing techniques, copula, and AI system have facilitated the field's better performance (Danish and Alam 2017; Kumar et al. 2015; Stamatatou et al. 2018).

Flood frequency analysis is one of the most common and popular studies among Hydraulic engineers and hydrologists. Pioneering work has been done by Jenkinson (1955), Hosking (1985) Fisher and Tippet (1928), Greenwood et al. (1979), Cunnane (1989a, b), Millington et al. (2011). There are two methods of flood frequency estimation—At site estimation and Regional. At the site frequency analysis has been done by Abdo et al. (2006), Rahman et al. (2014), Drissia et al. (2019), Leščešen and Dolinaj (2019), Ganamala and Sundar Kumar (2017), Kamal et al. (2017), Cassalho et al. (2018). Rahman et al. (2013) Karim and Chowdhury (1995). Regional flood frequency analysis using L-moments have been done by Noto and Loggia (2009), Parida et al. (1998),

Seckin et al. (2011), Kumar et al. (2003) Hussain and Pasha (2009). The present paper attempts to select the most suitable and reliable probability model of Baitarani River flood flow using three probability distribution model.

11.2 Materials and Methods

11.2.1 Study Area

The area of interest for the present study is Baitarani River which located in the state of Odisha and covering parts of Jharkhand. For purpose of better visualization, the area drained by Baitarani River named as Baitarani sub-basin is presented. It is located between 85°10′5″E to 86°20′30″E and 22°15′N to 20°55′30″N. Baitarani River is also called Baitarni or Vaitarani River. It is one of the major tributaries of Brahmani River (Fig. 11.1). The river originates from Gonasika village—Guptaganga hills of Kendujhar district in Odisha. The river has a length of about 365 km before draining out to the Bay of Bengal. Initially, the river flows northerly for about 80 km from the point of origin but it rotates at right angle upon reaching Champua. It then flows south-easterly, joining various tributaries from its left and right sides. It enters the floodplains at Anandapur before finally joining the Brahmani River at Dharma near Chandabali. Eventually, it drains out into the deltaic zone at Akhuapada and branches out into several distributaries into the Bay of Bengal.

11.2.2 Database

Daily discharge data from Central Water Commission published online by India-Water Resource Information System (WRIS) has been collected for Anandapur and Champua Station (Fig. 11.1). For Anandapur, the data is available for a period of 43 years (1976–2018) while for

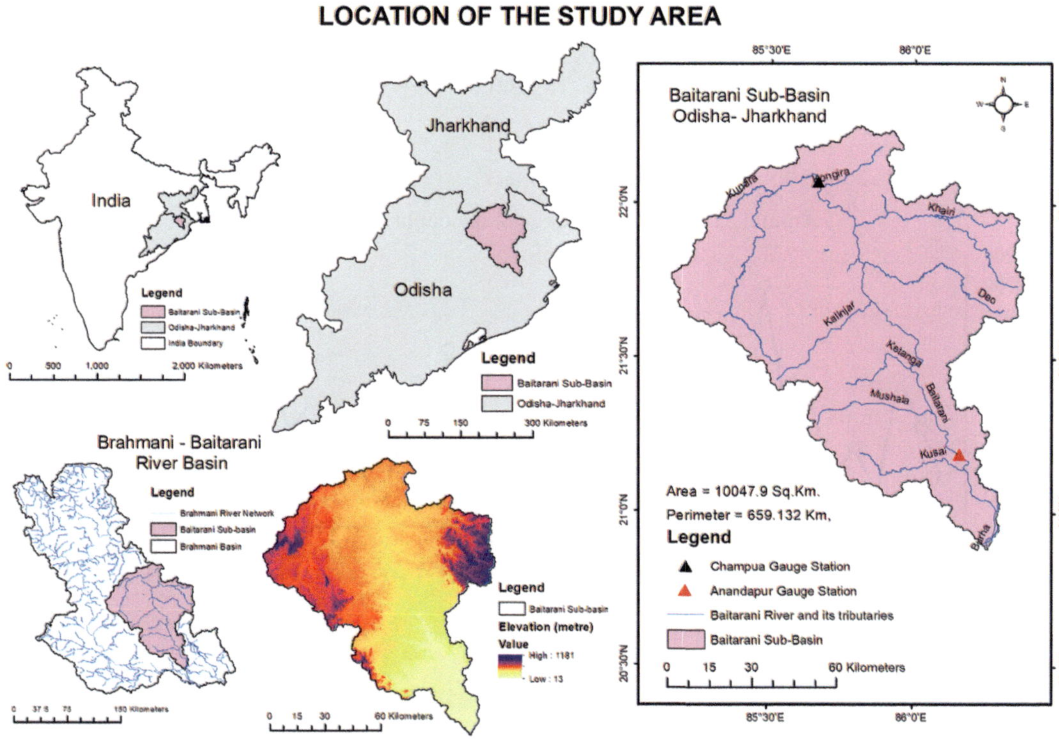

Fig. 11.1 Study area

Champua, it is available for 28 years (1991–2018). The largest of 365 days for each of the years is sorted out to form the annual maximum series (Gumbel 1941). The data is given in Table 11.1, presented in Figs. 11.2 and 11.3.

11.2.3 Methodology

In this study, three probability distributions: Generalized Extreme Value distribution (GEV), Gumbel distribution, also known as extreme

Table 11.1 Annual maximum daily discharge data for Anandapur and Champua station

Anandapur (discharge, Q in cumecs)						Champua (discharge, Q in cumecs)			
Year	Q	Year	Q	Year	Q	Year	Q	Year	Q
1976	2299.4	1990	2834.8	2004	3728.0	1991	1010.2	2005	920.8
1977	1567.7	1991	4760.0	2005	4105.6	1992	159.1	2006	420.7
1978	3253.6	1992	869.4	2006	3134.2	1993	479.5	2007	731.3
1979	4936.2	1993	4075.0	2007	5779.3	1994	652.8	2008	538.9
1980	1078.2	1994	4223.0	2008	6327.8	1995	403.5	2009	263.2
1981	2630.7	1995	5218.0	2009	1843.6	1996	503.2	2010	204.6
1982	1635.4	1996	5450.0	2010	567.7	1997	752.6	2011	1705.2
1983	4189.1	1997	5762.0	2011	7424.5	1998	343.0	2012	213.4
1984	3665.8	1998	1363.0	2012	750.0	1999	880.0	2013	550.8
1985	7974.0	1999	5856.0	2013	4840.0	2000	439.0	2014	1158.5
1986	1051.6	2000	2527.6	2014	7688.6	2001	499.7	2015	309.8
1987	869.4	2001	2787.9	2015	2381.9	2002	105.2	2016	325.1
1988	4293.6	2002	1030.1	2016	1434.6	2003	430.7	2017	1015.0
1989	5482.2	2003	1998.1	2017	2506.3	2004	363.0	2018	710.7
				2018	3321.0				

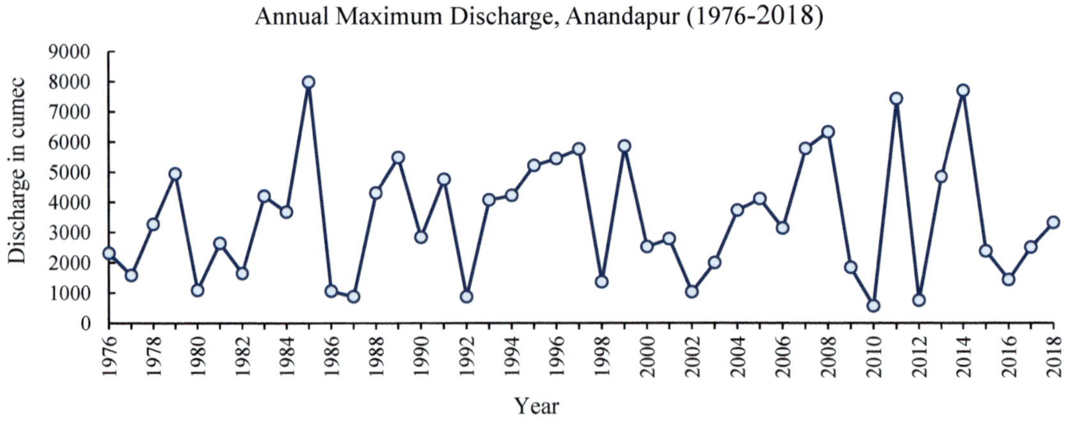

Fig. 11.2 Annual maximum discharge series, Anandapur

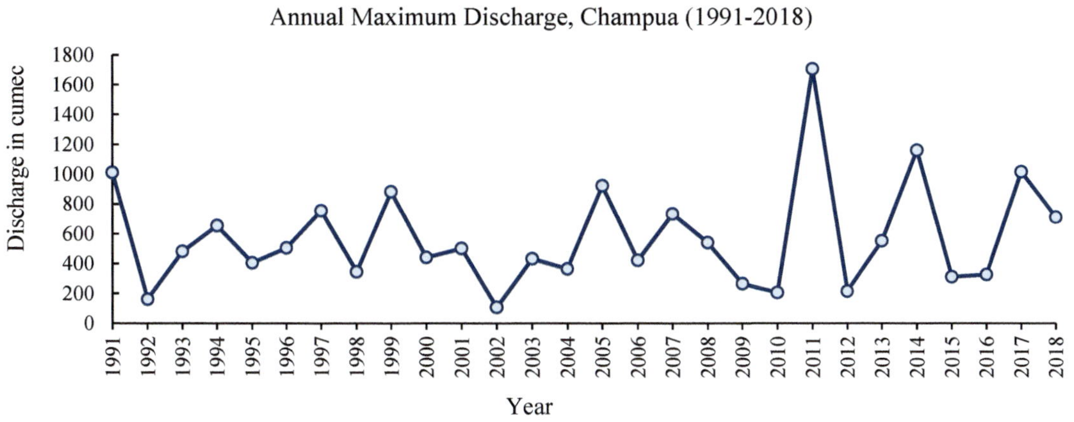

Fig. 11.3 Annual maximum discharge series, Champua

value type 1 distribution (EV1) and Log Pearson type 3 distribution (LP3) are analyzed for the stations. LP3 and GEV are distributions with three parameters, while EV uses 2 parameters. The methodological framework followed in this study is influenced by that of Millington et al. (2011) and Raghunath (2014). Gumbel distribution is calculated in two ways based on the method of parameter estimation is applied for comparison and named as Gumbel-1 and Gumbel-2. In this paper, L-Moments is used for estimation of parameter for GEV and Gumbel-1 while method of moments (MOM) is used in case of Gumbel-2 and LP3 which is adapted from Raghunath (2014).

L-moments are the modification of Probability weighted moments (PWMs) developed by Greenwood et al. (1979) that provide more accuracy and ease. L-moments are derived from the linear combinations of data which are arranged in increasing (ascending) order.

PWMs are required for the calculation of L-moments. The data are to be sorted in increasing order and ranked accordingly.

PWM are defined by Greenwood et al. (1979) as

$$M_{ijk} = E[x^i F^j (1-F)^k] = \int_0^1 x(F)^i F^j (1-F)^k \mathrm{d}F$$

where i, j, k are real numbers, F = Cumulative distribution function. If $j = k = 0$ and I is a non-negative integer then M_{100} represents the conventional moment of order I about the origin.

The following equations are to be calculated (Millington et al. 2011)

$$M_{100} = \frac{1}{N}\sum_{i=1}^{N} Q_i$$

$$M_{110} = \frac{1}{N}\sum_{i=1}^{N} \frac{(i-1)}{(N-1)} Q_i$$

$$M_{120} = \frac{1}{N}\sum_{i=1}^{N} \frac{(i-1)(i-2)}{(N-1)(N-2)} Q_i$$

$$M_{130} = \frac{1}{N}\sum_{i=1}^{N} \frac{(i-1)(i-2)(i-3)}{(N-1)(N-2)(N-3)} Q_i$$

where "i" is the rank of the value sorted in increasing order, N represents sample size and Q denotes the data value.

The L-moments equations are defined as (for details, see Hosking and Wallis 1997; Cunnane 1989a, b; Millington et al. 2011).

$\lambda_1 = M_{100}$.
$\lambda_2 = 2M_{110} - M_{100}$.
$\lambda_3 = 6M_{120} - 6M_{110} + M_{100}$.
$\lambda_4 = 20M_{120} - 30M_{120} + 12M_{110} - M_{100}$.
The moment ratios are calculated as follows:
$\theta_2 = \lambda_2/\lambda_1$ [L-CV].
$\theta_3 = \lambda_3/\lambda_2$ [L-Skewness].
$\theta_4 = \lambda_4/\lambda_2$ [L-Kurtosis].

(a) Generalized Extreme Value Distributions (After Millington et al. 2011).

GEV distributions was developed by Jenkinson (1959). It combines the three family of extreme value distribution: Gumbel (EV1), Fréchet (EV2) and Weibull (EV3). It has three parameters distribution: Location (δ), Scale (β) and Shape (ε). When shape parameter (ε) = 0, this is EV1, $\varepsilon > 0$, it becomes EV2 and when $\varepsilon < 0$, it is EV3.

The parameter of GEV as defined in Hosking and Wallis (1997) are calculated as

$$\varepsilon = 7.8590r + 2.9554r^2$$

$$\text{in which} \quad r = \frac{2}{3+\theta_3} - \frac{\ln(2)}{\ln(3)}$$

$$\beta = \frac{\lambda_2 \varepsilon}{(1-2^{-\varepsilon})\Gamma(1+\varepsilon)}$$

$$\delta = \lambda_1 - \beta\{1 - \Gamma(1+\varepsilon)\}/\varepsilon$$

where Γ is gamma function, δ = location; β = Scale; ε = shape parameter.

After the parameter estimation, the expected flood discharge can be calculated as

$$Q_T = \delta + \left(\frac{\beta}{\varepsilon}\right)\left\{1 - \left(-\log\left(\frac{T-1}{T}\right)\right)^{\varepsilon}\right\}$$

where T = return period in years.

Steps:

1. Arrange the data in ascending order.
2. Calculate the four PWMs.
3. Calculate the four L-Moments as given above.
4. Calculate the moment ratios and calculate the required parameter for the distribution using the equations given above.
5. For a desired return period (2, 5, 10, 25, 50, 100, 200 years), calculate the expected flood/Discharge value.

(b) Gumbel Distribution (After Millington et al. 2011).

Gumbel distribution has two parameters: Location (δ) and Scale (β).

$$\beta = \frac{\lambda_2}{\log(2)}$$

$$\delta = \lambda_1 - (\beta\zeta)$$

where ζ = 0.5772 (Euler's constant)

Discharge, Q_T is given by

$$Q_T = \delta + \beta y_T$$

where $y = -\ln[-\ln(1-\frac{1}{T}))$; T = Return period (Years).

(c) Gumbel Distribution (after Raghunath 2014).

As per the Gumbel theory of extreme values, the value of the discharge variate Q_T with return period T is given by the equation,

$$QT = \overline{Q} + K_{\alpha_{n-1}} \qquad (11.1)$$

where

$\overline{Q}$ is the mean of the discharge variate,

σ_n -1 Standard deviation of the sample which is given by $\sqrt{\frac{\sum(Q-\overline{Q})^2}{N-1}}$ where N is sample size;

K (Frequency Factor) = $\frac{y_T - \overline{y}_n}{S_n}$, where $\overline{y}_n$ = reduced variate, a function of return period T and expressed as $y_T = -\left[\ln . \ln \frac{T}{T-1}\right]$, S_n = reduced standard deviation. They are determined from Gumbel table. For N corresponding to 43, $\overline{y}_n = 0.5453$, $S_n = 1.1480$, and for $N = 28$, $\overline{y}_n = 0.5343$, $S_n = 1.1047$.

For the values of the expected flood as estimated by Gumbel method, confidence limits can be calculated. The confidence interval indicates the limits about the calculated value between which the true value can be said to lie with a specific probability based on sampling errors only.

For confidence probability c, the confidence interval of the variate Q_T is bounded by values Q_1 and Q_2 given by

$$Q_1 \text{ or } Q_2 = Q_T \pm f(c)S_e$$

where $F(c)$ is function of the confidence probability c determined by using the table of normal variate (Table 11.2).

S_e = probable error = $b\ (S_x)/\sqrt{N}$.

$B = \sqrt{1} + 1.3\ K + 1.1K^2$

where,

K frequency factor.
S_x Standard deviation of the sample.
N Sample size.

The steps involved in this method are:

Step 1: Collect the annual maximum data series and note the total sample size. Arrange the data in descending order of magnitude.
Step 2: Calculate the mean and standard deviation of the annual data series.
Step 3: Using Gumbel table on reduced mean and reduced standard deviation, determine the value of Y_n and S_n for N corresponding to 43 (Anandapur Station) and 28 (Champua Station).
Step 4: Calculate reduced variate, y.
Step 5: Calculate the required expected flood for the desired return period (2, 5, 10, 25, 50, 100 and 200 years).
Step 6: Calculate the confidence limits and plot it in the graph.

Probability of occurrence (P) of the given data series is plotted using Gumbel plotting position formula given by

$$P = 1 - e^{-e^{-y}}$$

where y is the reduced variate, return period (T) is the reciprocal of P.

(d) Log Pearson III Distribution (After Raghunath 2014).

This is one of the probability distributions developed by Karl Pearson popularized by US Water Resources Council. Here, the discharge variates are arranged in descending order and converted into logarithmic form and the transformed variates are analyzed. If "Q" represents the variate of the sample, then the transformed variate is represented by "Z" expressed as

$$Z = \log Q$$

For this Z series, the expected variate for any return period is given by

$$Z_T = \overline{Z} + K_z\alpha_z$$

where

K_Z frequency factor which is a function of T and C_s (Coefficient of Skew of Z)
σ_Z standard deviation of the Z variate series = $\sqrt{\sum(Z-\overline{Z})^2/(N-1)}$

$$C_s \frac{N\sum(Z-\overline{Z})^3}{(N-1)(N-2)(\alpha_z)^3}$$

$\overline{Z}$ mean of the Z values.
N sample size

After finding Z_T, the value of Q_T (Discharge for a specific return period) is given by:

$$Q_T = \text{antilog}\ (Z_T)$$

Goodness of Fit Test

GOF determine whether a probability distribution fit the input data. This is a statistical algorithm used to check the distance between the observed data and the expected outcome. Three goodness of fit namely "Kolmogorov–Smirnov test, Anderson–Darling and Chi-squared test" are

Table 11.2 Normal variate value

C in percent	80	90	95	99
$F(c)$	0.674	1.00	1.96	2.8

applied. The analysis is done using a software called EasyFit 5.6 Professional developed by MathWave Technologies. The formula for the test can be referred from Kamal et al. (2017).

11.3 Results and Discussion

A statistical examination of the discharge data (Table 11.3) at Anandapur station (1976–2018) shows an average discharge value of 3477.093 cumecs with a standard deviation of 2017.141 cumecs and a coefficient variation of 58.01%. At Champua, the mean discharge is 574.63 cumecs with a standard deviation of 354.79 cumecs and a coefficient of deviation of 61.74%. Historical maximum flows are observed during 1985, 1999, 2007 and 2008. As per the Odisha flood record, there was a flood in 1985 during August–September affecting 22 districts (http://dowrodisha.gov.in/). Super Cyclone of 1999 has affected the state and consequently affected the flood flow. In 2008 during the months of June and September, there was a recorded flood affecting 21 districts with 4.45 lakh hectares of the cropped area damaged, 651 breaches in rivers and 1276 breaches in canals. Estimating design flood is an important part of planning and management of floodplain and it is considered a precautionary measure that must be met. The return period can vary depending on the needs and requirements of the region under consideration. Table 11.3 tabulate the descriptive statistics of the observed data obtained from both the stations and Fig. 11.4 present the flood frequency plots between discharge and return period of the data.

The highest recorded flood flow at Anandapur of 7974.01 cumecs has a calculated return period of 11.04 years with a probability of occurrence of 9.05%. A flood flow of 567.65 cumecs has a return period of 1.18 years with 84.57% probability of occurrence. A median flood of 3253.6 cumecs has a return period of 2.12 years with a probability of occurrence of 46.87%. Based on Champua flood record, a high flood flow of 1705.20 cumecs has a calculated return period of 38.55 years with a probability of occurrence of 2.59%. A low flow of 105.22 cumecs is expected to have a return period of 1.13 years with high probability of occurrence of 88.07% whereas a median flood of 489.62 cumecs has a return period of 1.9 years with 52.30% probability of occurrence.

Tables 11.4 and 11.5 enumerate the result of the analysis for both the station. It shows the expected flood values using the three distribution models for a return period of 2 years, 5 years, 10 years, 25 years, 50 years, 100 years and 200 years.

From Table 11.4, it is observed that each of the models gave varying results with huge differences. At Anandapur station, for a return period of 2 years, GEV estimated an expected flood of 4647.1 cumecs, Gumbel-1 estimated 2665.5 cumecs, Gumbel-2 estimated 3032.8 cumecs and LP3 estimated 2902.1 cumecs. For 5 years return period, Gumbel-1 estimated the highest amount

Table 11.3 Descriptive statistics of the annual maximum series, Anandapur and Champua

Statistic	Anandapur	Champua
Sample size	43	28
Range	7406.4	1600
Mean	3477.1	574.63
Variance	4,068,900	125,880
Std. deviation	2017.1	354.79
Coef. of variation	0.58012	0.61743
Std. Error	307.61	67.05
Skewness	0.45829	1.3571
Excess Kurtosis	− 0.61316	2.4409

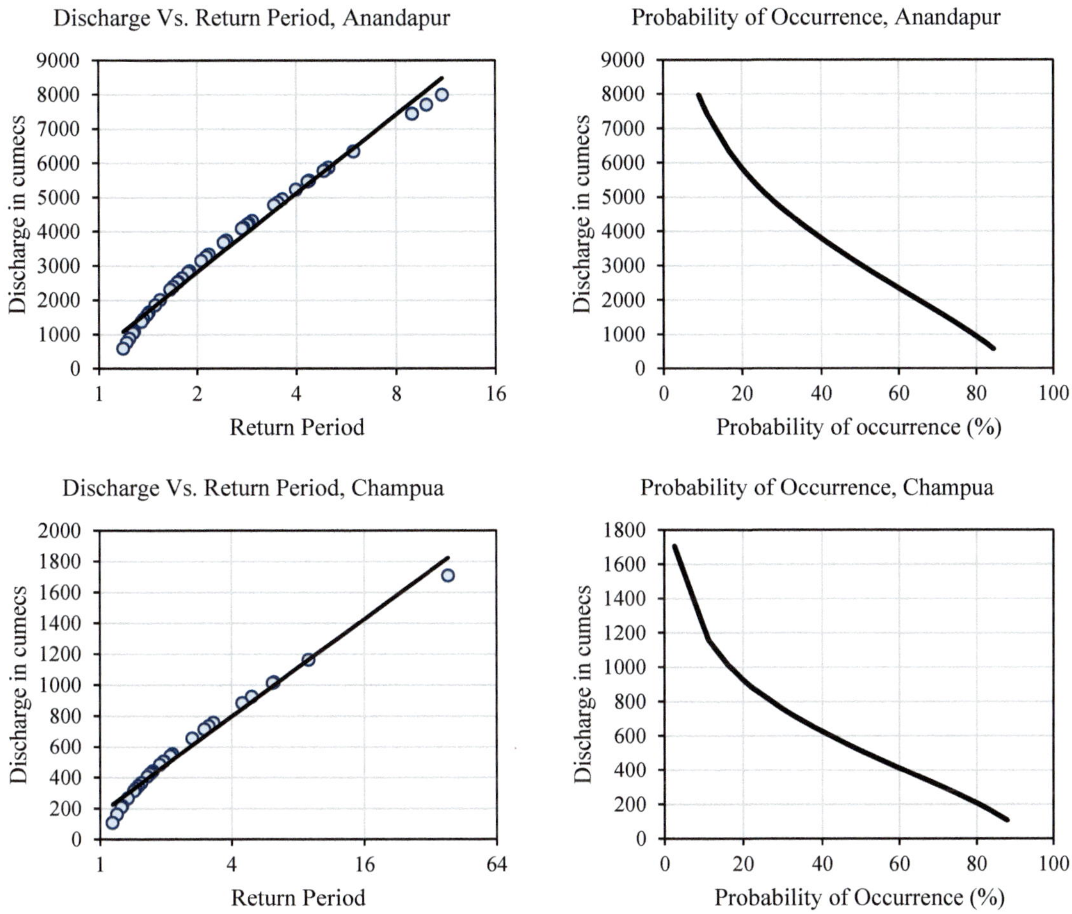

Fig. 11.4 Flood frequency plots, return period and probability of occurrence, Anandapur and Champua

of flood of about 7031.6 cumecs while GEV estimated a flood flow of 6381.9 cumecs, a difference of 649.9 cumecs from that of Gumbel-1. Gumbel-2 and LP3 estimated lower flood flow of 5849.3 cumecs and 5127 cumecs with a difference of 1182.3 cumecs and 1904.6 cumecs from that of Gumbel-1 which is a huge gap. From 10 years return period onward, we can observe that Gumbel-1 estimated higher expected flood, with 50 and100 years flood reaching up to discharge of 16,284.3 cumecs and 18,973.9 cumecs respectively. For 50 years, Gumbel-1 estimated the highest expected flood of 16,284.3 cumecs followed by Gumbel-2 with an estimation of 11,818 cumecs, a difference 4466.3 cumecs and LP3 estimated a flood flow of 10,993.1 cumecs with a difference of 5291.2 cumecs from that of Gumbel-1. GEV estimated the lowest expected flood for 50 years with a value of 9484.3 cumecs which is almost half of the estimate made by Gumbel-1. For 100 years, Gumbel-1 estimated a flood flow of 18,973.9 followed by Gumbel-2 with an estimate of 13,553 cumecs and LP3 (12,930.2 cumecs). The lowest estimate is given by GEV with an expected flood flow of 10,258.6 which is 8715.3 cumecs short to that of Gumbel-1.

A similar pattern of output is also observed in case of Champua. For 2 years return period, GEV estimated the highest flood flow of 717.3 cumecs followed by Gumbel-2 with an estimate of 513.5 (a difference of 203.8 cumecs) and LP3 with an estimate of 489.7 cumecs, having a difference of 227.6 cumecs from that of GEV

Table 11.4 Result showing the expected flood obtained using the three distribution models at Anandapur station

Return period (years)	Anandapur (expected flood in cumecs)			
	GEV*	Gumbel-1*	Gumbel-2**	LP3**
2	4647.1	2665.5	3032.8	2902.1
5	6381.9	7031.6	5849.3	5127.0
10	7429.9	9922.3	7714.0	6812.3
25	8649.4	13,574.8	10,070.1	9140.1
50	9484.3	16,284.3	11,818.0	10,993.1
100	10,258.6	18,973.9	13,553.0	12,930.2
200	10,979.6	21,653.7	15,281.7	14,966.9

* After Millington et al. (2011); ** After Raghunath (2014)

Table 11.5 Result showing the expected flood obtained using the three distribution models at Champua station

Return period (years)	Champua (expected flood in cumecs)			
	GEV*	Gumbel-1*	Gumbel-2**	LP3**
2	717.3	440.5	513.5	489.7
5	1059.4	1162.3	926.3	821.5
10	1312.2	1640.2	1199.6	1063.7
25	1665.2	2244.1	1544.9	1389.5
50	1953.5	2692.1	1801.0	1643.4
100	2264.3	3136.8	2055.3	1904.6
200	2600.3	3579.8	2308.6	2175.4

* After Millington et al. (2011); ** After Raghunath (2014)

estimate. The lowest estimate is given by Gumbel-1 having a value of 440.5 cumecs for 2 years return period. From 5 years onward, Gumbel-1 estimated the highest expected flood compared to the other models. It estimated a discharge of 1162.3 cumecs for 5 years while GEV which estimated the highest flood flow for 2 years period, estimated a lower value of 1059.4 cumecs followed by Gumbel-2 (926.3 cumecs) and LP3 (821.5 cumecs). For 50 years return period, Gumbel-1 estimated 2692.1 cumecs while GEV estimated 1953.5 cumecs having a difference of 738.9 cumecs from that of Gumbel-1. Gumbel-2 estimated 50 years expected flood of 1801 cumecs and LP3 estimated the lowest value of 1643.4 which makes a difference of 1048.7 cumecs from that of Gumbel-1. Gumbel-1 estimated a 100-year flood of 3136.8 cumecs, GEV gives 2264.3 cumecs, Gumbel-2 gives 2055.3 cumecs and LP3 gives the lowest flood flow of 1904.6 cumecs.

One interesting observation from the output is that even though GEV estimated exceptionally high expected flood flow for 2 years, it estimated a lower value for other return period comparatively. Whereas Gumbel-1 estimated a very low value for 2 years return period but continue to estimate exceptionally higher discharge value for all the other return periods. LP3 consistently estimated the lower values as compared to other models except for the 2 years period. Figures 11.5 and 11.6 gives the graphical representation of the expected flood obtained using the models.

As observed in the result above, the varying output shows that these models can lead to overestimation and underestimation of expected values from the true values. In order to minimize

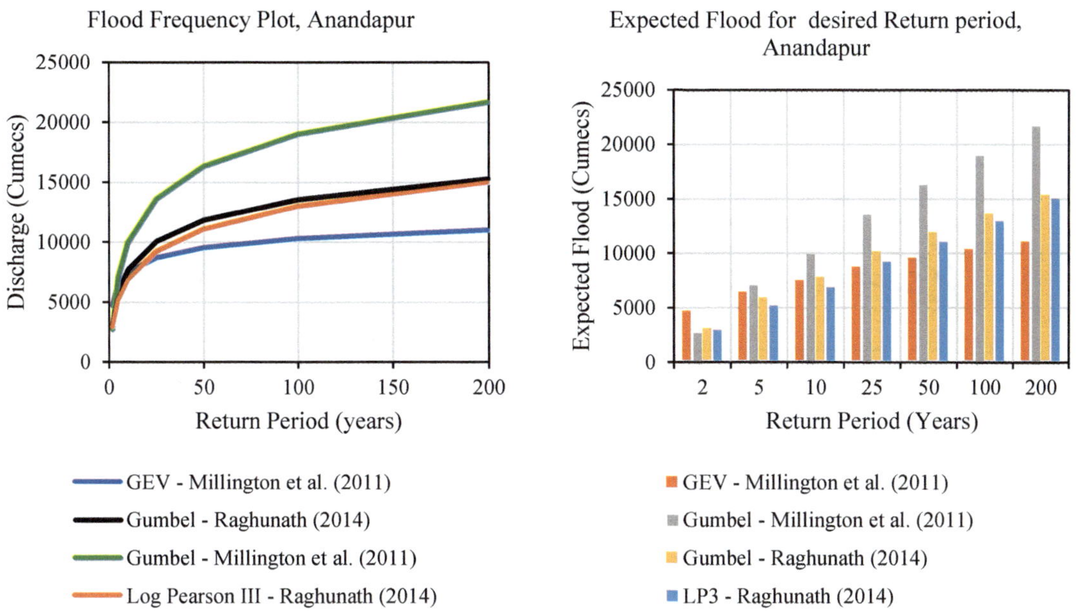

Fig. 11.5 Graph showing the flood frequency plot and estimated expected flood for the desired return period at Anandapur station, Baitarani sub-basin

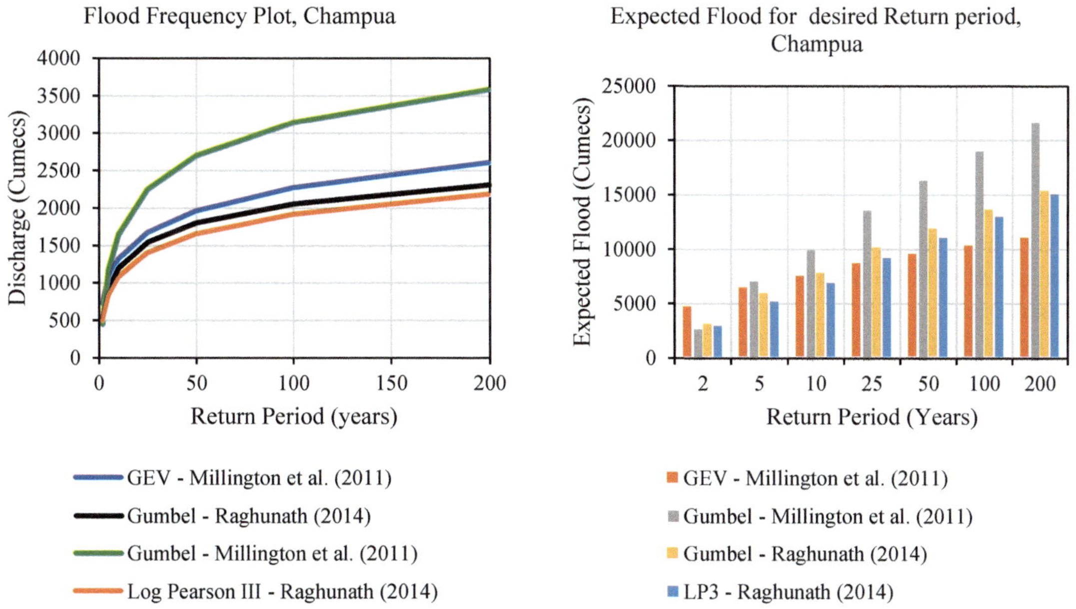

Fig. 11.6 Graph showing the flood frequency plot and estimated expected flood for the desired return period at Champua station, Baitarani sub-basin

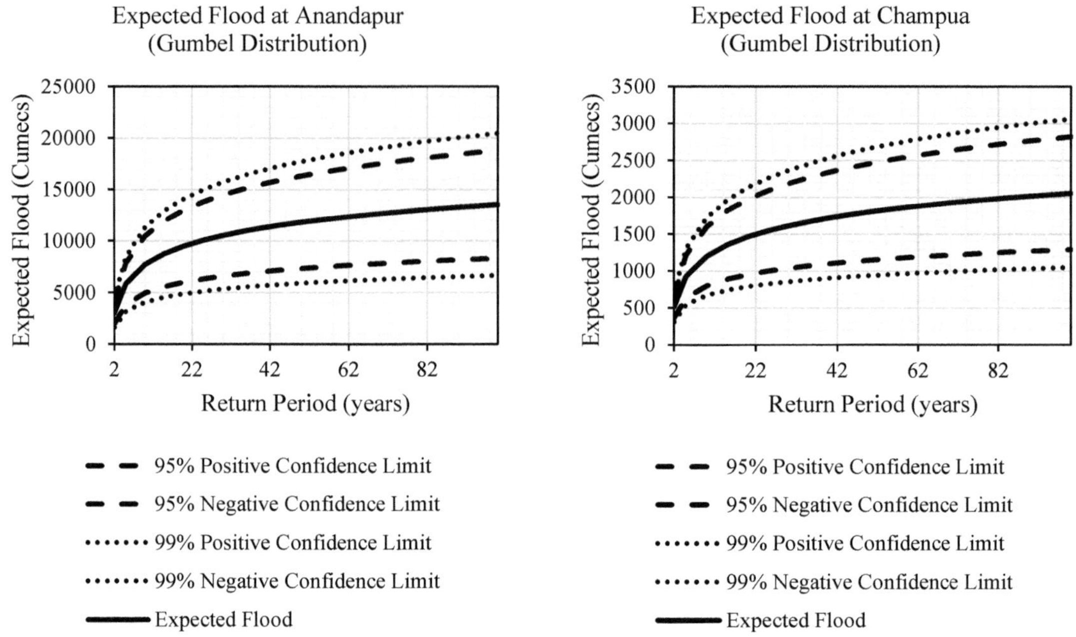

Fig. 11.7 Expected flood estimated by Gumbel distribution (after Raghunath 2014) along with confidence limits

such uncertainty, confidence limits are often calculated to highlight the likely range of values from the mean value. Figure 11.7 shows the plot of expected flood using Gumbel distribution (After Raghunath 2014) along with 95 and 90% confidence interval. The inconsistency and huge difference in the output results may be attributed to the differences in its methodological framework in terms of distribution, the method of parameter estimation chosen and the number of observations.

For purpose of finding the best suitable model, Goodness of Fit test is conducted. To test the model's fitting and examine its statistical significance of these models, the data has been subjected to three GOF tests, namely, "Anderson–Darling test, the Kolmogorov–Smirnov test and Chi-squared test" using Easyfit software. The models have been tested at the critical value of 0.05 (95% significance) level. The result is tabulated in Table 11.6. Based on the above results, a model is rejected or accepted compared to the critical value. The critical value of the Kolmogorov–Smirnov test at 0.05 is 0.14355, the Anderson–Darling test at 0.05 is 2.5018 and that of Chi-Squared is 12.592. If the test statistics are higher than the critical value, the model is rejected and vice versa. In the present study, none of the three models are rejected in all three tests. The ranking is based on the difference between the test statistic and the critical value. In all three models, it is observed that the ranking is consistent in case of Anandapur. GEV is the most suitable model for the data followed by Log Pearson 3 and Gumbel distribution.

However, for Champua station, the best fit model according to Chi-Square test is Gumbel followed by Log Pearson 3 and GEV while the other two tests predict GEV as the best fit model. From the overall observations, it can be said that GEV is the most suitable model for Baitarani based on the two station data. One limitation of model prediction is that the results are affected by the input factor like the number of observations, type of parameter and methods of parameter estimation, etc.

The lack of one definite clear model can often pose confusion and uncertainty which can affect the process of decision-making in flood control strategy. It is usually left on the physical

Table 11.6 Result of goodness of fit for both the stations

Goodness of fit—summary							
Anandapur	Distribution	Kolmogorov—Smirnov		Anderson—Darling		Chi-Squared	
		Statistic	Rank	Statistic	Rank	Statistic	Rank
1	Gen. Extreme Value	0.07161	1	0.31819	1	1.4453	1
2	Gumbel	0.09981	3	0.60967	3	2.6759	3
3	Log Pearson 3	0.09282	2	0.37258	2	1.5483	2
Champua							
1	Gen. Extreme Value	0.06873	1	0.11674	1	0.8871	3
2	Gumbel	0.10054	3	0.19719	3	0.68906	1
3	Log Pearson 3	0.07676	2	0.11757	2	0.76253	2

Calculated using EasyFit software by MathWave technologies

reasoning and user-discretion of the engineers or stakeholders who are involved in constructing the design flood. Software like Easyfit can be useful for finding the best model. A more comprehensive study can be undertaken with other multiparametric models like General Pareto, Log-Logistics, Weibull distribution, Copula, etc. or soft computational techniques for better estimation.

11.4 Conclusions

Baitarani River has been known for flooding and inundation. Due to its location in the direct route of southeast monsoon and seasonal cyclonic storms, the reason receives heavy rainfall and is frequently subjected to annual flooding. The present study aims to develop a suitable distribution model for the river basin to act as a guiding tool in developing hydraulic structures and planning flood hazard zones. The flood frequency analysis has been a widely and extensively used method for calculating expected flood for the specific return period. Notwithstanding its drawbacks, distribution models have proven to be an effective tool for flood management strategy. The present study analyzed three models for Baitarani River using the data available at two stations (Anandapur and Champua) for a period of 43 years and 28 years, respectively. The statistical analysis of the models accepted all the three models at a confidence level of 95%. From the GOF test, Generalized Extreme Value distribution comes out to be a better and reliable model than the other models. The study is helpful in understanding the importance of flood prediction using statistical methods which is the basic foundation for hydraulic engineers. The study highlights the varying results obtained using different models and the risk involved in choosing the best fit model for specific river basin. With the development of new techniques and advanced modeling, further studies are intended with higher-order multiparameter distributions to determine a more suitable and robust model for the same.

References

Abdo G, Sonbol M, Willems P (2006) Flood frequency analysis of the Eastern Nile Rivers. Friend/Nile 13621(April):1–13. http://www.unesco.org/fileadmin/MULTIMEDIA/FIELD/Cairo/pdf/REGIONAL_FLOOD_FREQUENCY_ANALYSIS.pdf

Cassalho F, Beskow S, de Mello CR, de Moura MM, Kerstner L, Ávila LF (2018) At-site flood frequency analysis coupled with multiparameter probability distributions. Water Resour Manage 32(1):285–300. https://doi.org/10.1007/s11269-017-1810-7

Cunnane C (1989a) Methods and merits of regional flood frequency analysis. J Hydrol 100(1–3):269–290. https://doi.org/10.1016/0022-1694(88)90188-6

Cunnane C (1989b) Statistical distributions for flood frequency analysis: World Meteorological Organization, Operational hydrology report No. 33 Secretariat of the World Meteorological Organization–No. 718, 61 p. plus appendixes (Issue 33)
Danish M, Alam J (2017) A soft computing approach to flood frequency analysis of river. Hydro-2017 International, L.D. College of Engineering Ahmadabad, India, pp 1743–1754
Drissia TK, Jothiprakash V, Anitha AB (2019) Flood Frequency analysis using L moments: a comparison between at-site and regional approach. Water Resour Manage 33(3):1013–1037. https://doi.org/10.1007/s11269-018-2162-7
Fisher RA, Tippett LHC (1928) Limiting forms of the frequency distribution of the largest or smallest member of a sample. Math Proc Cambridge Philos Soc 24(2):180–190. https://doi.org/10.1017/S0305004100015681
Foster HA (1924) Theoretical frequency curves and their application to engineering problem. Trans Am Soc Civ Eng 87(1):142–173. https://doi.org/10.1061/TACEAT.0003398
Ganamala K, Sundar Kumar P (2017) A case study on flood frequency analysis. International Journal of Civil Engineering and Technology 8(4):1762–1767
Greenwood JA, Landwehr JM, Matalas NC, Wallis JR (1979) Probability weighted moments: definition and relation to parameters of several distributions expressable in inverse form. Water Resour Res 15(5):1049–1054. https://doi.org/10.1029/WR015i005p01049
Gumbel EJ (1941) The return period of flood flows. Ann Math Stat 12(2):163–190
Hazen A (1914) Discussion on "flood flows" by W.E. Fuller. Transactions of the American Society of Civil Engineers 77:626–632
Hosking JRM (1985) Comment on "a correction for the bias of maximum likelihood estimates of Gumbel parameters", by Fiorentino and Gabrielle. J Hydrol 78 (3/4):393–396
Hussain Z, Pasha GR (2009) Regional flood frequency analysis of the seven sites of Punjab, Pakistan, using L-moments. Water Resour Manage 23(10):1917–1933. https://doi.org/10.1007/s11269-008-9360-7
Jenkinson AF (1955) The frequency distribution of the annual maximum (or minimum) values of meteorological elements. Q J R Meteorol Soc 81:158–171
Jenkinson AF (1969) Statistics of extremes. In: Estimation of maximum floods. In WMO No 233, TPI26 (Tech. Note No. 98)
Kamal V, Mukherjee S, Singh P, Sen R, Vishwakarma CA, Sajadi P, Asthana H, Rena V (2017) Flood frequency analysis of Ganga river at Haridwar and Garhmukteshwar. Appl Water Sci 7(4):1979–1986. https://doi.org/10.1007/s13201-016-0378-3
Karim AM, Chowdhury JU (1995) A comparison of four distributions used in flood frequency analysis in Bangladesh. Hydrol Sci J 40(1):55–66. https://doi.org/10.1080/02626669509491390
Kumar R, Chatterjee C, Kumar S, Lohani AK, Singh RD (2003) Development of regional flood frequency relationships using L-moments for middle Ganga plains subzone 1(f) of India. Water Resour Manage 17(4):243–257. https://doi.org/10.1023/A:1024770124523
Kumar R, Goel NK, Chatterjee C, Nayak PC (2015) Regional flood frequency analysis using soft computing techniques. Water Resour Manage 29(6):1965–1978. https://doi.org/10.1007/s11269-015-0922-1
Leščešen I, Dolinaj D (2019) Regional flood frequency analysis of the Pannonian Basin. Water (Switzerland), 11(2). https://doi.org/10.3390/w11020193
Millington N, Das S, Simonovic SP (2011) The comparison of GEV, Log-Pearson type 3 and Gumbel distributions in the upper Thames River watershed under global climate models (Issue 77). https://ir.lib.uwo.ca/cgi/viewcontent.cgi?article=1039&context=wrrr
Noto LV, La Loggia G (2009) Use of L-moments approach for regional flood frequency analysis in Sicily. Italy. Water Resources Management 23 (11):2207–2229. https://doi.org/10.1007/s11269-008-9378-x
Parida BP, Kachroo RK, Shrestha DB (1998) Regional flood frequency analysis of Mahi-Sabarmati basin (Subzone 3-a) using index flood procedure with L-Moments. Water Resour Manage 12(1):1–12. https://doi.org/10.1023/A:1007970800408
Raghunath HM (2014) Hydrology (Third). New Age International (P) Ltd.
Rahman AS, Rahman A, Zaman MA, Haddad K, Ahsan A, Imteaz M (2013) A study on selection of probability distributions for at-site flood frequency analysis in Australia. Nat Hazards 69:1803–1813. https://doi.org/10.1007/s11069-013-0775-y
Rahman A, Haddad K, Eslamian S (2014) Regional flood frequency analysis. Handbook of engineering hydrology: modeling, climate change, and variability, February, pp 451–469. https://doi.org/10.1201/b16683
Seckin N, Haktanir T, Yurtal R (2011) Flood frequency analysis of Turkey using L-moments method. Hydrol Process 25(April):3499–3505. https://doi.org/10.1002/hyp.8077
Stamatatou N, Vasiliades L, Loukas A (2018) Bivariate flood frequency analysis using Copulas. Proceedings 2 (11):635. https://doi.org/10.3390/proceedings2110635
Wu B, Goodridge JD (1976) Selection of frequency distributions for hydrologic frequency analysis. In Dept. of Water Resources, State of California, Sacramento

Application of Analytical Hierarchy Process (AHP) Method to Flood Risk Assessment at Sub-Himalayan Region Using Geospatial Data: A Case Study of Alipurduar District, West Bengal, India

12

Debasish Roy, Satyajit Das, Surajit Paul, and Surjapada Paul

Abstract

Floods are widespread natural calamities around the world, frequently resulting in fatalities as well as significant environmental and economic damage. For extracting precise vulnerability information to support flood remedial actions, a spatial vulnerability, risk, and susceptibility mapping technique including multi-criteria at the regional scale is required. This work provides a spatial multi-criteria-integrated method to flood vulnerability, risk, and susceptibility mapping using geospatial approaches at the regional scale. This suggested strategy was implemented in the Alipurduar district in West Bengal, India. Flood susceptibility, vulnerability, and flood risk assessment are the three components that were examined in this study. By integrating the susceptibility and vulnerability maps, the study created a flood risk map with 12 indicators for flood susceptibility and nine indicators for flood vulnerability. To facilitate the analytical hierarchy process, the requirements were translated into weighted spatial layers and normalized. Individual susceptibility and a weighted overlay approach were used to build vulnerability component maps, which were then utilized to build final flood risk maps. The produced maps were effective in determining the geographic extents and levels of susceptibility. Results reviled that the riverside area is more susceptible and the southern part of the district is more prone to flooding because of its proximity to the active waterway and low elevation. This approach was validated by AUC curve analysis using flood inventory data which were collected from Google map, flood recorded data available in the district disaster management handbook, and the Bhuvan web portal of ISRO. The results indicate that this method might be used to analyze the spatial variability of flood risk in flood vulnerable areas to build effective flood prevention plans and policies.

Keywords

Flood · Susceptibility · Vulnerability · Risk · AHP · Alipurduar

12.1 Introduction

Floods usually happen when a considerable amount of water exceeds and submerges typically dry land. Floods are common in tropical locations due to heavy rainfall or a storm surge

D. Roy (✉) · S. Das · S. Paul · S. Paul
Department of Geography and Applied Geography, University of North Bengal, Siliguri, West Bengal, India
e-mail: debasishroy176@gmail.com; rs_debasish@nbu.ac.in

J. Das and S. K. Bhattacharya (eds.), *Monitoring and Managing Multi-hazards*, GIScience and Geo-environmental Modelling, https://doi.org/10.1007/978-3-031-15377-8_12

from a tropical depression. Flood is the most devastating natural hazard among various hydro-meteorological events. These calamities frequently result in significant economic and environmental devastation, as well as the loss of life (Wang et al. 2011; Fernandez et al. 2016; Mojaddadi et al. 2017). According to a United Nations (UN) estimate, floods harmed 2.3 billion people globally between 1995 and 2015, killing 157,000 people (UNISDR 2015). Floods have cost the global economy roughly US $386 billion in the first 3 decades of the twentieth century (Wang et al. 2011). Several recent types of research have forecasted and assumed that climate change situations in the future significantly natural disasters will become more frequent and severe, such as floods (Chhen et al. 2015; Xenarios et al. 2016; Fang et al. 2016; Ntajal et al. 2017). Other causes, such as increased urbanization, economic development, and population growth, will exacerbate flood risk locations around the world (Mojaddadi et al. 2017). As a result, humans, wealth, and the environment will all be at risk for development. For reducing the effects of floods, disaster risk management strategies such as prevention and mitigation are recommended (Masuya et al. 2015). To create appropriate flood mitigation measures and information about important sensitive infrastructure and locations, it is necessary to identify the components that are responsible for vulnerability mapping. (Lee et al. 2017; Foudi et al. 2015). The purpose of the spatial vulnerability map is to show how dangerous it will affect humans, wealth, and the environment (Turner et al. 2003; Eckert et al. 2012; Rashid et al. 2013). Planners could also use the maps created through vulnerability assessments to establish innovative management strategies that focus on prevention and mitigation measures (Roy and Blaschke 2015; Rimba et al. 2017; Hoque et al. 2018). As a result, vulnerability and risk assessments can aid in the reduction of flood damage to humans, wealth, and the environment.

Several types of research have been conducted to create flood vulnerability maps using geospatial techniques with different approaches (Cozannet et al. 2013; Bhuiyan and Al Baky 2014). The detailed and precise vulnerability information is determined by the adequate and effective criterion selection, size, and components of vulnerability (Rashid 2013). Selecting appropriate criteria for each vulnerability component and their standard processing improves the trustworthiness of vulnerability information (Hoque et al. 2018). Other parameters, such as the local or regional scale of the study area, can influence the generation of detailed vulnerability data (Birkmann 2007; Fekete et al. 2010). Flood mitigation plans can be better prepared with detailed and precise vulnerable facts (Birkmann 2007). Another key challenge in vulnerability assessments is the selection of relevant components. The local community's coping abilities, as well as the surrounding environment and resources, play a vital role in protecting and minimizing flood effects (Hoque et al. 2018). As a result, it is critical to incorporate coping capability while assessing vulnerability to arrive at an outcome (Cutter et al. 2008).In the existing literature, few studies use coping capability in spatial vulnerability assessments employing multi-criteria-integrated geospatial methodologies at the microscale (Rana et al. 2018).

Being a floodplain area, various sections of the district are affected each year by the flood. Every monsoon, Alipurduar is prone to floods, particularly flash floods, which cause catastrophic damage to human dwellings, animals, and livelihood. In the peak of the monsoon season, road degradation and floods are particularly common in Alipurduar. The agricultural sector and the tea estate, which are also the district's main sources of income, have been severely harmed.

For acquiring geographical flood vulnerability information, a geospatial strategy combining remote sensing and spatial analysis is particularly successful (Mojaddadi et al. 2017; Ntajal et al. 2017). For planning and decision, remote sensing aids come for the gathering, analysis, and integration of a large number of datasets, whereas geospatial aids in the acquisition, assessment, and incorporation of numerous sets of data for planning and decision-making (Hoque et al. 2018). To combine multi-criteria for assessing geographical vulnerability, spatial decision-making procedures must use weighting and ranking. The analytical hierarchy process (AHP) is thought to be the best way for combining multicriteria for specific decision-making to generate spatially susceptible data (Wang et al. 2011; Roy and Blaschke 2015; Fernandez et al. 2016). The goal of this research is to develop and test a multi-criteria-integrated technique for assessing flood damage using AHP, combining data from the geographical analysis, GIS and data analysis, image remote sensing, data sets on coping skills, and validation data.

12.2 Study Area

This study was carried out in the Alipurduar district, West Bengal local administrative region (2789 km^2) in India's Eastern Himalayan foothills. Alipurduar is situated between latitudes 26° 25′ and 26° 51′ North and longitudes 89° 08′ and 89° 52′ E (Fig. 12.1). The area is dissected by many perennial rivers such as Torsa, Rydak, Sankosh, Kaljni, Jayanti, and Gadagar and it flows from the northern to the southern part of the district. The climate in this study area is tropical monsoon. The annual average temperature is 25°C. The monsoon season of the year gets a significant amount of rain. The yearly average rainfall is 3411 mm. The Indian state of West Bengal was flooded severely following heavy rains in July and August 2017. The geographical location, dense population, and all forms of resources are extremely vulnerable to flooding in areas of Alipurduar lowland (District Disaster Management Plan 2020). The residents in the study region have a low economic situation and a large number of people live in poverty. As a result, floods have a significant impact on socio-economic situations, and the area is triggered by heavy rainfall regularly and the destruction of water-holding wetlands (District Disaster Management Plan 2020).

12.3 History of Floods in the Sub-Himalayan Region of the Alipurduar District

Due to its geographical location, the region receives the most amount of rain in West Bengal, with an annual rainfall of 3753 mm and an average annual temperature of 24.33°C (Prokop and Walanus 2017). So, the selected study area receives maximum rainfall which has maximum chances of flood occurrence. Analysis of historical disasters is crucial for focusing on district risks. In the sub-Himalayan region, flood is a common phenomenon that occurs every year. The scale of the flood may vary in terms of the area flooded, the number of people harmed, and infrastructural damage. Every year, more or less all blocks are affected by the flood. Because of the role of the Himalayan foothills, many rivers have originated in the north and passed toward the south and the high rainfall pattern of the region affects changes in the river course and it may cause the flood. However, the intensity of rainfall is varying year-wise and so floods also vary, some of the year's floods have a devastating role that is genuinely disastrous are summarized in Table 12.1.

12.4 Geospatial Data Collection and Thematic Layer Preparation Technique

Large sets of spatial data obtained from a variety of sources in various formats, such as census data, satellite imagery, and weather data are all examples of geospatial data that are significantly used in this study. The source of data in detail has been provided in Table 12.2. For the assessment of flood risk in the sub-Himalayan region of Alipurduar district, a thorough literature review and expert viewpoints have been prioritized for the selection of the total nineteenth parameters in the study. Whereas, twelve indicators are connected to flood susceptibility factors viz. slope, elevation, TWI, TPI, NDVI,

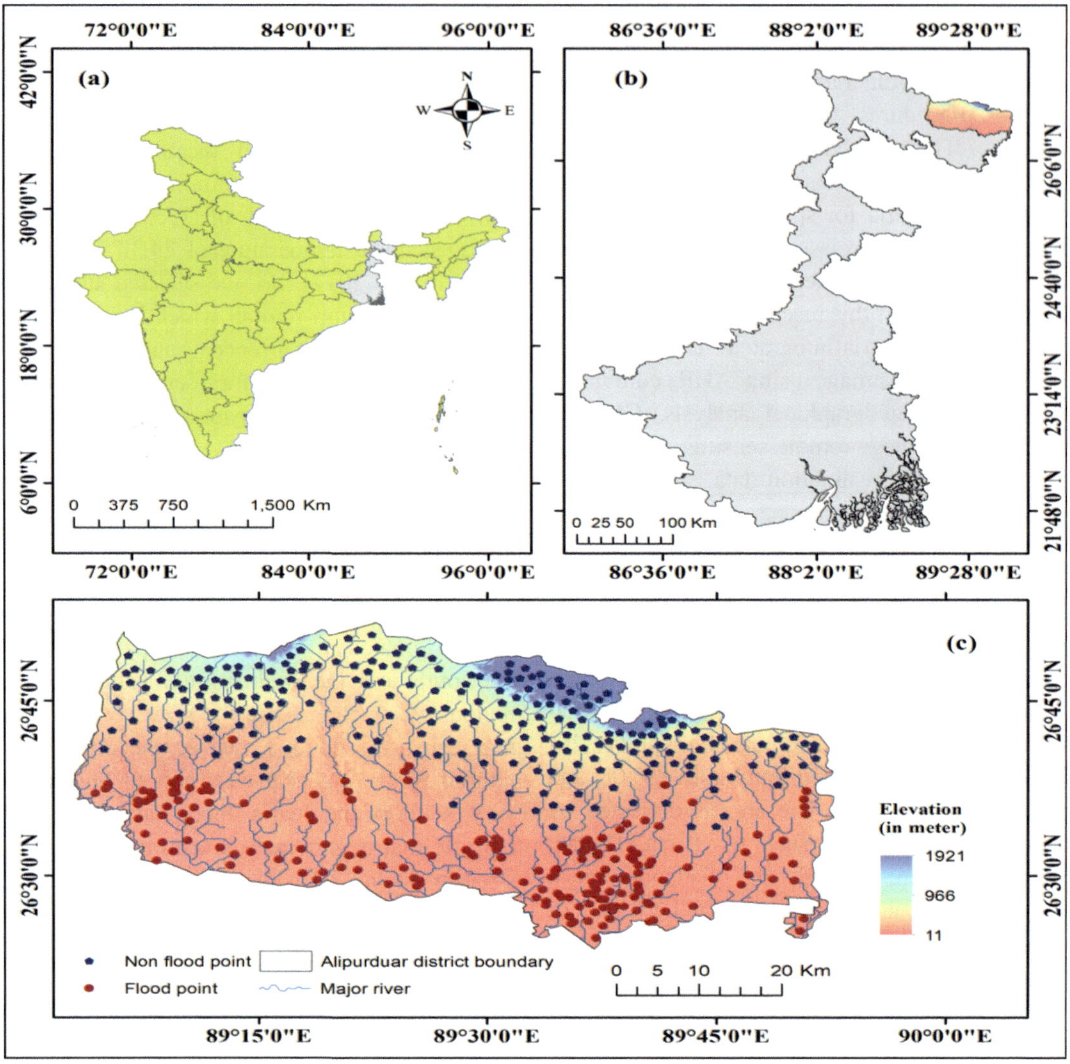

Fig. 12.1 Location map of the study **a** India, **b** West Bengal, and **c** Alipurduar district

MNDWI, drainage density, distance to river, MFI or RI (Rainfall intensity), geology, STI, and SPI. Subsequently, the remaining nine parameters are related to flood vulnerable factors viz. household density, child population <06 age, population density, literacy rate, employment rate, distance to flood shelters, distance to hospital, distance to road, and LULC. All of the thematic layers were created in the GIS platform.

In flood susceptibility factors preparation, ASTER-DEM data were presented to prepare elevation, slope, TWI, TPI, STI, and SPI. The data was downloaded with a 30 m spatial resolution map from NASA then the data was clipped with the study area. The data was processed for elevation map with fill in hydrology, subtool of spatial analyst tool and slope map prepared in surface sub-tool of spatial analyst tool and both the map were projected with UTM coordinate reference zone. TWI, STI, and SPI layers were prepared with DEM data through the process of flow direction, flow accumulation in hydrology subtool, and from math algebra subtool raster calculator used for raster slope (Tan slope, Radian

Table 12.1 Flood occurrences year and the level of effect in the sub-Himalayan region of Alipurduar district

Year	Affected level of flood in Alipurduar district
1991	Villages were affected, crops were damaged, and cattles were died
1993	Remarkable flood occurred due to heavy and continuous rainfall, roads disconnected, Torsa, Raidak flooded vast area, cattle ,crop , house, etc., were severely damaged, this occurrence left a deep mark to the local people who still remember the aftermath and trauma
1995	Villages were affected, crops were damaged, and houses besides the river bank were destroyed
1996	Heavy rain occurred for nearly seven consecutive days and the floodplain was inundated, people died, farmland affected, and houses were damaged
1998	Villages were affected, crops were damaged, and cattles were died
1999	Because of heavy rainfall in the last week of August, Torsa, Sankosh and Kaljani rives were flooded, people died, and farmland affected
2001	Flash flood occurred due to heavy rainfall, embankment collapsed, crops were damaged, and cattles were lost
2002	Several blocks, farmlands were damaged due to heavy rainfall and resultant flooding
2004	Crops damaged, cattle died and peoples were affected due to flood and pre-monsoonal calamities
2005	Villages were affected, crops were damaged, and houses were collapsed due to very heavy rainfall and flood.
2006	Villages were affected, crops were damaged, and cattles were died
2007	Uphill area of Dooars were flooded, houses were destroyed by mudslide
2010	Peoples, houses, and crops were affected due to flood
2011	Peoples, houses, and crops were affected due to flood
2012	Peoples, houses, and crops were affected due to flood
2013	Peoples, houses, and crops were affected due to flood
2014	Peoples, houses, and crops were affected due to flood
2015	Peoples, houses, and crops were affected due to flood
2016	Peoples, houses, and crops were affected due to flood
2017	Massive flood occurred due to incessant rainfall. While the death toll was not high, damage to properties like fish pond, road, culvert, bridge, etc., were huge
2018	Some people lose their shelter, some died due to thunderstruck
2019	Damage of crop and various public properties like road, culvert, embankment, etc.

Source District Disaster Management Plan report (2020)

slope) in ArcGIS platform to create the layers. The land facet corridor tool was used to create a topographic position index thematic layer. Landsat 8 OLI data are collected to prepare the NDVI and MNDWI mapping. After radiometric correction of 30 m spatial resolution satellite image, the raster calculator tool was initiated for mapping. After that, line density tool and Euclidian distance, the tool was used for the preparation of drainage density and distance to river layer. Subsequently, 35 years (1985–2020) gridded very high resolution ($0.25° \times 0.25°$) data has been collected from the Indian meteorological department (IMD) and extracted from a study area to prepare rainfall intensity or modified fourier index mapping using IDW interpolation technique. Subsequently, the geological unit was represented in the study area which was gathered from the GSI and is available in the Bhukosh server.

In flood vulnerability factors preparation, census data 2011 was performed to prepare population density, household density, no. of the child population, literacy rate, and employment rate in the ArcGIS platform. The demographic

Table 12.2 Sources and details of flood indicators used for flood risk assessment

Sl no	Flood indicators	Data type	Details	Source
Flood susceptibility dimensions				
1	Elevation	ASTER GDEM, GeoTIFF format	Spatial resolution 30 m, 1° × 1° tiles	https://earthdata.nasa.gov
2	Slope	ASTER GDEM, GeoTIFF format	Spatial resolution 30 m, 1°× 1° tiles	https://earthdata.nasa.gov
3	TWI(Topographic wetness index)	ASTER GDEM, GeoTIFF format	Spatial resolution 30 m, 1° × 1° tiles	https://earthdata.nasa.gov
4	TPI(Topographic position index)	ASTER GDEM, GeoTIFF format	Spatial resolution 30 m, 1° × 1° tiles	https://earthdata.nasa.gov
5	NDVI (Normalized difference vegetation index)	Landsat 8 OLI/TIRS Collection 2	Spatial resolution 30 m, WRS Path138 WRS Row 042	https://earthexplorer.usgs.gov
6	MNDWI (Modified normalized differe -nce water index)	Landsat 8 OLI/TIRS Collection 2	Spatial resolution 30 m, WRS Path138 WRS Row042	https://earthexplorer.usgs.gov
7	Drainage density	ESRI, Shapefile	GloRic (Global River Classification),version 1.0	https://www.hydrosheds.org
8	Distance to river	ESRI, Shapefile	GloRic (Global River Classification),version 1.0	https://www.hydrosheds.org
9	MFI (Modified Fourier index)	High spatial resolution (0.25°× 0.25°)	IMD Daily gridded rainfall data	https://www.imdpune.gov.in
10	Geology	Shapefile	Geological survey of India (GSI) data	https://bhukosh.gsi.gov.in
11	STI (Sediment transport index)	ASTER GDEM, GeoTIFF format	Spatial resolution 30m, 1° × 1° tiles	https://earthdata.nasa.gov
12	SPI (Stream power index)	ASTER GDEM, GeoTIFF format	Spatial resolution 30m, 1° × 1° tiles	https://earthdata.nasa.gov
Flood Vulnerability dimensions				
1	Population density	Attribute data	Population enumeration data, census of India, 2011	https://censusindia.gov.in
2	Household density	Attribute data	Population enumeration data, census of India, 2011	https://censusindia.gov.in
3	Child population <06 age	Attribute data	Population enumeration data, census of India, 2011	https://censusindia.gov.in
4	Literacy rate	Attribute data	Population enumeration data, census of India, 2011	https://censusindia.gov.in
5	Employment rate	Attribute data	Population enumeration data, census of India, 2011	https://censusindia.gov.in
6	Distance to flood shelters	Flood shelter locations, point data	District Disaster Management Authority Office of the District Magistrate Alipurduar, West Bengal	https://alipurduar.gov.in

(continued)

Table 12.2 (continued)

Sl no	Flood indicators	Data type	Details	Source
7	Distance to hospital	Shapefile	Open street map data	https://www.openstreetmap.org
8	Distance to road	Shapefile	Open street map data	https://www.openstreetmap.org
9	LULC	GeoTIFF format, tiles scene 45R	ESRI (10 m resolution map), earth's land surface 2020	https://livingatlas.arcgis.com

data was formed in the attribute table after rectifying the data processed to create thematic layers in the ArcGIS platform. After that, distance to flood shelters thematic layer was created using the Euclidiean tool, and the location point of coordination was arranged from the district disaster management authority office Alipurduar, 2020 handbook. Similarly, point and line data of hospitals and roads respectively are available in the open street map server, and the layers were prepared using the Euclidian distance technique in the ArcGIS platform. And lastly, for the LULC thematic layer, ESRI provides 10 m resolution earth's surface land use land cover data in GeoTIFF format, the study area was clipped for the process of modifying the zonation with help of ESRI enlisted categories in the ArcGIS environment.

As a result, all thematic layers were rasterized and transformed to UTM projection zone 45N with a 30 m resolution cell size, and raster layers were categorized and presented for weighting using the AHP approach.

12.5 Identification of Multicollinearity Problems in Variables

The variance inflation factors (VIF) and tolerance method were applied to identify the collinearity problem between the 12 conditioning components of flood susceptibility and nine conditioning factors of flood vulnerability in this model assessment. The model's prediction accuracy will be reduced if there is linear collinearity between the conditioning factors (Rahmati et al. 2016). There are several methods used for the multicollinearity test, in this present study VIF method is performed to analyze the model. For the accuracy assessment of this test, tolerance should be more than 0.1, and the percentage of VIF should be less than 10. If there are two factors affected by this problem then one should be removed from the model assessment. The following formula is used to calculate the VIF

$$\text{Tolerance of the } i\text{th predictor variable } (T_i) = 1 - R_i^2$$

$$\text{VIF of the } j\text{th predictor variable } \left(\text{VIF}_j\right) = 1/T_i$$

where R_i^2 represents the coefficient of determination of the regression equation.

In this study, 1000 random points were selected using create random points tool in ArcGIS for the multicollinearity test. After extraction of multi-value point data, SPSS software was used to analyze the linear regression with collinearity diagnostic statistics. Collinearity statistics (Table 12.3), for the flood susceptibility dimensions, show that STI has the highest VIF percentage (4.911) and MNDWI has the lowest VIF percentage (1.017). Subsequently, in the flood vulnerability dimensions, it shows that population density has the highest VIF percentage (7.269), and the employment rate has the lowest VIF percentage (1.024). Therefore, the result reveals that all parameters used in susceptibility, as well as vulnerability, are significantly standard. It is indicating that each parameter is individually important for the

Table 12.3 Multicollinearity diagnosis test for flood susceptibility and vulnerability indicators in sub-Himalayan region Alipurduar district

Susceptibility factors			Vulnerability factors		
Dimensions	Tolerance	VIF	Dimensions	Tolerance	VIF
Elevation	0.598	1.671	Population density	0.138	7.269
Slope	0.714	1.401	Household density	0.134	7.443
TWI	0.786	1.273	Child pop <06 age	0.139	7.183
TPI	0.827	1.210	Literacy rate	0.858	1.166
NDVI	0.912	1.096	Employment rate	0.977	1.024
MNDWI	0.984	1.017	Distance to flood shelters	0.621	1.611
Drainage density	0.635	1.575	Distance to hospital	0.612	1.635
Distance to river	0.643	1.556	Distance to road	0.805	1.242
MFI	0.854	1.171	LULC	0.897	1.114
Geology	0.868	1.575			
STI	0.204	4.911			
SPI	0.221	4.515			

assessment of flood risk and it asserts that there is no high correlation among the conditioning factors.

12.6 Normalized Weight Assignment Using Multi-Criteria Decision-Based AHP Method

Multi-criteria decision analysis (MCDA) is a standard technique for resolving spatially relevant problems as well as natural hazard predictions. This technique provides a geospatial framework to perform various conditioning variables into a hierarchical structure (Duleba and Moslem 2019). AHP method is an important criteria-based weighted technique method in the decision-making process by assigning multiple parameters on basis of expert opinion. This AHP method is popularly used for the estimation of expected prediction analysis in various research studies which was introduced by Saaty (1980). To predict the estimation, this method has to gather some indicators and assign priority hierarchically to minimize the complexity in decision-making. For the judgment of each criterion for risk assessment, the priority criteria were assigned by Saaty's relative important scale.

In the present research work, the AHP method is considered for the delineation of flood risk estimation and all the important criteria of flood susceptibility and vulnerability were performed. This method is a basic criteria-based subjective method and the indicators were assigned respective weights by AHP. There are so many decision-making methods to decide the specific goal, viz., TOPSIS, WASPAS, VIKOR, and AHP, etc. AHP method is well-known for the assessment of flood hazards and vulnerability as a promising technique considered in reliable method (Souissi et al. 2019; Das 2020; Saha and Agrawal 2020; Roy et al. 2021). It is difficult to assign weight to each parameter to estimate the outcome of the flood risk zonation. Therefore, relative importance has been allocated to each criterion with a comparative matrix according to Saaty's 1–9 scale (Table 12.4). Saaty's scale is considering only non-zero value, i.e. the minimum value of 1 represents the equally important and the maximum value of nine indicates the extreme importance of the pair-wise judgment. The process of the AHP method calculation is divided into four sequential stages, viz., (i) pair-wise comparison matrix evaluation, (ii) weight

Table 12.4 Saaty's relative importance scale value

Less important				Equal	More important			
Extremely	Very strongly	Strongly	Moderately	1	Moderately	Strongly	Very strongly	Extremely
1/9	1/7	1/5	1/3		3	5	7	9

2, 4, 6, and 8 are the intermediate values between the two adjacent judgments in a pair-wise comparison matrix

assignment, (iii) weight normalization criteria, as well as a consistency check (Das et al. 2017; Ghosh and Ghosal 2020; Roy et al. 2021). To check the consistency of the model following steps are generated (Tables 12.5, 12.6, 12.7, 12.8).

Step (i): Pair-wise comparison matrix is the initial step of the model. The comparison of each indicator has been developed using Saaty's numerical scale. The total number of matrix elements is calculated using the formula $n(n-1)/2$, where n means the total number of indicators allocated in susceptibility zonation and vulnerability zonation. The following equation is performed for susceptibility zonation and vulnerability zonation pair-wise comparison matrix in Tables 12.5 and 12.9.

$$P_x = \begin{bmatrix} x_{11} & x_{12} & x_{13} & \cdots & x_{1n} \\ x_{21} & x_{22} & x_{23} & \cdots & x_{2n} \\ \vdots & \vdots & \vdots & \vdots & \vdots \\ \vdots & \vdots & \vdots & \vdots & \vdots \\ x_{n1} & x_{n2} & x_{n3} & \cdots & x_{nn} \end{bmatrix} \tag{12.1}$$

where P_x denotes the pair-wise comparison matrix, x_{nn} is the factors of P_x.

Step (ii): Weight normalization of the criteria has been calculated by the geometrical mean of an individual row of the indicators. The weight of the criteria has been assigned in Tables 12.6 and 12.10 with the below equation.

$$\mathrm{NW} = \left(\frac{GM}{\sum_{n-1}^{N} GM_n}\right) \tag{12.2}$$

Table 12.5 Pair-wise comparison matrix of twelve thematic layers for flood susceptibility

	EL	SL	DD	DR	TWI	MNDWI	RI	NDVI	GL	TPI	STI	SPI
EL	1.00	4.00	3.00	3.00	4.00	3.00	3.00	5.00	6.00	4.00	6.00	6.00
SL	0.25	1.00	3.00	3.00	3.00	4.00	2.00	5.00	5.00	6.00	6.00	5.00
DD	0.33	0.33	1.00	2.00	5.00	5.00	3.00	4.00	5.00	6.00	6.00	7.00
DR	0.33	0.33	0.50	1.00	3.00	4.00	3.00	4.00	3.00	4.00	3.00	4.00
TWI	0.25	0.33	0.20	0.33	1.00	3.00	2.00	3.00	4.00	4.00	4.00	3.00
MNDWI	0.33	0.25	0.20	0.25	0.33	1.00	2.00	2.00	3.00	3.00	2.00	3.00
RI	0.33	0.50	0.33	0.33	0.50	0.50	1.00	3.00	2.00	2.00	2.00	3.00
NDVI	0.20	0.20	0.25	0.25	0.33	0.50	0.33	1.00	3.00	2.00	2.00	3.00
GL	0.17	0.20	0.20	0.33	0.25	0.33	0.50	0.33	1.00	2.00	2.00	2.00
TPI	0.25	0.17	0.17	0.25	0.25	0.33	0.50	0.50	0.50	1.00	2.00	1.00
STI	0.17	0.17	0.17	0.33	0.25	0.50	0.50	0.50	0.50	0.50	1.00	2.00
SPI	0.17	0.20	0.14	0.25	0.33	0.33	0.33	0.33	0.50	1.00	0.50	1.00

Where EL = elevation, SL= slope, DD = drainage density, TWI = topographic wetness index, MNDWI= modified normalized different water index, RI = rainfall intensity, NDVI= normalized difference vegetation index, GL = Geology, TPI= topographic position index, STI = sediment transportation index, and SPI = stream power index

Table 12.6 Normalized pair-wise comparison matrix of twelve thematic layers for flood susceptibility

	EL	SL	DD	DR	TWI	MNDWI	RI	NDVI	GL	TPI	STI	SPI	Weight
EL	0.26	0.52	0.33	0.26	0.22	0.13	0.17	0.17	0.18	0.11	0.16	0.15	0.22
SL	0.07	0.13	0.33	0.26	0.16	0.18	0.11	0.17	0.15	0.17	0.16	0.13	0.17
DD	0.09	0.04	0.11	0.18	0.27	0.22	0.17	0.14	0.15	0.17	0.16	0.18	0.16
DR	0.09	0.04	0.05	0.09	0.16	0.18	0.17	0.14	0.09	0.11	0.08	0.10	0.11
TWI	0.07	0.04	0.02	0.03	0.05	0.13	0.11	0.10	0.12	0.11	0.11	0.08	0.08
MNDWI	0.09	0.03	0.02	0.02	0.02	0.04	0.11	0.07	0.09	0.08	0.05	0.08	0.06
RI	0.09	0.07	0.04	0.03	0.03	0.02	0.06	0.10	0.06	0.06	0.05	0.08	0.06
NDVI	0.05	0.03	0.03	0.02	0.02	0.02	0.02	0.03	0.09	0.06	0.05	0.08	0.04
GL	0.04	0.03	0.02	0.03	0.01	0.01	0.03	0.01	0.03	0.06	0.05	0.05	0.03
TPI	0.07	0.02	0.02	0.02	0.01	0.01	0.03	0.02	0.01	0.03	0.05	0.03	0.03
STI	0.04	0.02	0.02	0.03	0.01	0.02	0.03	0.02	0.01	0.01	0.03	0.05	0.03
SPI	0.04	0.03	0.02	0.02	0.02	0.01	0.02	0.01	0.01	0.03	0.01	0.03	0.02

Where EL = elevation, SL= slope, DD = drainage density, TWI = topographic wetness index, MNDWI= modified normalized different water index, RI = rainfall intensity, NDVI= normalized difference vegetation index, GL = Geology, TPI= topographic position index, STI= sediment transportation index, and SPI = stream power index

where NW is indicating the estimation of normalized weights, and GM_n is the geometric mean calculation of the nth row in P_x.

Step (iii): in step three, the geometrical mean is multiplied with each assigned importance value in the matrix table individually for the calculation of $\lambda_{\max}$. The following equation was used to determine the consistency index (CI) value.

$$\mathrm{CI} = \frac{(\lambda_{\max} - n)}{n - 1} \tag{12.3}$$

where $\lambda_{\max}$ is denoted principle eigenvalue, and n denotes the number of parameters in total.

Step (iv): The consistency ratio (CR) is derived using the following equation to validate the AHP approach (Basak et al. 2021)

$$\mathrm{CR} = \frac{\mathrm{CI}}{\mathrm{RI}} \tag{12.4}$$

where consistency ratio (CR) is calculated by dividing the consistency index (CI) and random index (RI).

Random index (RI) values were given by Saaty (1980) to determine the consistency of the Pair-wise comparison matrix. In the study, the RI value (Table 12.7) for twelve flood susceptibility and nine flood vulnerability parameters are 1.48 and 1.45, respectively. For this model, to perform the consistency ratio value should be <0.10 or <10% to keep working on the model. In this research study, the calculated consistency ratios 0.074 and 0.066 were assigned accordingly in both susceptibility and vulnerability indicators (Table 12.8) which are adequate to be accepted in the model performance for the flood risk management in the Alipurduar district.

12.7 Flood Susceptibility, Vulnerability, and Risk Zonation Map Preparation

The AHP approach was used to assign indicators weight depending on the expert's view and self-investigation, and also this approach was used to assign each sub-criteria weight of the indications. After weight was assigned Arc GIS platform was used to measure both the susceptibility and vulnerability zonation mapping. The following formulas were used to delineate the susceptibility and vulnerability mapping (Malik et al. 2020)

Table 12.7 Random consistency index value

n	1	2	3	4	5	6	7	8	9	10	11	12	13	14	15
RCI	0.00	0.00	0.58	0.90	1.12	1.24	1.32	1.41	1.45	1.49	1.51	1.48	1.56	1.57	1.59

Table 12.8 AHP model summery

Flood susceptibility						Flood vulnerability					
λ_{max}	N	RI	CI	CR	Consistency	λ_{max}	N	RI	CI	CR	Consistency
13.204	12	1.48	0.109	0.074	CR<0.1(yes)	9.857	9	1.45	0.107	0.073	CR<0.1(yes)

Table 12.9 Pair-wise comparison matrix of nine thematic layers for flood vulnerability

	PD	HD	CP	LR	ER	DFS	DH	DR	LULC
PD	1.000	2.000	4.000	3.000	4.000	3.000	4.000	3.000	5.000
HD	0.500	1.000	2.000	3.000	3.000	4.000	4.000	5.000	5.000
CP	0.250	0.500	1.000	3.000	3.000	2.000	3.000	2.000	3.000
LR	0.333	0.333	0.333	1.000	2.000	0.333	0.500	3.000	3.000
ER	0.250	0.333	0.333	0.500	1.000	0.333	0.333	0.500	0.333
DFS	0.333	0.250	0.500	3.000	3.000	1.000	1.000	3.000	4.000
DH	0.250	0.250	0.333	2.000	3.000	1.000	1.000	2.000	3.000
DR	0.333	0.200	0.500	0.333	2.000	0.333	0.500	1.000	2.000
LULC	0.200	0.200	0.333	0.333	3.000	0.250	0.333	0.500	1.000

Where PD = population density, HD = Household density, CP = Child population <06 age, LR= Literacy rate, ER = Employment rate, DFS= Distance to flood shelters, DH= Distance to hospital, DR = Distance to road, and LULC= Land use landcover

$$FSI = \sum_{i=1}^{n} \left(W_i^F \times R_i^F\right) \quad (12.5)$$

$$FVI = \sum_{i=1}^{n} \left(W_i^V \times R_i^V\right) \quad (12.6)$$

where FSI and FVI stand for flood susceptibility and vulnerability indexes, respectively, with n denoting the number of components and W_i and R_i denoting the weight and rank of the indicators.

Flood risk assessment, which includes geoenvironmental hazards and socio-economic aspects, is a critical task for mitigating and managing floods. Flood risk is determined by the relationship between flood hazard and flood susceptibility. Therefore, flood risk mapping in the sub-Himalayan region of the Alipurduar district has been calculated by-product of FSI and FVI using a raster calculator in the ArcGIS platform. For the creation of flood risk index (FRI) mapping, the following equation was used.

$$FRI = FSI \times FVI \quad (12.7)$$

12.8 Flood Inventory Map

A flood inventory is the crucial mapping that was prepared to predict the future flood risk event in the sub-Himalayan flooding region. Historical details of flood events are required in inventory mapping to predict future risk assessment. This map is containing the spatial distribution of inundations. A flood inventory map can be created using a variety of different techniques that

Table 12.10 Normalized pair-wise comparison matrix of nine thematic layers for flood vulnerability

	PD	HD	CP	LR	ER	DFS	DH	DR	LULC	Weight
PD	0.290	0.395	0.429	0.186	0.167	0.245	0.273	0.150	0.190	0.258
HD	0.145	0.197	0.214	0.186	0.125	0.327	0.273	0.250	0.190	0.212
CP	0.072	0.099	0.107	0.186	0.125	0.163	0.205	0.100	0.114	0.130
LR	0.097	0.066	0.036	0.062	0.083	0.027	0.034	0.150	0.114	0.074
ER	0.072	0.066	0.036	0.031	0.042	0.027	0.023	0.025	0.013	0.037
DFS	0.097	0.049	0.054	0.186	0.125	0.082	0.068	0.150	0.152	0.107
DH	0.072	0.049	0.036	0.124	0.125	0.082	0.068	0.100	0.114	0.086
DR	0.097	0.039	0.054	0.021	0.083	0.027	0.034	0.050	0.076	0.053
LULC	0.058	0.039	0.036	0.021	0.125	0.020	0.023	0.025	0.038	0.043

Where PD = population density, HD = household density, CP = child population <06 age, LR= literacy rate, ER = employment rate, DFS= distance to flood shelters, DH= distance to hospital, DR = distance to road, and LULC= land use landcover

are dependent on various factors such as data availability, historical flood records, and access to geoenvironmental factors (Khosravi et al. 2019; Roy et al. 2021). In the present study of the sub-Himalayan Alipurduar district, 201 flood points were identified for the preparation of a flood inventory map, as well as 246 non-flood location points were identified. These location points were acquired from Google maps, flood recorded data available in the district disaster management handbook, and the Bhuvan web portal of ISRO. This flooding and non-flooding point data are also required for crossvalidation.

12.9 Flood Susceptibility Indicators

After literature review and expert opinion, twelve indicators (Fig. 12.2) were chosen for the flood susceptibility zonation. Therefore, the elaboration of the susceptibility indicators has been mapped with the help of ArcGIS software, and the calculation is summarized in Table 12.11.

12.9.1 Elevation

Elevation plays an important role as it has an immense impact on flood risk mapping in the study area, The GDEM was used to create an elevation outline for the whole study area, revealing that the lowest elevations are found in the district's southwestern and southeastern parts (<122 m), those areas have maximum chances to get affected by the high magnitude of inundation. Therefore, more weight was given to the areas having the lowest elevation (<122 m). Figure 12.2a shows the elevation of the Alipurduar district which is categorized into five classes.

12.9.2 Slope

The slope is another important criterion for the assessment of flood hazards in the entire study area. A large area of flat terrain with a very gentle slope is flooded for a long time, whereas a moderate-to-high slope allows the flood water to flow away easily. By using ASTER GDEM a slope map has been prepared and it's revealed that from middle to the southern section of the study area is confined with a gentle slope (0°–4°), while the northern part has a steep slope (ranging from 10° to 67°). Figure 12.2b shows the slope of the Alipurduar Himalayan foothill region which is categorized into five classes.

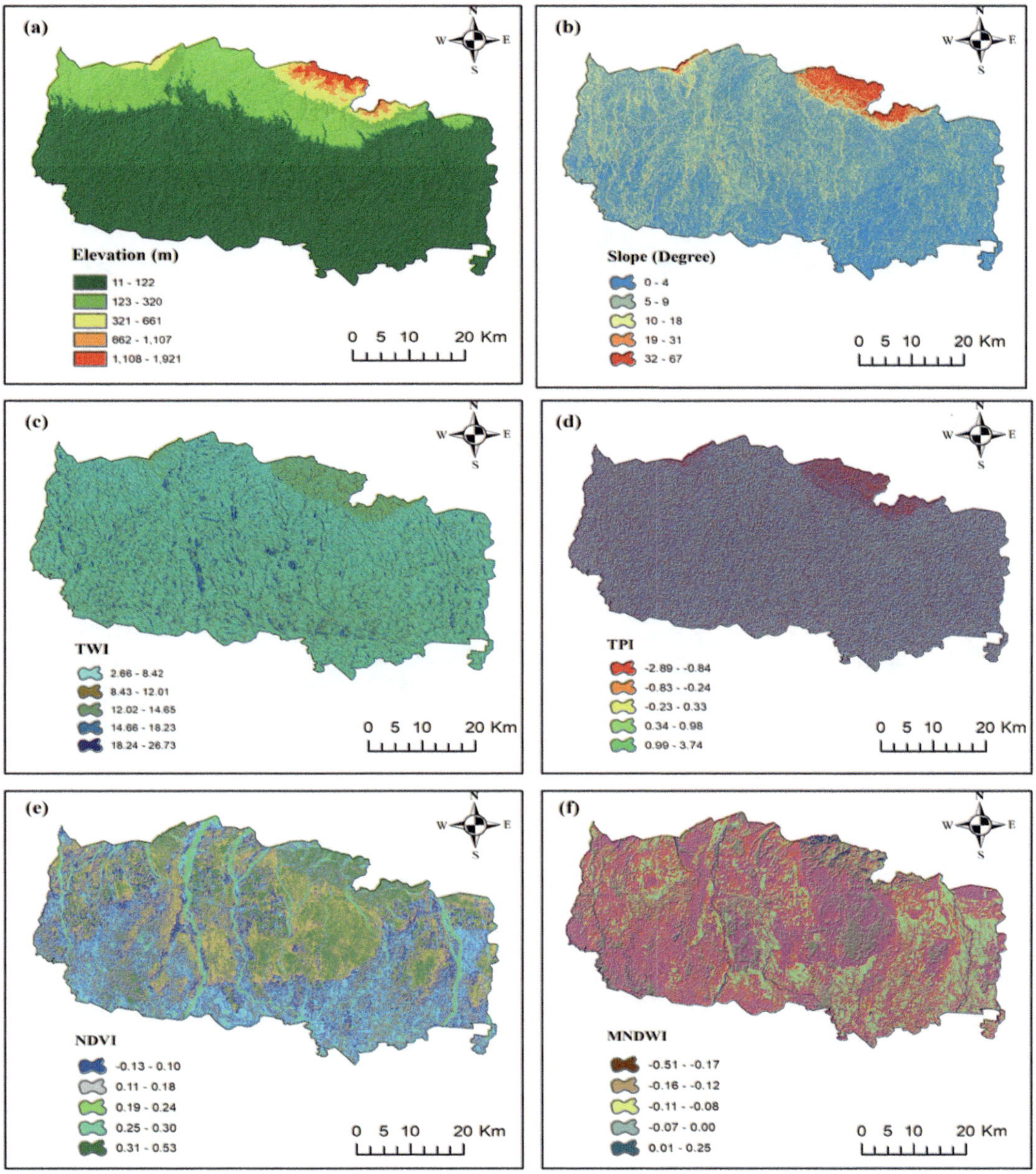

Fig. 12.2 Flood susceptibility indicators, **a** elevation, **b** slope, **c** TWI, **d** TPI, **e** NDVI, **f** MNDWI, **g** drainage density (DD), **h** distance to river, **i** RI, **j** geology, **k** STI, **l** SPI

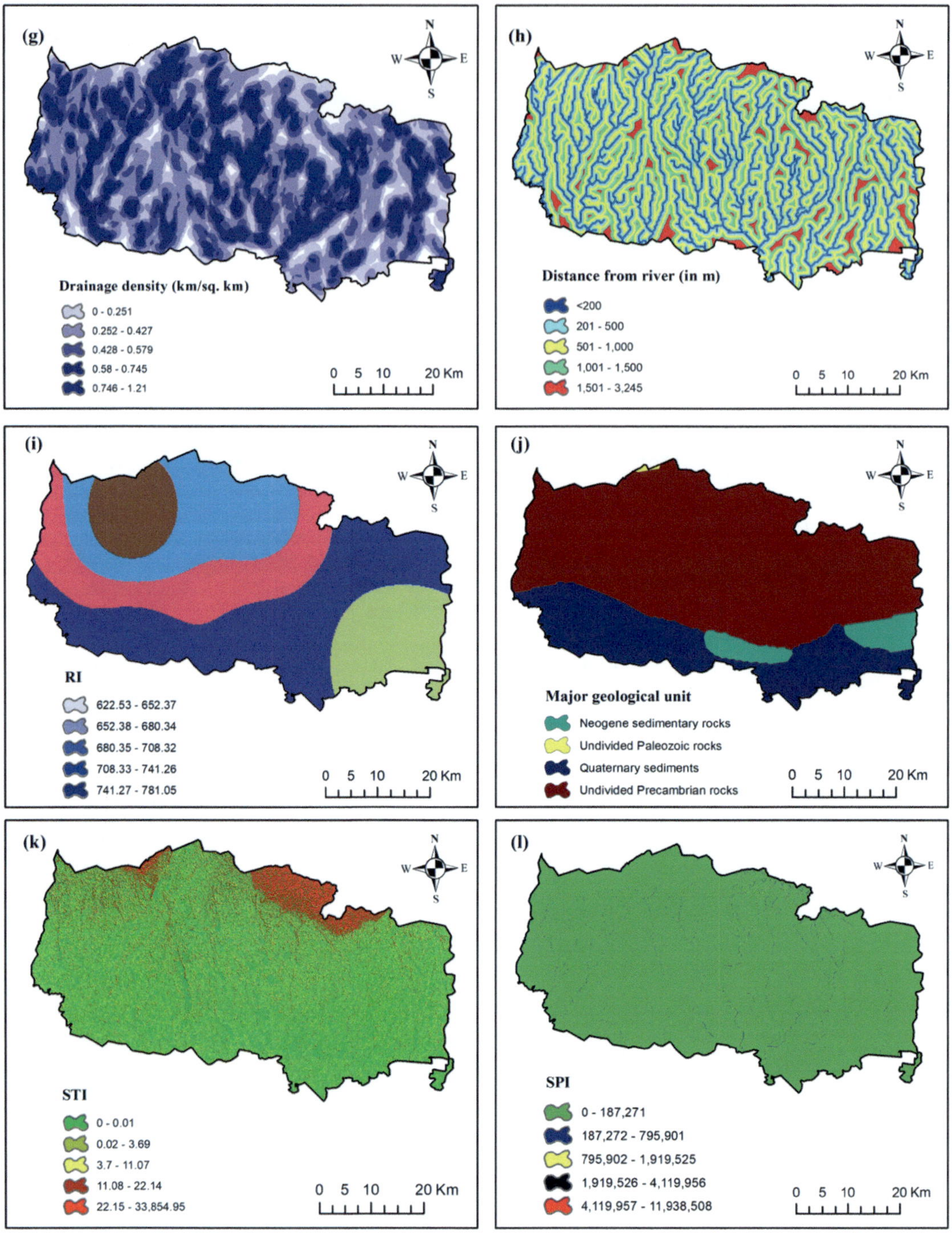

Fig. 12.2 (continued)

Table 12.11 Summary of criteria, AHP weight, class range, level of the flood, area in sq. km, area in %, and rating of each sub criteria for evaluation of flood susceptibility zonation

Sl no.	Parameters	AHP Weight	Reclass class	Class range	Flood level	Area in km^2	Area in %	Rating
1	Elevation	0.223	1	11–122	Very high	2114.02	75.40	0.489
			2	123–320	High	558.46	19.92	0.237
			3	321–661	Medium	68.75	2.45	0.142
			4	662–1107	Low	40.85	1.46	0.084
			5	1108–1921	Very low	21.74	0.78	0.047
2	Slope	0.169	1	0–4	Very high	1340.98	47.92	0.483
			2	5–9	High	1025.33	36.64	0.260
			3	10–17	Medium	305.06	10.90	0.140
			4	18–31	Low	79.60	2.84	0.073
			5	32–67	Very low	47.56	1.70	0.045
3	DD	0.156	1	0–0.251	Very low	388.92	14.21	0.042
			2	0.252–0.427	Low	846.23	30.92	0.089
			3	0.428–0.579	Medium	704.46	25.74	0.167
			4	0.58–0.745	High	551.46	20.15	0.229
			5	0.746–1.21	Very high	245.35	8.97	0.473
4	DR	0.109	1	<200	Very high	485.21	17.61	0.509
			2	201–500	High	739.77	26.85	0.218
			3	501–1000	Medium	954.90	34.66	0.131
			4	1001–1500	Low	441.78	16.03	0.093
			5	1501–3245	Very low	133.68	4.85	0.049
5	TWI	0.082	1	2.66–8.42	Very low	1078.94	38.48	0.041
			2	8.43–12.01	Low	622.66	22.21	0.096
			3	12.02–14.65	Medium	608.83	21.71	0.162
			4	14.66–18.23	High	403.33	14.39	0.244
			5	18.24–26.73	Very high	90.03	3.21	0.456
6	MNDWI	0.059	1	−0.513 to −0.17	Very low	426.30	15.20	0.041
			2	−0.169 to −0.122	Low	736.56	26.27	0.104
			3	−0.121 to −0.077	Medium	1063.27	37.92	0.165
			4	−0.076 to 0.003	High	528.24	18.84	0.249
			5	0.004–0.248	Very high	49.34	1.76	0.441
7	RI	0.056	1	623–652	Very low	376.82	13.44	0.040
			2	653–680	Low	990.53	35.32	0.088
			3	681–708	Medium	585.41	20.88	0.155
			4	709–741	High	601.45	21.45	0.223
			5	742–781	Very high	250.11	8.92	0.494

(continued)

Table 12.11 (continued)

Sl no.	Parameters	AHP Weight	Reclass class	Class range	Flood level	Area in km^2	Area in %	Rating
8	NDVI	0.041	1	−0.133 to 0.01	Very high	17.79	0.63	0.508
			2	0.011–0.14	High	368.94	13.16	0.239
			3	0.141–0.22	Medium	748.86	26.71	0.117
			4	0.221–0.35	Low	1557.76	55.56	0.093
			5	0.351–0.529	Very low	110.37	3.94	0.043
9	Geology	0.032	1	*N*	High	151.58	5.41	0.244
			2	pC	Medium	1977.78	70.54	0.112
			3	Pz	Very low	3.55	0.13	0.075
			4	Q	Very high	670.79	23.93	0.568
10	TPI	0.027	1	−2.89 to −0.84	Very high	358.51	12.86	0.500
			2	−0.83 to −0.24	High	699.29	25.08	0.225
			3	−0.23 to 0.33	Medium	851.80	30.55	0.134
			4	0.34–0.98	Low	622.00	22.31	0.096
			5	0.99–3.74	Very low	256.71	9.21	0.044
11	STI	0.025	1	0.00–0.10	Very high	1476.64	52.66	0.487
			2	0.11–3.69	High	571.12	20.37	0.225
			3	3.70–11.07	Medium	387.65	13.83	0.140
			4	11.08–22.14	Low	135.72	4.84	0.096
			5	22.15–33,854.95	Very low	232.71	8.30	0.052
12	SPI	0.021	1	0–187,271	Very high	2794.78	99.68	0.530
			2	187,272–795,901	High	6.99	0.25	0.190
			3	795,902–1,919,525	Medium	1.61	0.06	0.134
			4	1,919,526–4,119,956	Low	0.41	0.01	0.093
			5	4,119,957–11,938,508	Very low	0.044	0.002	0.053

12.9.3 Topographic Wetness Index (TWI)

Topographic wetness index (TWI) also called compound topographic index (CTI) which depicts the spatial wetness land of the area introduced by Beven and Kirkby (1979), Choudhury et al. (2022). The impact of wetness on topography, and wet levels of the land can generate runoffs that could be determined as flood risk (Rahmati et al. 2016; Haghizadeh et al. 2017). The wetness of a region is measured with a high TWI value. Hence, the high TWI value zone refers high risk of inundation. A raster calculator tool was used to determine TWI using digital elevation model. The following equation was used in the raster calculator tool to prepare TWI.

$$\mathrm{TWI} = \mathrm{In}(A_S/\tan\beta) \qquad (12.8)$$

where A_s = upslope area and β = slope gradient (degree).

The study area is confined to very low to the low category of TWI value (ranging from 2.66 to 12.01), and approximately 18% area represents high to very high wetness in the study area, Fig. 12.2c.

12.9.4 Topographic Position Index (TPI)

The topographic position index (TPI) was calculated by using raster DEM. It is the altitude difference between the elevation value of the cell and the average altitude of the neighboring cell. A positive value means the high altitude and a negative value means the lower altitude. In this study, area topographic position index categorized into five classes in Fig. 12.2d, viz., (−2.89)–(−0.84), (−83)–(−0.24), (−0.23)–0.33, 0.34–0.98, and 0.99–3.74. TPI value near zero means the topography is either the flat area or mid-slope area, which is more susceptible to flood.

12.9.5 Normalized Difference Vegetation Index (NDVI)

For predicting flood susceptibility normalized difference vegetation index (NDVI) is another important factor. Generally, non-vegetated features such as water logging areas, snow cover, barren land, and many other features that come under lower NDVI value (negative or close to 0) may cause the flood susceptible zone. Shrub and grassland come under moderate values, while forests come in high values (0.6–0.8) (Tehrany et al. 2013; Khosravi et al. 2019). The study area covered mostly low-to-medium value of the NDVI index area, i.e., mixed forest, a small amount of shrub and grassland which is considerably affecting in flood susceptible condition. Figure 12.2e shows NDVI of the Alipurduar district which is categorized into five classes.

12.9.6 Modified Normalized Difference Water Index (MNDWI)

Modified normalized difference water index (MNDWI) has been prepared using SWIR bands collected from Landsat 8 satellite imagery. For the flood susceptibility assessment, MNDWI is an important indicator. MNDWI expressed by the following equation.

$$\text{MNDWI} = \frac{\text{Green} - \text{SWIR}}{\text{Green} + \text{SWIR}} \tag{12.9}$$

Here, the SWIR band refers the band 6 of the Landsat 8 image.

In NWDI, water information is found contaminated due to built-up land and gives positive values therefore, to overcome the problem SWIR bands were used in modified NDWI. MNDWI was categorized into five classes in this study, viz., (−0.51)–(−0.17), (−0.16)–(−0.12), (−0.11)–(−0.08), (−0.07)–(0.00, and 0.01–0.25.). Figure 12.2f shows the MNDWI of the Alipurduar district which is categorized into five classes.

12.9.7 Drainage Density (DD)

The length of the river inside a basin divided by the basin's area represents drainage density (Saha et al. 2022). Drainage density is the most important indicator to delineate the flood susceptibility map. In the present study, drainage density value ranges 0.25–1.21 km/km^2. Low drainage density means the low level of flooding condition, whereas high drainage density means the high level of flood condition. In this study, about 30% area is considered high to very high drainage density. Figure 12.2g shows the drainage density of the Alipurduar Himalayan foothill region which is categorized into five classes.

12.9.8 Distance from the River

Another important factor to estimate the flood susceptibility is the distance to river. Distance up to 0.5 km from the rivers, people, and property is in severe danger. Therefore, the distance below 0.5 km from the rivers is usually categorized as the most susceptible area, while the distance more than 1.5 km is usually categorized as the least susceptible. Settlements with a distance of 200–1000 m from the river are mostly seen in the study area, which depicts higher chances of flood. Locations within 90 m of rivers are more vulnerable (Pradhan et al. 2009; Samanta and Pal 2011). Five classes ranging from less than 250 m to more than 1500 m were used here for the present study. Figure 12.2h shows, distance to river which is categorized into five classes.

12.9.9 Rainfall Intensity (RI/MFI)

Heavy rainfall and flooding have a positive relationship. The frequency of rainfall increases in the study area from July to September (i.e., the rainy season). Surface runoff from heavy rains is highly important for predicting flash floods, which are very fast, flashy, and have a short duration (Samanta and Pal 2011). The rainfall intensity index was used as a triggering element for the flood susceptibility study since a higher intensity indicates adequate rainfall, while a less intensity of rainfall indicates the deficiency of rainfall and increases the chance of drought. In the study area, the intensity and geographical mapping were calculated using the average and observed yearly rainfall of the Alipurduardistrict from 1985 to 2020. The modified fournier index approach was used to create the rainfall intensity map (Tiwari et al. 2020). Figure 12.2i shows the rainfall intensity of the Alipurduar district which is categorized into five classes.

12.9.10 Geology

The study area has been divided into four geological types. The paleozoic rocks and quaternary sediments are located in a high-elevation location that is rarely flooded during the monsoon season. As a result, it was given the lowest weight. Flood plain deposits are found in places, where flooding is a common occurrence. This was assigned the uppermost priority, followed by the classes of sand with silts and clays. Figure 12.2j shows the geology of the Alipurduar district which is categorized into five classes.

12.9.11 Sediment Transportation Index (STI)

The STI index, which is used to quantify landscape erosion, takes into consideration of flow convergence and divergence, which influences the occurrence of floods. The areas with more runoff have more sediment transport and are less prone to flooding. Based on the calculation, an STI map was arranged and classified into five categories in Fig. 12.2k and out of them, the study area confined the very high (0.11–3.69) to high (3.70–11.07) category of STI value.

12.9.12 Stream Power Index (SPI)

Stream power index (SPI) measures the erosion force of water flow. Moore et al. provide the formula for calculating SPI (Moore et al. 1991).

$$\mathrm{SPI} = (\alpha X \tan \beta) \tag{12.10}$$

where α shows the area of the basin and *tan* β is the angle of slope in degrees at that center (Jaafari et al. 2014). The stream power index (SPI) map followed five categories of class in Fig. 12.2l, viz, 0–187,271, 187,272–795,901, 795,902–1,919,525, 1,919,526–4,119,956, and 4,119,957–11,938,508. The majority of the rivers in the study area have a very high and high category of stream power (i.e., 0–795,901).

12.10 Flood Vulnerability Indicators

In this study, nine indicators (Fig. 12.3) were chosen for the flood vulnerability zonation. All the vulnerability indicators have been mapped in ArcGIS software, and the calculation has been summarized in Table 12.12.

12.10.1 Population Density

The most important variable in evaluating social vulnerability is population density as people are at the epicenter of tragedy and vulnerability. As a result, the population density indicator received more weight. At every level of the decision-making process, higher emphasis was given to high population concentration, while less priority was given to those areas having lower density. However, the population density of the Alipurduar district is categorized into five classes in Fig. 12.3a.

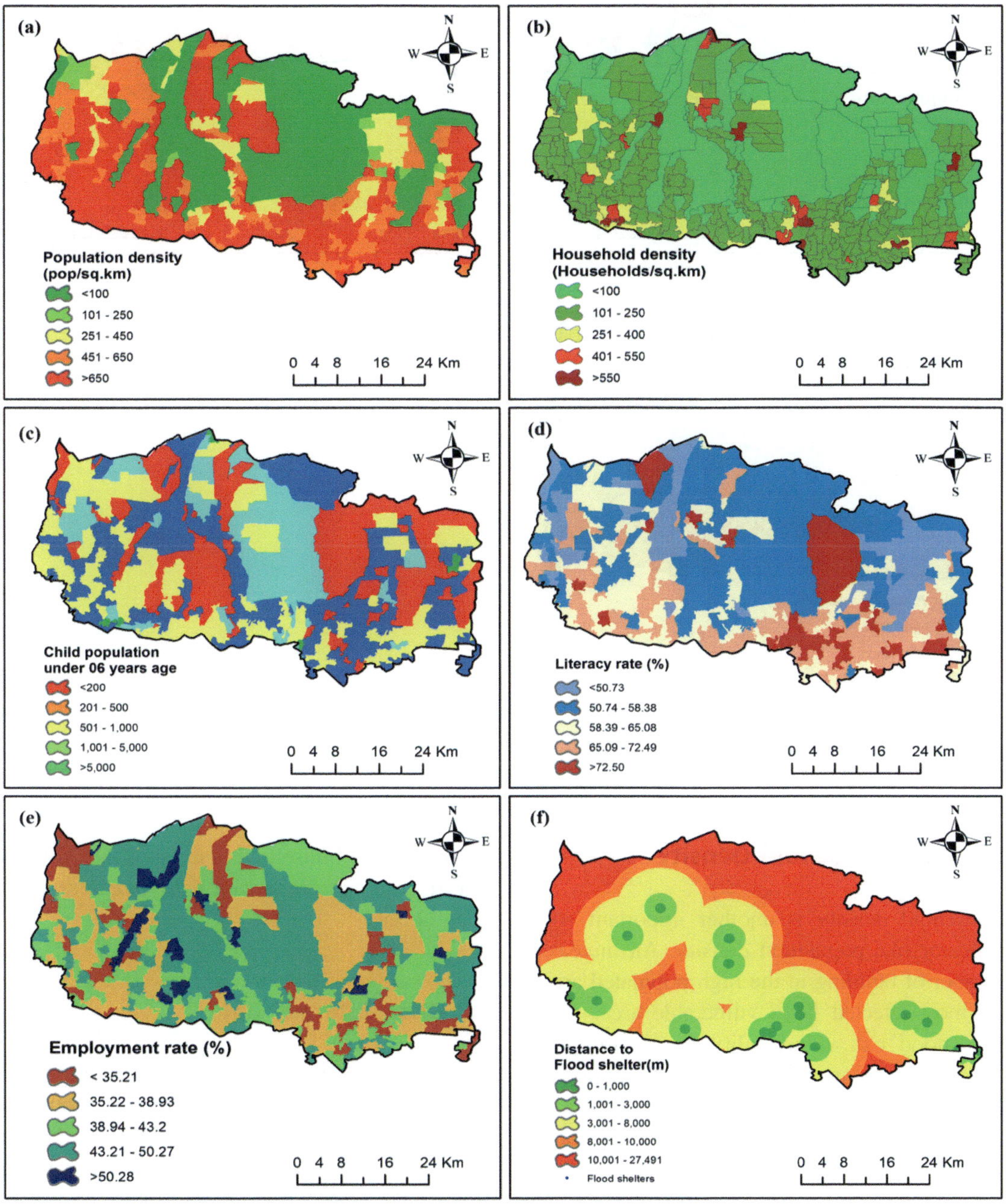

Fig. 12.3 Flood vulnerability indicators, **a** population density, **b** household density, **c** child population <06 aged, **d** literacy rate, **e** employment rate, **f** distance to flood shelters, **g** distance to hospital, **h** distance to road, **i** LULC

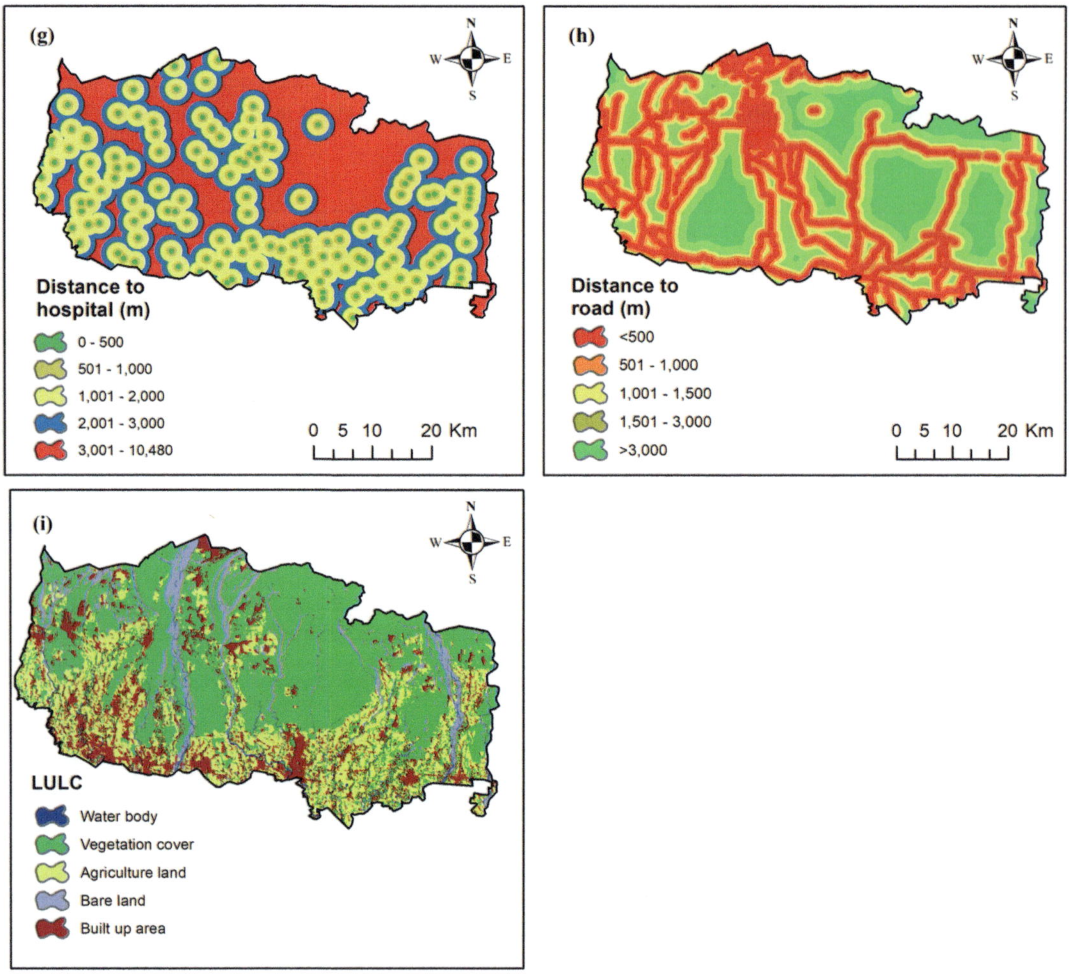

Fig. 12.3 (continued)

12.10.2 Household Density

Household density is another important social indicator like population density. An increasing number of building in the high populated area is the most affected flood vulnerable area. Therefore, flood risk vulnerability has a strong and close relationship with household density that influences the severity of risk (Ghosh and Kar 2018). As the southern and middle part of the district has a higher number of households than the northern part, therefore, the damage and casualty are maxima in the south of this area. However, the household density of the Alipurduar district is categorized into five classes in Fig. 12.3b.

12.10.3 Literacy Rate

The literacy rate in the study was calculated based on the total population. Illiterate people are highly vulnerable to floods, whereas literate people less vulnerable in the case of a disaster. In this study, literacy rate of the Alipurduar district can be classified into five levels in Fig. 12.3d.

Table 12.12 Summary of criteria, AHP weight, class range, level of the flood, area in sq. km, area in %, and rating of each sub criteria for evaluation of flood vulnerability zonation

Sl no.	Parameters	AHP Weight	Class range	Flood level	Area in km^2	Area in %	Rating
1	population Density	0.270	< 100	Very low	1052.16	37.65	0.059
			101–250	Low	62.75	2.24	0.092
			251–450	Moderate	247.67	8.86	0.150
			451–650	High	450.71	16.13	0.222
			>650	Very high	981.64	35.12	0.477
2	Household Density	0.216	<100	Very low	1369.44	49.00	0.066
			101–250	Low	1191.08	42.62	0.100
			251–400	Moderate	130.20	4.66	0.152
			401–550	High	55.74	1.99	0.207
			>550	Very high	48.46	1.73	0.474
3	Child population aged < 06	0.130	<200	Very low	726.31	25.99	0.054
			201–500	Low	910.86	32.59	0.099
			501–1000	Moderate	679.14	24.30	0.145
			1001–5000	High	463.72	16.59	0.208
			>5000	Very high	14.89	0.53	0.495
4	Literacy Rate	0.071	<50.73	Very high	383.35	13.72	0.398
			50.74–58.38	High	1133.18	40.54	0.267
			58.39–65.08	Moderate	509.62	18.23	0.157
			65.09–72.49	Low	483.08	17.28	0.124
			>72.50	Very low	285.70	10.22	0.053
5	Employment Rate	0.038	<35.21	Very high	284.34	10.17	0.403
			35.22–38.93	High	708.83	25.36	0.253
			38.94–43.2	Moderate	676.23	24.19	0.183
			43.21–50.27	Low	1019.78	36.49	0.113
			>50.28	Very low	105.75	3.78	0.049
6	Distance to flood shelter	0.102	0–1000	Very low	48.46	1.73	0.050
			1001–3000	Low	301.87	10.77	0.076
			3001–8000	Moderate	1153.25	41.13	0.121
			8001–10000	High	362.54	12.93	0.246
			10001–27491	Very high	937.59	33.44	0.507
7	Distance to hospital	0.081	0–500	Very low	111.72	3.98	0.069
			501–1000	Low	303.02	10.81	0.104
			1001–2000	Moderate	839.06	29.93	0.152
			2001–3000	High	668.11	23.83	0.230
			3001–10,480	Very high	881.80	31.45	0.446

(continued)

Table 12.12 (continued)

Sl no.	Parameters	AHP Weight	Class range	Flood level	Area in km^2	Area in %	Rating
8	Distance to road	0.051	<500	Very low	763.66	27.24	0.065
			501–1000	Low	536.53	19.14	0.110
			1001–1500	Moderate	377.65	13.47	0.158
			1501–3000	High	642.98	22.93	0.214
			>3000	Very high	482.89	17.22	0.453
9	LULC	0.041	Water body	Very low	427.10	1.69	0.072
			Vegetation cover	Moderate	13748.89	54.49	0.139
			Agriculture land	High	5914.18	23.44	0.204
			Bare land	Low	2137.06	8.47	0.094
			Built up area	Very high	3006.18	11.91	0.491

As a result, the high literacy population is weighted as low vulnerability, and the low literacy population is weighted as highly vulnerable.

12.10.4 Employment Rate

The employment rate was computed by dividing the total number of people in the study area by the number of people participating in various secondary and tertiary activities. The economic background of those who are impacted by floods is critical, as they will require financial assistance during a flood. Hence, in this study, the majority of residents have the lowest employment rate (i.e., below 40%). Figure 12.3e, shows the employment rate of the Alipurduar district which is categorized into five classes.

12.10.5 Flood Shelter Distance

Residents are forced to abandon their homes during floods due to the possibility of casualties. As a result, the exiled Publicis are weaker and more susceptible during the flood because their safety, health, and cleanliness are all threatened (Roy et al. 2020). The distance from the shelter has been classified into five groups in Fig. 12.3f, as the distance between flood shelters raises, so does the vulnerability to floods and the risk of being flooded.

12.10.6 Distance to Hospital

When there are casualties, a suitable number of beds in hospitals and medically skilled as well as technical employees are required for effective hazard response (Chen et al. 2013, Roy et al. 2020). The study region is divided into five levels in Fig. 12.3g, i.e., the distances from hospitals are indicated as follows: the farther away from medical services, the greater the risk of flooding.

12.10.7 Distance to Road

Highways were flooded amid heavy rains, causing major connection and accessibility issues (Das 2020). Furthermore, the availability of major highways is crucial during floods, especially in terms of relief efforts, because all rescue and help are only accessible via major national and state highways (Ghosh and Kar 2018). Asian highway (AH02), National Highway (31-C), and Major

Table 12.13 Descriptive statistics of single parameter sensitivity analysis of flood susceptibility

Thematic layers	Empirical weight (%)	Effective weight in %			
		Min	Max	Mean	SD
EL	22	4.839	61.318	34.608	8.251
SL	17	2.202	54.147	21.305	7.429
DD	16	1.705	43.915	9.312	5.598
DR	11	1.531	53.655	8.470	5.278
TWI	8	0.775	32.823	3.781	2.915
MNDWI	6	0.553	27.663	3.713	2.342
RI	6	0.493	28.567	3.672	3.139
NDVI	4	0.407	23.221	2.463	1.751
GL	3	0.784	9.368	2.684	2.083
TPI	3	0.284	16.767	1.950	1.521
STI	3	0.298	15.868	3.156	1.770
SPI	2	0.317	16.901	4.376	1.366

State highways pass through the Alipurduar district, which has a very crucial role in terms of relief and rescue efforts during flood hazards. Figure 12.3h shows distance to road in Alipurduar district which is categorized into five classes.

12.10.8 Land Use and Land Cover

Because anthropological practices such as deforestation and urbanization have a direct impact on environmental disasters, and regions with compact communities are more prone to floods, land use is a critical component in flood vulnerability mapping (Komolafe et al. 2018). The five forms of land use and land cover statistics include water bodies, vegetation cover, agricultural land, barren land, and built-up area. Because the possible effects of floods are high for the built-up area and agricultural land category in terms of life and economic loss due to floods, floods have a significant impact on rural people and the local economy. This is one of the reasons why the built-up and agriculture category have the most weight (Dandapat and Panda 2017). Figure 12.3i shows land use and land cover of the Alipurduar district which is categorized into five classes.

12.10.9 Sensitivity Analysis

The AHP method is a subjective method, and the weights of this method have been calculated based on the expert's opinion and sometimes prior knowledge of the author. Therefore, it is difficult to understand the standard weight of the individual parameter used in the assessment. Furthermore, the AHP method has some limitations to take the number of factors for the model assessment; therefore, it also seems to be uncertain of individual weight on the final output layer. Because lots of factors are neglected for the model, may be neglected factors have some potential to change the estimation. Therefore, the effective weight of each model has to be needed for the model validation.

In this study, a single parameter sensitivity assessment was performed to evaluate the individual effective weight compared with the AHP model empirical weight. This measurement was developed to evaluate the real weight effect on final index mapping. The effective weight (W) of flood susceptibility and vulnerability (Tables 12.13 and 12.14) factors were calculated for each AHP model parameter using the following equation.

Table 12.14 Descriptive statistics of single parameter sensitivity analysis of flood vulnerability

Thematic layers	Empirical weight (%)	Effective weight in %			
		Min	Max	Mean	SD
PD	27	5.859	77.584	35.714	21.230
HD	22	5.247	64.723	12.371	4.109
CP	13	2.052	38.286	9.539	4.398
LR	7	1.225	32.466	10.483	6.207
ER	4	0.632	19.439	4.615	2.260
DFS	10	1.695	53.444	16.930	11.803
DH	8	1.565	41.256	13.899	8.618
DR	5	0.894	26.148	6.405	4.950
LULC	4	0.850	25.288	4.888	2.523

$$W = \frac{P_r \cdot P_w}{\text{FSI}} \times 100 \quad (12.11)$$

$$W = \frac{P_r \cdot P_w}{FVI} \times 100 \quad (12.12)$$

where FSI=flood susceptibility index and FVI=-flood vulnerability index, W = effective weight of each parameter, and P_r and P_w are the rating value and weight for each factor.

The operation indicates variation between empirical weight and effective weight in both susceptibility and vulnerability assessment. In the case of susceptibility indexing (Table 12.3), elevation and the slope are the most important factor as it 34.60% and 21.30% effective weight, respectively, which are greater than empirical weight. TPI is the minimum significant factor for the susceptibility zonation mapping as it is 1.59% effective weight. SPI is also an important factor as its effective weight (4.376%) is higher than the theoretical weight (2%). In flood vulnerability indexing (Table 12.4), population density is the most significant in model assessment as to its effective weight of 35.71%. Subsequently, literacy rate, employment rate, distance to flood shelters, distance to households, distance to roads, and LULC (10.48, 4.61, 16.93, 13.89, 6.40, and 4.88%) have the higher effective weight than the theoretical weight (7, 4, 10, 8, 5, 4%). Whether, the theoretical weights are assigned maximum in household density and child population, which are 22% and 13%, respectively.

12.11 Results and Discussion

12.11.1 Spatial Variation of Flood Susceptibility in the Alipurduar District

In this work, an AHP weight was used to create a flood susceptibility map, and 12 parameters were chosen for the mapping. As there are many research articles which are having fewer factors for the flood susceptibility zonation mapping. However, it also reveals that field-based surveys, availability and acquiring spatial data, and expert-based opinion are also important for the accuracy of rating in susceptibility zonation. The susceptibility zonation map (Table 12.15 and Fig. 12.4) was classified into five categories in the natural breaks method. The susceptibility zonation map of the sub-Himalayan Alipurduar district has revealed that the 977.56 km^2 which is 35% of the total area is captured with highly susceptible to flood. About 137.33 km^2 area (4.92%) is covered with a very low flood susceptibility zone in the northern part of the study area, the region is high in elevation, slope, etc. About 395.48 km^2 area (14.18%) is occupied by a low susceptibility zone in the southern region of the very low susceptibility, it is also in the northern part of the study area, this region is also in high altitude, high slope, low drainage density also in far away from the river, so that the region

Table 12.15 Distribution of flood susceptibility, vulnerability and risk of Alipurduar district

Level of flood	Flood susceptibility		Flood vulnerability		Flood risk	
	Area in km^2	Area in %	Area in km^2	Area in %	Area in km^2	Area in %
Very low	137.33	4.92	584.68	20.94	607.37	21.83
Low	395.48	14.18	856.29	30.67	889.39	31.97
Moderate	859.35	30.80	461.95	16.54	601.83	21.63
High	977.56	35.04	732.23	26.22	532.65	19.15
Very high	420.25	15.06	157.04	5.62	150.77	5.42

is low susceptible in flood. 859.35 km^2 area (30.80%) is the moderate region, and the covered area of 420.25 km^2 (15.06%) is the very high susceptibility zone or flood-prone zone, situated in the inter-fluvial part of the region along with the river basin of the Himalayan foothill region. Therefore, it is clear that the region with high altitude, maximum elevation, low slope, TWI, and other induces factors may cause the highly susceptible flood-prone zone.

12.11.2 Spatial Variation of Flood Vulnerability in Alipurduar District

In the present study, the flood vulnerability index was prepared based on the AHP weight in the ArcGIS platform. In this study, nine important parameters were selected on basis of an extensive literature review, and expert opinion to prepare vulnerability mapping. This map is prepared in a combination of village-level socio-economic data. In this study, the vulnerable map (Fig. 12.4) is prepared the combination of indicators viz. population density, household density, child population, literacy rate, employment rate, flood shelters, distance to hospital, road, and LULC. Within the above-stated indicators, population density refers to the highest weighted value in the AHP method, and LULC has referred to as the least Weightage in AHP model performance on final vulnerability mapping in the study area. The vulnerability map was categorized into five classes (Table 12.15) using the natural break method in ArcGIS, viz., very low vulnerability zone, low vulnerability zone, moderate vulnerability zone, high vulnerability zone, and very high vulnerability zone in Himalayan foothill Alipurduar district. In the vulnerable map, it is revealed that the 732.23 km^2 (26.22%) are covered under high vulnerable to flood, hence the area indicating high population density, high household density, etc., low and very low classes have covered about 50% of vulnerable area combinedly, it reveals a middle and northern portion of the study area has less vulnerable due to low population and household density, child population, and better performance of hospitals and accessibility of road supply, etc. Subsequently, only 5% area is under very high vulnerability in this region. Furthermore, low literacy, insufficient flood shelter rescues, and low road hospital performance can further increase the vulnerability prospects in the entire sub-Himalayan region of the Alipurduar district.

12.11.3 Spatial Distribution of Flood Risk

A flood risk map (Fig. 12.4) is an integration of flood susceptibility and flood vulnerability index which was prepared in ArcGIS using the raster calculator tool. The flood risk zonation map has been categorized into five classes (Table 12.15), viz., very low and low-risk zone, moderate, high, and very high-risk zone developed in this study. In the risk map, low and very low categories have covered 50% area of the Alipurduar district, and the rest of the 50% area covered in every block (Madarihat, Falakata, kalchini, Alipurduar I, Alipurduar II, kumargram) of the study

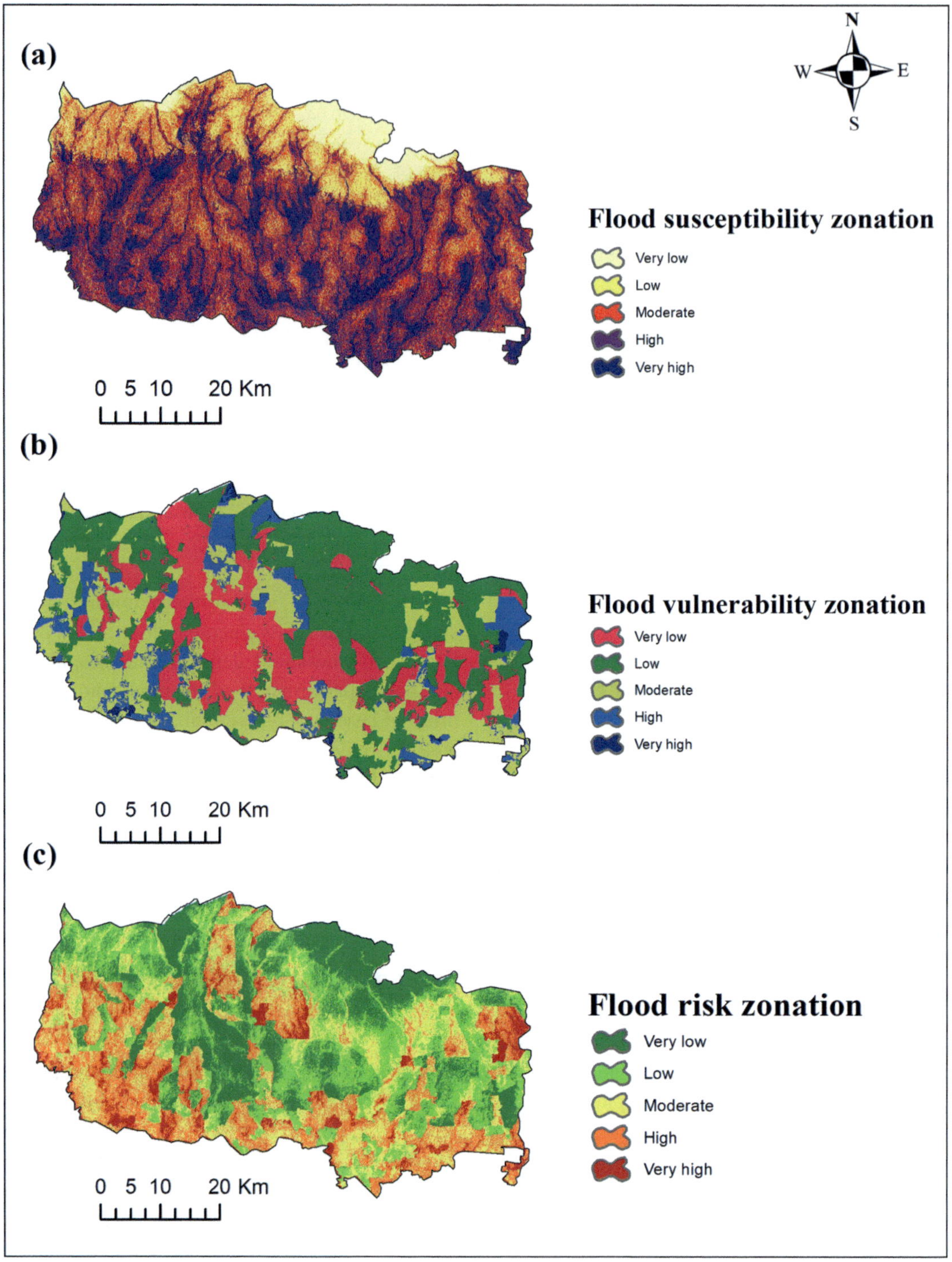

Fig. 12.4 Final results of **a** flood susceptibility map, **b** vulnerability map, **c** flood risk map of the Alipurduar foothill region based on multi-criteria-based analytical hierarchy process (AHP)

area. Categorically, 21.63%, 19.15%, and 5.42% risk inundate areas are observed in moderately, high, and very high respectively in the study area. Low and very low inundate risks are noticed in high altitude, sloppy, low population, and household density regions. Moderate risk has been found in agricultural land, moderate annual rainfall, low literacy, and employment rate region of the study area, and high and very high flood risk have found in the heavy rainfall, river proximity, and low elevation portion of the study area.

The study area is covered by an extensive river network. The major rivers are Tista, Kaljani, Raidak, Jayanti, and Sankosh which originate in the Himalayas and flow toward the south. These rivers become highly erratic during the heavy rainfall, and consequently, many people shifted their dwellers. Due to heavy rainfall, all subsequent rivers have become flooded and created a massive sheet of water which increases the intensity of flood risk level. As the entire sub-Himalayan Alipurduar district is almost flat except for the northern part of Alipurduar, it increases the flood risk of stagnant floodwater over lengthy periods. As a result of the high population density and continued expansion of towns along the river banks, the situation became more hazardous. Therefore, it is clear that the low elevated region near the river bank indicates high vulnerability to flood whereas it is found that the high altitude interfluves region has moderate low risk vulnerable to flood. However, the entire Alipurduar Himalayan foothill region is vulnerable as a result, and flood management and planning should be emphasized.

12.12 Model Validation

Model validation is one of the most important tasks after developing any model to ensure that the outcome is valid. In this study multi-criterion, GIS-based AHP model was developed for the flood risk assessment, after the development of the model historical recorded data was used to validate the model using the AUC (Area under the curve) technique. However, there are many techniques to validate the data, using AUC for the simplicity of the graph and the high accuracy estimation. For the AUC accuracy, it can be categorized into five classes viz. >0.9 means

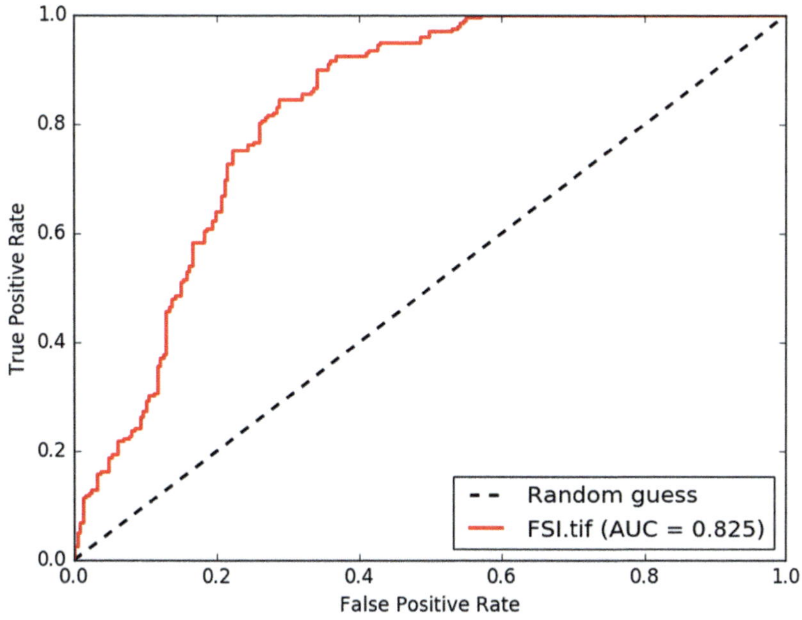

Fig. 12.5 Flood risk assessment by AHP model, validated using the area under curve (AUC)

excellent accuracy, 0.8–0.9 means good validation, 0.7–0.8 means the model is simply accepted, 0.5–0.7 means considerable level, and <0.5 AUC value means the model performance is not accepted for the outcome, which is considerably invalid for prediction estimation. According to this research study (Fig. 12.5), and AUC (area under the curve) value is 0.825 which means the model is 82.5% accurate and is considered an excellent outcome for the model. The high AUC value (0.825) indicates that flood mapping in the sub-Himalayan Alipurduar district is highly accurate, which could be helpful for flood management, control, and mitigation in the future.

12.13 Conclusions

The current study employs a multi-criterion statistical approach to investigate the primary components and triggering elements behind flood hazard, flood susceptibility, and flood risk in a quantitative manner. Flooding has had a significant impact on the Himalayan foothills over time. The destruction brought by floods underlined the importance of risk assessment to gain a better knowledge of the factors that produce flood hazards. The study shows how an analytical hierarchy process (AHP) based technique to flood hazard assessments can improve existing risk assessment approaches. The paper also recommends employing AHP-driven weights to evaluate flood hazard and susceptibility, as well as including socio-economic elements alongside physical ones. The goal of the study is to link hydrological and geomorphic elements to flood hazard intensity, as well as population and their socio-economic, and infrastructure aspects to susceptibility. Both are needed to create risk intensity classifications, which will aid in the identification of priority locations for efficient budgeting and important concern. The findings clearly show that the bottom, low-height plain regions, which encompass almost the entire districts' northwestern and southwestern parts are particularly vulnerable due to a high possibility of flood hazard and disorder, population concentration, agricultural expansion, and insufficient resilience capacity to minimize hazardous scenarios. So, the study's findings may be useful in identifying elements that promote resilience, which may then be included in the future management of flood hazards and planning decisions. This type of research can also be conducted at the district and basin levels, allowing for the production of a more comprehensive risk map and a more accurate assessment of riverine flood risk.

References

Basak A, Das J, Rahman ATM, Pham QB (2021) An integrated approach for delineating and characterizing groundwater depletion hotspots in a coastal state of India. Journal of the Geological Society of India 97(11):1429–1440. https://doi.org/10.1007/s12594-021-1883-z

Beven KJ, Kirkby MJ (1979) A physically based, variable contributing area model of basin hydrology/Un modèle à base physique de zone d'appel variable de l'hydrologie du bassin versant. Hydrol Sci J 24(1):43–69

Bhuiyan SR, Al Baky A (2014) Digital elevation-based flood hazard and vulnerability study at various return periods in Sirajganj SadarUpazila, Bangladesh. International Journal of Disaster Risk Reduction 10:48–58

Birkmann J (2007) Risk and vulnerability indicators at different scales: applicability, usefulness and policy implications. Environmental Hazards 7(1):20–31

Chen Y, Huang C, Ticehurst C, Merrin L, Thew P (2013) An evaluation of MODIS daily and 8-day composite products for floodplain and wetland inundation mapping. Wetlands 33(5):823–835

Chhen H, Ito Y, Sawamukai M, Tokunaga T (2015) Flood hazard assessment in the Kujukuri plain of Chiba prefecture, Japan, based on GIS and multicriteria decision analysis. Natural Hazards 78(1):105–120

Choudhury S, Basak A, Biswas S, Das J (2022) Flash flood susceptibility mapping using GIS-based AHP method. In: Pradhan B, Shit PK, Bhunia GS, Adhikary PP, Pourghasemi HR (eds) Spatial modelling of flood risk and flood hazards. GIScience and Geo-Environmental Modelling. Springer, Cham. https://doi.org/10.1007/978-3-030-94544-2_8

Cozannet GL, Garcin M, Bulteau T, Mirgon C, Yates ML, Méndez M, Baills A, Idier D, Oliveros C (2013) An AHP-derived method for mapping the physical vulnerability of coastal areas at regional scales. Natural Hazards and Earth System Sciences 13(5):1209–1227

Cutter SL, Barnes L, Berry M, Burton C, Evans E, Tate E, Webb J (2008) A place-based model for understanding community resilience to natural disasters. Global Environmental Change 18(4):598–606

Dandapat K, Panda GK (2017) Flood vulnerability analysis and risk assessment using analytical hierarchy process. Modeling Earth Systems and Environment 3 (4):1627–1646

Das J, Gayen A, Saha S, Bhattacharya SK (2017) Modelling of alternative crops suitability to tobacco based on analytical hierarchy process in Dinhata subdivision of Koch Bihar district, West Bengal. Modeling Earth Systems and Environment 3(4):1571–1587. https://doi.org/10.1007/s40808-017-0392-y

Das S (2020) Flood susceptibility mapping of the Western Ghat coastal belt using multi-source geospatial data and analytical hierarchy process (AHP). Remote Sensing Applications: Society and Environment 20:100379

District Disaster Management Plan (2020) A publication of the district disaster management authority, Alipurduar, March 2020

Duleba S, Moslem S (2019) Examining Pareto optimality in analytic hierarchy process on real data: an application in public transport service development. Expert Systems with Applications 116:21–30

Eckert S, Jelinek R, Zeug G, Krausmann E (2012) Remote sensing-based assessment of tsunami vulnerability and risk in Alexandria. Egypt. Applied Geography 32(2):714–723

Fang X, Pomeroy JW (2016) Impact of antecedent conditions on simulations of a flood in a mountain-head water basin. Hydrol Process 30(16):2754–2772

Fekete A, Damm M, Birkmann J (2010) Scales as a challenge for vulnerability assessment. Natural Hazards 55(3):729–747

Fernandez P, Mourato S, Moreira M (2016) Social vulnerability assessment of flood risk using GIS-based multicriteria decision analysis. A case study of Vila Nova de Gaia (Portugal). Geomatics, Natural Hazards and Risk 7(4):1367–1389

Foudi S, Osés-Eraso N, Tamayo I (2015) Integrated spatial flood risk assessment: the case of Zaragoza. Land Use Policy 42:278–292

Ghosh M, Ghosal S (2020) Climate change vulnerability of rural households in flood-prone areas of Himalayan foothills, West Bengal, India. Environ Dev Sustain 23:2570–2595

Ghosh A, Kar SK (2018) Application of analytical hierarchy process (AHP) for flood risk assessment: a case study in Malda district of West Bengal. India. Natural Hazards 94(1):349–368

Haghizadeh A, Siahkamari S, Haghiabi AH, Rahmati O (2017) Forecasting flood-prone areas usingShannon's entropy model. J Earth Syst Sci 126(3):1–11

Hoque MAA, Phinn S, Roelfsema C, Childs I (2018) Modelling tropical cyclone risks for present and future climate change scenarios using geospatial techniques. International Journal of Digital Earth 11 (3):246–263

Jaafari A, Najafi A, Pourghasemi HR, Rezaeian J, Sattarian A (2014) GIS-based frequency ratio and index of entropy models for landslide susceptibility assessment in the Caspian Forest, northern Iran. International Journal of Environmental Science and Technology 11(4):909–926

Khosravi K, Shahabi H, Pham BT, Adamowski J, Shirzadi A, Pradhan B, Dou J, Ly H-B, Gróf G, Ho HL, Hong H, Chapi K, Prakash I (2019) A comparative assessment of flood susceptibility modeling using multi-criteria decision-making analysis and machine learning methods. Journal of Hydrology 573:311–323

Komolafe AA, Herath S, Avtar R (2018) Methodology to assess potential flood damages in urban areas under the influence of climate change. Natural Hazards Review 19(2):05018001

Lee S, Kim JC, Jung HS, Lee MJ, Lee S (2017) Spatial prediction of flood susceptibility using random-forest and boosted-tree models in Seoul metropolitan city, Korea. Geomatics, Natural Hazards and Risk 8 (2):1185–1203

Malik S, Pal SC, Chowdhur I, Chakrabortty R, Roy P, Das B (2020) Prediction of highly flood prone areas by GIS based heuristic and statistical model in a monsoon dominated region of Bengal Basin. Remote Sens Appl Soc Environ 19:100343

Masuya A, Dewan A, Corner RJ (2015) Population evacuation: evaluating spatial distribution of flood shelters and vulnerable residential units in Dhaka with geographic information systems. Natural Hazards 78 (3):1859–1882

Mojaddadi H, Pradhan B, Nampak H, Ahmad N, Ghazali AHB (2017) Ensemble machine-learning-based geospatial approach for flood risk assessment using multi-sensor remote-sensing data and GIS. Geomatics, Natural Hazards and Risk 8(2):1080–1102

Moore ID, Grayson RB, Ladson AR (1991) Digital terrain modelling: a review of hydrological, geomorphological, and biological applications. Hydrological Processes 5(1):3–30

Ntajal J, Lamptey BL, Mahamadou IB, Nyarko BK (2017) Flood disaster risk mapping in the lower Mono river basin in Togo, West Africa. International Journal of Disaster Risk Reduction 23:93–103

Pradhan B, Shafiee M, Pirasteh S (2009) Maximum flood prone area mapping using RADARSAT images and GIS: Kelantan River basin. International Journal of Geoinformaticss 5(2):11–23

Prokop P, Walanus A (2017) Impact of the Darjeeling-Bhutan Himalayan front on rainfall hazard pattern. Natural Hazards 89(1):387–404

Rahmati O, Zeinivand H, Besharat M (2016) Flood hazard zoning in Yasooj region, Iran, using GIS and multi-criteria decision analysis. Geomatics, Natural Hazards and Risk 7(3):1000–1017

Rashid AM (2013) Understanding vulnerability and risks. Disaster risk reduction approaches in Bangladesh. Springer, Tokyo, pp 23–43

Rimba AB, Setiawati MD, Sambah AB, Miura F (2017) Physical flood vulnerability mapping applying geospatial techniques in Okazaki City, Aichi Prefecture. Japan. Urban Science 1(1):7

Roy DC, Blaschke T (2015) Spatial vulnerability assessment of floods in the coastal regions of Bangladesh. Geomatics, Natural Hazards and Risk 6(1):21–44

Roy P, Pal SC, Chakrabortty R, Chowdhuri I, Malik S, Das B (2020) Threats of climate and land use change on future flood susceptibility. J Cleaner Prod 272:122757

Roy S, Bose A, Chowdhury IR (2021) Flood risk assessment using geospatial data and multi-criteria decision approach: a study from historically active flood-prone region of Himalayan foothill India. Arabian Journal of Geosciences 14(11):1–25

Saaty TL (1980) The analytic hierarchy process. McGraw-Hill, New York, pp 20–25

Saha AK, Agrawal S (2020) Mapping and assessment of flood risk in Prayagraj district, India: a GIS and remote sensing study. Nanotechnology for Environmental Engineering 5(2):1–18

Saha S, Das J, Mandal T (2022) Investigation of the watershed hydro-morphologic characteristics through the morphometric analysis: a study on Rayeng basin in Darjeeling Himalaya. Environmental Challenges 100463. https://doi.org/10.1016/j.envc.2022.100463

Samanta S, Pal M (2011) Fuzzy tolerance graphs. International Journal of Latest Trends in Mathematics 1(2):57–67

Souissi D, Zouhri L, Hammami S, Msaddek MH, Zghibi A, Dlala M (2019) GIS based MCDM–AHP modeling for flood susceptibility mapping of arid areas, southeastern Tunisia. Geocarto Int 35:991–1017

Tehrany MS, Pradhan B, Jebur MN (2013) Spatial prediction of flood susceptible areas using rule-based decision tree (DT) and a novel ensemble bivariate and multivariate statistical model in GIS. Journal of Hydrology 504:69–79

Tiwari V, Kumar V, Matin MA, Thapa A, Ellenburg WL, Gupta N, Thapa S (2020) Flood inundation mapping-Kerala 2018; harnessing the power of SAR, automatic threshold detection method and google earth engine. Plos One 15(8):e0237324

Turner BL, Kasperson RE, Matson PA, McCarthy JJ, Corell RW, Christensen L, Schiller A (2003) A framework for vulnerability analysis in sustainability science. Proceedings of the National Academy of Sciences 100(14):8074–8079

UNISDR C (2015) The human cost of natural disasters: a global perspective

Wang Y, Li Z, Tang Z, Zeng G (2011) A GIS-based spatial multi-criteria approach for flood risk assessment in the Dongting Lake Region, Hunan. Central China. Water Resources Management 25(13):3465–3484

Xenarios S, Nemes A, Sarker GW, Sekhar NU (2016) Assessing vulnerability to climate change: are communities in flood-prone areas in Bangladesh more vulnerable than those in drought-prone areas? Water Resources and Rural Development 7:1–19

GIS-Based Landslide Susceptibility Mapping: A Case Study from Kegalle District, Sri Lanka

13

Kupendrathas Guhananth, Pattiyage I. A. Gomes, and H. D. Abeysiriwardana

Abstract

A landslide occurs when a piece of rock, a piece of earth, or a pile of debris slides down a slope. There are a few key geological and hydrological factors that influence the occurrence of landslides. However, these factors do not provide an equal contribution to landslide susceptibility. In this study, Analytical Hierarchy Process (AHP) was used to identify landslide susceptible areas in Kegalle district, Sri Lanka, where several past landslides occurred. This research analyzed the contributing parameters of landslides such as slope, aspect, soil class, lithology, rainfall, land use, distance to roads, and distance to streams. The AHP gave an acceptable (i.e., ≤ 1) Consistency Ratio (CR) of 0.032. The final landslide susceptibility model showed a 71% level accuracy with the area under the curve value of 0.705. About 4% of the entire Kegalle district was identified as very highly susceptible, while approximately 14% of the study area is showing high susceptibility. Moderate and Low susceptibility zones cover about 39% and 26% of the study area, respectively. Nearly 16% of the study area belongs to the very low susceptible zone. Soon after the study was finalized, a landslide occurred in the Dombemeda area, which was identified as a very high susceptible area by the model; though unfortunate, it verified the model. This showed the importance of liaising with relevant authorities - since the developed model can be used as an aid for landslide preparedness and mitigation.

Keywords

Analytical hierarchy process · GIS · Landslides · Landslide susceptibility mapping

13.1 Introduction

A natural disaster is a complex, destructive event beyond the control of humans. The loss of human life and property damages due to natural disasters occurring worldwide is rapidly increasing (Kumar and Anbalagan 2016), and this is the case for landslides too (Petley et al. 2005). Landslides are displacements of soil, rock, and debris down a slope, that occur suddenly without warning, creating immense socio-economic and environmental impacts (Bloechl and Braun 2005;

K. Guhananth · P. I. A. Gomes (✉) ·
H. D. Abeysiriwardana
Department of Civil Engineering, Faculty of Engineering, Sri Lanka Institute of Information Technology, Malabe, Sri Lanka
e-mail: ayantha.g@sliit.lk

K. Guhananth
Department of Civil Engineering, Faculty of Engineering, University of Jaffna, Jaffna, Sri Lanka

J. Das and S. K. Bhattacharya (eds.), *Monitoring and Managing Multi-hazards*,
GIScience and Geo-environmental Modelling, https://doi.org/10.1007/978-3-031-15377-8_13

Guzzetti et al. 2012). Landslides can occur anyplace in the world due to natural and/or man-made reasons. The impacts vary from one landslide to another and are governed by the landslides' magnitude and aerial extent (Kumar and Anbalagan 2016). In contrast to the popular perspective that landslides always cause negative impacts, landslides can have some positive impacts as well (Gomes et al. 2020a) (e.g., exposing underlying stable soil layers assisting the environment to reach a stable condition).

Many landslides are triggered by natural physical, geological, and morphological causes, but some are due to anthropogenic activities (Singh et al. 2017) such as slope alterations and vegetation removal (Yaa'cob et al. 2020). Evaluating the relationship between landslide triggering factors and past occurrences of landslides in order to predict the likelihood of a landslide occurring on a certain site or area in the future is known as landslide susceptibility (Kumar and Anbalagan 2016; Quan and Lee 2012). A variety of approaches, such as fuzzy logic, neural network, logistic regression, and Analytic Hierarchy Process (AHP), can be used in landslide susceptibility studies to find above-mentioned relationship (Hong et al. 2017). With the recent development in computer-based technology, Geographical Information Systems (GIS) and Remote Sensing (RS) technologies have become effective tools in landslide susceptibility mapping (Abeysiriwardana and Gomes 2022). RS and GIS support the delineation of landslide-prone areas (Yaa'cob et al. 2020; Sarkar and Kanungo 2004) with or without a large amount of field data. Nevertheless, Gomes et al. (2020a) showed the importance of field data (e.g., vegetation) in understanding landslides, and Abeysiriwardana and Gomes (2022) successfully incorporated those data in landslide susceptibility modeling via logistic regression. The general practice in landslide susceptibility mapping is to classify the study area ranging from low to high based on the likelihood of occurring landslides (Guzzetti et al. 2012).

Landslides are one of the natural disasters that have been frequently occurring since a few decades back in Sri Lanka (Hemasinghe et al. 2018). There have been incidents of landslides occurring in a few more regions where landslides were not common too (Gomes et al. 2020a). Recent landslides in Sri Lanka could be attributed to high-intensity low-frequency rainfalls (Ratnayake and Herath 2005; Kaleel and Reeza 2017; Nawagamuwa and Perera 2017) and anthropogenic disturbances (Gomes et al. 2020a). The frequency and the magnitude of landslides in Sri Lanka are likely to be on the increase due to climate variability and anthropogenic disturbances due to the growing population (Abeysiriwardana and Gomes 2022). However, the landslide susceptibility analyses conducted in Sri Lanka are still scant and/or unavailable in the public domain. Thus, this study aimed at mapping landslide susceptibility of one of the landslides prone districts in Sri Lanka, as a case study incorporating AHP approach and GIS tools.

13.2 Materials and Methods

13.2.1 Study Area

The Kegalle district, which covers an area of 1693 km^2, is one of Sri Lanka's largest semi-urban districts. The location of the study area (Fig. 13.1) is in the wet zone of Sri Lanka, which receives mean annual rainfall ranges from 2000 to 3000 mm. Annual temperature in the range of 25–30 °C. The Kegalle district witnessed several major landslides including the Aranayake landslide in 2016, killing over 130 people and 75 homes were destroyed along the landslide's path (Dang et al. 2019).

13.2.2 Landslide Causative Factors

Eight factors that could make an area fundamentally predisposed to landslides (slope angle, aspect, soil type, lithology, land use, rainfall, distance to roads, and distance to streams) were selected for this study based on the literature review and are referred to as causative factors hereafter (Fig. 13.2 shows thematic maps obtained via ArcGIS). Since slope governs the shear strength of the soil/sediment, the slope can

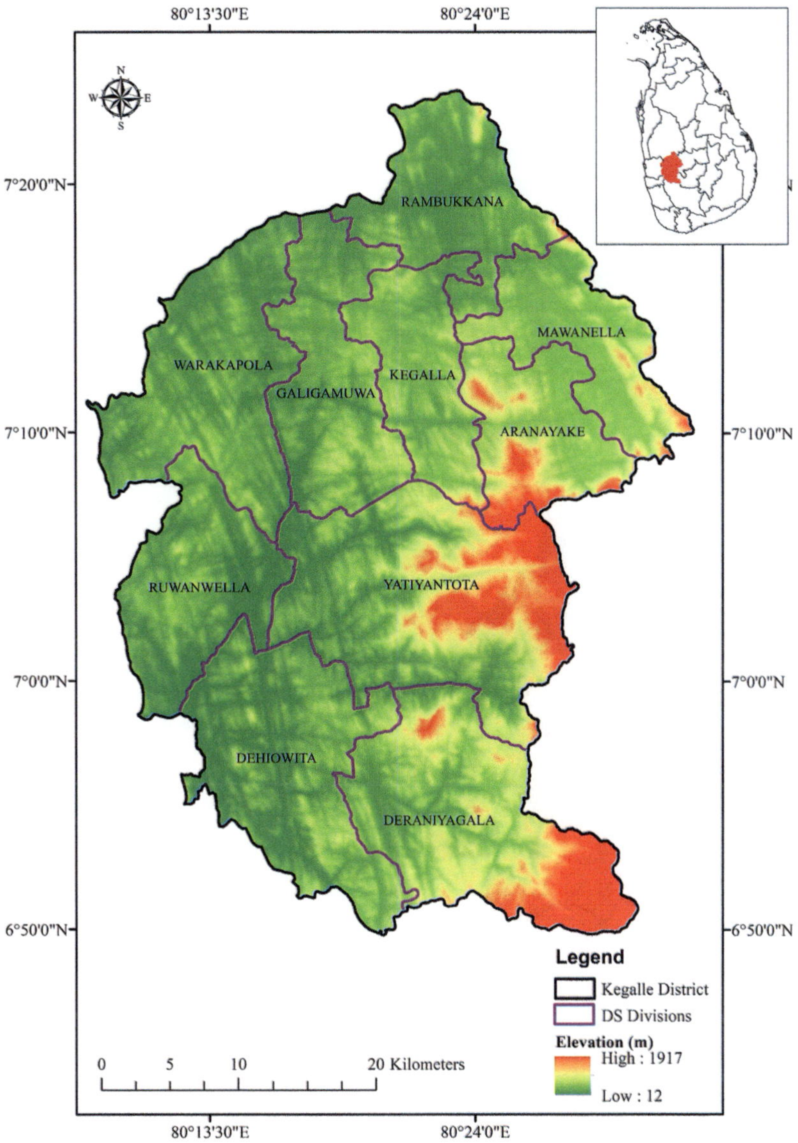

Fig. 13.1 Location of study area with general topographic features

be considered as one of the key factors that trigger landslides (Yalcin 2008). In general, the study area had a mix of mild slope (<10°) and steep slope (40°–50°) areas (Fig. 13.2a). Slope aspect, the direction of maximum slope of the terrain (Fig. 13.2b) is another one of the widely considered causative factors by many researchers, because some parameters related to aspect such as exposure to sunlight, drying winds, rainfall (degree of saturation), and discontinuities that may determine the occurrence of landslides could be attributed to aspect (Yalcin 2008).

Slope stability can be affected by soil physical characteristics such as particle size and pore dispersion in the soil matrix (Lee 2005; Schuster and Highland 2007). More than 80% of the study area is Orthic Acrisols soil, whereas the remainder is Ferric Acrisols (Fig. 13.2c). Geomorphological characteristics (created by geological processes) of an area have a major influence on landslides and are determined by the lithologic properties of that area (Yalcin 2008; Persichillo et al. 2017). As illustrated in Fig. 13.2d, the different types of lithological

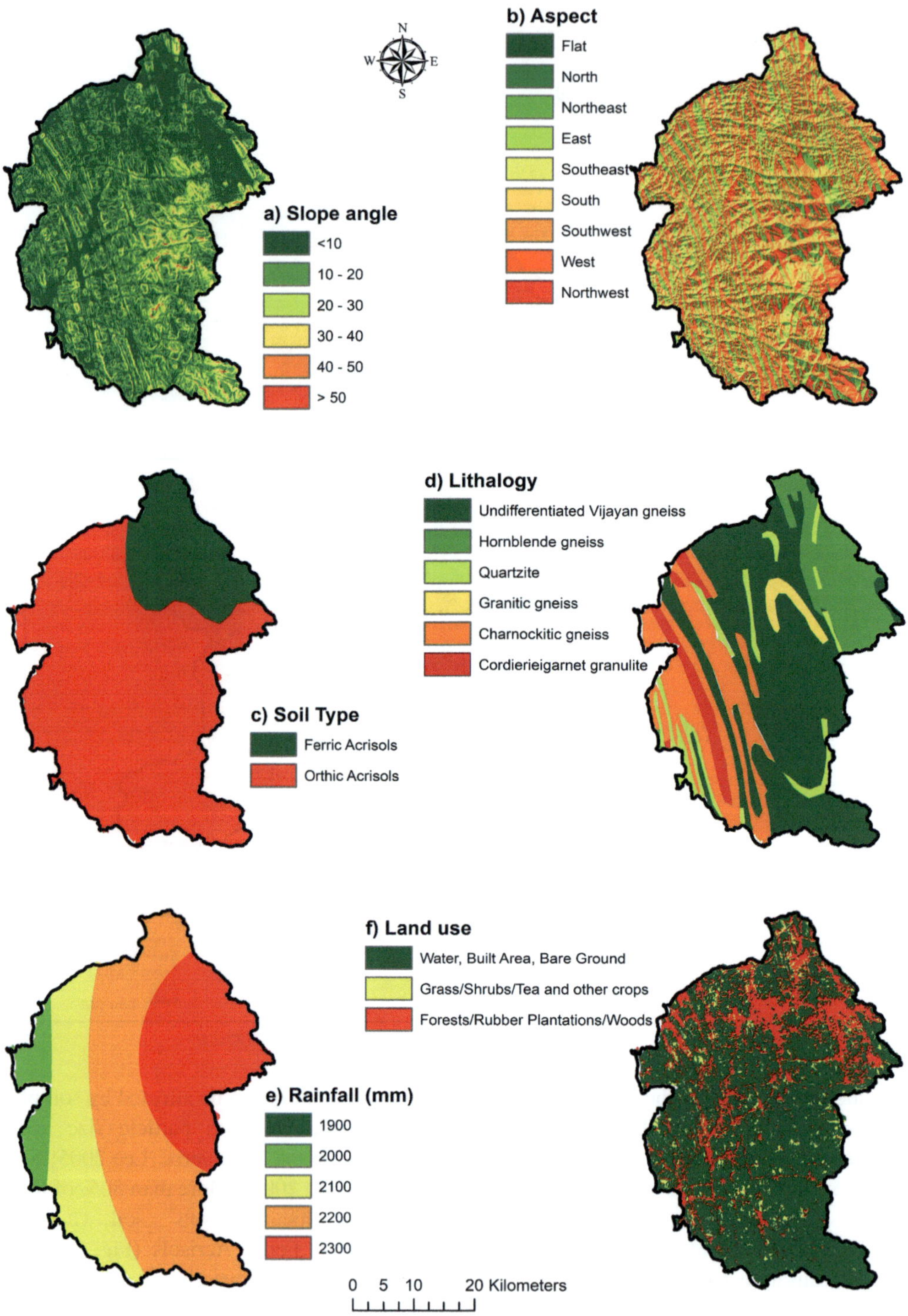

Fig. 13.2 Thematic layers of causative factors; **a** slope **b** aspect **c** soil class **d** lithology **e** precipitation **f** land use **g** distance to roads **h** distance to streams

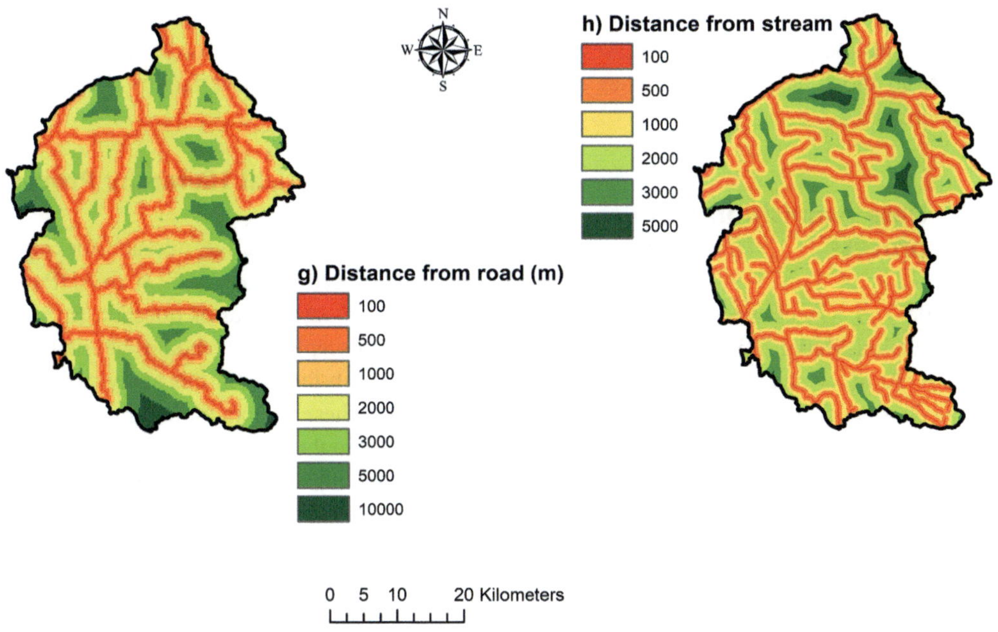

Fig. 13.2 (continued)

groupings in the study area were divided into seven basic classifications, namely, Biotic gneiss, Cordierieigarnet granulite, Charnockitic gneiss, Granitic gneiss, Quartzite, Hornblende gneiss, and undifferentiated Vijayan gneiss.

Many studies have also considered land use as one of the major causative factors of landslide. Human activities, such as settlements, agriculture, and transportation in hilly terrains can affect the slope stability (Jayasinghe et al. 2017). As shown in Fig. 13.2f, the South-Eastern quarter is mostly occupied by vegetation, whereas the remaining three quarters are a mix of residential areas and vegetation, an indication of human settlements and home gardens (homesteads).

Rainfall governs the soil pressure by increasing pore water pressure in soil. Rainfall also aids the mobility of soil mass on slopes (Panagos et al. 2017) . Most landslides are caused by the exceedance of a threshold of rainfall (Yalcin 2008). Spatial distribution of rainfall over the Kegalle district is shown in Fig. 13.2e. The observed highest yearly rainfall in the study area for the past 10 years was 2250 mm.

Roads change the topography of an area and reduce the shear strength of the slope's toe (Ayalew and Yamagishi 2005). In addition, the imperviousness of roads promotes water infiltration into slopes adjacent to roads increasing soil saturation leading to unstable slopes. Therefore, distance from roads can have significant influence on the slope stability of an area. Road buffer (Fig. 13.2g) had a close surrogation with the built areas thematic map other than the northern and southern tips. It should be noted that the northern tip has a densely built area. Streams could enhance the likelihood of landslides by increasing soil moisture, hence reducing the stability of the slope (Yalcin 2008). The fact that over 80% of any location is only 100 m away from a stream/river (Fig. 13.2h) is understandable considering the hilly geography and the influence of monsoonal rains. Such conditions result in a dense headwater stream network

(Gomes et al. 2020b). The influence of roads and streams in landslide phenomena decreases as the distance from those and the area under consideration increases.

13.2.3 Data Acquisition and Pre-process

Slope angle and aspect were derived from the Digital Elevation Model (DEM) with a resolution of 30 m × 30 m, which was downloaded from the United States Geological Survey (Earth-Explorer) website. FAO/UNESCO Soil Map of the World was utilized to extract the soil type map of the study area. Lithology and land use maps of Sri Lanka (scale 1:10,000) were obtained from the Department of Survey and the Department of Land use and Policy Planning of Sri Lanka. Annual rainfall data from eight rainfall gauging stations within and around the Kegalle district for the past 10 years were obtained from the Department of Meteorology Sri Lanka. Averaged rainfall values at each station were interpolated using Inverse Distance Weighted (IDW) method in ArcGIS to obtain rainfall distribution thematic layer (30 m × 30 m). Road network and stream network maps of Sri Lanka were obtained from the Department of Sri Lanka. Using the road and stream network vector files, buffer zones around each feature were generated. Buffer zones indicate the distance from each feature. Then all the vector files (soil type, land use, lithology, road, and stream buffer) were converted into raster files with 30 m × 30 m grids, in order to generate thematic layers of causative factors (Fig. 13.2).

13.2.4 Analytical Hierarchy Process

AHP is a multi-objective decision-making technique introduced by Saaty (1977). In AHP method, firstly, the research goal or question (in this case landslide susceptibility) should be first broken-down into sub-factors (in this study causative factors). Then these factors are arranged in a hierarchical order in the form of a matrix which contains the number of rows and columns equal to the number of factors (Basak et al. 2021). Afterward, the factors are systematically evaluated by comparing them to each other, two at a time, based on their relative importance in each other, commonly known as a pairwise comparison (Saaty 1977). In the pairwise comparison, the factors are assigned weights on a 9-point ordinal scale (Das et al. 2017), where 1 denotes equal importance of the two factors, and 9 denotes one factor is more significant than others. Table 13.1 lists the preferences indicated by the other digits in the scale. In the pairwise comparison matrix, weights along the diagonal are equal to 1, the weights recorded above the diagonal vary between 1 and 9, and conversely and the weights recorded below the diagonal varies between the reciprocals 1/2 and 1/9, indicating that a row factor is important than the corresponding columns factors. Once the matrix is built, the relative weights of the factors are derived by calculating the maximum

Table 13.1 Preferences indicated by each digit in the 9-point ordinal scale

Preference/Scale	Degree of preference	Remarks
1	Equally	Factors inherit equal contribution
3	Moderately	One factor moderately favored over other
5	Strongly	Judgment strongly favors one over other
7	Very strongly	One factor very strongly favored over other
9	Extremely	One factor favored over others in the highest degree
2, 4, 6, 8	Intermediate	Compensation between weighs 1, 3, 5, 7 and 9
Reciprocals	Opposite	Refers inverse comparison

or principal eigenvector of the matrix. Finally, the consistency ratio (CR) (Eq. 13.1) is calculated to check whether there is adequate consistency. If the CR < 0.1, then the pair-wise comparison is considered to be adequately consistent. However, if CR $\geq$ 0.1, then the pair-wise comparison should be repeated until CR is <0.1 (Saaty 1977). Where CI is the consistency index and RI is the random inconsistency index. CI can be computed by Eq. 13.2, where $\lambda_{\max}$ is the maximum eigenvalue and n is the number of factors.

$$CR = \frac{CI}{RI} \tag{13.1}$$

$$CI = \frac{\lambda_{\max} - 1}{n - 1} \tag{13.2}$$

Pairwise comparison is influenced by the decision maker's opinions and assumptions. In this study, the weights assigned in pairwise comparison (Table 13.2), were determined by sending questionnaires to different experts and through a literature review.

13.2.5 Validation of the Susceptibility Model

The landslide susceptibility model (LSM) of the study area was developed based on AHP results. The prediction accuracy of the LSM was analyzed using the Area Under the Curve (AUC) value of the Receiver Operating Characteristic Curve (ROC). The ROC curve plots TPR (Eq. 13.3) as a function of False Positive Rate (FPR) (Eq. 13.4) (Abeysiriwardana and Gomes 2022).

$$TPR = \frac{TP}{TP + FN} \tag{13.3}$$

$$FPR = \frac{FP}{TN + FP} \tag{13.4}$$

where TP is a true positive, TN is a true negative, FP is a false positive, and FN is a false negative. To plot ROC, firstly a landslide inventory was created. In the study area, over 40 landslide prone locations have been identified—including the major eight historical landslides which resulted in significant loss of lives and property. These locations were included in the landslide inventory as landslide points, while an approximately similar number of locations that have not recorded historical landslides were taken as non-landslide points in the landslide inventory. ROC for this study was then generated (Fig. 13.3) and AUC was obtained in ArcGIS. Model performances can be considered good if AUC is greater than 0.7—better if closer to 1 (Abeysiriwardana and Gomes 2022).

Table 13.2 Pairwise comparison matrix

Factor	Slope	Aspect	Soil	Lithology	Land use	Rainfall	Distance to roads	Distance to streams
Slope	1	3	4	5	6	7	8	8
Aspect	0.33	1	3	4	5	6	7	7
Soil	0.25	0.33	1	1	3	4	5	5
Lithology	0.20	0.25	1.00	1	3	4	6	6
Land use	0.17	0.20	0.33	0.33	1	5	6	6
Rainfall	0.14	0.17	0.25	0.25	0.20	1	3	3
Distance to roads	0.13	0.14	0.20	0.17	0.17	0.33	1	1
Distance to streams	0.13	0.14	0.20	0.17	0.17	0.33	1.00	1

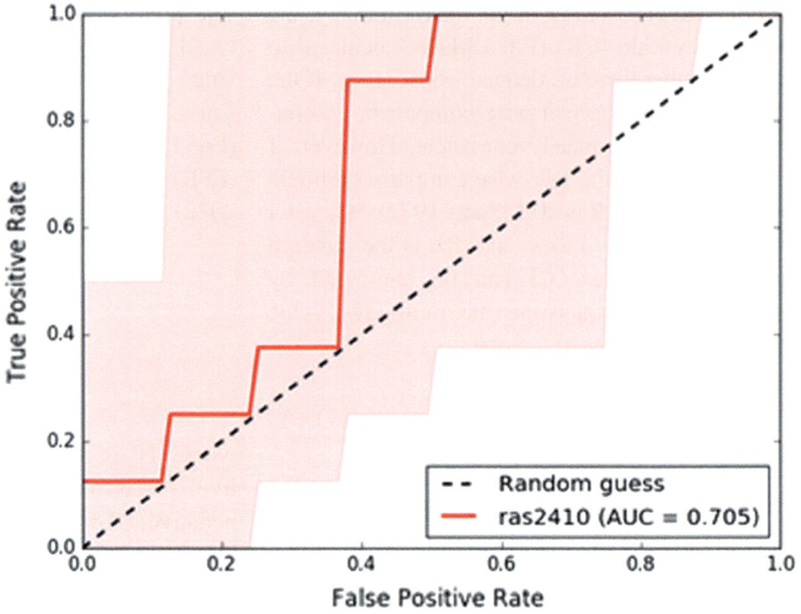

Fig. 13.3 Receiver operating curve

13.2.6 Landslide Susceptibility Modeling

All eight thematic layers of causative factors were reclassified into different subclasses using ArcGIS raster classification tool. The subclasses were ranked considering whether the causative factor's contributions to landslide occurrence is proportionate or inversely proportionate (Table 13.3), so that 1 indicates the least probable contribution and the highest digit indicates the highest probable contribution. Then the reclassified thematic layers were overlaid in ArcGIS based on the LSM to generate landslide susceptibility map of the study area, which was finally classified into five classes ranging from very high susceptibility to very low susceptibility.

13.3 Results and Discussion

Table 13.4 displays a summary of eigenvector calculations and final weightages for eight causative factors. The study gave a CR value of 0.032, thus indicating the pairwise comparison is acceptable. The analysis demonstrated that the AHP LSM has an AUC of 0.705 (Fig. 13.3) and prediction accuracy of 71%, thus proving the model is acceptable to be used in real-world applications.

Landslide susceptibility model (LSM) developed for the study area is given in Eq. 13.5, where S is the slope angle, A is the aspect, Sl is the soil type, L is the lithology, LU is land use, RF is rainfall, d_r is the distance to roads, and d_s is the distance to streams.

$$\begin{aligned}\text{LSM} = {} & 0.354S + 0.229A + 0.116\text{Sl} + 0.118L + 0.091\text{LU} \\ & + 0.043\text{RF} + 0.024d_r + 0.024d_s\end{aligned} \tag{13.5}$$

Figure 13.4 shows the landslide susceptibility map with zones corresponding to susceptibilities ranging from very low, to very high. A landslide was recorded on November 9, 2021, in the Dombemeda area in the Rambukkana District Secretariat (DS) division with four fatalities. In fact, this was an area predicted as very highly susceptible by our model, which was finalized in July 2021. This landslide occurrence further validated the prediction power of the model, indicating that the model can be applied for landslide susceptibility analysis in Kegalle

Table 13.3 The causative factors and relative ranks assigned to subclasses

Major factor	Subclasses	Rank
Slope	<10°	1
	10°–20°	2
	20°–30°	3
	30°–40°	4
	40°–50°	5
	>50°	6
Aspect	Flat	1
	North	2
	Northeast	3
	East	4
	Southeast	5
	South	6
	Southwest	7
	West	8
	Northwest	9
Soil class	Orthic Acrisols	1
	Ferric Acrisols	2
Lithology	Cordierieigarnet granulite	6
	Charnockitic gneiss	5
	Granitic gneiss	4
	Quartzite	3
	Hornblende gneiss	2
	Undifferentiated Vijayan gneiss	1
Mean annual rainfall (mm)	1900	1
	2000	2
	2100	3
	2200	4
	2300	5
Land use	Water/built area/bare ground	3
	Grass/shrubs/tea and other crops	2
	Forests/rubber plantations/woody area	1
Distance from streams (m)	100	6
	100–500	5
	500–1000	4
	1000–2000	3
	2000–3000	2
	3000–5000	1

(continued)

Table 13.3 (continued)

Major factor	Subclasses	Rank
Distance from roads (m)	100	7
	100–500	6
	500–1000	5
	1000–2000	4
	2000–3000	3
	3000–5000	2
	5000–10,000	1

Table 13.4 Normalized pairwise comparison matrix (Judgment matrix)

Factor	*S*	*A*	Sl	*L*	LU	*RF*	d_r	d_s	Priority vector	Weight (*W*) (%)
S	0.427	0.573	0.401	0.420	0.324	0.253	0.216	0.216	0.354	35
A	0.142	0.191	0.301	0.336	0.270	0.217	0.189	0.189	0.229	22
Sl	0.107	0.064	0.100	0.084	0.162	0.145	0.135	0.135	0.116	12
L	0.085	0.048	0.100	0.084	0.162	0.145	0.162	0.162	0.118	13
LU	0.071	0.038	0.033	0.028	0.054	0.181	0.162	0.162	0.091	9
RF	0.061	0.032	0.025	0.021	0.011	0.036	0.081	0.081	0.043	5
d_r	0.053	0.027	0.020	0.014	0.009	0.012	0.027	0.027	0.024	2
d_s	0.053	0.027	0.020	0.014	0.009	0.012	0.027	0.027	0.024	2

where *S*—Slope, *A*—Aspect, Sl—Soil class, *L*—Lithology, LU—Land use, RF—Rainfall, d_r—Distance to roads, and d_s—Distance to streams

District with high confidence. Also, this is an indication that proper communication between researchers such as us and government agencies is a must so that the government can take precautionary action (it should be noted government agencies probably have done similar risk assessments, mainly by in-situ assessments covering all landslide prone areas).

As for the share of landslide susceptibility classes (Table 13.5), the lowest share was recorded in the very high (4%) and very low (16%) classes. The moderate and low susceptible zones account for 39% and 28% of the study area, respectively. About 14% of the total study area has been marked as highly susceptible. In fact, 18% of the territory of Kegalle District is characterized by very high to high landslide susceptibility, while 83% of the District is identified as very low to moderate landslide susceptible areas.

Furthermore, the landslide map was intersected with the land use map to investigate the correspondence of susceptibility classes with land use categories. The share of each DS division and each land use type of their total areas corresponding to individual landslide susceptibility class are given in Tables 13.6 and 13.7, respectively. Slightly more than 6% of Aranayaka, Dehiowita, and Ruwanwella DS divisions have been identified as very high susceptible areas, while 17% of Dehiowita, and Ruwanwella DS divisions have been marked as high susceptible areas, indicating that the two DS divisions should be given extra attention in preparing for future landslide phenomena.

However, the high-risk areas should be prioritized in any disaster management plan. Thus, the study paid special focus to residential and agricultural areas.

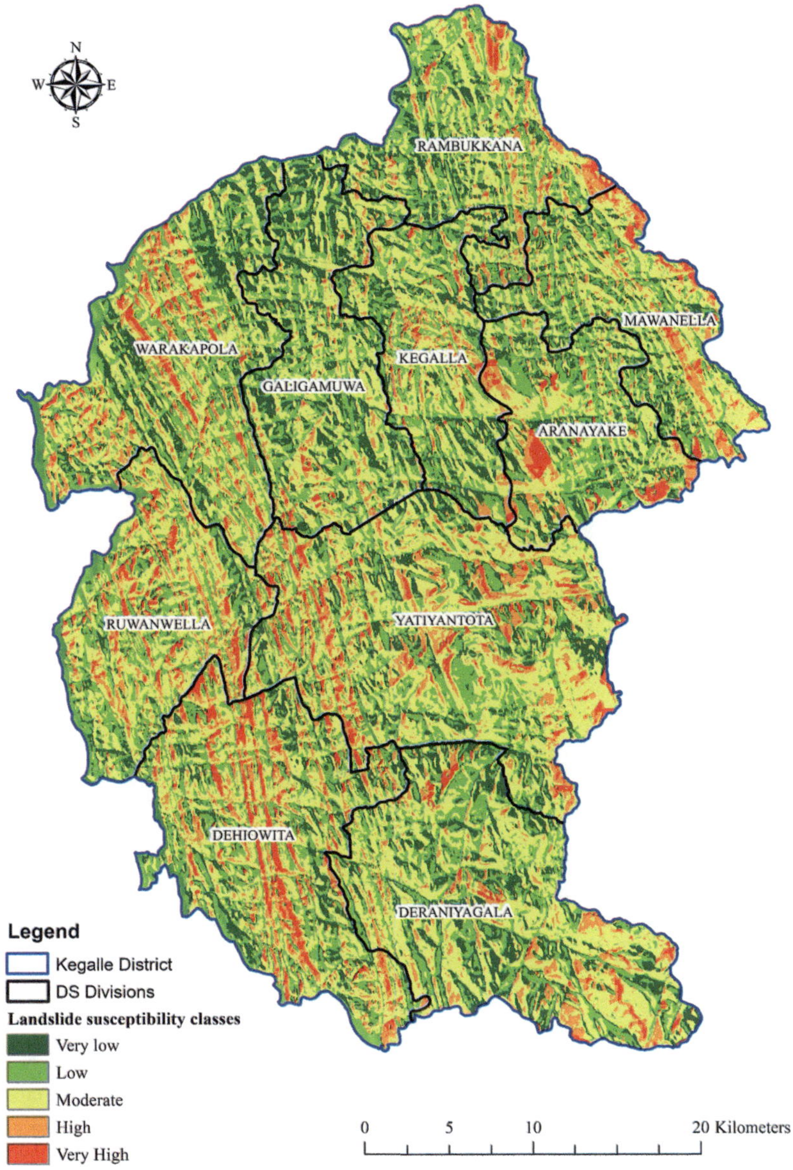

Fig. 13.4 Landslide susceptibility map of Kegalle district

Table 13.5 Area coverage of flood susceptibility classes

Susceptibility class	Area (km^2)	% Area
Very low	254.78	16
Low	451.67	28
Moderate	632.48	39
High	219.43	14
Very high	65.19	4
Total study area	1623.54	

Table 13.6 Area share of DS divisions corresponding to the individual landslide susceptibility class

Name of the DSD	Total area of DSD (km^2)	% Area of each DS division that belongs to each susceptibility class				
		Very low	Low	Medium	High	Very high
Rambukkana	130.18	16	30	38	12	3
Mawanella	114.95	14	27	39	15	4
Aranayake	114.07	16	30	35	13	6
Kegalla	108.82	18	31	37	11	2
Galigamuwa	148.58	25	29	39	7	1
Warakapola	184.48	16	29	37	14	4
Ruwanwella	131.46	8	30	40	17	6
Yatiyantota	275.09	13	25	42	15	5
Deraniyagala	217.62	19	26	42	11	2
Dehiowita	198.28	12	27	36	17	8

Table 13.7 Area share of each LU type corresponding to the individual landslide susceptibility class

Land use type	Total area of LU type (km^2)	Area % of the study area	% of each LU type that corresponds to individual susceptibility classes				
			Very low	Low	Medium	High	Very high
Built area/Homestead	317.70	20	12	26	40	16	6
Forests/rubber plantations/woody area	1169.27	72	17	28	38	13	4
Tea and other crops	33.51	2	15	24	47	12	2
Grass/shrubs/water bodies/bare lands	103.06	6	14	28	41	13	5

Table 13.8 Composition of susceptibility classes in terms of LU type

Susceptibility class	Built area/Homestead	Forests/Rubber plantations/Woody area	Tea and other crops	Grass/Shrubs
Very low	15	80	2	3
Low	18	77	2	3
Medium	20	74	3	3
High	24	71	2	3
Very high	27	68	1	4

As per Table 13.7, the total area of homestead land cover type, corresponding to the high and very high susceptibility class was 16% and 6%, respectively. As for the agricultural areas (tea and rubber), their share in the high susceptibility class is about 13%, while in the case of the very high susceptibility, they are around 3–4%. Table 13.8 shows the land use wise composition of each landslide susceptibility class. About 27%, 24%, and 20% of the very high, high, and moderate susceptibility, respectively, comprise residential settlements. More than 65% of the high and very high susceptibility classes are covered by forest/rubber plantation areas. The aforementioned statistics emphasized the need for proper preparedness for future landslide events, as a significant share of residential and agricultural areas have been identified as high to very high susceptible areas, indicating the socio-economic threat in case of future landslide events.

Notwithstanding the high accuracy, as the AHP method produces biased outputs based on decision maker's understanding and opinions, the model could be further improved by employing statistical methods, which are not biased, to establish relationship between causative factors and landslide occurrences.

13.4 Conclusion

Considering the fatalities and economic losses, it is obvious that a prediction of landslides is vital. Nevertheless, due to a large number of associated factors, their spatial–temporal combinations, and difficulties associated with quantifying the magnitude of factors, a 100% accurate or even close to 100% accurate model is impossible. In that regard, the model we developed using AHP can be considered a strong one since its accuracy is 71%. This is particularly advantageous since we used a standard set of landslide prediction factors, where data can be easily collected.

An occurrence of a landslide soon after the study, gave a chance to validate the model since it happened in one of the highly susceptible areas demarcated by the model. The developed landslide susceptibility map can be used as an aid to prevent future landslide hazards in the Kegalle district by decision-makers as well as geospatial scientists. It is recommended that the changes in geographical and geological characteristics of the mentioned DS divisions and the likelihood of hazards in the future be studied. Also, as a future study, we recommend incorporating further breakdown and fine-tuning of factors to make the prediction more accurate. As an example, instead of the generic factor of built areas, the type of built environment that shows the stress to the environment (e.g., construction with shallow foundations creates less disturbance to the ground than construction that involves basement construction) should be considered. Nevertheless, we would like to conclude that a model that incorporates a standard set of variables can also be used to predict landslide susceptibility with good accuracy. Thus, until fine-tuned super accurate models come, such should be used in disaster prevention and mitigation.

Acknowledgements This research was funded by a research grant by Faculty of Graduate Studies and Research, Sri Lanka Institute of Information Technology (FGSR/RG/FE/2021/12).

References

Abeysiriwardana HD, Gomes PIA (2022) Integrating vegetation indices and geo-environmental factors in GIS-based landslide-susceptibility mapping: using logistic regression. J Mt Sci 19(2):477–492. https://doi.org/10.1007/s11629-021-6988-8

Ayalew L, Yamagishi H (2005) The application of GIS-based logistic regression for landslide susceptibility mapping in the Kakuda-Yahiko Mountains, Central Japan. Geomorphology 65(1–2):15–31. https://doi.org/10.1016/j.geomorph.2004.06.010

Basak A, Das J, Rahman ATM, Pham QB (2021) An integrated approach for delineating and characterizing groundwater depletion hotspots in a coastal state of India. J Geol Soc India 97(11):1429–1440. https://doi.org/10.1007/s12594-021-1883-z

Bloechl A, Braun B (2005) Economic assessment of landslide risks in the Swabian Alb, Germany-research framework and first results of homeowners' and experts' surveys. Nat Hazard 5(3):389–396. https://doi.org/10.5194/nhess-5-389-2005

Dang K, Sassa K, Konagai K, Karunawardena A, Bandara RM, Hirota K, Tan Q, Ha ND (2019) Recent rainfall-induced rapid and long-traveling landslide on 17 May 2016 in Aranayaka, Kagelle district, Sri Lanka. Landslides16(1):155–164. https://doi.org/10.1007/s10346-018-1089-7

Das J, Gayen A, Saha S, Bhattacharya SK (2017) Modelling of alternative crops suitability to tobacco based on analytical hierarchy process in Dinhata subdivision of Koch Bihar district, West Bengal. Model Earth Syst Environ 3(4):1571–1587. https://doi.org/10.1007/s40808-017-0392-y

Gomes PIA, Aththanayake U, Deng W, Li A, Zhao W, Jayathilaka T (2020a) Ecological fragmentation two years after a major landslide: correlations between vegetation indices and geo-environmental factors. Ecol Eng 153:105914. https://doi.org/10.1016/j.ecoleng.2020.105914

Gomes PIA, Wai OW, Dehini GK (2020b) Vegetation dynamics of ephemeral and perennial streams in mountainous headwater catchments. J Mt Sci 17 (7):1684–1695. https://doi.org/10.1007/s11629-017-4640-4

Guzzetti F, Mondini AC, Cardinali M, Fiorucci F, Santangelo M, Chang KT (2012) Landslide inventory maps: New tools for an old problem. Earth Sci Rev 112(1–2):42–66. https://doi.org/10.1016/j.earscirev.2012.02.001

Hemasinghe H, Rangali RS, Deshapriya NL, Samarakoon L (2018) Landslide susceptibility mapping using logistic regression model (a case study in Badulla District, Sri Lanka). Procedia Eng 212:1046–1053. https://doi.org/10.1016/j.proeng.2018.01.135

Hong H, Pradhan B, Sameen MI, Chen W, Xu C (2017) Spatial prediction of rotational landslide using geographically weighted regression, logistic regression, and support vector machine models in Xing Guo area (China). Geomat Nat Haz Risk 8(2):1997–2022. https://doi.org/10.1080/19475705.2017.1403974

https://doi.org/10.1007/s10346-018-1089-7

https://doi.org/10.1016/j.ecoleng.2020.105914

https://doi.org/10.1016/j.geomorph.2006.07.019

https://doi.org/10.1016/j.scitotenv.2016.09.125

https://doi.org/10.1088/1755-1315/540/1/012084

https://doi.org/10.2113/gseegeosci.13.1.25

Jayasinghe GJMSR, Wijekoon P, Gunatilake J (2017) Landslide susceptibility assessment using statistical models: a case study in Badulla district, Sri Lanka. Ceylon J Sci 46(4):26–41. https://doi.org/10.4038/cjs.v46i4.7466

Kaleel MIM, Reeza MJ (2017) The impact of landslide on environment and socio-economy: GIS based study on Badulla district in Sri Lanka. World Sci News 88 (2):69–84. https://dx.doi.org/10.2139/ssrn.3455589

Kumar R, Anbalagan R (2016) Landslide susceptibility mapping using analytical hierarchy process (AHP) in Tehri reservoir rim region, Uttarakhand. J Geol Soc India 87(3):271–286. https://doi.org/10.1007/s12594-016-0395-8

Lee SARO (2005) Application of logistic regression model and its validation for landslide susceptibility mapping using GIS and remote sensing data. Int J Remote Sens 26(7):1477–1491. https://doi.org/10.1080/01431160412331331012

Nawagamuwa UP, Perera LP (2017, May) Recommending rainfall thresholds for landslides in Sri Lanka. In: Workshop on world landslide forum. Springer, Cham, pp 267–272. https://doi.org/10.1007/978-3-319-53483-1_31

Quan HC, Lee BG (2012) GIS-based landslide susceptibility mapping using analytic hierarchy process and artificial neural network in Jeju (Korea). KSCE J Civ Eng 16(7):1258–1266. https://doi.org/10.1007/s12205-012-1242-0

Panagos P, Borrelli P, Meusburger K, Yu B, Klik A, Jae Lim K, Yang JE, Ni J, Miao C, Chattopadhyay N, Sadeghi SH (2017) Global rainfall erosivity assessment based on high-temporal resolution rainfall records. Scien Rep 7(1):1–12.https://doi.org/10.1016/j.catena.2015.09.015

Persichillo MG, Bordoni M, Meisina C (2017) The role of land use changes in the distribution of shallow landslides. Sci Total Environ 574:924–937. https://doi.org/10.1016/j.scitotenv.2016.09.125

Petley DN, Dunning SA, Rosser NJ (2005) The analysis of global landslide risk through the creation of a database of worldwide landslide fatalities. In: Landslide risk management. CRC Press, pp 377–384. https://doi.org/10.1201/9781439833711-18

Ratnayake U, Herath S (2005) Changing rainfall and its impact on landslides in Sri Lanka. J Mt Sci 2(3):218–224. https://doi.org/10.1007/bf02973195

Sarkar S, Kanungo DP (2004) An integrated approach for landslide susceptibility mapping using remote sensing and GIS. Photogram Eng Remote Sens 70(5):617–625. https://doi.org/10.14358/pers.70.5.617

Saaty TL (1977) A scaling method for priorities in hierarchical structures. J Math Psychol 15(3):234–281. https://doi.org/10.1016/0022-2496(77)90033-5

Schuster RL, Highland LM (2007) Overview of the effects of mass wasting on the natural environment. Environ Eng Geosci 13(1):25–44. https://doi.org/10.2113/gseegeosci.13.1.25

Singh A, Singh T, Nagarajan R (2017, Nov) Impact of landslides on environment. In: Euro-Mediterranean conference for environmental integration. Springer, Cham, pp 1869–1871. https://doi.org/10.1007/978-3-319-70548-4_540

Yaa'cob N, Abd Rashid ZNA, Tajudin N, Kassim M (2020, July) Landslide possibilities using remote sensing and geographical information system (GIS). IOP Conf Ser: Earth Environ Sci 540(1):012084. https://doi.org/10.1088/1755-1315/540/1/012084

Yalcin A (2008) GIS-based landslide susceptibility mapping using analytical hierarchy process and bivariate statistics in Ardesen (Turkey): comparisons of results and confirmations. Catena 72(1):1–12. https://doi.org/10.1016/j.catena.2007.01.003

Landslide Susceptibility Evaluation and Analysis: A Review on Articles Published During 2000 to 2020

14

Jonmenjoy Barman, David Durjoy Lal Soren, and Brototi Biswas

Abstract

Landslides are man-induced or natural triggering natural hazards with large-scale environmental and socioeconomic impact. From the last decade, landslide susceptibility zonation, using quantitative and qualitative techniques, is an interesting area of interest among scholars. The purpose of the present study was to analyze and review the trend of the published articles, methodologies adopted, and the area of study of the published articles during 2000 to 2020. The result of the review revealed that among the various methodologies adopted, machine learning and logistic regression were the maximum implemented, and south and southeast Asian countries are the most landslide-prone areas. The development of remote sensing and GIS has played a significant role in data gathering, analysis, visualization, and identification of landslide susceptible zones for proper monitoring. Knowledge-based study like the geographically weighted overlay method is much applicable in the northeastern states of India. Researchers emphasize on slope angle, topographic wetness index, and land use/land cover as important conditioning factors in landslide occurrence. The study would be helpful for the researchers to choose study areas, methodologies, and preferable journals for publishing their research.

Keywords

Landslide · Susceptibility · Machine learning · RS · GIS

14.1 Introduction

Landslide or landslip is a geological hazard that commonly happens in mountainous areas due to high energy and instability of mass (Roy and Saha 2019). The landslide has a complex mechanism, mainly driven by geomorphology, topography, geology, and seismic events (Pourghasemi et al. 2018). These factors are broadly divided into two groups such as causative factors and triggering factors. Factors like lithology, slope angle, altitude, aspect, faults, land use, drainage density, and soil are known as causative factors, while human interventions, earthquakes, precipitations are known as triggering factors (Pourghasemi et al. 2018). Hundreds of billions of properties have been damaged by landslides as well affecting about 1.5 million people in the world (Sterlacchini 2011; Chen and Chen 2021). Susceptibility, hazard,

J. Barman · D. D. L. Soren (✉) · B. Biswas
Department of Geography and RM, Mizoram University, Aizawl 796004, India
e-mail: devid.dls.king@gmail.com

J. Das and S. K. Bhattacharya (eds.), *Monitoring and Managing Multi-hazards*,
GIScience and Geo-environmental Modelling, https://doi.org/10.1007/978-3-031-15377-8_14

Table 14.1 Approaches of landslides study

Models	Features	Details
Physical based	Quantitative	Emphasis on landslide failure mechanism
Knowledge based	Qualitative	Emphasis on landslide conditioning factors and their weights
Data based	Qualitative	Emphasis on geoenvironmental characteristics of landslide

and risk mapping are the three steps for landslide analysis. Terminologies such as landslide susceptibility mapping, landslide susceptibility assessment had newly arrived in the literature regarding landslide. Landslide susceptibility mapping is the potentiality of spatial landslide occurrence of known slope slide, a set of given geoenvironmental conditions including historical landslide sites mapping (Guzzetti et al. 2006; Merghadi et al. 2020), while landslide susceptibility assessment is temporal and spatial perdition of landslide and landslide susceptibility map preparation (Chen and Chen 2021). Analysis of landslides is a hierarchical process consisting of susceptibility, possibility, and risk as expressed in formulas (14.1–14.3) (Lee and Min 2001).

$$\text{Susceptibility} = f(\text{Landslide-related factors, Landslide}) \tag{14.1}$$

$$\text{Possibility} = f(\text{impact factors, susceptibility}) \tag{14.2}$$

$$\text{Risk} = f(\text{Damageable objects, possibility}) \tag{14.3}$$

Although natural hazard like landslide cannot be fully mitigated, a suitable understanding of scientific methodologies could be an important tool for reducing vulnerability (Pourghasemi et al. 2018). Methods for working landslide susceptibility mainly can be divided into three types: physical-based, knowledge-based, and data-based methods (Table 14.1) (Huang and Zhao 2018). Identification of susceptible zones is the first step to mitigate any hazard. During the past decade, satellite data-based landslide susceptibility modeling has increased due to development in the field of remote sensing & GIS (Huang and Zhao 2018). To analysis landslide susceptibility, a total of 201 relevant published articles during the last 20 years (2000–2020) were categorized according to publishing year, models and methodologies used, study area, etc. Although a number of literature reviews on landslide susceptibility were attempted by Segoni et al. (2018), Budimir et al. (2015), Aleotti and Chowdhury (1999) and Kanungo et al. (2012), however the present study is different from the previous since the previous studies focused on a specific model or region, while the present study is an overall study on landslide susceptibility evolution and analysis.

14.2 Materials and Methods

At the onset, a total of 201 articles published during 2000 to 2020 were collected from Google Scholar using searching keyword landslide susceptibility. Articles were then categorized according to their publication year, i.e., from 2000 to 2020. A quantitative and qualitative literature review has been done to understand the temporal and spatial changes. A database was then created for all the review papers categorizing them into years of publication, methodologies, name of journals, and study area.

14.3 Results and Discussion

14.3.1 Temporal Trend of Published Articles

Out of the total of 201 review papers from 2000 to 2020, highest number of articles were published in 2019 (14%) followed by 2020 (11%), 2018 and 2010 (7%), 2012 (7%), and 2016 (6%) details in Fig. 14.1.

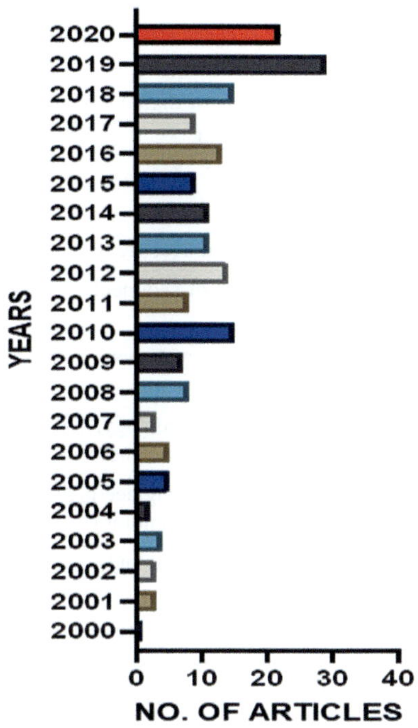

Fig. 14.1 Temporal trend of journals publication

14.3.2 Spatial Trend of Published Articles

The 201 research articles taken into consideration encompassed 46 countries. Almost 55% of the articles represents works from 8 countries, with India ranking first (19%) followed by China (10%), South Korea (6%), and Iran (5%), respectively. Furthermore, it was found that most of the authors had taken their study area from the southeast and south Asian monsoonal climatic region (India, China, and South Korea); revealing rainfall is an important triggering factor for landslides details in Fig. 14.2.

14.3.3 The Trend of Publishing Journals

From 2000 to 2020, a total of 79 journals have published articles on landslide susceptibility. *Geomorphology, Natural hazard, Engineering geology, Environmental earth science,* and *Landslide* have published more than 10 articles each and account for 38% of the total published articles. *Computer and geoscience, Journal of the Geological Society of India, Environmental geology, Natural hazard and earth system science, Catena, Environmental modeling and software, Geocarto international, Bulletin of engineering geology,* and *The Environment* were the most trending journals giving priorities about publishing articles on landslide susceptibility (Table 14.2).

Fig. 14.2 Spatial distribution of study areas

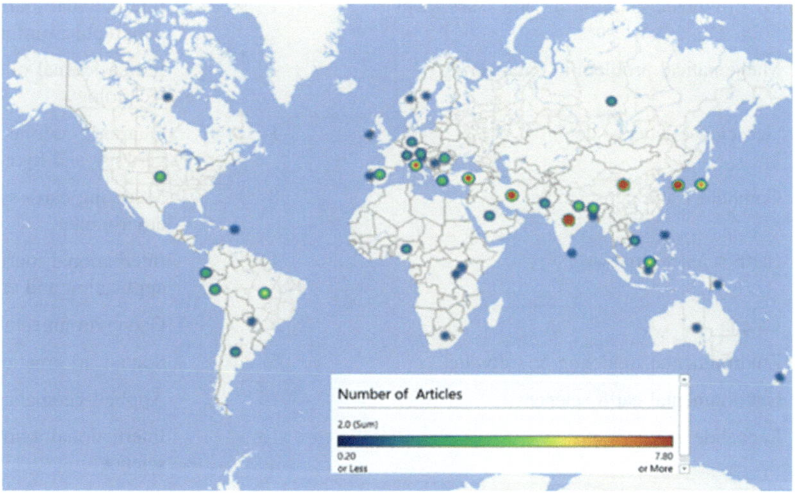

Table 14.2 Published articles in different journals

Journal	Number	Journal	Number
Geomorphology	16	International journal of digital earth	1
Environmental geology	4	Civil Engineering and environmental system	1
Natural hazard	19	Forest	1
Engineering geology	12	Journal of Zhejiang university	1
Earth surface processes and landforms: the journal of the British geomorphological research group	1	International journal of sediment research	1
International Journal of remote sensing	1	Journal of the society of remote sensing	2
Science of total environment	1	International journal of geology, earth and environmental sciences	1
Engineering with computers	1	International journal of engineering sciences and research technology	2
Journal of the geological society of India	3	Environment, development and sustainability	4
Environmental disasters	1	International journal of engineering and technical research	1
Entropy	1	Nature, environment and pollution technology	1
Palynology	1	Thematic journal of geography	1
Geoscience journal	1	Geotechnical and geological engineering	2
Journal of soils and sediments	1	Journal of applied geophysics	1
Natural hazard and earth system sciences	4	Sustainability	1
Canadian geotechnical journal	1	Theoretical and applied climatology	1
Geojournal	1	Innovative infrastructure solution	1
Annals of GIS	1	Remote sensing of environment	1
Science China earth sciences	1	Spatial information research	1
Environmental monitoring and assessment	1	International journal of geographical information science	1
Mathematical problem in engineering	1	Indian journal of science and technology	1
Geoscience letters	1	Quarterly journal of engineering geology and hydrology	1
Computers and geosciences	8	Modeling earth system and environment	2
Earth science reviews	1	International journal of computer application and engineering	1
Catena	5	Geoenvironmental disasters	1
Environmental modeling & software	5	Spatial information research	1
Environmental earth sciences	15	Applied geomatics	1
Landslide	13	International journal of disaster risk science	1
Mathematical geosciences	1	Advance civil engineering	1

(continued)

Table 14.2 (continued)

Journal	Number	Journal	Number
Hamburger Beitragezur Physischen Geographie Und Landschaftsokologie	1	Water	1
Big earth data	1	Geoscience	2
Geocarto international	6	In landslide science for a safer geoenvironment	2
Bulletin of engineering geology and the environment	5	Arabian journal of geoscience	3
International journal of applied earth observation and geoinformation	1	The Egyptian journal of remote sensing and space science	1
Procedia engineering	2	In transportation soil engineering in cold region	1
Remote sensing	2	Journal of maps	1
Journal of mountain science	2	Expert system with application	1
Remote sensing and multi-criteria decision analysis: in-publication for urban planning and development	1	In 2020 seventh international conference on edemocracy and government	1
In spatial modeling in GIS and R for earth and environmental sciences	1		

14.3.4 Trending of Methodology

A total of 77 types of methodologies have been used by the authors; among them, logistic regression ranked first (31) (Kincal et al. 2009; Samia et al. 2018) followed by the frequency ratio (28) (Mind'je et al. 2019; Silalahi et al. 2019), support vector machine (21) (Marjanović et al. 2011), artificial neural networks (20) (Bragagnolo et al. 2020), GIS weighted overlay (18) (Pachuau 2019), and analytical hierarchy processes (17) (Ahmed 2015; Basu and Pal 2017), respectively. Of the various articles, authors used single methodology in 92 research works, while the rest of the works were comparative analysis among various methodologies (Fig. 14.3). Some of the works focused on validation checking, chorology of landslide susceptibility, correlation between topography and geological structure for landslide susceptibility, etc. Among single methodology used, maximum authors have chosen GIS weighted overlay (15). Intergrade methodologies like neuro-fuzzy and fuzzy-AHP have taken a footprint on recent trending methodologies as well as machine learning methodologies. Support vector machine is the highest trending machine learning method followed by artificial neural networks (20) and random forest, which covered 72% of the total articles as represented in Table 14.3.

Except the traditional methodologies, some researchers have discussed about the correlation between geological structure and topography with landslide susceptibility, geomorphological and structural features extraction of landslide susceptibility.

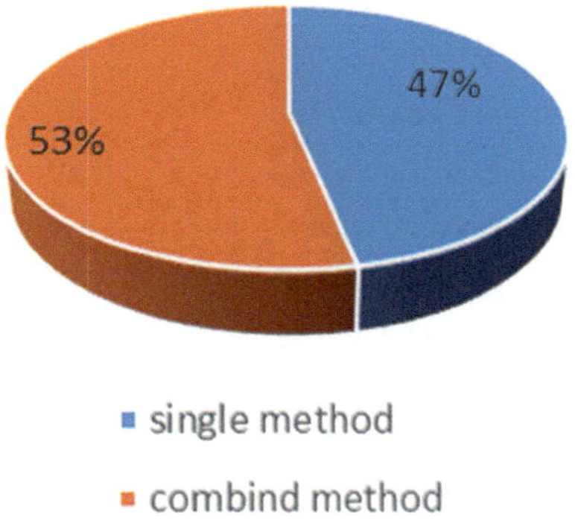

Fig. 14.3 Methodology involved for analysis

Table 14.3 Methodologies emphasized by the researchers

Name of method	Number of articles	Name of method	Number of articles
Frequency ratio	28	Temporal probability	1
Keefer method	1	Convolutional neural network	2
Harp and Noble method	1	Recurrent neural network	1
Analytical hierarchy method	17	TRIGRS	1
GIS Weighted overlay	18	Neuro-fuzzy	1
Weight of evidence	16	AdaBoost	2
Information value	10	Satellite sar interferometry	1
Index of entropy	9	Fuzzy-Gama	2
Yule index	1	Factor analysis	1
Distance distribution analysis	1	Landslide rupture hypothesis	1
Probability of landslide occurrence	1	Stability index mapping	1
Binary logistic regression	3	Support vector machine	21
Bayesian theory	7	Decision tree	11
SINMAP model	3	Multi-layer perception neural network	1
Adaptive neuro-fuzzy interface system	2	Kernel logistic regression	1
Random forest	15	Artificial neural network	20
Fuzzy membership	9	Reduced error pruning trees	1
Relative effect	1	Bagging	5
Logistic regression	31	Multi-boost	4
Evidential belief function	3	Rotation forest	6
Classification and regression tree	1	Random subspace	3
GPS and GRP	1	Dempster-Shafer	4
RSAGA package	1	Stability index mapping	1
Quantitative heuristic method	1	Certainty factor	6
Landslide density	1	Statistical index	6
Like hood ratio	4	Fuzzy-AHP	2
Discriminant analysis	5	Kernel logistic regression	1
Slope model	1	Classification and regression tree	1
Alternative decision tree	2	Logistic model tree	1
Weighted linear combination	1	Empirical conditional probability	3
Ordered weighted averaging	1	Newmark's method	1
Landslide relative frequency	1	Naïve Bayes	1
Forest canopy density model	1	MLP neural networks	1
Landslide susceptibility index	1	Functional tree	1
General linear model	2	Structural and geomorphological feature extraction	1
Multi-layer perception	1	Shallow landslide stability (SHALSTAB)	2

(continued)

Table 14.3 (continued)

Name of method	Number of articles	Name of method	Number of articles
Geographically weighted regression	2	Rainfall I-D threshold	1
Interactive back analysis and sensitivity	1	Root cohesion	1
Spatial regression	1	–	–

14.3.5 Authors

Of the 201 articles considered for review, 10 papers were mono-authored, and the rest have poly authorship. Most of the authors published one paper, while only 6 authors published above 5 papers, as viewed in Fig. 14.4b. Lee S has the highest number of published papers (20) with co-authors.

14.4 Summary of Reviews

The work Landslide Susceptibility Evaluation and Analysis conducted from 2000 to 2020 around the world. Various methods applied to evaluate landslide susceptibility evaluation, Broeckx et. al. (2019), Ayalew and Yamagishi (2005) used logistic regression. The result was very satisfactory occupied with 0.94% accuracy level. Frequency ratio method also widely used with attending accuracy level of 0.73% (Jana et al. 2019), to assess the landslide slope is argued as most influencing factor then other (Mind'je et al. 2019). Singh et al. (2020) conducted one study in Indian sub-continent; there investigation shows that less vegetated and barren land is most influencing indicator than TWI, lineament density, geomorphology, and slope, respectively. Artificial neural network (ANN) also widely used method to landslide assessment with commendable accuracy level (AUC= 0.84%) (Chauhan et al. 2010). Landslide study also conducted with one single method

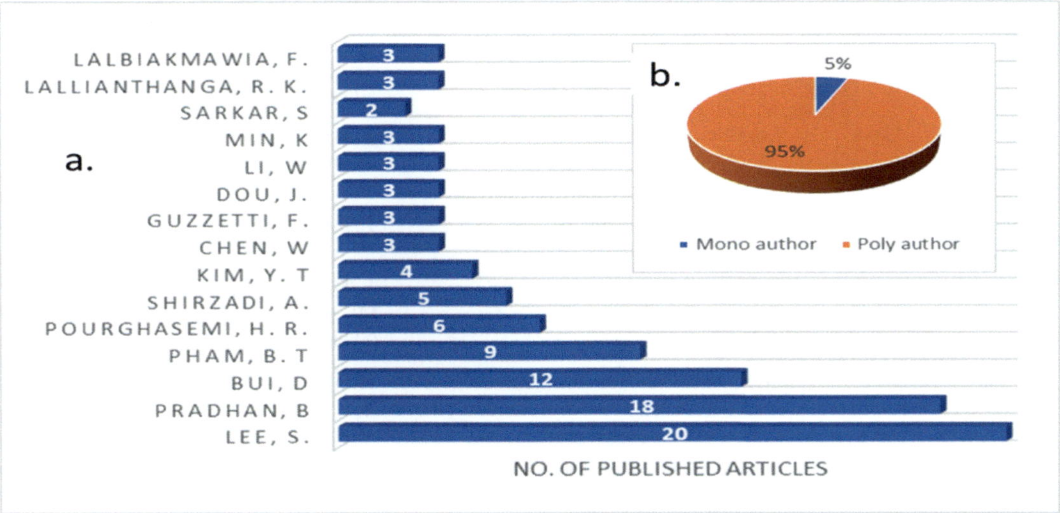

Fig. 14.4 **a** Top 15 authors and their published articles on landslide susceptibility. **b** Number of authors for collaboration

(Cárdenas and Mera 2016; Neuhäuser and Terhorst 2007; Ayalew and Yamagishi 2005) as well as used combined methodologies (Chen et al. 2017; Shirzadi et al. 2018; Pham et al. 2016). The detailed of the study area, methodology with findings referred in Table 14.4.

14.5 Conclusion

Landslide susceptibility is an emerging study in literacy number of methodologies also increased day by day. These growths happen due to the development of remote sensing and GIS.

Table 14.4 Brief summary of review

Author	Study area	Methods	Findings
Broeckx et al. (2019)	Uganda	Logistic regression	1. AUC= 0.94
			2. Landslide susceptibility is a significant indicator of LMR
			3. Landslide is a major sediment source in Africa
Mind'je et al. (2019)	Rwanda	Frequency ratio	1. AUC= 84.6%
			2. Slope is the most influencing factor (17.23%) followed by land use land cover
Singh et al. (2020)	India	Information value, index of entropy	1. LULC (sparse vegetation and barren land) rank first for a landslide in this region followed by TWI, lineament density, geomorphology, and slope respectively
Cárdenas and Mera (2016)	Ecuador	Yule coefficient	1. Result found that slope facing north and northeast was the highest chance of landslide occurrence
Neuhäuser and Terhorst (2007)	Germany	Weight of evidence	1. Result indicated that the slope between 11° to 26° with colluvial soil layer was an indicator of slope instability
Mondal and Mandal (2019)	India	Index of entropy	1.ROC= 78.2%
			2. Method resulted that soil types were the highest causative factor whereas NDVI was the least factor
Ayalew and Yamagishi (2005)	Japan	Logistic regression	1. Road network found the major causative factor of landslide in this study area
Chauhan et al. (2010)	India	Artificial neural network	1.AUC= 0.84%
Chen et al. (2017)	China	Evidential belief function, certainty factor, frequency ratio	1. The success rates of the EBF, CF, and FR models are 0.8038, 0.7924, and 0.8088, respectively
Dias and Gunathilake (2014)	Sri Lanka	WAA, SINMAP	1. The factor of safety ranging from 0 to >1.5
Dunning et al. (2009)	Bhutan	Structural and geomorphological feature extraction	1. Nam Ling Landslide is indicative of deeper-seated deformation and is not purely a function of road construction

(continued)

Table 14.4 (continued)

Author	Study area	Methods	Findings
Shirzadi et al. (2018)	Iran	Alternating decision tree, bagging, multiboost, random subspace, rotation forest	1. Study resulted TWI and slope angle was most triggering factors of landslide
Pham et al. (2016)	India	Support vector machine, linear regression, Fisher's linear discriminant analysis, Bayesian network, Naïve Bayes	1. Among five models, SVM model played the best performance
Jana et al. (2019)	Papua New Guinea	Frequency ratio	AUC= 0.73%
Roy and Saha (2019)	India	Fuzzy- LNRF, Fuzzy- AHP	AUC= 91% and 90%, respectively

Landslide susceptibility zonation is the first step to landslide hazard mitigation, management, and measures. In the present study, we reviewed 201 published articles to understand the spatial and temporal changes. The number of publishing articles increase 7.19 times per year during the period 2000–2020; it is increased quite less by about 4.1 times per year during the period 2000–2009. The overall increase has happened due to the availability of digital information and the development of machine learning methodologies. Logistic regression, frequency ratio, analytical hierarchy method, GIS overlay weighted, information value, the weight of evidence, and recent methodologies like support vector machine, artificial neural networks, random forest, and decision tree were widely used methodologies that reflect their high accuracy. Review articles were covered across 46 countries over the world among them most of the countries fall in the southeast and East Asian countries that also indicated the high landslide-prone area. A lot of journals have given priority to publishing articles on landslide susceptibility and associates works; most of the articles are published under science direct, Elsevier, and Springer base journals like Geomorphology, Natural Hazard, Engineering geology, Environmental earth science, and landslide published more than 10 papers. Although landslide is an unprotectable natural hazard, we can reduce risk magnitude by proper management and awareness.

References

Ahmed B (2015) Landslide susceptibility mapping using multi-criteria evaluation techniques in Chittagong metropolitan area, Bangladesh. Landslides 12 (6):1077–1095

Aleotti P, Chowdhury R (1999) Landslide hazard assessment: summary review and new perspectives. Bull Eng Geol Env 58(1):21–44

Ayalew L, Yamagishi H (2005) The application of GIS-based logistic regression for landslide susceptibility mapping in the Kakuda-Yahiko mountains, Central Japan. Geomorphology 65(1–2):15–31

Basu T, Pal S (2017) Exploring landslide susceptible zones by analytic hierarchy process (AHP) for the Gish river basin, West Bengal, India. Spatial Inf Res 25(5):665–675

Bragagnolo L, da Silva RV, Grzybowski JMV (2020) Landslide susceptibility mapping with r. landslide: a free open-source GIS-integrated tool based on artificial neural networks. Environ Modell Softw 123:104565

Broeckx J, Maertens M, Isabirye M, Vanmaercke M, Namazzi B, Deckers J, Tamale J, Jacobs L, Thiery W, Kervyn M, Vranken L (2019) Landslide susceptibility and mobilization rates in the Mount Elgon region, Uganda. Landslides 16(3):571–584

Budimir MEA, Atkinson PM, Lewis HG (2015) A systematic review of landslide probability mapping using logistic regression. Landslides 12(3):419–436

Cárdenas NY, Mera EE (2016) Landslide susceptibility analysis using remote sensing and GIS in the western Ecuadorian Andes. Nat Hazards 81(3):1829–1859

Chauhan S, Sharma M, Arora MK, Gupta NK (2010) Landslide susceptibility zonation through ratings derived from artificial neural network. Int J Appl Earth Obs Geoinf 12(5):340–350

Chen W, Xie X, Wang J, Pradhan B, Hong H, Bui DT, Duan Z, Ma J (2017) A comparative study of logistic model tree, random forest, and classification and

regression tree models for spatial prediction of landslide susceptibility. Catena 151:147–160
Chen X, Chen W (2021) GIS-based landslide susceptibility assessment using optimized hybrid machine learning methods. Catena 196:104833
Dias AV, Gunathilake AAJK (2014) Evaluation of sensitivity of the Waa and SINMap models (static) for landslide susceptibility risk mapping in Sri Lanka. In: Landslide science for a safer geoenvironment. Springer, Cham, pp 167–173
Dunning SA, Massey CI, Rosser NJ (2009) Structural and geomorphological features of landslides in the Bhutan Himalaya derived from terrestrial laser scanning. Geomorphology 103(1):17–29
Guzzetti F, Reichenbach P, Ardizzone F, Cardinali M, Galli M (2006) Estimating the quality of landslide susceptibility models. Geomorphology 81(1–2):166–184
Huang Y, Zhao L (2018) Review on landslide susceptibility mapping using support vector machines. Catena 165:520–529
Jana SK, Sekac T, Pal DK (2019) Geo-spatial approach with frequency ratio method in landslide susceptibility mapping in the Busu River catchment, Papua New Guinea. Spatial Inf Res 27(1):49–62
Kanungo DP, Arora MK, Sarkar S, Gupta RP (2012) Landslide susceptibility zonation (LSZ) mapping–a review. J South Asia Disaster Stud 2
Kıncal C, Akgun A, Koca MY (2009) Landslide susceptibility assessment in the Izmir (West Anatolia, Turkey) city center and its near vicinity by the logistic regression method. Environ Earth Sci 59(4):745–756
Lee S, Min K (2001) Statistical analysis of landslide susceptibility at Yongin, Korea. Environ Geol 40 (9):1095–1113
Marjanović M, Kovačević M, Bajat B, Voženílek V (2011) Landslide susceptibility assessment using SVM machine learning algorithm. Eng Geol 123 (3):225–234
Merghadi A, Yunus AP, Dou J, Whiteley J, ThaiPham B, Bui DT, Avtar R, Abderrahmane B (2020). Machine learning methods for landslide susceptibility studies: a comparative overview of algorithm performance. Earth-Sci Rev:103225
Mind'je R, Li L, Nsengiyumva JB, Mupenzi C, Nyesheja EM, Kayumba PM, Gasirabo A, Hakorimana E (2019). Landslide susceptibility and influencing factors analysis in Rwanda. Environ Dev Sustain:1–28
Mondal S, Mandal S (2019) Landslide susceptibility mapping of Darjeeling Himalaya, India using index of entropy (IOE) model. Appl Geomatics 11(2):129–146
Neuhäuser B, Terhorst B (2007) Landslide susceptibility assessment using "weights-of-evidence" applied to a study area at the Jurassic escarpment (SW-Germany). Geomorphology 86(1–2):12–24
Pachuau L (2019) Zonation of landslide susceptibility and risk assessment in Serchhip town, Mizoram. J Indian Soc Remote Sens 47(9):1587–1597
Pham BT, Pradhan B, Bui DT, Prakash I, Dholakia MB (2016) A comparative study of different machine learning methods for landslide susceptibility assessment: a case study of Uttarakhand area (India). Environ Model Softw 84:240–250
Pourghasemi HR, Yansari ZT, Panagos P, Pradhan B (2018) Analysis and evaluation of landslide susceptibility: a review on articles published during 2005–2016 (periods of 2005–2012 and 2013–2016). Arab J Geosci 11(9):1–12
Roy J, Saha S (2019) Landslide susceptibility mapping using knowledge driven statistical models in Darjeeling district, West Bengal, India. Geoenviron Disasters 6(1):1–18
Samia J, Temme A, Bregt AK, Wallinga J, Stuiver J, Guzzetti F, Ardizzone F, Rossi M (2018) Implementing landslide path dependency in landslide susceptibility modelling. Landslides 15(11):2129–2144
Segoni S, Tofani V, Rosi A, Catani F, Casagli N (2018) Combination of rainfall thresholds and susceptibility maps for dynamic landslide hazard assessment at regional scale. Front Earth Sci 6:85
Shirzadi A, Soliamani K, Habibnejhad M, Kavian A, Chapi K, Shahabi H, Chen W, Khosravi K, Thai Pham B, Pradhan B, Ahmad A (2018) Novel GIS based machine learning algorithms for shallow landslide susceptibility mapping. Sensors 18(11):3777
Silalahi FES, Arifianti Y, Hidayat F (2019) Landslide susceptibility assessment using frequency ratio model in Bogor, West Java, Indonesia. Geosci Lett 6(1):1–17
Singh P, Sharma A, Sur U, Rai PK (2020) Comparative landslide susceptibility assessment using statistical information value and index of entropy model in Bhanupali-Beri region, Himachal Pradesh, India. Environ Dev Sustain:1–18
Sterlacchini S, Ballabio C, Blahut J, Masetti M, Sorichetta A (2011) Spatial agreement of predicted patterns in landslide susceptibility maps. Geomorphology 125(1):51–61

Assessment of the Social Impact of Arsenicosis Through Groundwater Arsenic Poisoning in Malda District

15

Debapriya Poddar, Sarbari Mukhopadhyay, and Jayanta Das

Abstract

Arsenic (As) concentration in groundwater is a significant environmental issue for the different parts of the Ganga basin. The social status of the inhabitants of Malda district has been profoundly distressed by groundwater arsenic toxicity as several symptoms and diseases have been stimulated by the widespread arsenic exposure. Moreover, the systematic study on this vital issue is scarce for the study area and West Bengal. Therefore, the present study has been conducted to depict the social impact of arsenicosis due to arsenic contamination in the groundwater of Malda district of West Bengal. The results show that the most prominent social hazards are difficulties in getting married to arsenicosis victims' women, the problem of social instability, the dominance of the dowry system, preferences to remain unidentified, and the mental stress of women, etc. A mean vulnerability score of the social hazards index (0.65) suggests the arsenicosis victims have endured an assortment of the high level of different social vulnerability due to arsenicosis illness. To overcome the social hazards, increase the study of the environmental education and knowledge of empowerment as it helps to overcome the traditional culture, ostracism, and some superstitious problems related to arsenicosis.

Keywords

Arsenic · Groundwater contamination · Arsenicosis · Social impact

15.1 Introduction

Water, a fundamental element of life, is caused to death when contaminated. So, the water quality should be adequately sustained (Yadav et al. 2012, 2015). Generally, the groundwater is more dangerous to use when it is highly contaminated due to its close interaction with the various minerals such as Fluoride, Salt, Iron, Arsenic, etc., present in the aquifers (Chakraborty et al. 2011). In 1984 groundwater arsenic contamination was discovered in the lower Ganga Plain of West Bengal (Garai et al. 1984). Arsenic is a ubiquitous element in the earth's crust. Still, less

D. Poddar (✉)
Department of Geography and Applied Geography, University of North Bengal, Darjeeling 734013, India
e-mail: debapriyapoddar2011@gmail.com

S. Mukhopadhyay
Department of Geography, Maynaguri College, Maynaguri, Jalpaiguri 735224, India

J. Das
Department of Geography, Rampurhat College, Rampurhat, Birbhum 731224, India
e-mail: jayanta.daas@gmail.com

J. Das and S. K. Bhattacharya (eds.), *Monitoring and Managing Multi-hazards*,
GIScience and Geo-environmental Modelling, https://doi.org/10.1007/978-3-031-15377-8_15

than 1% of most rocks, coals, and soils constitute this element (Alam et al. 2002), generally released from minerals and some geogenic sources. Human activities have also caused extensive groundwater arsenic contamination in different parts of the world (Rahman et al. 2018; Smith et al. 1998). Due to the significant level of arsenic concentration, groundwater contamination is also a crucial environmental and social issue in Southeast Asian countries.

Extensive groundwater arsenic contamination has affected the highly populated and comparatively low-lying area of Bhagirathi–Ganga delta plain in the southern part of West Bengal (Nickson et al. 1998). Generally, arsenic has naturally released in flood plain sediments (Kinnibugh and Smedley 2001; Acharyya et al. 2005). In West Bengal, the arsenic-polluted areas are primarily located in the east of the Bhagirathi River (Nickson et al. 1998). The severe arsenic-contaminated districts of West Bengal are Malda, Murshidabad, Nadia, North 24 Parganas, South 24 Parganas, Burdwan, Hoogly, Howrah, and Kolkata (Rana 2013; Das 2013, 2015; Farooq et al. 2011). The intensity of the arsenic-caused health effects in the Malda district is too high, and it is the worst-hit arsenic-affected district of West Bengal. The arsenic-affected blocks of Malda district are Kaliachak-I, Kaliachak-II, Kaliachak-III, Manikchak, and English Bazar, Ratua–I, and Ratua–II. So, the arsenic contamination has been considered a localized threat for the inhabitants of the Malda district, where 12 lakh populations of 229 villages of seven blocks of the district and about 696,822 populations are afflicting this problem which is really of considerable concern.

Subsequently, the social status of the inhabitants has been profoundly distressed by such groundwater arsenic toxicity as several symptoms and diseases have been stimulated by the widespread arsenic exposure to human beings (Rahman et al. 2018; Chatterjee et al. 2010). Accordingly, it has generated some societal problems (Ahmed et al. 2011). Furthermore, the hazardous impact of arsenic poisoning has significantly influenced the victims' social lives (Hassan et al. 2005) and ultimately distorted the populations' social structure (Chowdhury et al. 2006). Arsenic contamination has acquired an enormous social impact on its victims in the arsenic-prone region of the study area. Finally, arsenicosis illness triggers social nuisance. Hence, the present chapter has explored the social impact of arsenic toxicity on peoples' daily lives. Moreover, this chapter has focused on the social hazards and the social risk of arsenicosis illness. So, the principal objective of this research work is to find out the social impact of arsenicosis on the inhabitants of the study area.

15.2 Database and Methodology

15.2.1 Study Area

Malda district has been chosen as the study area being located within the latitudes of 24°30′ N to 25°32′ N and longitude of 87°48′ E to 88°30′ E, covering an area of 3566.17 km^2 (Fig. 15.1). The Malda district is divided into15 blocks. Among them, seven blocks are contaminated, having beyond the permissible limit of arsenic concentration in groundwater (0.05 mg/l, Indian standard). Kaliachak-I, Kaliachak-II, Kaliachak-III, Manikchak, and English Bazar are severely affected, whereas, Ratua–I and Ratua–II are less affected. In these blocks' the maximum concentration of arsenic in a shallow tube well varies between 0.072 and 0.929 mg/l (PHE, Malda). The rest of the blocks bear an insignificant amount of arsenic (Fig. 15.2). Moreover, a considerable number of tube wells were contaminated with the presence of arsenic. The water supply depends on groundwater (Madhavan and Subramanian 2006) extracted from the shallow aquifers. In general, the district is a low plain area with a general slope from north to south. It comprises alluvial soil originating from the different rivers like Ganga, Kalindi, Tangon, Punarbhaba, and Mahananda. The Ganga is the principal river of the Malda district. This river moves through the western part of the study area, flowing through the block of Manikchak, Kaliachak-II, and Kaliachak-III. The elevated arsenic occurrences in the groundwater were found in alluvial sands, concentrated away from the Ganga margin (Purkait and Mukherjee 2008). Additionally, the Holocene sediments coming from the Himalayan Mountain have actively contaminated those areas through which the Ganga River flows actively, and arsenic originated as a groundwater contaminant (Madhavan and Subramanian 2006). The economy is purely rural as most of the work depends on primary activities being the principal source of livelihood.

15.2.2 Database

The present study is based on primary data sources. Primary data have been collected from the arsenic-affected areas of the study area through a survey method using a questionnaire. Multi-stage sampling is proposed here to find the study areas. In the first stage, identify the affected blocks and villages with a high level of arsenic concentration. These data have been collected from secondary sources. Sample villages have been identified randomly using the random table. The sample size among the seven arsenic concentrated blocks has been calculated using G*Power 3.1.9.7 software. Using this method, the estimated sample villages is 42. All seven blocks have been covered for sampling. For the purposes, simple random sampling with a non-proportionate technique has been adopted due to variations of arsenicosis patients among the villages. After that, calculated the N/n ratio in which N is the total number of inhabited villages. At the same time, n is the total number of sample villages, i.e., 1657 (including the inhabited municipal area of English Bazar block)/42 = 39.45. Using this relationship, one sample village represents every 40 villages of Malda district. After selecting the sample villages, arsenic-affected households were identified. Then, the desired sample size was calculated using Cochran's method (Cochran 1963), as regards the minimum sample size, the approach adopted wherein among all the observations pertaining to various variables lowest prevalence was anticipated at 25%, required minimum sample size came to be 288 at 95% level of confidence. The following formula has been used

$$n = \frac{Z^2pq}{e^2} \qquad (15.1)$$

n = Sample size
Z = Z value found in the Z table at a given confidence interval
p = Estimated proportion of an attribute that is present in the population
$q = 1 - p$
e = Desired level of precision.

To overcome the unavoidable, no response rate of about 4% of the sample respondent is proposed to cover a sample of 300 instead of 288.

15.2.3 Methods

Descriptive statistics of arsenic concentration in groundwater such as Median, Mean, Variance, and Standard deviation have been calculated using SPSS statistical analytics. MS Office Excel, SPSS 15, and R 3.4.3 were used to calculate these statistics.

The social impact of arsenicosis is assessed through a scaling technique. Likert five-point scales are applied here to collect the data regarding the level of such effects related to arsenicosis illness as well as arsenic contamination. The social variables will be measured on a Likert scale, and scores will assign for each statement.

Social vulnerability scores have also been calculated through the scaling technique, consolidated efficiently through a Social Hazards index. The Social hazards index measures the vulnerable condition of arsenicosis patients in social aspects. In each of the vulnerabilities of arsenicosis patients, 22 questions were asked to recognize the social impacts of the victims of arsenicosis patients. Social vulnerabilities are

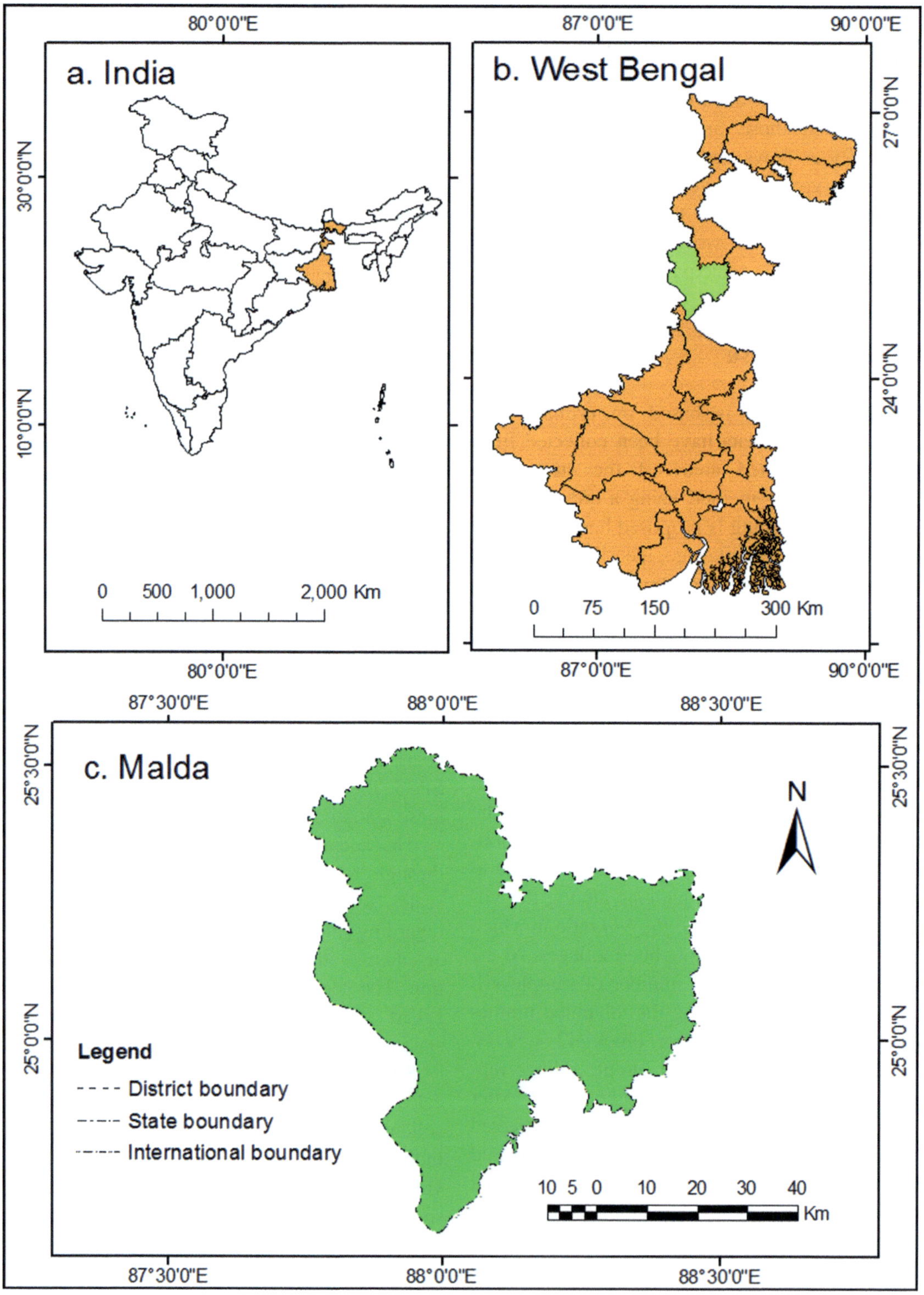

Fig. 15.1 Location map of the study area

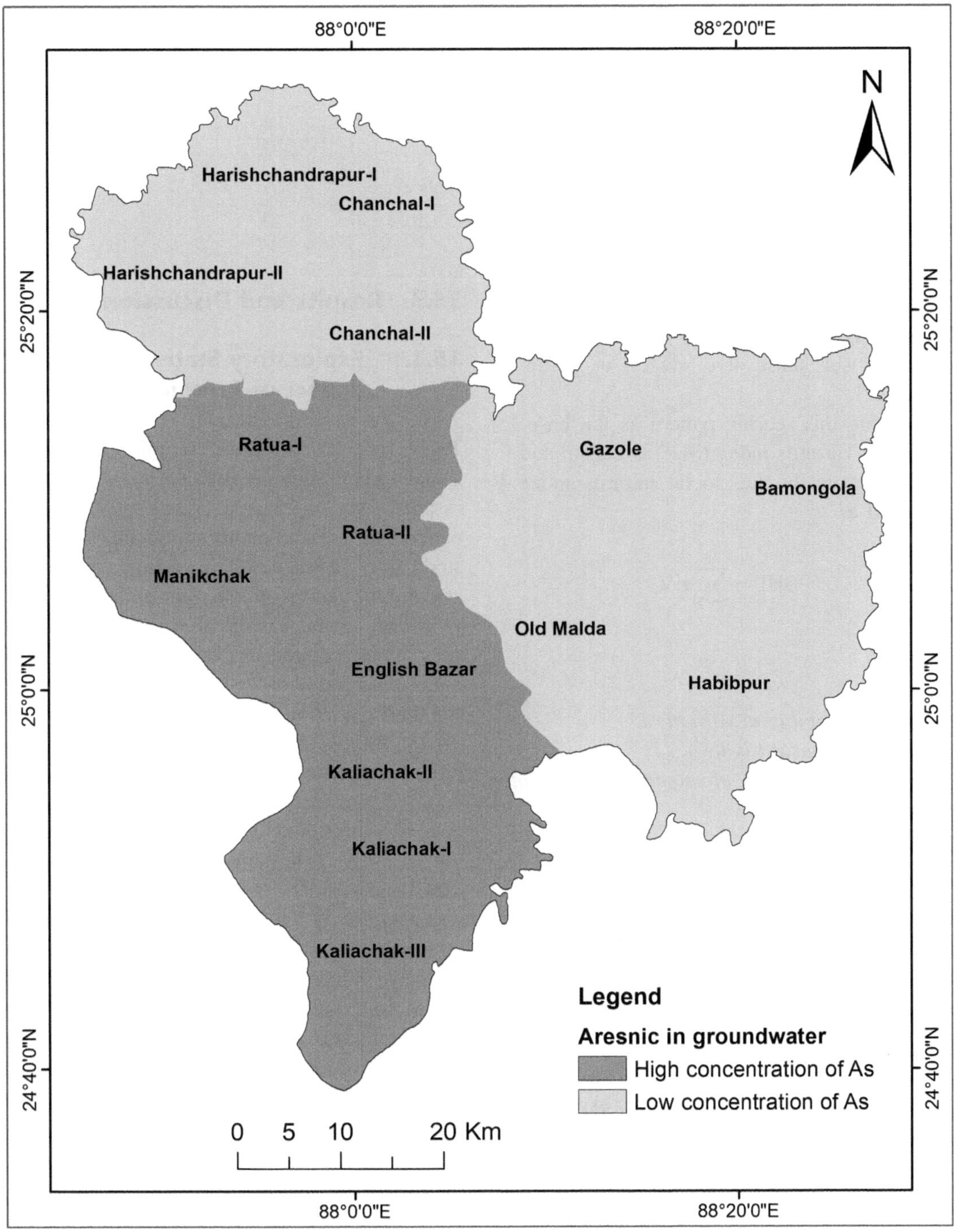

Fig. 15.2 Location of high and low arsenic concentrated blocks of Malda district

Table 15.1 Five stages of social vulnerabilities

Sl. No.	Vulnerability	Scores
1	Strongly disagree	1
2	Disagree	2
3	Neutral	3
4	Agree	4
5	Strongly agree	5

assessed at five stages. They are 1. Strongly disagree 2. Disagree 3. Neutral 4. Agree 5. Strongly agree. The scoring pattern of the vulnerability for these five stages is given in Table 15.1.

Keeping this scoring pattern as the base for the Social Hazards index (SHI) is constituted.

Mathematically, the social hazards index is presented as

$$\mathrm{SHI} = \sum_{i=1}^{n} \mathrm{EV}_i$$

where

EV_i = *Vulnerability in variable i.*
SHI = *Social hazard index.*
$i = 1, \ldots, n$ *or Number of vulnerability variables included in the index.*
$n = 22$.

The following six independent socio-economic variables are chosen to conduct an ANCOVA test. These are family size (X_1), age (X_2), monthly household income (X_3), marital status (X_4), education status (X_5), and occupation (X_6). And the dependent variables of different social impacts Y were selected. In this study, a one-factor ANOVA attempts to explore the relationship between an impact and a variable by measuring the effect of the social background of the arsenicosis inhabitants on the perceived social consequences of arsenicosis disease. These socio-economic variables have significantly influenced the severity of the social impact of arsenicosis diseases.

15.3 Results and Discussion

15.3.1 Exploratory Statistics of Social Impact of Arsenicosis

Table 15.2 exhibits the descriptive statistics of social hazards due to arsenicosis in the sample respondents of Malda district. Arsenicosis patients have faced enormous social risks in the study area. Difficulties in getting married to arsenicosis victims women are among the most significant social hazards due to arsenicosis illness, have acquired the first rank with a maximum mean value of 4.82. At the same time, its' standard deviation is 0.78, and its skewness is −4.48. Moreover, the problem of social instability has got the second rank as its mean value is 4.48, and the standard deviation is 0.80. Simultaneously, the third rank has been achieved by the social hazards of the dowry system with a mean and standard deviation of 4.29 and 0.45, respectively. While its skewness is 0.93. Concurrently, the social problem of patients who prefer to remain unidentified has obtained the fourth rank with a mean and standard deviation of 4.09 and 0.75, respectively. The next rank has been achieved by the social hazards of "women face mental stress" with a mean of 4.06. Though the arsenicosis patients are treated as "untouchable" as this social stigma has obtained 15 ranks but, it got the highest standard deviation with skewness of −0.05, and kurtosis is −1.53. At the same time, the standard deviation of the social hazards varies from 0.45 to 1.55, with an average value of 1.1. The maximum value is found in

Table 15.2 Descriptive statistics of different social hazards due to arsenicosis in the sample respondents of Malda district

Sl. No.	Variable	Mean	SD	Skewness	Kurtosis	Rank
1	Social instability	4.48	0.8	−1.75	3.32	2
2	Forced to leave the village	2.95	1.19	0.6	−1.36	17
3	Migration due to arsenic threat	2.94	1.06	0.52	−1.26	18
4	Arsenic patients are treated as untouchable	3	1.55	−0.05	−1.53	15
5	Relatives discontinued visiting arsenic-affected villages	3	1.03	−0.27	−1.17	16
6	School of dropout	3.34	0.95	−0.52	−1.34	14
7	Child labor due to arsenicosis	3.64	1.07	−0.62	−0.38	9
8	Lack of cooperation	3.87	0.92	−0.44	-0.62	7
9	Refused water collection from the neighbors	3.47	0.93	−0.64	0.42	12
10	Problem of selling cultivated products	2.16	0.65	0.28	0.73	25
11	The affected are avoided in social activities	2.81	0.82	−0.67	0.69	19
12	Suffering from depression	3.63	1.15	−0.41	−0.58	10
13	Patients prefer to remain unidentified	4.09	0.75	−1.15	1.93	4
14	Loss of faith	3.49	1.16	−0.19	−0.65	11
15	Parents are suffering from depression, thinking of social exclusion due to the arsenic threat	2.76	1.34	0.22	−0.87	20
16	Not able to educate children	4.06	1.27	−1.48	1.08	6
17	Spending money on treatment is a waste	3.37	1.53	−0.53	−1.29	13
18	Arsenic patients often try committing suicide	2.17	0.93	0.19	−0.78	24
19	Woman faces the mental stress due to arsenicosis related problem	4.06	0.94	−0.89	0.62	5
20	Women are finding it difficult to get married	4.82	0.78	−4.48	18.69	1
21	Dowry	4.29	0.45	0.93	−1.14	3
22	Women's are debarred from social function	2.53	0.84	0.78	0.87	23

Source Household survey, 2019

arsenic patients treated as untouchable as significant social stigma. In contrast, a minimum value is found in increasing the dowry system. At the same time, it has acquired the third rank (Fig. 15.3).

The result of the ANCOVA test (*F*) is displayed in Table 15.3. Except for the problem of loss of faith, all kinds of social vulnerability are significantly influenced by the concerned socio-economic variables since the respective "F" statistics of these impacts are significant at 95% significance level, where $p \leq 0.0001$.

15.3.2 Social Instability

"Arsenicosis" as a threat of groundwater arsenic contamination causes extensive social instability in human life. Generally, arsenicosis sufferers have faced some incredible difficulties allied to social instability (Rahman et al. 2018). The study of Brinkel et al. (2009) has described arsenicosis victims of Bangladesh have also suffered from the problem of social instability. Simultaneously, social instability has produced considerable social problems in arsenic vulnerable regions of

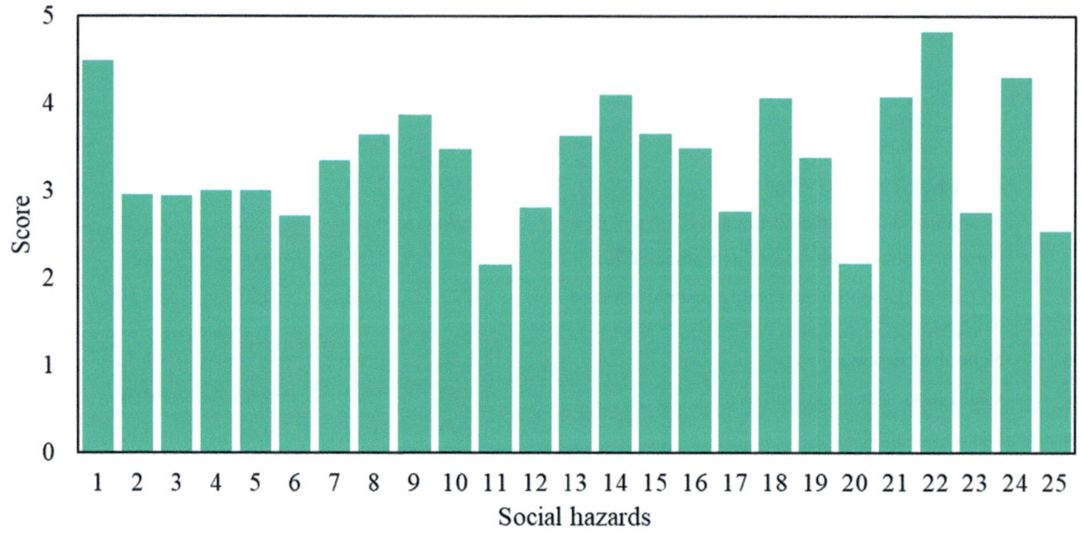

Fig. 15.3 Different social hazards due to arsenicosis

Malda district in diverse ways. Such instability, a significant social impact of arsenicosis illness is influenced by socio-economic variables such as family size, age, monthly household income, marital status, education status, and occupation of the arsenicosis patients.

On the other hand, most of the sample respondents (62.67%) strongly agreed with the concept that social instability originated through arsenicosis illness. Only 1% of study participants strongly disagreed with this concept, while 8.33% of respondents are neutral about such impact of arsenicosis.

Some significant evidence of social instability that arsenicosis victims of Malda district have experienced is discussed below.

15.3.2.1 Forced to Leave the Village

"Arsenicosis" sufferers of the study area have often been forced to leave the villages, raising unsafe circumstances in society. Such instability is highly varied with some socio-economic variables as mentioned above. When they were asked whether the arsenicosis patients were forced to leave villages, the majority of the respondents (59.67%) disagreed with such impact, whereas 14% of study participants strongly disagreed with this. In contrast, 26.33% of respondents agreed with this question. There is a significant effect of the family size, age, monthly household income, marital status, education status, and occupation on the enforcement of the victims to leave the villages.

15.3.2.2 Migration Due to Arsenicosis Illness

The incidence of migration frequently ensues among the arsenicosis victims in the study area as some unstable conditions arise in the individuals' social life. Arsenicosis patients have practiced such movement mainly for the following issues. Some have migrated to consume the required arsenic-free safe water to conquer such arsenicosis disease. Some of them have intended to migrate to defeat the negligence of neighborhoods they desired to depart from such unstable conditions. Such social impact is considerably controlled by some variables such as family size, age, monthly household income, marital status, education status, and occupation status of the sufferers. The maximum proportion of the respondents (51.67%) has disagreed with the question of whether arsenicosis patients have practiced an occurrence of migration. Whereas about 29.67% of respondents have acquired a contrary view, and they are agreed with such impact. About 8% of respondents are strongly agreed with this social impact. At the same time,

Table 15.3 The level of respondents' responses concerning the different social impacts of arsenicosis

Sl. No.	Variable	Strongly disagree	Disagree	Neutral	Agree	Strongly agree	*F*
1	Social instability	1	1.67	8.33	26.33	62.67	2.92
2	Forced to leave the village	14	59.67	0	26.33	0	2.06
3	Migration due to arsenic threat	0	51.67	10.67	29.67	8	1.8
4	Arsenic patients are treated as untouchable	27	15.33	12.33	21.33	24	18.3
5	Relatives discontinued visiting arsenic-affected villages	6.67	31.33	19	41.33	1.67	5
6	School of dropout	0	30.33	8	58.67	3	6.7
7	Child labor due to arsenicosis	3.33	14.67	17.67	43	21.33	3
8	Lack of cooperation	5.67	9	22.33	41.67	21.33	5.02
9	Refused water collection from the neighbors	4.33	8	34	43.67	10	4.84
10	Problem of selling cultivated products	12.67	60.33	26	0.67	0.33	3.94
11	The affected are avoided in social activities	10.33	12.67	63.67	12.33	1	4.13
12	Suffering from depression	5	9.33	33	23.33	29.33	1.67
13	Patients prefer to remain unidentified	0	7	3	64	26	2.94
14	Loss of faith	6	9	42.67	15	27.33	1.07
15	Parents are suffering from depression thinking of social exclusion due to the arsenic threat	25.33	11.67	41.33	5.33	16.33	3.14
16	Not able to educate children	11	2.33	4.33	34.33	48	4.95
17	Spending money on treatment is a waste	21.33	12.33	2.33	35.67	28.33	8.05
18	Arsenic patients often try committing suicide	29.67	29.67	35.67	4.33	0.67	3.46
19	Woman faces the mental stress due to arsenicosis related problem	2	2.33	21.67	35.33	38.67	2.17
20	Women are finding it difficult to get married	3.67	0	1	1	94.33	7.12
21	Dowry	0	0	0	71	29	2.87
22	Women's are debarred from social function	6	48.33	35.33	7	3.33	1.94

Source Household survey, 2019

approximately 10.67% of study participants have attained a neutral approach.

15.3.2.3 Untouchability

The term "untouchability" is associated with arsenicosis patients as this concept has emerged in some unstable conditions in society (Hassan et al. 2005). Generally, such patients are being neglected as they have been treated as untouchable, based on some superstitious belief. About 27% of sample respondents have strongly disagreed with this belief, whereas nearly same percentage (24%) of respondents have strongly agreed. On the other hand, only 12.33% of sample respondents have practiced neutral behavior with this perception, and 15.33% disagreed with untouchability (Table 15.3). Some variables usually influence such perceptions about arsenicosis. The relationship between family size, age, monthly household income, marital status, education status, occupation (independent variables), and beliefs of untouchability about arsenicosis patients is statistically significant. In Bangladesh, this type of social impact of arsenicosis was also detected by Chowdhury et al. (2006).

15.3.2.4 Relatives Discontinued Visiting Arsenic-Affected Villages

In the Malda district, some study participants have experienced relatives being discontinued from visiting arsenic-affected villages due to arsenicosis disease. Consequently, it has raised instability in social life as considered as an imperative social impact of arsenicosis illness. The majority of the respondents (41.33%) agree with this view, while a tiny proportion (1.67%) of participants strongly agree with this vulnerability. On the contrary, about 31.33% of the study participants disagree with this consequence. Only 6.67% of respondents are strongly disagreed with this assessment. Besides this, about 19% of the study participants are neutral with such social instability. The relationship between concerned socio-economic variables and this social problem is statistically significant (Table 15.3).

15.3.2.5 Increasing Rate of Dropout

The social stigma of "untouchability" toward clinically ill arsenicosis children often debars them from continued schooling. On the flip side, children from families with arsenicosis ailment of the earning members end up shouldering the cited responsibility of being the breading earners for the family, adding to an increased rate of dropouts. This situation is precarious for such children. On the other hand, such social impact is usually varied with some socio-economic variables. In other words, a significant effect of these socio-economic variables such as family size, age, monthly household income, marital status, education status, and occupation is found on the occurrences of school dropout due to arsenicosis illness. The majority of the respondents (58.67%) agree with whether arsenicosis illness is responsible for school dropouts. In contrast, only 3% of participants are strongly agreed with this subject. At the same time, about 30.33% of sample respondents have given disagreed attitudes toward this quarry.

15.3.2.6 Child Labor

One of the most considerable social instability caused by arsenicosis ailments is the production of child labor in society. The concerned disease takes a substantial toll on the lives of children under 14 by aiding liability of becoming the earning members of their families as their parents are shorn of being victims of arsenicosis, thereby pushing the former group into child laboring. About 43% of sample respondents in the Malda district have agreed with this concept. In contrast, only a very small proportion of the sample respondents (3.33%) strongly disagreed with this view that child labor is really formed through arsenicosis illness. On the other hand, about 17.67% of study participants have neutral attitudes toward this observation. Such social impact of arsenicosis is highly influenced by these socio-economic variables such as family size, age, monthly household income, marital status, education status, and occupation.

15.3.2.7 Noncooperation

Arsenicosis victims have suffered from the lack of cooperation in society in various aspects as it is indeed a big issue for raising social instability. Most of the study participants are agreed (41.67%) with this view, whereas; a very small proportion (5.67%) of the study participants are strongly disagreed (41.67%). At the same time, about 22.33% of the sample respondents have acquired a neutral approach as they cannot decide whether the patients with arsenicosis have suffered from a lack of cooperation. On the other hand, 21.33% of participants are strongly agreed with this social problem. Besides this, some socio-economic variables have influenced the severity of non-cooperation. A significant association existed between the concerned socio-economic variables and the problem of noncooperation due to arsenicosis ailments.

15.3.2.8 Refused to Water Collection

In Malda district, the arsenicosis victims have abstained from water collection from their neighbor's tube wells and ponds; this is truly a severe social problem as well as the social impact of arsenicosis illness. Such social impact is also identified in Bangladesh by Chowdhury et al. (2006) in their research work. Though most of the respondents have experienced arsenicosis, patients have refused to collect water as well as they agree with this observation, whereas 34% of sample respondents have neutral approaches toward this societal problem. At the same time, a tiny proportion of the sample respondents have carried out a contradictory view as they strongly disagreed with such kind of social impact (Table 15.3).

15.3.2.9 Problem of Selling Cultivated Products

The farmers with arsenicosis illness have faced trouble to selling their cultivated products based on untouchability. This societal problem has generally provided an unstable condition for the victims. On the other hand, such impact of arsenicosis is varied with different socio-economic variables. Besides this, about 60.33% of sample respondents disagree about social instability. In comparison, 12.67% of study participants are strongly disagreed. On the other hand, about 26% of sample victims are neutral about this problem. At the same time, about 0.67% of sample respondents agree, and 0.33% strongly agree with this social vulnerability.

15.3.2.10 Avoided Social Activities

Arsenicosis patients have frequently been shunned in social activities as this kind of social hazard has assembled some unstable conditions in the society in the Malda district. They have quite often been debarred from social participation and social work. But this societal problem is highly influenced by several variables such as family size, age, monthly household income, marital status, education status, and occupation status of the respondents. Most of the sample respondents (63.67%) have acquired neutral feelings toward the social violence of abandonment to participation in social activities as a tremendous social instability. Concurrently, 12.67% of sample respondents disagree with this opinion. Surprisingly, nearly the same proportion (12.33%) of the sample respondents agreed with this impact. On the other hand, about 10.33% of the study participants strongly disagreed with such consequences, whereas only 1% of sample respondents strongly agreed.

15.3.3 Ostracism and Depression

Ostracism is the curse for arsenicosis patients (Rakib et al. 2015). In Malda district, a significant number of arsenicosis patients have suffered from this social vulnerability. According to the observational status, a considerable number of arsenicosis victims in Bangladesh were also facing the problem of ostracism, as reported by Ahmed et al. (2007). Though 29.33% of study participants strongly agreed with the belief that arsenicosis patients have carried out the problem of ostracism, and 23.33% of sample respondents agreed with this. Unexpectedly, 33% of study participants have a neutral mentality concerning the arsenicosis patients suffering from ostracism. But this social impact is influenced by some socio-economic variables such as education status, occupational status, etc.

15.3.3.1 Preference of Remain Unidentified

A further social instability as well as the social impact of arsenicosis ailments is the preference of arsenicosis patients to remain unidentified themselves to evade some social vulnerability. A similar problem has noticed in Bangladesh by Hassan et al. in 2005. Such social instability is responsible for generating the victims' hesitation regarding their illness. On the other hand, some of them are not intended to disclose their problem for fear of noncooperation (Ahmed et al. 2007). There is prevailed a significant relationship between the concerned socio-economic variables (family size, age, monthly household income, marital status, education status, and occupation status) and this social crisis. A large proportion of the study participants (64%) agreed that arsenicosis patients preferred to stay unidentified themselves. At the same time, about 26% of respondents strongly agreed, and 7% of participants disagreed with such a question. But there are, none of the respondents strongly disagreed with this query. But, only 3% of sample respondents have practiced a neutral approach to this view.

15.3.3.2 Loss of Faith

Arsenicosis patients have experienced the problem of loss of faith, which is one of the most significant social impacts of arsenicosis illness. Most of the respondents (42.67%) had practiced neutral approaches with such impact. At the same time, the percentage of the sample respondents who are strongly agreed with this belief is 27.33%. Concurrently, about 9% of respondents disagree with this view, and 65% strongly disagree. This social impact does not influence by these concerned socio-economic variables (family size, age, monthly household income, marital status, education status, and occupation).

15.3.3.3 Parent's Depression

Sometimes, the parents with arsenicosis diseases in the Malda district have experienced the dilemma of depression regarding different issues. Such kinds of problems naturally arise unstable conditions in their social life. This problem also has a significant social impact on arsenicosis patients. Only 5.33% of sample respondents have believed and agreed with the view of the emergence of depression among the parents through social exclusion due to arsenic threat. At the same time, the percentage of the respondents with strongly agreed with such a view is 16.33%. On the contradiction, about 11.67% of the study participants disagree, and 25.33% strongly disagree with such social impact. Besides this, the percentage of the respondents with neutral attitudes toward such a view is 41.33%.

15.3.3.4 Hamper the Child Education

Arsenicosis diseases hinder child education as it has emerged as an unstable situation in society. A large proportion of the respondents (48%) have acquired a maximum score of “5” since study participants strongly agreed with this social impact as such victims have obstructed to continuing their child's education. At the same time, the percentage of the respondents who have obtained a score of “4” agreed with this problem is 34.33%. On the other hand, about 11% of the sample respondents are strongly disagreed, whereas only 2.33% of the respondents disagreed with this social nuisance of arsenicosis ailments. Family size, age, monthly household income, marital status, education status, and occupation (socio-economic variables) significantly impact this social problem (hamper the child's education).

15.3.3.5 Spending Money on Treatment is a Waste of Opinion

While the respondents who agree that spending money on treatment purposes is waste cover a maximum proportion of the respondents (35.33%), whereas 2.33% of respondents have attained neutral attitudes with such estimation. At the same time, about 21.33% of sample respondents have acquired a score of “1” as they strongly disagreed with this perception. Whereas 12.33% of respondents have provided a contrary view and disagreed with this query. Such social impact differs with family size, age, monthly household income, marital status, education status, and occupation status of the respondents.

15.3.3.6 Tendency to Suicide

In several cases, arsenicosis patients lead a bleak life as they cannot involve any common social interaction. Consequently, they become depressed and attempt to commit suicide. There was a significant effect of family size, age, monthly household income, marital status, education status, and occupation on the tendency to suicide due to such problem. The majority of the study participants (35.67%) have acquired a score of “3” as they have practiced neutral attitudes toward this approach. At the same time, an equal proportion of the sample respondents have obtained a 1 and 2 scores as they strongly disagreed and disagreed with this social and mental problem. On the contrary, only 0.67% of respondents strongly agreed. The respondents who have acquired a score of “4” as they are agreed is 4.33% (Table 15.3).

15.3.4 Impact on Women

Females patients in Malda district are the worst victims of arsenicosis as they are a significant vulnerable group in society. The difficulties of women are manifested in different forms. Some of them are neglected by their husbands, whereas; some are divorced or separated. Such kind of impact is also noticed in Bangladesh by Rahman et al. (2018).

15.3.4.1 Mental Stress

The problem of mental stress due to arsenicosis illness has generated a substantial social impact on a woman. However, such a result is naturally influenced by some variables mentioned above. The maximum proportion of the study participants (38.67%) strongly agreed with this question. On the other hand, the percentage of the respondents who agreed with such social instability is 35.33%. About 21.67% of respondents have practiced neutral attitudes toward this view. At the same time, small proportions of the respondents who agree and strongly disagree with this question are 2.33% and 2%, respectively.

15.3.4.2 Denied of Marriage

Difficulties in getting married manifested in various manners. Such as, a woman with such diseases has faced some social problems principally related to pre-marital and post-marital relationships. Since people become indisposed to establishing marital rapport with arsenicosis victims' families, which ultimately results in significant social hazards in Malda district. Chowdhury et al. (2006) have also found such a problem in Bangladesh. The highest percentage of the sample respondents (94.33%) has acquired a score of "5" as they strongly agreed with the view that arsenicosis women patients have suffered from the problem of getting married (Table 15.3). Surprisingly, there was not a single respondent who disagreed with this view. Still, only 3.67% of sample respondents are strongly disagreed. A significant association has existed between the concerned socio-economic variables (family size, age, monthly household income, marital status, education status, and occupation status) and difficulties in getting married as social instability due to arsenicosis diseases.

15.3.4.3 Dowry

Dowry is imperative to social violence in society. Moreover, arsenicosis is accountable for enhancing this societal problem (Rahman et al. 2018). It is tremendously difficult for the parents to arrange the marriage for their young girl suffering from arsenicosis without offering a considerable dowry. There was prevailed a significant effect of the family size, age, monthly household income, marital status, education status, and occupation on the growing quantity of dowry. The vulnerability of such social impact is as widespread as none of the respondents have acquired disagree, strongly disagree, or neutral attitudes toward this view. On the other hand, all the respondents have similar experiences with this social violence. About 71% of study participants agreed with this hazard. At the same time, about 29% of the sample participants strongly agreed with this impact (Table 15.3).

15.3.4.4 Debarred from the Social Function

In the preference of social stigma of untouchability, arsenicosis women often debar from social function. In the Malda district, such social vulnerability has prevailed on a minor scale. The majority of the respondents (48.33%) disagree with the question of whether women arsenicosis sufferers are debarred from social function? At the same time, 6% of sample respondents are strongly disagreed with this question. On the other hand, the respondents with neutral attitudes are 35.33%. On the contrary, about 7% of the sample respondents agreed with such an impact. In comparison, 3.33% of the respondents have acquired a maximum score of "5" as they strongly agreed with this question. The relationship between these socio-economic variables like family size, age, monthly household income, marital status, education status, occupation, and the avoidance of the social function of the arsenicosis woman patient is statistically significant.

Table 15.4 Level of social vulnerability in the sample respondents of Malda district

Level	Social index	Number	Percentage (%)	Mean score
Low vulnerability	<0.50	82	27.33	0.46
Medium vulnerability	0.50–0.61	146	48.67	0.58
High vulnerability	>0.61	72	24.00	0.65

Source Household survey, 2019

15.3.5 Social Vulnerability Index

Based on the level of vulnerability, the respondents are categorized into three categories as

(1) Respondents with a high level of vulnerability
(2) Respondents with a medium level of vulnerability
(3) Respondents with a low level of vulnerability.

In order to group the different categories of respondents into three categories, the mean score has been used.

Table 15.4 exhibits the level of social vulnerability of arsenicosis patients. About 146 (48.67%) respondents have faced a medium level of vulnerability with a mean score of 0.58. On the other hand, 82 (27.33%) respondents have experienced a low level of vulnerability with a mean score of 0.46. At the same time, 72 (24%) respondents belong to a high level of vulnerability group. Though the percentage of the respondents with a low level of vulnerability is more (27.33%) than a high level (24.33%) but a mean vulnerability score of the respondents with a high level of vulnerability is more (0.65%) than a low level of vulnerability group (0.46%).

15.4 Conclusion

The problem of arsenicosis is a vast and severe burden to the society of Malda district and the country itself. Since some significant social hazards have emerged in the study area's arsenic-affected region. Such difficulty has elicited several unpredicted social harms. Simultaneously, arsenicosis has caused the fallout of the social structure. Such patients have experienced severe negative social impacts such as social uncertainty, social injustice, social isolation, problematic family issues, breaking marital relationships, untouchability, social stigma, etc. The most significant social problems in the study area are difficulties in getting married to arsenicosis victims women, the issue of social instability, the dominance of the dowry system, preferences to remain unidentified, and the mental stress of women, etc. Furthermore, a mean vulnerability score of the Social hazards index (0.65%) suggests that respondents with a high level of social vulnerability are more. Hence, the arsenicosis victims have endured an assortment of high levels of different social vulnerability due to arsenicosis illness. Due to the improper knowledge regarding arsenicosis, many victims are not permitted to participate in the social program. Therefore, arsenicosis patients with visible symptoms have been shunned in the community. Perhaps the worst social disaster is yet to come, though it can be avoided if some suitable strategies are assumed and a systematic plan can solve this social threat.

Arsenicosis patients should contact social services and some sanitary authorities to decrease the exclusion of sick individuals from society and others. To overcome the social hazards, improve the quality of social life by improving some social indicators. Such as increasing the literacy rate and studying environmental education and knowledge empowerment to achieve sustainability helps to overcome the traditional culture, ostracism, and some superstitious problems. At the same time, the community members and law enforcement authorities should construct some laws to prevent separation and ostracism as it may help to improve the mental health of the arsenic victims. On the other hand, some rehabilitation programs for arsenicosis patients, especially women, are needed. Several community activities such as group discussions and community education regarding arsenicosis should be organized to reduce the social discrimination against the patients. After all, psychosocial support should be needed to overcome the social crisis. At the same time, in the arsenic hot-spot areas of Malda district, a preventive campaign should be organized to raise awareness, judge the scale and severity of such social problems, and plan for prevention or mitigation. During the study, we perceived an enormous social impact on its victims, and eventually, arsenicosis illness triggered several social nuisances.

References

Acharyya SK, Shah BA, Ashyiya ID, Pandey Y (2005) Arsenic contamination in groundwater from parts of Ambagarh-Chowki block, Chhattisgarh, India: source and release mechanism. Environ Geol 49(1):130–148

Ahmad SA, Sayed MH, Khan MH, Karim MN, Haque MA, Bhuiyan MS, & Faruquee MH (2007) Sociocultural aspects of arsenicosis in Bangladesh: a community perspective. J Environ Sci Health, Part A, 42(12):1945–1958

Ahmed AM, Alam MJB, Ahmed AM (2011) Evaluation of socio-economic impact of arsenic contamination in Bangladesh. J Toxicol Environ Health Sci 3(10):298–307

Alam MGM, Allinson G, Stagnitti F, Tanaka A, Westbrooke M (2002) Arsenic contamination in Bangladesh groundwater: a major environmental and social disaster. Int J Environ Health Res 12(3):235–253

Brinkel J, Khan MH, Kraemer A (2009) A systematic review of arsenic exposure and its social and mental health effects with special reference to Bangladesh. Int J Environ Res Public Health 6(5):1609–1619

Chakraborti D, Das B, Murrill MT (2011) Examining India's groundwater quality management. Environ Sci Technol 45(1):27–33

Chatterjee D, Halder D, Majumder S, Biswas A, Nath B, Bhattacharya P, Bhowmick S, Mukherjee-Goswami A, Saha D, Hazra R, Maity PB (2010) Assessment of arsenic exposure from groundwater and rice in Bengal delta region, West Bengal, India. Water Res 44 (19):5803–5812

Chowdhury MA, Uddin MT, Ahmed MF, Ali MA, Rasul SM, Hoque MA, Alam R, Sharmin R, Uddin SM, Islam MS (2006) Collapse of socio-economic base of Bangladesh by arsenic contamination in groundwater

Cochran WG (1963) Sampling techniques, 2nd Ed., New Work, John Wileyand Sons, Inc., Contingent Valuation Method. Resources for the Future, Washington, DC, p. 82

Das A (2013) Socio economic and gender aspects of arsenicosis—a case study in rural West Bengal. J Humanities Soc Sci 13(7):74–83

Das A (2015) Ground water arsenic contamination—a study of major arsenic affected districts of West Bengal. Int Jr Sci Res 4(6):2993–2996

Farooq SH, Chandrasekharam D, Norra S et al (2011) Temporal variations in arsenic concentration in the groundwater of Murshidabad district, West Bengal, India. Environ Earth Sci 62(2):223–232

Garai R, Chakraborty AK, Dey SB et al (1984) Chronic arsenic poisoning from tube-well water. J Indian Med Assoc 82(1):34

Hassan MM, Atkins PJ, Dunn CE (2005) Social implications of arsenic poisoning in Bangladesh. Soc Sci Med 61(10):2201–2211

Kinniburgh DG, Smedley PL (2001) Arsenic contamination of groundwater in Bangladesh. Vol 2: final report. British Geological Survey

Madhavan N, Subramanian V (2006) Factors affecting arsenic concentration in groundwater in West Bengal. Environ Chem Lett 4(2):79–82

Nickson R, Mc Arthur J et al (1998) Arsenic poisoning of Bangladesh groundwater. Nature 395(6700):338–338

Purkait B, Mukherjee A (2008) Geostatistical analysis of arsenic concentration in the groundwater of Malda district of West Bengal, India. Front Earth Sci China 2 (3):292–301

Rakib MA, Elahi M, Akter MS, Hossain MB, Hossain SM & Huda ME (2015) Arsenic poisoning and social crisis in Bangladesh. Adv Res:1–10

Rahman MA, Rahman A, Khan MZK, Renzaho AM (2018) Human health risks and socio-economic perspectives of arsenic exposure in Bangladesh: a scoping review. Ecotoxicol Environ Saf 150:335–343

Rana MJ (2013) Arsenic contamination of West Bengal, with reference to Malda district. Int J Scien Res 2 (2):166–169

Smith ERG, Naidu R, Alston AM (1998) Arsenic in the soil environment. Doctoral dissertation, Academic Press

Yadav IC, Singh S, Devi NL, Mohan D, Pahari M, Tater PS, Shakya BM (2012) Spatial distribution of arsenic in groundwater of southern Nepal. In: Reviews of environmental contamination and toxicology, vol 218. Springer, Boston, MA, pp 125–140

Yadav IC, Devi NL, Singh S (2015) Spatial and temporal variation in arsenic in the groundwater of upstream of Ganges river basin, Nepal. Environ Earth Sci 73 (3):1265–1279

16 Groundwater Depletion Zonation Using Geospatial Technique and TOPSIS in Raipur District, Chhattisgarh, India

Pooja Gupta, Sanjay Tignath, Dhananjay Kathal, Subhashis Choudhury, Koyel Mukherjee, and Jayanta Das

Abstract

Identifying and visualizing groundwater depletion zonation is crucial for the scientific management of precious groundwater resources. Therefore, the main objective of this present study is to delineate groundwater depletion zonation (GDZ) using Technique for Order Preference by Similarity to an Ideal Solution (TOPSIS) integrated with GIS. The 13 relevant groundwater depletion parameters, i.e., NDBI, GMIS, groundwater trend and magnitudes, cropping intensity, irrigated area, population density, RAI in wet and dry seasons, and VCI were considered. This study also evaluated rainfall and groundwater trends and magnitudes using non-parametric Mann–Kendall (MK) and Sen's slope (Q) estimator tests. From the MK result, groundwater trend in wet season ranges −4.11 to 3.05 and dry season ranges −2.24 to 3.28, whereas Sen's slope in wet season −0.11 to 0.07 and dry season −0.11 to 0.39. The groundwater depletion zonation showed that 23.09, 49.90, 19.02, 793 and 0.06 areas were delineated as very low, low, medium, high, and very high susceptibility zones. Results also showed that middle portions of this district face high to very high groundwater depletion. So, a proper sustainable groundwater management plan is needed to replenish this precious natural resource.

Keywords

Ground water · Stress zonation · Multicriteria decision making · Geographical information system · Raipur

16.1 Introduction

Ground water is fresh water, found in cracks and spaces of soil in the subsurface layer. It is a gift of nature which is used in drinking, irrigation, etc. (Basak et al. 2021). It gives security to farmers and others consumers and is essential for

P. Gupta (✉) · S. Tignath
Department of Geology, Government Science College, Jabalpur, Madhya Pradesh, India
e-mail: pooja.geo@gmail.com

S. Tignath
e-mail: tignathsanjay@gmail.com

D. Kathal
Plant Pathology, College of Agriculture, Powarkheda, India
e-mail: dkathal@jnkvv.org

S. Choudhury
Centre for Himalayan Studies, University of North Bengal, Darjeeling, West Bengal 734013, India
e-mail: rs_subhashis@nbu.ac.in

K. Mukherjee · J. Das
Department of Geography, Rampurhat College, Birbhum, West Bengal 731224, India
e-mail: koyel.mukherjee.suri@gmail.com

J. Das
e-mail: jayanta.daas@gmail.com

J. Das and S. K. Bhattacharya (eds.), *Monitoring and Managing Multi-hazards*,
GIScience and Geo-environmental Modelling, https://doi.org/10.1007/978-3-031-15377-8_16

maintaining drought resilience across India. It is very much significant to get the success of the Green Revolution. It is observed that around 0.6 million people in India are facing acute water shortages, specifically during the summer. In arid and semiarid regions, the condition becomes more critical where the inhabitants purely depend on subsurface water because of the uncertainty of rainfall (Al-Juaidi et al. 2018; Arabameri et al. 2019). Subsurface water or groundwater is measured as an important natural resource in regions where the surface water is inadequate. Even many areas with high rainfall and surface water still depend on ground water (Choudhury et al. 2021). In this way, this natural resource regulates industrial, urban, and domestic sectors.

The increasing demand for groundwater has generated an acute water shortage in some areas. As a result, fresh water is taken from the ground at a higher rate than the recharge rate in an aquifer (Mukherjee and Singh 2012). It is clear that both the quality and quantity of ground water are deteriorated and depleted. For example, a large number of aquifers in Chhattisgarh are being over exploited. Side by side, global warming, clearance of vegetation, lack of water conservation techniques, etc. have a negative impact on ground water storage (Gain et al. 2012; Shao et al. 2018). Continuously growing population and irregular rainfall are the main factors for ground water depletion in this area. Local people including farmers of this area are the main ground water users, but they are not concerned about the necessary information released by the related department.

As ground water depletion depends on different factors, thorough research is needed to collect and gather data. Among different current techniques, and MCDM is one of them. It is the simplest and upgraded approach (Wang et al. 2019; Joerin et al. 2001; Sahoo et al. 2021; Zolekar and Bhagat 2015). The Technique for Order Preference by Similarity to an Ideal Solution (TOPSIS) is used in MCDM analysis. This statistical technique ideally can represent the groundwater depletion susceptible zones. In the TOPSIS method, the optimal solution can be achieved by assessing the difference between the estimated and actual solution (Wang et al. 2019, Chung et al. 2017). This procedure depicts the importance of each parameter used in research work. That is why it has been used in different research works like ground water recharge zonation (Al-Abadi et al. 2020), etc.

Different associated factors related to the fluctuations in groundwater levels are also analyzed here. TOPSIS method is useful to assess groundwater depletion zoning on the basis of hydrological, climatological, and geological factors. The main goals of the study are to (1) examine the depth of groundwater, (2) verify the role of associated driven forces on groundwater level fluctuation, (3) measure the spatial groundwater depletion zone through the TOPSIS method for identifying the groundwater depletion in the Raipur district and also to evaluate the performance level of the TOPSIS method.

16.2 Study Area

The study was conducted over 12 meteorological stations of Raipur district lying between latitudes 21° 4′ 0″ N and 21° 28′ 0″ N and longitude 81° 36′ 0″ E to 82° 0′ 0″ E (Fig. 16.1). The observed area is surrounded via the Bilaspur district within the north, Baster district and a part of the state of Orissa within the south, Raigarh district and a part of Orissa nation in the east and Durg district within the west. The benchmark of the study area is about 347 m above mean sea level. Mahanadi is the principal river of district Raipur. The study area is encompassed by the north Balodabajar district and south Dhamtari district. The study area covers an area of 2902 km^2. The district occupies the south-eastern part of the higher Mahanadi valley and the bordering hills in the south and the east. From the hydro-climatological elements, the yearly average rainfall over Raipur is about 1250–1300 mm. greater than 90% of the whole annual rainfall takes place in the month of June–October (Swain et al. 2015).

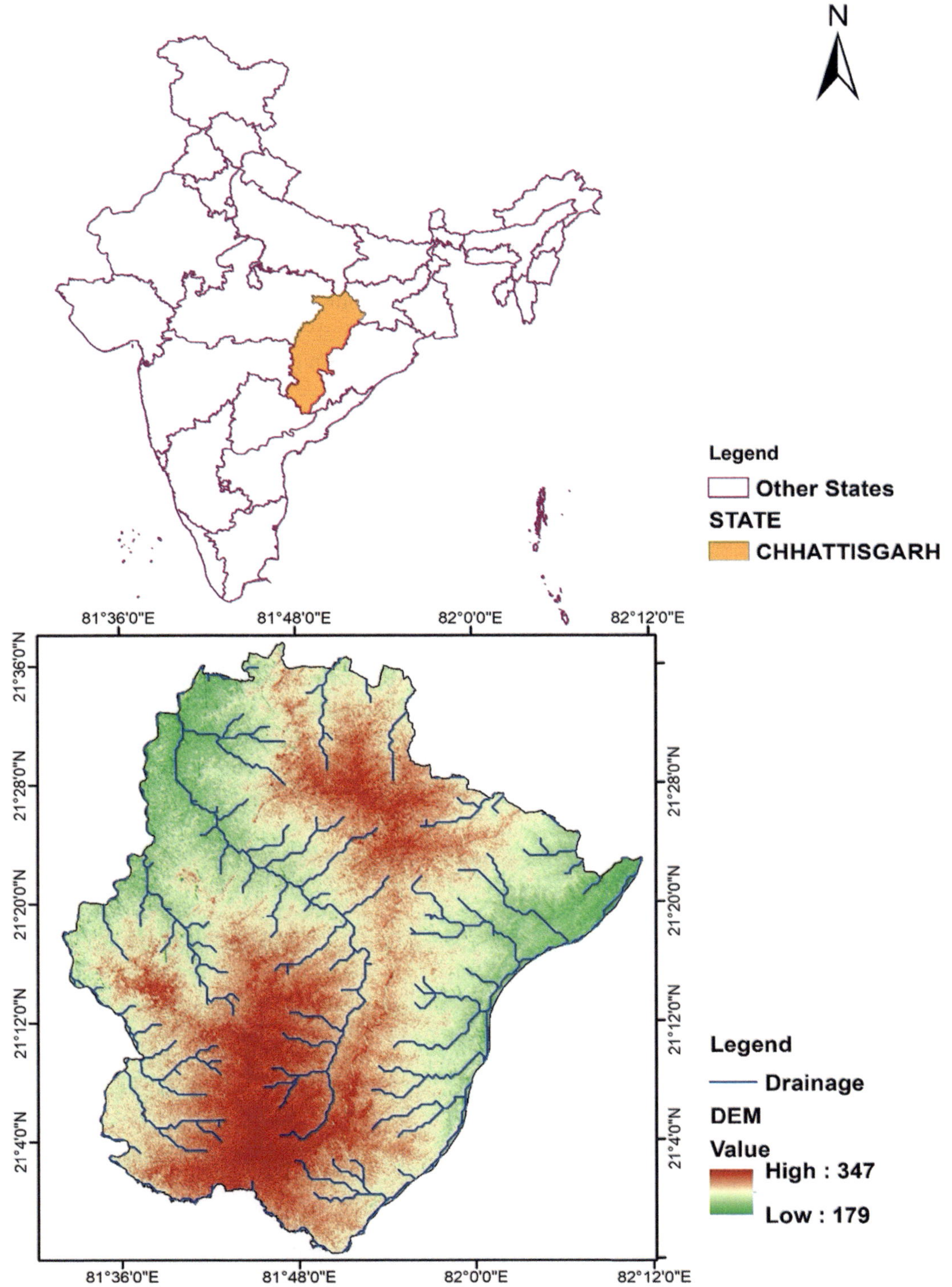

Fig. 16.1 Location map of the study area

16.3 Database and Methodology

16.3.1 Database

The data used in this work includes the Raipur district from 1996 to 2018. The statistics was gathered from CGWB, 2011. These facts furnished the GWD Monsoon, Sen's slope Monsoon, GWD post-monsoon Rabi, and Sen's slope put up-monsoon-Rabi. Rainfall gridded day-by-day data gathered from the India Metrological Department (IMD), Pune (www.imdpune.gov.in), and modified to month-to-month and annual records by the use of Microsoft excel 2019 to create an annual rainfall map by the usage of QGIS, and R software program. SPSS software program used for multicollinearity test of dataset in a particular manner.

Population facts are gathered from Census of India and processed; evapotranspiration information is accrued from Figshare.com and also processed; cropping intensity and irrigation records extracted from the internet site of Chhattisgarh agriculture department and processed. ArcGIS 10.5 version and R 3.5.1 version software program have been used for mapping and facts reading functions, while inverse distance weighted (IDW) approach turned into used for interpolation of factors values. Information related to measuring evaporation has been collected and processed by the FAO Penman–Monteith method (FAO-PM).

We used the groundwater records for 23 years (1996–2018) from the monitoring networks of the central ground water board (CGWB, India). For the explore to specializing in Raipur district (Fig. 16.1), the data consists of recorded groundwater facts in January, August, and November for the duration from 1996 to 2018 from monitoring wells (boreholes, dug well). Due to economic limitations and absence of availability, CGWB has taken measurements of GWD four times a year yet. There are missing remark wells which have been changed at one-of-a-kind factors in time, the time series dataset is asymmetrical in nature. So, many studies have hired unique screening movements not most effective to choose wells with most observation (Ashoka et al. 2017; Bhanja et al. 2016), however moreover with admire to the depth (deep as opposed to shallows) of wells as their trend's variety appreciably (Hora et al. 2019).

16.3.2 Methodology

16.3.2.1 Rainfall Anomaly Index (RAI)

Rainfall anomaly index (RAI) was presented by Van Rooy in 1965. It is the same as to decile index and it is a positioned-based drought index that incorporative a ranking practice to delegate scales to precipitation variances either positive or negative (Keyantash and Dracup 2002 and Raziei 2021). The rainfall anomaly index is figured as follows

$$\mathrm{RAI} = +3\frac{P_i - \overline{P}}{\overline{M} - \overline{P}} \tag{16.1}$$

In this equation, P_i is uniform precipitation, $\overline{P}$ is mean of precipitation, $\overline{M}$ is mean of 10 extremes, and ± 3 is the precede used to limit the upper and lower bounds of differences (Aksoy and Haralick 2001, Khansalari et al. 2018, Keyantash and Dracup 2002, Van Rooy 1965).

16.3.2.2 Cropping Intensity

Cropping intensity is stated as the proportion of gross cropped area to the net cropped area. The agriculture boom may increase both by bringing more land below cultivation or by using growing manufacturing of crops and cropping intensity of each combination (Deshmukh and Tanaji 2017). Cropping intensity is computed as follows

$$\text{Cropping Intensity Index} = \frac{\text{Gross cropped area}}{\text{Net shown area}} \times 100 \tag{16.2}$$

16.3.2.3 Irrigation Intensity

Irrigation is fundamentally the artificial application of water to overcome shortages in rainfall for rising crops. Irrigation is very vital to overcoming the basin's problems of agriculture. Irregular, uncertain, and uneven rainfall in time and space is not sufficient for crop production. The irrigation intensity means the percentage ratio between the net areas irrigated to the net area shown.

$$\text{Irrigation Intensity} = \frac{\text{Net irrigated area}}{\text{Net shown area}} \times 100 \quad (16.3)$$

16.3.2.4 GMIS

For the target year 2010, the Landsat Global man-made Impervious Surface (GMIS) dataset contains a global estimate of fractional impervious cover obtained from the Global Land Survey (GLS) Landsat dataset. The GMIS data collection has two parts: a global percent of impermeable cover and per pixel-related impervious cover uncertainty. The dataset is anticipated to include a wide range of topics, with the goal of better considerate the impact of man-made surfaces on climate and the environment on a continental to global scale (Brown et al. 2017). GMIS data is raster data that represents the built-up area density of the selected ranges between 0 and 200.

16.3.2.5 Normalized Difference Built-Up Index (NDBI)

The built-up area and unadorned soil replicate more SWIR than NIR. In the case of the greenie surface, the reflection of NIR is higher than SWIR band. For better results, we used the Built-up index. The study of urban pattern using NDBI and NDVI is known as built-up index. The built-up index indicating built-up and barren land with only a maximum positive value which allowing BU to map the area automatically.

$$\text{BU} = \frac{\text{NDBI}}{\text{NDVI}} \quad (16.4)$$

NDBI calculation is a modest way to derive. NDBI can be calculated by the formula given below

$$\text{NDBI} = \frac{(\text{SWIR} - \text{NIR})}{(\text{SWIR} + \text{NIR})} \quad (16.5)$$

$$\text{For Landsat 8 data NDBI} = \frac{(\text{Band 6} - \text{Band 5})}{(\text{Band 6} - \text{Band 5})} \quad (16.6)$$

16.3.3 GWD Data

16.3.3.1 Mann–Kendall Test

The Mann–Kendall (Mann 1945; Kendall 1975; Das and Bhattacharya 2018; Das et al. 2019) test statics (S) of the series $x_1, x_2, x_3, \ldots,$ and x_n, are calculated by using the below equation:

$$S = \sum_{k=1}^{n-1} \sum_{j=k+1}^{n} \text{sign}(x_j - x_k) \quad (16.7)$$

where n is the quantity of information points, sign significant the signum work, x_j, and x_k speaks to the data points of time j, and k

$$\text{sign}(x_j - x_k) = \begin{cases} +1 & \text{if } x_j - x_k > 0 \\ 0 & \text{if } x_j - x_k = 0 \\ -1 & \text{if } x_j - x_k < 0 \end{cases} \quad (16.8)$$

The variance of S, Var(S) is computed by the following equation:

$$\text{Var}(S) = \frac{1}{18}\left\{n(n-1)(2n+5) - \sum_{p=1}^{g} t_p(t_p - 1)(2t_p + 5)\right\} \quad (16.9)$$

where "g" represents the number of binding groups which indicates a set of sample data with similar values and t_p indicates the extent of pth binding number.

The estimation of (S) and Var(S) is used to estimate process the standardized test measurement Z as taken after the following equation (Pearson and Hartley 1966).

$$Z = \begin{cases} \frac{(S-1)}{\sqrt{\text{Var}(S)}}, & \text{if } S > 0 \\ 0, & \text{if } S = 0 \\ \frac{(S+1)}{\sqrt{\text{Var}(S)}}, & \text{if } S < 0 \end{cases} \quad (16.10)$$

16.3.3.2 Sen's Slope Estimator

To discover the precise quantity of slope (exchange/unit time) in a hydro-meteorological time series facts, Sen's slope (Sen 1968) estimator is a final preference. To achieve an estimation of slope Q, the slopes of N pairs of information are considered by means of the underneath equation (Das et al. 2021).

$$Q_i = \frac{x_k - x_j}{k - j}, \quad i = 1, 2, \ldots, N, \ k > j \quad (16.11)$$

where x_k and x_j are the value of data at time, j, and Q_i is the median slope, respectively.

16.3.3.3 Population Density

Population density is calculated formula by

$$\mathrm{Pd} = \frac{\mathrm{NP}}{\mathrm{TA}} \quad (16.12)$$

where Pd is density of population, NP is the total number population, and TA is the total land area covered by the population. TA is basically either square kilometer or square miles.

16.3.3.4 Vegetation Condition Index (VCI)

The normalized difference vegetation index (NDVI) is a degree of the power of plant life. Its much resultant based on the identified radiometric properties of vegetation, the use of near-infrared (NIR) and seen (pink) radiation

$$\mathrm{NDVI} = \frac{\mathrm{Visible} - \mathrm{NIR}}{\mathrm{Visible} + \mathrm{NIR}} \quad (16.13)$$

Because when daylight hits a plant, utmost of the red wavelength of the perceptible part of the spectrum 0.4–0.7 μm is fascinated by the leaf chlorophyll, a natural phenomenon when the leaf cell structure reflects most NIR radiation 0.7–1.1 μm (Deering 1975).

NDVI values vary from −1 to +1, with values signifying maximum no-green vegetation, and values close to +1 indicate maximum possible plant density.

Landsat 8 data download from USGS Earth Explorer site vegetation condition index (Pettorelli et al. 2005) calculated the established formula.

The VCI is calculated as

$$\mathrm{VCI}_i = 100\frac{\mathrm{NDVI}_i - \mathrm{NDVI}_{\mathrm{min}}}{\mathrm{NDVI}_{\mathrm{max}} - \mathrm{NDVI}_{\mathrm{min}}} \quad (16.14)$$

where NDVI_i is the flattened NDVI and NDVI-max and NDVImin are the absolute maximum and minimum NDVI, respectively (Steven and Srinivasan 2009)

16.3.4 Multicollinearity

The multicollinearity is a regression model to check the multicollinearity of various thematic layers and here one parameter has to take an independent and the remaining other thematic layers are considered dependent variable is highly correlated with other thematic layers. Through this multicollinearity test, it can be easily predicted one input parameter is linearly from another thematic layer of this produced model (for instance, slope and elevation) which will outcome is not a negligible amount of appropriateness in the produced model result of GDZ. Hence, it is necessary to run this model of regression to check the multicollinearity of the selected thematic layers. Less than 0.10 of a tolerance and variance inflation factor (VIF) of greater than 10 suggest problems of multicollinearity (Basak et al. 2021). The tolerance (T_i) and variance inflation factor (VIF_i) were calculated using following equations

$$\text{Tolerance of the } i\text{th predictor variable}(T_i) = 1 - R_i^2 \quad (16.15)$$

$$\text{VIF of the } i\text{th predictor variable}(\mathrm{VIF}_i) = 1/T_i \quad (16.16)$$

In this assessment, those parameters have to be removed which has a tolerance value less than 0.10 and VIF amount greater than 10. To identify

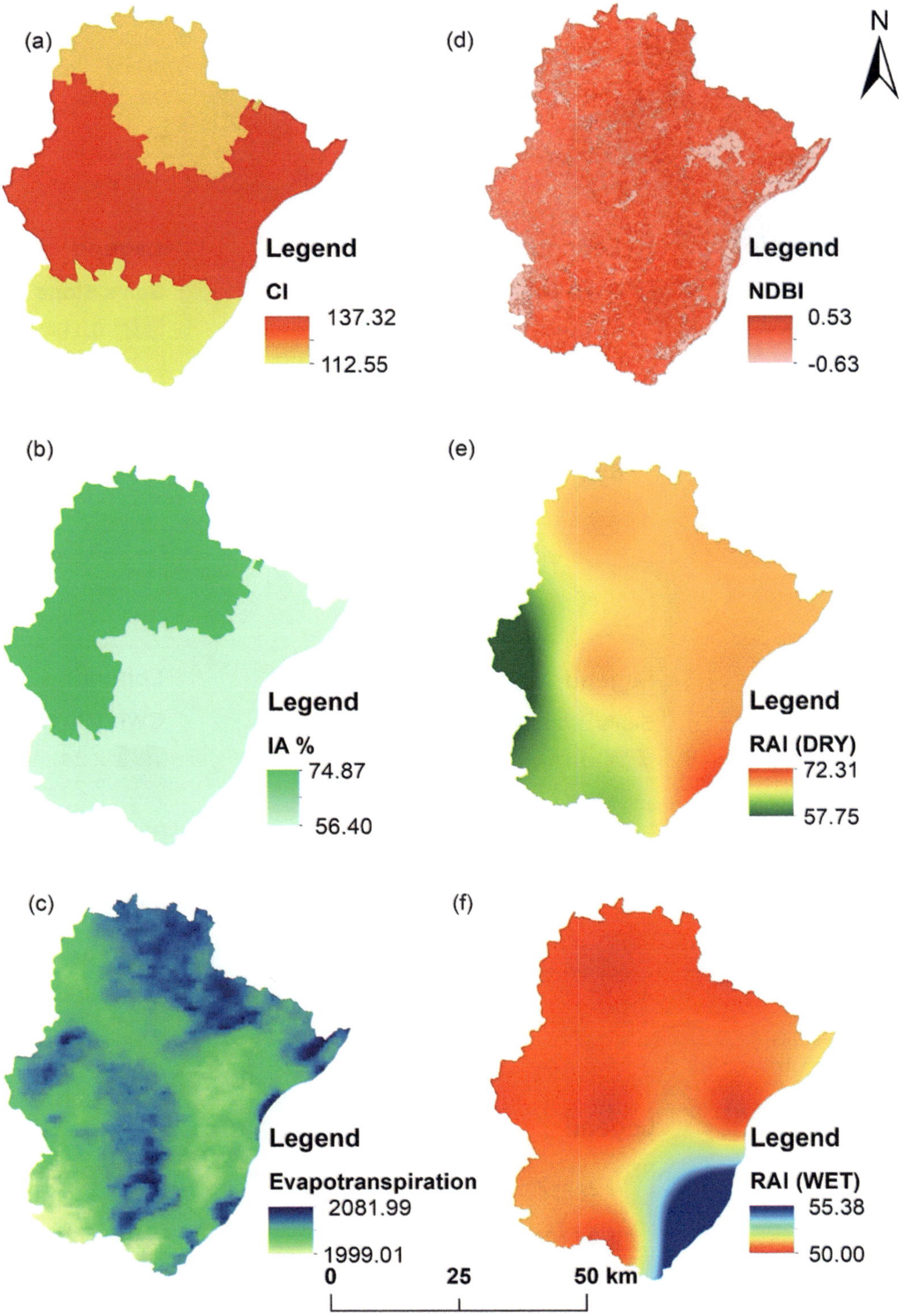

Fig. 16.2 **a** Cropping intensity, **b** irrigation area, **c** evapotranspiration, **d** normalized difference built-up area, **e** rainfall anomaly index dry, **f** rainfall anomaly index wet

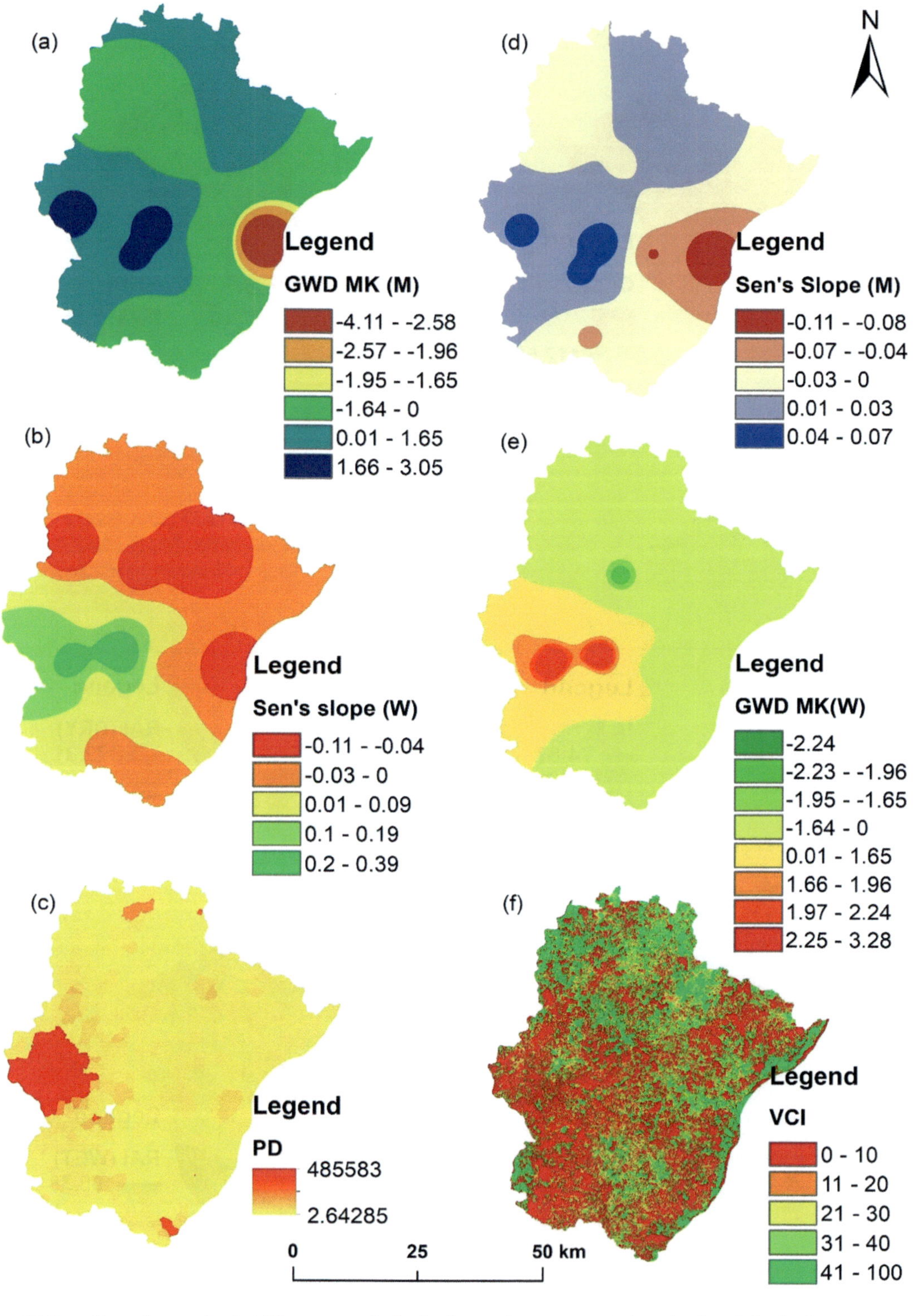

Fig. 16.3 **a** Groundwater depth MK monsoon, **b** Sen's slope winter, **c** population density, **d** Sen's slope monsoon, **e** groundwater depth M winter, **f** vegetation condition index

the problem of multicollinearity within the parameters, 1000 points have been randomly created within this study space with the help of sampling tool of the data management tools package in the Arc GIS platform. The data was extracted from all thematic layers one by one using multivalue to points of extraction tool of the spatial analyst tools package in the ArcGIS platform and then with the help of SPSS software (version-26) this test was done.

16.3.5 Method of TOPSIS

We count on the rankings of alternatives and weigh represented by means of numerical information. The problem is said by means of a single choice-maker within the conventional approach for order desire by similarity to a super answer method. While there are a couple of decision-makers, complexity happens due to the fact the desired solution has to be agreed upon by means of numerous interest agencies, each of which has a different set of goals. Sections 16.1 and a pair systematically outline the conventional TOPSIS algorithm for a single decision-maker and group decision-making, respectively (MCDM via applying the TOPSIS technique to crisp and interval records).

16.3.5.1 Estimate the Normalized Decision Matrix

This step transmutes the shape of characteristic dimensions into non-dimensional features, allowing comparisons throughout criteria. Due to the fact, several necessities are usually measured in several units, and the rankings in the evolution matrix x must be altered to a normalized scale the standardization of values can be finished with the aid of using manner of one in every of some of seemed standardized formulation. A number of the regularly used techniques of calculating the standardization value

$$n_{ij} = \frac{x_{ij}}{\sqrt{\sum_{i=1}^{m} x_{ij}^2}} \tag{16.17}$$

$$n_{ij} = \frac{x_{ij}}{\max_i x_{ij}} \tag{16.18}$$

And analyze weight normalized decision matrix.

The weighted normalized value V_{ij} is considered in the given formula

$$V_{ij} = \overline{X}_{ij} \times Wj \tag{16.19}$$

where W_j is the weight of the jth criterion

$$\sum_{j=1}^{n} W_j = 1 \tag{16.20}$$

Determine negative ideal solution and the positive ideal.

The positive solution is the maximum benefit standard and the cost standard, while the negative ideal solution maximizes the cost standard and lowers the benefit standard.

Positive ideal explanation A^+ has the form

$$\begin{aligned} A^+ &= \left(v_1^+, v_2^+, \ldots, v_n^+\right) \\ &= \left(\left(\max_i v_{ij} \,|\mathrm{j} \in \mathrm{I}\right), \left(\min_i v_{ij} |\mathrm{j} \in \mathrm{J}\right)\right) \end{aligned} \tag{16.21}$$

and negative ideal explanation A^- has the form

$$\begin{aligned} A^- &= \left(v_1^+, v_2^+, \ldots, v_n^+\right) \\ &= \left(\left(\max_i v_{ij} \,|\mathrm{j} \in \mathrm{I}\right), \left(\min_i v_{ij} |\mathrm{j} \in \mathrm{J}\right)\right) \end{aligned} \tag{16.22}$$

where "i is associated with advantage criteria and j with the cost criteria $i = 1,\ldots,m,\ j = 1\ldots,n$". "Analyze the separation measures from the positive ideal solution and the negative ideal solution."

Some distance metrics may be implemented in the TOPSIS method. The parting of each substitute from an effective perfect solution is given below

$$S_i^+ = \left(\sum_{j=1}^{n}\left(v_{ij} - v_j^+\right)^p\right)^{\frac{1}{p}}, \quad i = 1, 2, \ldots, m \tag{16.23}$$

The separation of every opportunity from the terrible ideal answer is given as

$$S_i^- = \left(\sum_{j=1}^{n}\left(v_{ij} - v_j^+\right)^p\right)^{\frac{1}{p}}, \quad i = 1, 2, \ldots, m \tag{16.24}$$

Calculate the relative closeness to the positive ideal solution.

The relative closeness of the *i*th alternative S_j with respect to S^+ is defined as

$$P_i = \frac{S_i^-}{S_i^- + S_i^+} \tag{16.25}$$

whereas $0 \leq P_i \leq 1, \quad i = 1, 2, \ldots, m.$

16.4 Results and Discussion

16.4.1 Multicollinearity

The multicollinearity statistical analysis result was illustrated in Table 16.1. The VIF value is <10 and the tolerance value >0.1 for each thematic layer revealed that there is no collinearity found among the 11 parameters which was selected for this GDZ assessment. The results also suggest that there is no uncertainties found in the multicollinearity condition. The highest VIF (9.34) was found in the groundwater trend in monsoon season, followed by rainfall trend in the winter season, Sen's slope of rainfall in monsoon season, Sen's slope of groundwater in winter season, irrigated area, cropping intensity, RAI in dry season, RAI in wet season, evapotranspiration, GMIS, NDBI, and VCI, respectively. On the other hand, the tolerance value is more than 0.1 for the parameter of GMIS, cropping intensity, irrigated area, evapotranspiration, NDBI, rainfall anomaly index dry, RAI in wet season, VCI, groundwater trend in monsoon season, Sen's slope of groundwater in the winter season, rainfall trend in the winter season, Sen's slope of rainfall in monsoon season.

16.4.2 Causative Factors of Groundwater Depletion

Table 16.2 shows that the study area's rainfall anomaly index in pre-monsoon season ranges between 56.67 and 73.33 and rainfall anomaly index in monsoon season varies between 50 and 56.67% in the year. Northern, eastern and north-eastern parts of the study area occupied higher value of the RAI in dry season, whereas lower value of RAI in dry season observed in the south and south-western parts of the study area (Fig. 16.2e). On the other hand, higher value of RAI in wet season is found in the marginal south-eastern corner of the study area, whereas rest of the area (northern, eastern, western, south-western, and central) occupied lower value of the RAI in wet season (Fig. 16.2f). It indicates that the GWD mostly occurred in the south and south-western parts of the study area.

Compound growth rate in cropping intensity shown in the Table 16.2 is 137.31; it shows that higher cropping intensity and 122.55 of cropping intensity shows less cropping intensity in the study area.

North-western irrigation area of the study area shown (Fig. 16.2b) high intensity 74.86 and in the south-eastern part shown less intensity 56.39 (Deshmukh and Tanaji 2017), which indicates more GWD in the study area.

In the study area, evapotranspiration ranges from 2081 to 1999. The higher evapotranspiration in the study area covers north-eastern and a few parts of the south-west but lowest evapotranspiration in the study area cover mostly the south-western region (Fig. 16.2c), which designates high GWD in the area where comparatively higher evapotranspiration is observed. (FAO 1998) (Fick and Hijmans n.d.; UNEP 1997; Walter et al. 2000).

Normalized difference built-up index value lies between −1 and +1. Positive values shown built-up areas, whereas negative value represent water bodies. Low NDBI value designates vegetation cover (Tek 2018). In the present study,

Table 16.1 Multicollinearity diagnosis test (tolerance and VIF) for conditioning factors used in TOPSIS

Constant	Tolerance	VIF
GMIS	0.79	1.26
Cropping intensity	0.41	2.43
Irrigated area	0.34	2.93
Evapotranspiration	0.69	1.43
NDBI	0.82	1.21
Rainfall anomaly index in dry season	0.44	2.24
Rainfall anomaly index in wet season	0.52	1.91
Groundwater trend in monsoon season	0.14	9.34
Sen's slope of groundwater in winter season	0.19	6.07
Rainfall trend in winter season	0.17	7.24
Sen's slope of rainfall in monsoon season	0.26	5.63
VCI	0.85	1.16

Table 16.2 Rainfall anomaly index (%) in year

Station	RAI in dry season	RAI in wet season
1	66.67	50.00
2	73.33	56.67
3	56.67	50.00
4	70.00	50.00
5	70.00	50.00
6	70.00	50.00
7	70.00	50.00

built-up index varies between 0.06 and 0.50 in the study area, which shows higher GWD in the study area (Fig. 16.2d).

Mann–Kendall (MK) test shows that groundwater MK monsoon value ranges between −4.11 and 3.05 and Mann–Kendall (MK) winter value ranges between −2.24 and 3.28. Table 16.3 represents groundwater data during 1996–2018 of groundwater MK test result of GWD of Z-test of Mann–Kendall (ZM) winter (Fig. 16.3a), GWD Sen's winter (Fig. 16.3b), GWD ZM monsoon (Fig. 16.3d) and GDW Sen's monsoon (Fig. 16.3e). Here, Sen's slope of winter value ranges between −0.11 and 0.39 and Sen's slope monsoon value varies −0.11 to 0.07. Here is direct relationship between different MK value with GWD. Higher value indicates higher GWD and vice-versa (Table 16.4).

Table 16.3 Percentage of irrigated area and cropping intensity value

Sl. No.	Irrigated area (%)	Cropping intensity
1	56.39	137.31
2	56.67	112.55
3	65.14	119.99
4	74.86	119.69

In the present study, VCI value ranges between 0 and 100. It is observed from the study that the south, south-western, and north-eastern parts of the study area had the lower VCI value and central, north-western, and eastern parts of the study area occupied higher VCI value (Fig. 16.3f). It shows that the higher GWD in the area has the lower VCI value (Mandal et al. 2021).

16.5 TPOSIS Model

GWD is the most essential element for determining the spatial volume of a strain junction and the strain applied to the surrounding groundwater system. In general, the distance of the upper boundary of the saturation area from the groundwater level is taken into account by the intensity of the groundwater (Mandal et al.

Table 16.4 Groundwater data (1996–2018)

Station code	GWD ZM winter	GWD Sen's winter	GWD ZM monsoon	GWD Sen's monsoon
S1	−1.30	−0.04	−0.95	−0.05
S2	−1.04	−0.04	−1.00	−0.02
S3	−1.59	−0.09	−4.11	−0.11
S4	−0.26	−0.04	−0.81	−0.08
S5	3.28	0.27	1.06	0.03
S6	−0.99	−0.10	−0.70	−0.03
S7	0.46	0.03	2.12	0.04
S8	3.13	0.39	3.05	0.07
S9	1.71	0.14	2.54	0.05
S10	0	0	0	0
S11	−2.24	−0.08	−1.14	−0.01
S12	−0.43	−0.11	0.99	0.03

2021). Broadly, higher GWD illustrates the scarcity of groundwater availability, and even a decrease in intensity suggests a depression in the aquatic system. Furthermore, better GWD has intensified strain in aquatic systems. GWD has improved significantly in all seasons at some point in terms of length. The existing study considered thirteen significant parameters to create the GWD of the surrounding test.

In the Technique for Order of Preference by Similarity to Ideal Solution (TOPSIS) method, for each test point were calculated by the Si+ and Si− values. Here Si+ values ranges between minimum 0.000 and maximum 0.020 are calculated to signify positive ideal explanation and for Si– minimum 0.000 and maximum 0.019 are calculated to characterize negative ideal explanation. Finally, TOPSIS values are calculated to determine the area of groundwater depletion as shown in Table 16.5.

For GWD (Slope and ZM each) parameter values for rainy, and winter seasons, RAI (each dry and wet season), and NDBI, CI, PD, and for irrigated locations. Ultimately, the reputation of groundwater depletion becomes the calculation for each drilling well, where lower values indicate dire conditions and higher values indicate better groundwater conditions. The obtained TOPSIS results are classified into five aesthetics, namely very high (value < 0.10%), high, (value 0.10–10%), medium (value 10–20%), low (value 20–30%). And very small regions (>30%) (Table 16.5 and Figs. 16.4a, b). As indicated in Fig. 16.4, approximately 0.06% of the area is found under very high or supercritical zones; 7.99% of the area is concentrated in subsupercritical zones, especially in the middle of the location. Also, 19.02% of the regions were identified as critical sector as well as medium region, extending from extra north to south region. However, 23.09% of the area was below the subcritical sector or lower region, with relevant and intense north–south elements scattered throughout the study area. About half the space (49.90%) found in a great replenishment area or very few GWDs has become this extra-discovery experiment, distributed over the entire study location (Fig. 16.4a, b) (Fig. 16.5).

Table 16.5 Groundwater depletion via TOPSIS for Raipur district

Sl. No.	Status	Area (km^2)	Area (%)
1	Very low	670.07	23.09
2	Low	1448.49	49.9
3	Medium	552.2	19.02
4	High	230.06	7.93
5	Very high	1.74	0.06

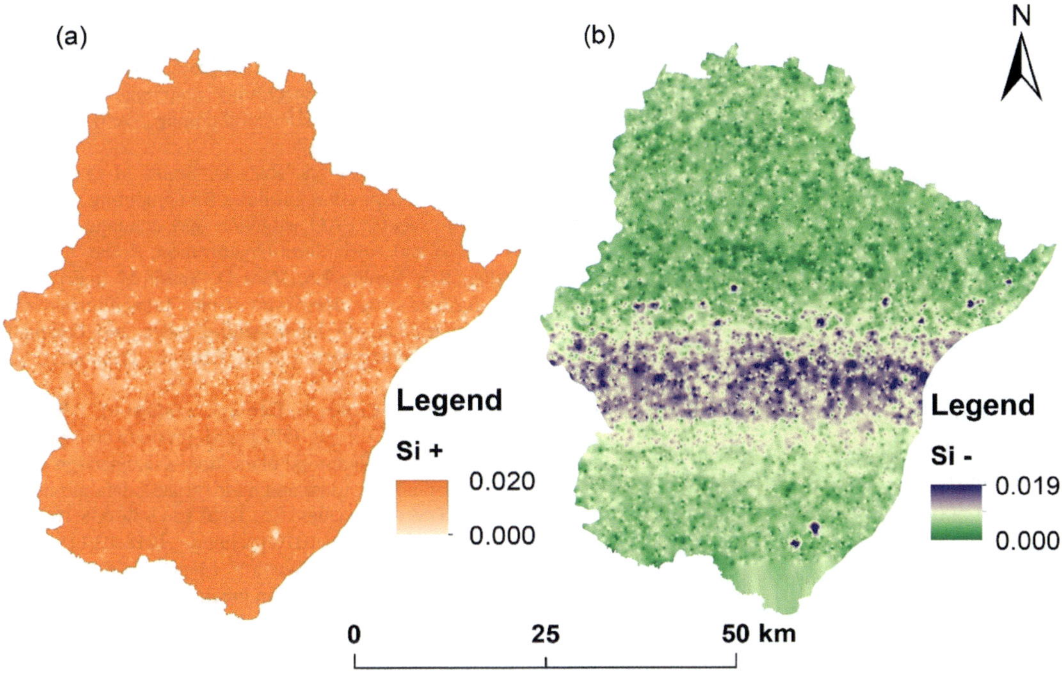

Fig. 16.4 Criteria of positive and negative impact of TOPSIS **a** S-Positive **b** S-Negative

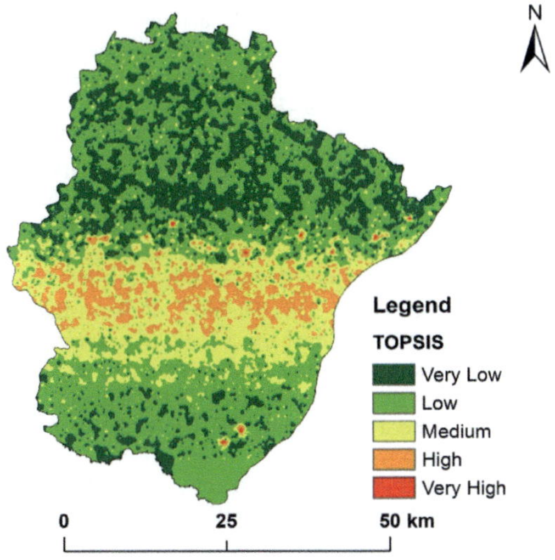

Fig. 16.5 Groundwater depletion zone of Raipur district via TOPSIS model

16.6 Conclusion

From the above discussion, it can be concluded that the groundwater which is a significant natural source of water, is facing risk of depletion. It is mainly consumed by agricultural populace. With the increasing rate of population, a large amount of ground water withdrawal is also accelerated. The present study reveals how ground water depletion zonation is related to the changing trend of rainfall, cropping pattern, irrigation, NDVI, VCI etc. with time in this study area. It is facing ground water depletion problem for over usage of water in agriculture and household purpose which stimulates lessening rainfall tendency. They are the leading driving pressure for groundwater drainage in the supercritical and critical sections.

Due to deteriorating agricultural diversity, the demand for water has increased in a unique range. In the rainy (wet) and ice (dry) seasons, a speedy decrease in groundwater level or an increase in GWD is due to irreversible abstraction rather than natural regeneration. Consequently, the practice of actual groundwater use and higher irrigation water control needs to be logically integrated with nearby sources and climatic circumstances. Participation of the network in surface water control and groundwater monitoring can play an important role in efficient use of resources together with waste reduction. Assessment of crop water requirement for soil moisture reduction and water conservation for future resources and development of water conservation for irrigation system is required. Therefore, an integrated, participatory watershed control software is similarly needed to harness sustainable water resources. It is much needed to reduce moisture loss from the soil and retain soil water. This is why crop water requirement assessment and higher water conservation irrigation methods need to be higher. Above all, an integrated watershed management program is necessary to use water through a scientific way.

16.7 Data Availability Statement

Data utilized in the study is obtained from both open and government agencies. The Landsat 8 download from USGS. Earth explores and extracts NDBI, data downloaded from IMD, Pune which is processed in Arc10.5, Census data, 2011 collected from the website. And groundwater data was downloaded from the website of https://indiawris.gov.in/wris/.

References

Aksoy S, Haralick RM (2001) Feature normalization and likelihood-based similarity measures for image retrieval. Pattern Recogn Lett 22(5):563–582

Al-Abadi AM, Ghalib HB, Al-Mohammdawi JA (2020) Delineation of Groundwater Recharge Zones in Ali Al-Gharbi District, Southern Iraq Using Multi-criteria Decision-making Model and GIS. J Geovisualization Spat Anal 4(1):1–12.https://doi.org/10.1007/s41651-020-00054-7

Al-Juaidi AE, Nassar AM, Al-Juaidi OE (2018) Evaluation of flood susceptibility mapping using logistic regression and GIS conditioning factors. Arab J Geosci 11(24):1–10

Allen RG, Pereira LS, Raes D, Smith M (1998) Crop evapotranspiration guidlines for computing crop water requirments. FAO irrigation and drainage paper 56. Food and Agriculture Organization, Rome

Allen RG, Pereira LS, Raes D, Smith M (n.d.) https://figshare.com. Retrieved from https://figshare.com/articles/dataset/Global_Aridity_Index_and_Potential_Evapotranspiration_ET0_Climate_Database_v2/7504448

Arabameri A, Rezaei K, Cerda A, Lombardo L, Rodrigo-Comino J (2019) GIS-based groundwater potential mapping in Shahroud plain, Iran: a comparison among statistical (bivariate and multivariate), data mining and MCDM approaches. Sci Total Environ 658:160–177. https://doi.org/10.1016/j.scitotenv.2018.12.115

Ashoka A, Gleeson T, Wada Y, Mishra V (2017) Relative contribution of monsoon precipitation and pumping to changes in groundwater storage in India. Nat Geosci 10:117–190

Basak A, Das J, Rahman ATMS, Pham QB (2021) An integrated approach for delineating and characterizing groundwater depletion hotspots in a coastal state of India. J Geol Soc India 97:1429–1440. https://doi.org/10.1007/s12594-021-1883-z

Bhanja S, Mukhrjee A, Saha D, Velicogna I, Famiglietti J (2016) Validation of grace based groundwater storage anomaly using in-situ groundwater level measurment in India. J Hydrol:729–738

Brown B, Nuberg I, Llewellyn R (2017) Negative evaluation of conservation agriculture: perspectives from African smallholder farmers. Int J Agric Sustain 15(4):467–481

CGWB (n.d.) Dynamic groundwater resource of India (as on March, 2009, CGWB, New delhi), 2011

Choudhury M, Jyethi DS, Dutta J, Purkayastha SP, Deb D, Das R, Bhattacharyya D (2021) Investigation of groundwater and soil quality near to a municipal waste disposal site in Silchar, Assam, India. Int J Energy Water Resour. https://doi.org/10.1007/s42108-021-00117-5

Chung ES, Abdulai PJ, Park H, Kim Y, Ahn SR, Kim SJ (2017) Multicriteria assessment of spatial robust water resource vulnerability using the TOPSIS method coupled with objective and subjective weights in the Han river basin. Sustainability (Switzerland) 9(1):29. https://doi.org/10.3390/su9010029

Deering DW (1975) Measuring "forage production" of grazing units from Landsat MSS data. In: Proceedings of the tenth international symposium of remote sensing of the environment, pp 1169–1198

Das J, Bhattacharya S (2018). Trend analysis of long-term climate parameter in Dinhata of Koch Bihar district, West Bengal. Spatial Inf Res:1–10

Das J, Mandal T, Saha P (2019) Spatial-temporal trend and change point detection of winter tempreture of North Bengal, India. Spatial Inf Res:411–424

Das J, Mandal T, Rahman AS, Saha P (2021) Spatio-temporal characterization of rainfall in Bangladesh: an innovative trend and discrete wavelet transformation approaches. Theoret Appl Climatol 143(3):1557–1579. https://doi.org/10.1007/s00704-020-03508-6

Deshmukh M, Tanaji VS (2017) Cropping intensity Index and irrigation intensity in India. North Asian Int Res J Soc Sci Human 3(2)

Fick SE, Hijmans RJ (n.d.) Worldclim 2: new 1-km spatial resolution climate surface for global land areas. Int J Climatol

Gain AK, Giupponi C, Renaud FG (2012) Climate change adaptation and vulnerability assessment of water resources systems in developing countries: a generalized framework and a feasibility study in Bangladesh. Water 4(2):345–346. https://doi.org/10.3390/w4020345

Hora T, Shrinivasan V, Basu N (2019) The groundwater recovery paradox in south India. Geophys Res Lett

Joerin F, Thériault M, Musy A (2001) Using GIS and outranking multicriteria analysis for land-use suitability assessment. Int J Geogr Inf Sci 15(2):153–174

Kendall MG (1975) Rank correlation methods. Griffin London United Kingdom

Keyantash J, Dracup JA (2002) The quantification of drought: an evaluation of drought indices. Bull Am Meteor Soc 83(8):1167–1180

Khansalari S, Raziei T, Mohebalhojeh AR, Ahmadi-Givi F (2018) Moderate to heavy cold-weather precipitation occurrences in Tehran and the associated circulation types. Theoret Appl Climatol 131(3):985–1003

Mandal T, Saha S, Das J, Sarkar A (2021). Groundwater depletion susceptibility zonation using TOPSIS model in Bhagirathi river basin, India. Model Earth Syst Environ:1–21

Mann H (1945) Non parametric tests against trend. Econometrica 13:245–259

Mukherjee P, Singh CK (2012) Delineation of groundwater potential zones in arid region of India—a remote sensing and GIS approach. Water Resour Manag 26 (9):2643–2672. https://doi.org/10.1007/s11269-012-0038-9

Pearson E, Hartley H (1966) Biometrika tables for statisticians, vol. 1, 3rd ed. Cambridge University Press

Pettorelli N, Vik JO, Mysterud A, Gaillard JM, Tucker CJ, Stenseth NC (2005) Using the satellite-derived NDVI to assess ecological responses to environmental change. Trends Ecol Evol 20(9):503–510

Raziei T (2021) Revisiting the rainfall anomaly index to serve as a simplified standardized precipitation index. J Hydrol 602:126761

Sahoo S, Chakrabotry S, Pham QB, Sharifi E, Saad SS, Matej V et al (2021) Recognition of district-wise groundwaterstress zones using the GLDAS-2 cathcment land surface model during lean season in the Indian state of West bengal. Springer Acta Geophysica

Sen P (1968) Estimate of the regression coefficient based on Kendall's tau. J Am Stat Asso 63:1379–1389

Shao G, Guan Y, Zhang D, Yu B, Zhu J (2018) The impacts of climate variability and land use change on streamflow in the Hailiutu river basin. Water 10 (6):814. https://doi.org/10.3390/w10060814

Steven MQ, Srinivasan G (2009) Evaluting the utility of the vegetation condition index (VCI) for monitoring meterological drough in Texas. Elsevier

Swain S, Verma M, Verma MK (2015) Statistical trend analysis of monthly rainfall for Raipur District, Chhattisgarh. Int J Adv Eng Res Studi/IV/II/Jan–March, 87, 89

Tek BK (2018) NDvi, NDBi and NDWI calculation using Landsat 7, 8. Researchget, 1–7

UNEP (1997) World atlas of desertification. United Nations Environment programme

Van Rooy MP (1965) A rainfall anomally index independent of time and space, Notos

Walter I et al (2000) ASCE's standardized refrence evapotranspiration equation. In: Proceedings of 4th national irrigation symposium, ASAE, Phoenix, A.Z

Wang Y, Qiu M, Shi L, Xu D, Liu T, Qu X (2019) A GIS-based model of potential groundwater yield zonation for a sandstone aquifer based on the EWM and TOPSIS methods. In: IMWA "Mine water: technological and ecological challenges", pp 387–393

Zolekar RB, Bhagat VS (2015) Multi-criteria land suitability analysis for agriculture in hilly zone: remote sensing and GIS approach. Comput Electron Agric 118:300–321

Part II

Multi-hazards Management

17 Terrain Sensitivity Guided and People's Perception Based Risk Area Management of the Hills of Darjeeling District, India

Sudip Kumar Bhattacharya

Abstract

The hills of the Darjeeling district having litho-geomorphologically complex and diversified landform, have been subjected to rapid human agglomeration since the post-independence time with establishment of ordered Government and hill tourism in the panoramic sites of the Darjeeling hills. Human pressure on the landscape in the hills of Darjeeling district has given rise different magnitudes of Terrain Sensitivity and inhabitants of the hills equally perceived human activities on the landscape as the cause of different degrees of 'Risk Area occurrences' which are hazardous in different degrees for safe human habitation. Therefore, the present paper takes an attempt to identify and propose management strategies for the different risk areas for human habitation guided by Terrain Sensitivities appeared on the landscape and based on People's Perception that mandate various degrees of fragility associated with landforms with human impact or use. Result reveals four categories of risk areas associated with the hills of Darjeeling district for which appropriate management strategies have been proposed.

Keywords

Human pressure on the landscape · Terrain sensitivity · People's perception · Risk Area occurrence · Management strategies

17.1 Introduction

The panoramic hills of the Darjeeling district experienced quite rapid human agglomeration with a motive of grabbing lands following the establishment of ordered Government after independence as well as flourishing hill tourism over the geologically complex (Mallet 1874; Pawde and Saha 1982; Acharyya et al. 2017) and geomorphologically form-process interacted diversified landscape. As a consequence, the region has undergone deforestation (Bhattacharya 2009) with gradual expansion of settled areas and arable lands. Hill tourism, moreover, has become a very attractive recreational medium for the outside visitors in the panoramic and scenic sites in the hills which have added impetuous on rapid agglomeration of population and accelerated expansion of urban areas and business induced settlement growth in the hills of Darjeeling district. All these resulted dwindling

S. K. Bhattacharya (✉)
Department of Geography and Applied Geography, University of North Bengal, Siliguri, West Bengal 734013, India
e-mail: skbhatt2002@yahoo.co.in

J. Das and S. K. Bhattacharya (eds.), *Monitoring and Managing Multi-hazards*, GIScience and Geo-environmental Modelling, https://doi.org/10.1007/978-3-031-15377-8_17

of forest and expansion of concrete areas and substantially contributed the change in the natural setup of this region with increase of built environment. This change of natural set up and human pressure on the geologically and geomorphologically complex landscape is seen to have significant bearing upon the stability of this region (Bhattacharya 2016) because human pressure has been randomly exerted upon the landscape irrespective of the degree of tolerance the landscape can endure. Such human pressure is manifested by frequent occurrence of devastating landslides almost every year after rain as well as soil erosion in the present time (Bhattacharya 1999, 2002, 2012a, b, 2014; Ghosh et al. 2009; Ghosh 2011).

In the present time, the great challenge of the human being is to look into the consequences on nature resulted from the activities done by human itself. Many important works have been carried out by the different experts all over the world on the activities of human being on the nature and alteration of nature and related consequences (Bathrellos et al. 2012, 2013; Bhattacharya 2012a, b; Papadopoulou-Vrynioti et al. 2013). Over exercise by human on the nature has become consequent boomerang (Bhattacharya 1993) in many ways like temperature increase in the atmosphere, increasing contamination and pollution in the biosphere, human pressure on the landscape and the resultant increase in soil erosion, slope failure, and many more Bhattacharya (2014, 2016). Therefore, appraisal of the human activities on nature is necessary. Terrain Sensitivity assessment is such a way to measure and perceive terrain's degree of ability to sustain the pressure of land use developed over a region (Bhattacharya 2018, 2021). With terrain sensitivity appraisal it is also necessary to see what local people think over the status of the area they live in so that both spatial terrain sensitivity status and human perceived terrain's condition can pave a way of environmental planning for sustainable habitation.

17.2 Area of Study

The hills of the Darjeeling district after being separated from Kalimpong part is extended from 26° 46″ 32′ to 27° 13″ 13′ N and 87° 59″ 42′ to 88° 28″ 53′. The hills of Darjeeling district has been delineated by sharp break of slope from the contiguous undulating Terai plains. The entire hill region of the Darjeeling district covers an area of 1206.3 Km^2 bounded by Nepal on the West, Kalimpong district on the East, Sikkim on the North and Terai undulating plains on the South (Fig. 17.1).

17.3 Environmental Planning

Environmental planning is targeted to save an area from further deterioration and to recover the area to a certain extent from the damage what has already been done. Environmental planning has attracted both physical and human geographers since a long time as it provides the statutory instruments for changing land use designations, for instance protecting national parks from certain types of development. It is also a key forum for mediating between those with different views of the future for particular areas, most notably with regard to whether or not to allow development on a particular piece of land and, if allowed, what kind of development should be permitted (Haughton 2017).

The terrain sensitivity map (Fig. 17.2) of the hills of Darjeeling district gives a clear picture of the spatial variation of sensitivity ranging from high to stable zones resulted from different degrees of human pressure on the landscape for which some kinds of damage related to the stability over the different magnitudes of sensitivity is inevitable and higher sensitive zones are in no way congenial for further development. Investigation has depicted (Bhattacharya 2016, 2021) that the hills of the Darjeeling district, has been subjected to steady development by the human

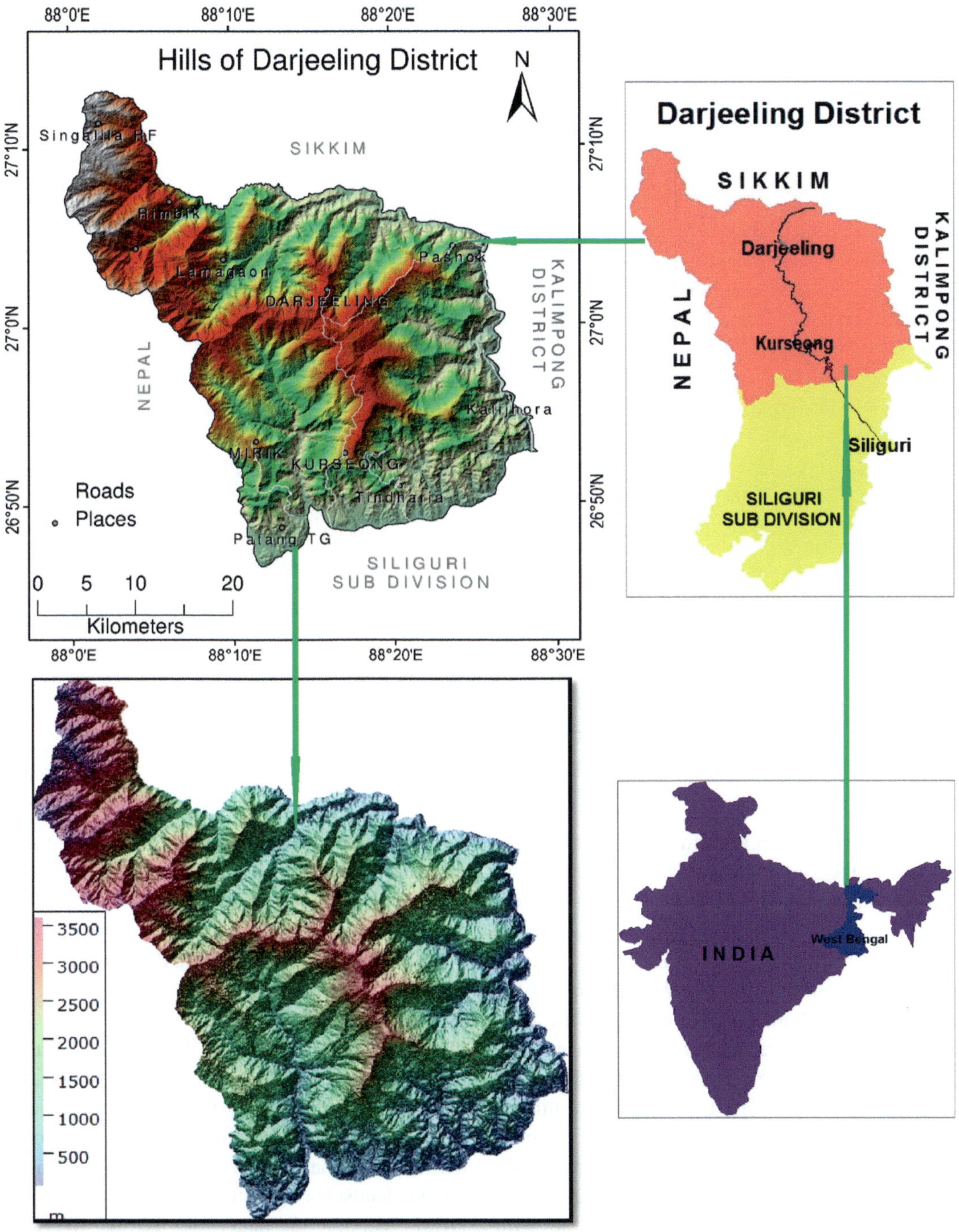

Fig. 17.1 Location map of the hills of Darjeeling district

being for which human induced pressure has converted congenitally better tolerable zones to a zone of increased terrain sensitivity where immediate planning and management is necessary. Where the terrain parameters being considerably weak, made the area least tolerable;

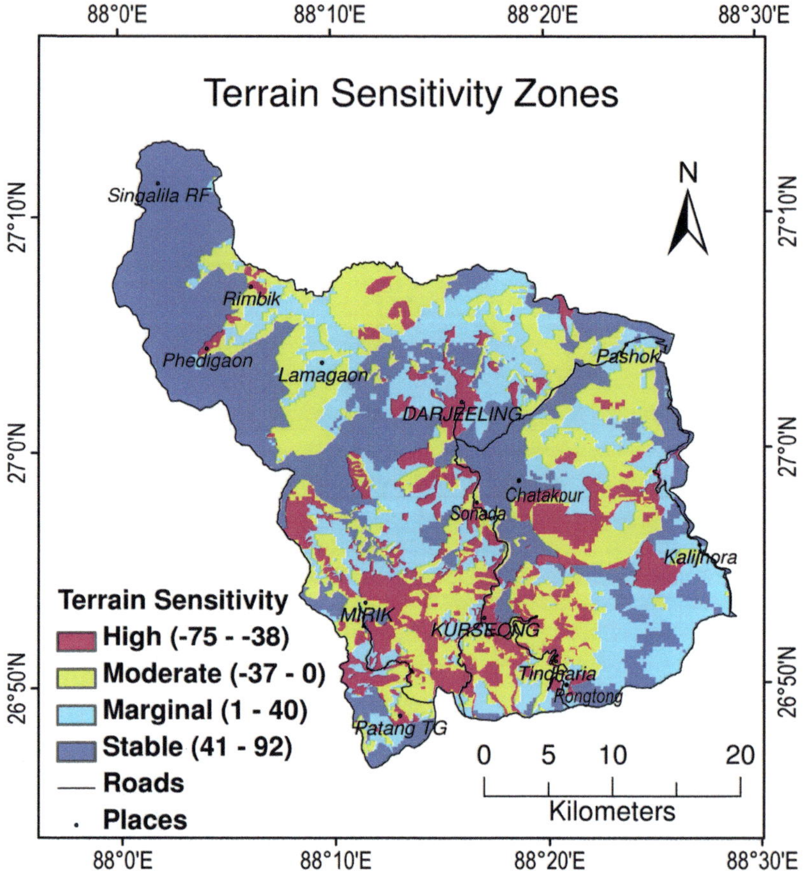

Fig. 17.2 Terrain Sensitivity map of the hills of Darjeeling. *Source* Bhattacharya (2021)

human activities like settlement growth, improper use of the land for cultivation, deforestation, and conversion of land in to scattered tree fields, converted the area into highly sensitive terrain for human habitation (Bhattacharya 2021). So, it is the crucial time to think over how much we will spread our hands over the hills of the Darjeeling district where many areas are in no way congenial for further development. Moreover, what will be the strategy to recoup (with utmost effort) the damage that has already been done on different terrain sensitivity zones in study area?

Therefore, the terrain sensitivity guided environmental planning can be a way out. Along with this the investigator also has taken an attempt to see the humans' perception on the 'built environment' and their grown up need for planning. Both humans' perception and spatial terrain sensitivity, therefore, will be used for a well thought planning and management for the sustainable use of the terrain of the study area.

17.4 Materials and Methods

The map of terrain sensitivity zones (Fig. 17.2) is the product of Proportional Value-added Rating (PVR) algorithm-based Terrain Tolerance (TT) appraisal through multi-criteria decision analysis (MCDA) followed by a deductive method of human influence (pressure) estimation on the landscape (Bhattacharya 2021). ASTER DEM Version 3 data for terrain tolerance appraisal and SOI topographical maps (RF = 1:50,000) and LISS-IV satellite images (5.8 m resolution) for estimation of human influence on the landscape, i.e., land use factors have been used along with thorough field investigation.

Perception study has been carried out to see what is the opinion of the people about an area; whether the locality or the concerned terrain has been changed to a different condition due to peoples influence over the terrain through their developmental activities and for which the area need a specific level of planning and management. The perception study has been carried out by different experts in the different field of study (McDonald 2012, Ghosh and Debajit 2012). For this study a questionnaire has been prepared. The questionnaire has been framed in such a way that all the important aspects of built environment and need of planning and management have been covered. The term 'built environment' refers to "the human-made surroundings that provide the setting for human activity, ranging in scale from buildings and parks or green space to neighbourhoods and cities that can often include their supporting infrastructure, such as water supply or energy networks. The built environment is a material, spatial, and cultural product of human labour that combines physical elements and energy in forms for living, working, and playing" (Kaklauskas and Gudauskas 2016). It has been defined as "the human-made space in which people live, work, and recreate on a day-to-day basis" (Wikipedia 2021). Therefore, the built environment and its significance as perceived by human being has been used as a tool to evaluate the status of environment of an area to compare with the achieved Terrain Sensitivity of different grades (Fig. 17.2). This comparison will also help in validating the spatial magnitudes of Terrain sensitivity of the study area. Finally, the humans perception on the **need of planning and management** of selected localities will be checked with various magnitude of terrain sensitivity zone to propose some comprehensive management measures to restore the hills of Darjeeling district from further damage and for sustainable use of the terrain.

The material used for this purpose is a close-ended questionnaire having thoughtfully chosen questions for both the built environment and need of planning and management. Survey has been carried out on reasonably conscious people at least have completed their secondary education. Scaled questions have been answered by those people of selected localities ticking on the appropriate ratings of the questions in the given Questionnaire as 1, 2, 3, and 4. At least five numbers of answered questionnaires have been collected from each locality to get the opinion of the respondents for both the built environment of the locality and need of planning and management of the concerned area. On the basis of these the Tables 17.1 and 17.2 have been prepared where the maximum number of similar opinions given by the respondents of each question has been considered as the right answer for that question having certain assigned rating value. Finally, analysis has been done on the basis of the data obtained in the Tables (17.1 and 17.2) and its cartographic representation by column and pie diagrams in respect of the ratings of each question (Figs. 17.3 and 17.4).

17.5 Findings and Discussion

17.5.1 People's Perception on the Built Environment

As mentioned above that built environment is the human-made condition in the natural set up of an area or region for developing supporting infrastructure for living; a material, spatial and cultural product with supported amenities for living. It is in one sentence-a human-made space in the name of development to achieve ease and comfort in living. So, it is, on the other hand, the human built energy on the nature compromising with the ecological balance to a certain extent in a particular area or region. So, with the increase of built environment it can be expected that the alteration of natural set up of an area will lead to a kind of deterioration in the nature if balanced with human activities will remain in stable condition or if not balanced will invite instability threshold inevitably. Therefore, the assessment of built environment, the way the inhabitants or the people of the area or region perceive, has been considered as an important parameter in which a set of questions have been put carefully

Table 17.1 Built environment ratings

Questions	Darjeeling	Kurseong	Mirik	Tindharia	Pasok	Rimbik	Sonada	KaliJhora	Patang	Rangtong	Singalila	Phedi Gaon	Chatakpur
1	3	3	4	3	2	1	2	3	2	2	1	1	1
2	4	4	4	4	3	3	3	2	2	1	1	1	1
3	4	4	3	4	3	3	2	2	2	2	2	3	2
4	4	4	4	4	3	3	3	2	2	2	1	1	1
5	4	4	4	4	3	3	3	3	3	3	1	1	1
6	4	4	4	4	2	2	3	2	2	2	1	1	1
7	4	4	3	3	1	1	2	3	1	1	1	1	1
8	2	2	3	3	3	3	3	2	4	4	4	3	3
9	3	3	3	3	3	3	3	2	4	4	4	2	2
10	4	4	4	4	3	3	3	2	2	2	1	2	2
11	4	3	3	3	2	3	2	1	2	3	2	2	2
12	3	3	3	2	2	1	2	2	2	2	1	1	1
13	2	2	2	4	2	1	2	2	3	2	1	1	1
14	4	4	4	3	3	2	2	2	2	2	1	1	1
15	4	4	4	4	3	3	3	2	2	2	2	2	2

Table 17.2 Need of planning and management ratings

Questions	Darjeeling	Kurseong	Mirik	Tindharia	Pasok	Rimbik	Sonada	KaliJhora	Patang	Rangtong	Singalila	Phedi Gaon	Chatakpur
1	3	2	4	4	3	2	2	2	2	3	1	1	1
2	4	2	4	4	3	3	3	3	2	2	1	2	2
3	4	3	3	3	2	2	3	3	1	2	1	1	1
4	4	3	4	3	3	2	2	2	2	2	1	2	1
5	3	3	3	4	3	3	3	2	2	2	2	2	2
6	4	4	4	4	3	3	2	2	2	2	2	2	2
7	4	4	3	3	2	2	3	2	1	2	1	1	1
8	4	4	4	4	3	3	3	3	2	2	2	2	2
9	4	4	3	4	3	3	3	2	2	4	2	1	2
10	2	4	4	4	3	3	3	2	3	4	2	2	2
11	2	4	4	4	3	3	2	2	2	3	2	2	2
12	4	4	3	4	2	3	2	2	2	2	1	1	1
13	3	4	4	4	2	3	2	2	2	2	1	1	1
14	4	4	3	4	2	2	2	2	2	2	1	1	1
15	4	4	4	4	3	3	3	2	2	2	1	1	1

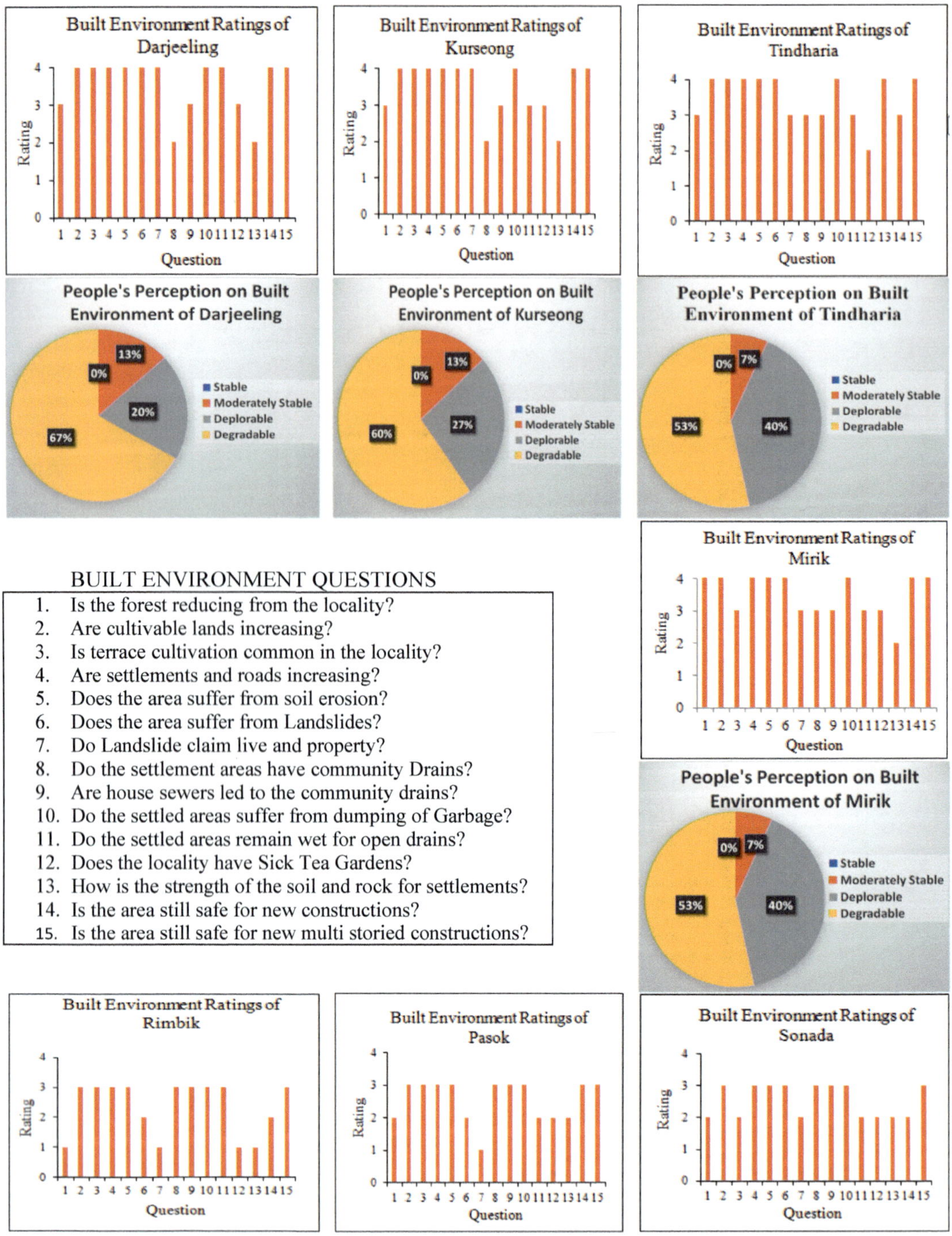

Fig. 17.3 Built environment perception study

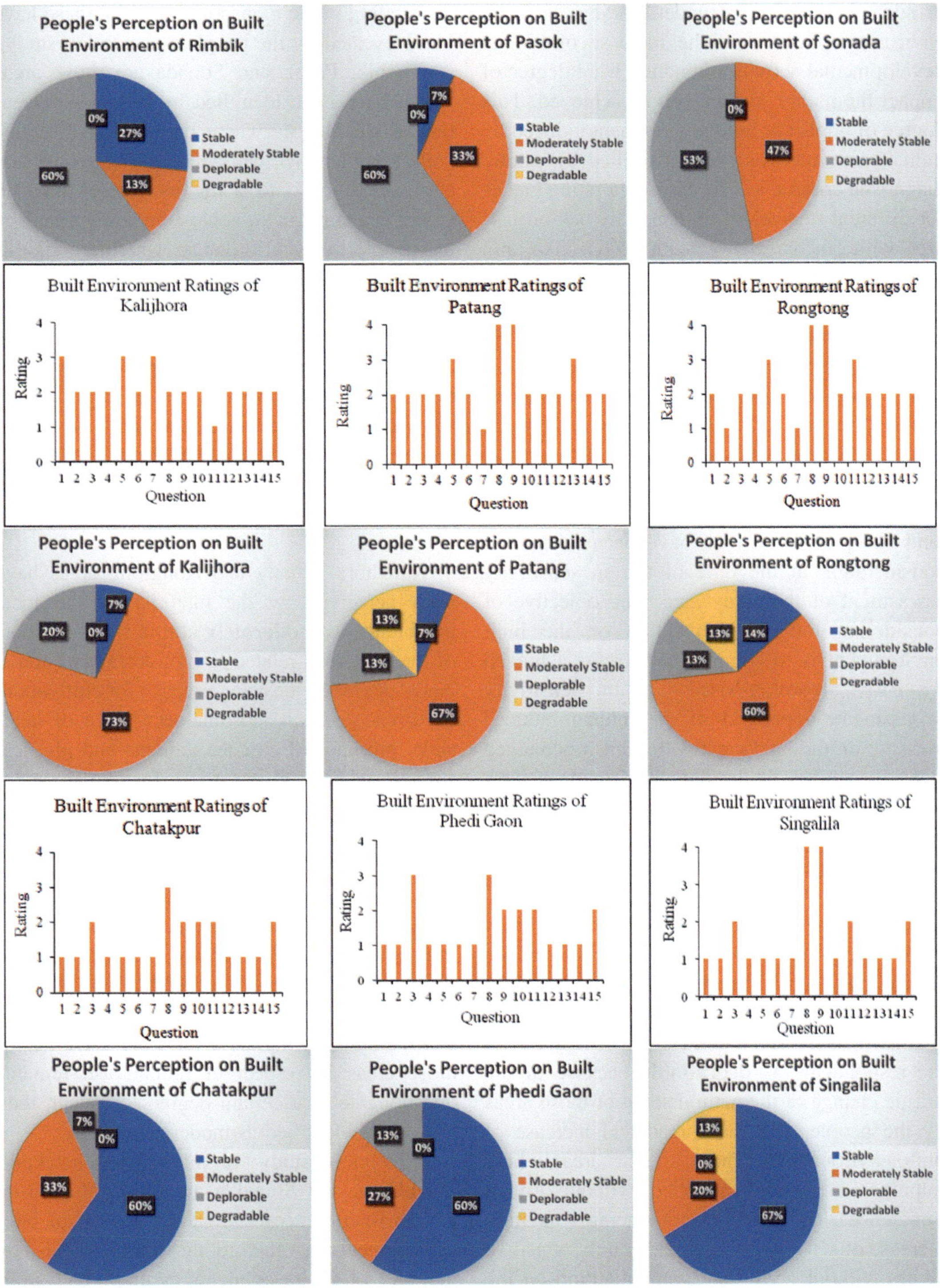

Fig. 17.3 (continued)

in such a manner (for understanding the built environment) that with the increase of human developmental activities in nature what degree of impact the nature sustain, can be achieved. This degree of impact of built environment induced by human activities has been quantified by setting diagnostic criteria, i.e., by assigning rating values for different degrees of human activities on nature with increasing generalization like no, slowly, steadily, rapidly or no, rare, few, and common as relevant to the each question for different areas. Fifteen selected questions have been put forward to cover up the aspects of human activities on nature like deforestation, increase of arable lands, nature of cultivation practice etc., as well as human-made material, spatial, and cultural space in the form of establishing, settlements. These are nothing but the land use practices which have been taken into consideration at the time of terrain sensitivity assessment of the study area. The objective of considering human's perception on the built environment is to compare the results achieved by PVR algorithm-based Terrain Sensitivity Assessment, a product of the human induced pressure on the landscape with human perceived state of stability of the built environment. This can validate how much the geospatial-data based terrain sensitivity assessment has resemblance with human perceived degree of stability related to built environment of the study area.

Figure 17.3 shows the result of the built environment perception study of the hills of the Darjeeling district where it is clear that the localities like Darjeeling, Mirik, Kurseong and Tindharia have been identified by the majority of the respondents as **degradable** because of the drastic change of the natural setup of those areas by the people with the spread and increase of built environments. These are the areas where improper cultivation practice without giving impotence of scientific terracing method, increase of settlements and roads on the sensitive slopes without giving adequate attention on the constructional rules have invited severe soil erosion and increased the magnitude and frequency of landslide occurrences (in many years devastating in the monsoon time). All these have been revealed by the people's perception study.

Rimbik, Pasok and Sonada are those areas which have been identified by people's perception as **deplorable** where human activities like cultivation practice without adopting proper scientific method like terracing, increase of road and settlement constructions have been identified by the people living in the areas which are significantly against the physical stability of these localities. The human practices like dumping garbage in the settled areas (Fig. 17.4e) and the dearth of community drains to pass sewer water too, give indications that these localities are subjected to soil erosion and landslide occurrences, and most of these have been rated by the local people as 3 for which those areas have been graded as deplorable from environmental points of view.

Kalizhora, Patang and Rongtong areas have been perceived by the majority of the local inhabitants as **moderately stable** zones as most of the questions related to the increase of arable lands, degree of terracing practice for cultivation, occurrence of soil erosion, occurrence of landslide, presence of sick tea gardens and physical strength of the soil and rocks for human settlements commonly have yielded rating value 2 in respect to the alteration of natural environment and thereby the growth rate of the built environment. There are some questions, in those localities like rate of deforestation, lack of community drains, answered by the local people which indicate human interference on the landscape has left a few worrying signature on the lack of human awareness for which stability of those areas in respect of human use remains questionable to a certain degree for which they have been identified as moderately stable.

Perception study reveals that Chatakpur, PhediGaon and Singalila are mostly free from human interferences and steady growth of built environment as the rating value given by majority of the respondents as one which indicates degree of human energy in the natural ecosystem is almost in a balanced state although there are some negative signs like lack of

a: Congested Built Environment in Kurseong

b: Congested Built Environment in Darjeeling town

c: Highrising Building in Darjeeling Hills

d: A part of Sick Tea Garden in High Altitude

e: Garbage dump and wetness behind buildings

f: Flakey and Friable rocks in the Daling Thrust

g: Skeletal soil on the oversteepened hill slipe

h: Ability of Amliso to anchor rocky surface

Fig. 17.4 Relevant photos of the study area

community drains and unmanaged sewer. As the human has not overpowered the ecological setup of these localities they have been considered as **stable**.

When compared the human mandates, regarding the conditions of these localities, with the various zones of terrain sensitivity resulted from human pressure on the landscape, it reveals that high, moderate, marginal and stable terrain sensitivity zones have fairly good resemblance with human perceived conditions of those localities, since the **degradable** localities like Darjeeling, Kurseong, Mirik and Tindharia; the **deplorable** localities like Rimbik, Pasok, Sonada; **moderately stable** areas like Kalizora, Patang and Rongtong and **stable** locations like Chatakpur, Phedi Gaon and Singalila almost correspond with high, moderate, marginal and stable terrain sensitivity zones (Fig. 17.2), respectively. So, the People's Perception adequately validates that PVR algorithm-based assessment of the spatial distribution of terrain sensitivity zones are significant.

17.5.2 People's Perception on the Need of Planning and Management

Human's perception about their need of planning and management for their localities have also been studied and the following results have been obtained (Fig. 17.5):

In the areas of Darjeeling, Kurseong, Mirik and Tindharia **very high** level of planning and management have been proposed by the majority of the respondents who feel strong to high necessity of tree and grass plantation for retaining slope stability along with high demand of terracing practice on the cultivated fields in association with strong to high need of mulching practice on the crop fields to protect the slopes. No respondent encouraged building or multi-storied building construction over all these localities and in these localities majority of the respondents demanded maintenance of drainage system in the settled areas with dense and high rising buildings and addition of community drains for free passage of water from house sewers except only in Darjeeling town area where present status of draining system with community drains to lead all the house sewers is better being the head quarter of the district. Most of the respondents of these localities also felt high need of landslide, soil erosion managements since they felt these areas are environmentally highly fragile.

Rimbik, Pasok and Sonada areas have been perceived by majority of the local respondents as the places which need **high** level of planning and management for restoration of the stability of these areas as in the opinion of the respondents these are the places which are adequately fragile from environmental points of view. These areas strongly need plantation of slope stabilizing grasses and/or trees, mulching and terracing on the crop fields as well as moderate to strong level of landslide and soil erosion management as perceived by the local inhabitants. People are strongly against the construction of multi-storied buildings, they feel necessity of garbage removal from the settlement areas by community effort as well as community drains to lead house sewer in their locality.

People's perception about the Kalijhora, Patang and Rongtong denotes that these areas are partially fragile from environmental points of view where **moderate** level of management and planning are needed keeping attention on the maintenance of ground cover, mulching and terracing on the cultivable fields. People also need moderate level of garbage removal by community effort, improvement of drainage system in the settled areas along with addition of community drain in the locality for free passage of sewer water so that the ground surface of the settled areas will be free from unwanted wetness. Most interesting feature of these locations in common is that the need of management and planning on the basis of the majority of the aspects taken into consideration, proposes moderate level, but there are some aspects which need even higher level of management as reflected in the people's perception too.

Need of planning and management for Chatakpur, Phedi Gaon and Singalila as perceived by

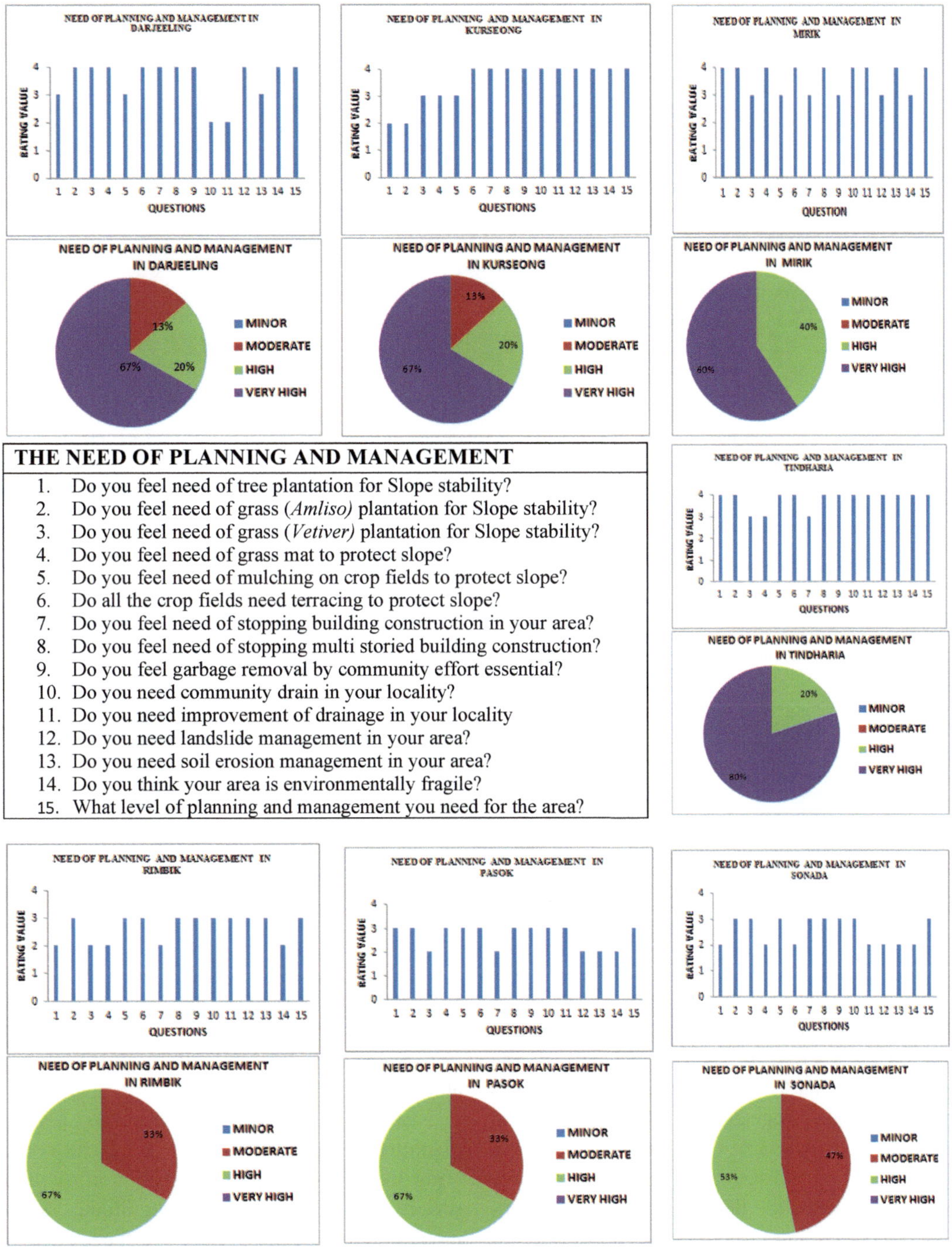

Fig. 17.5 Need of planning and management perception study

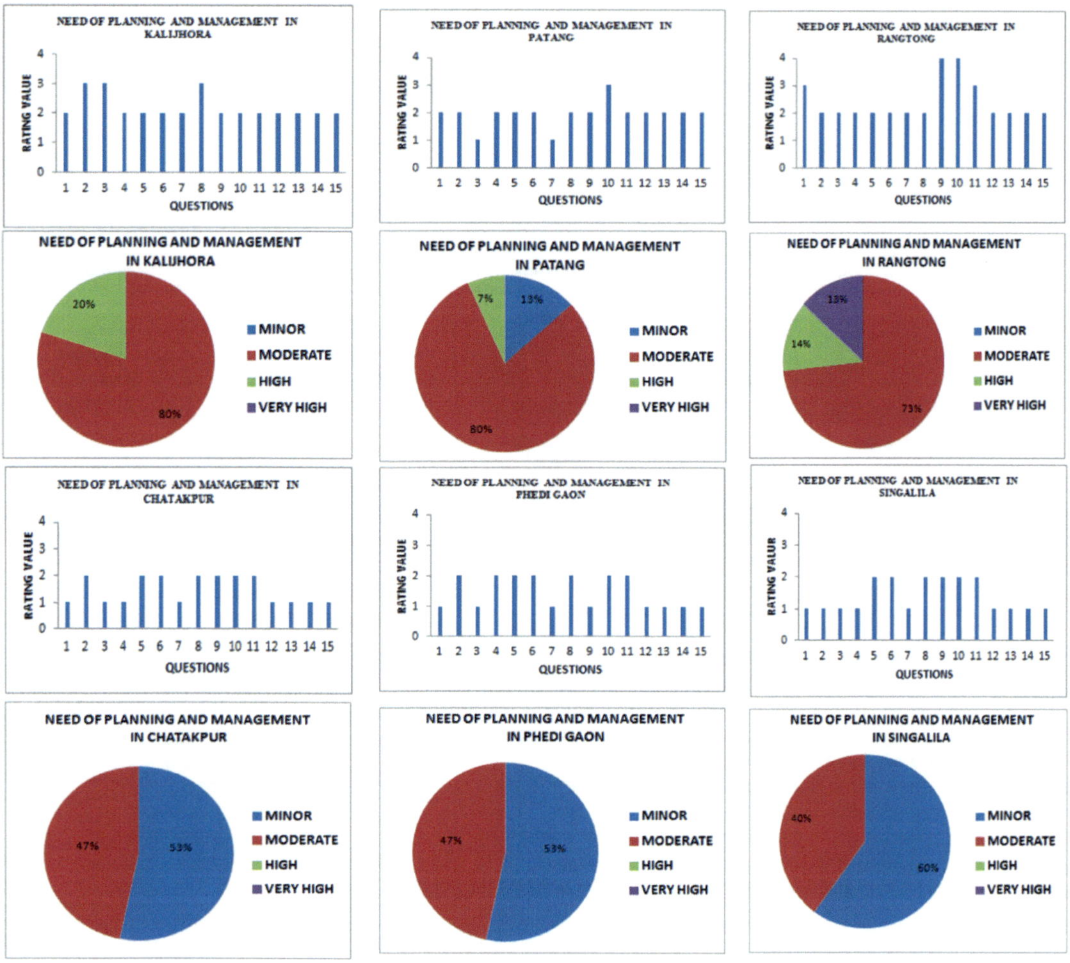

Fig. 17.5 (continued)

the people of the locality is **minor** since necessity of increasing vegetation cover by planting trees on these areas are felt by majority of the local respondents as negligible since these areas have adequate vegetation. Majority of the respondents do not have the opinion of stopping further building construction, nor do they have any urge of landslide and soil erosion management for these areas, since these areas are not environmentally fragile in their opinion in the present state. However, plantations of grass, mulching and terracing on the crop fields and stopping multi-storied buildings in the localities have their top priority to keep the balance of nature with the built environment of these areas.

Keeping the peoples' perceived need of planning and management in mind and considering the assessed spatial distribution of terrain sensitivities having varied magnitudes, a well thought management plan is necessary for restoration of the environmental balance of the study area. Therefore, the following management plans have been proposed:

17.5.3 Environmental Planning and Management for the Study Area

A plan for management is intended for elevating the condition of an environment (be it built environment) from a state where the stability of the area will become free from vulnerability risk. The human activities on the landscape has exerted pressure on the study area as understood from terrain sensitivity assessment as well as from people's perception for which it is high time to think over a well thought management plan so that the built environment will have a balance with the natural setup of the area concerned. The following management strategies have been proposed:

As apparent from the Figs. 17.2, 17.3 and Table 17.3 that the terrains of high sensitivity is actually the human perceived degradable zones where human activities have flourished as an incidence mostly on geologically softer rocks and disturbed zones (Fig. 17.6) of Lesser Himalayan Sequence (LHS) having Palaeoproterozoic and Permian sedimentary formations and on the Miocene rocks of lower Siwalik; adequately deformed (have become flakey and friable) by the effect of the Daling Thrust (DT) (Fig. 17.4f), one of the most important tectonic zone between Daling and Damuda groups as well as by the Main Central Thrust (MCT) and Main Boundary Thrust (MBT). The thrust induced flakey and friable nature of the relatively harder phylitic rocks of Daling series, Permo-carboniferous softer Damuda rocks of Lower Gondwana and foreland molasse sediments (sandstone/shale) of the Siwalik group with thrust caused fold effect (Kellett et al. 2014; Ghosh et al. 2016; Acharyya et al. 2017) altogether being deeply entrenched by the turbulent rivers have given rise high dissection and ruggedness over this part of the study area and thereby a kind of physical instability over which high congestion of concrete and multi-storied

Table 17.3 Recommended strategies for environmental management plan

Terrain sensitivity, human perceived terrain status and grade	Present built environment and landuse status	Recommended management strategies and human perceived need of planning and management
High sensitivity, degradable zones (Grade IV, Very high risk area)	i. Highly congested concrete settlements with many multi-storied structures in the towns and semi-town areas ii. Closely scattered concrete structures and buildings in the settled areas iii. Fallow or bare land surfaces iv. Row crops on the un-terraced land surfaces v. Sick tea garden areas	**Very high** i. Zones should be strictly prohibited for further settlement growth and multi-storied building constructions ii. Un-vegetated fallow or bare lands to be planted by community efforts with intense root anchoring fodder and fuel wood plants for local consumption and with profit earning Amlisho or Broom Grass (Thysanolaena maxima). Another grass species Vetiver (*Vetiveria zizanioides* L.) which is also economically prospective and very effective in soil and water conservation to be planted with equal importance iii. Row cropping practice without terracing should be restricted by enforcing rules iv. In the open spaces of the settled areas social forestry program to be initiated from community level v. Improvement of sick tea gardens to be brought under the notice of garden authority by local administration

(continued)

Table 17.3 (continued)

Terrain sensitivity, human perceived terrain status and grade	Present built environment and landuse status	Recommended management strategies and human perceived need of planning and management
Moderate sensitivity, deplorable zones (Grade III, High risk area)	i. Congested settlements in urban and semi-urbanized areas almost with concrete structures ii. Vast closely scattered settlement areas iii. Tea garden, row crop cultivation areas iv. Small forest patch v. Large area under scattered tree fields	**High** i. Further settlement growth on the congested areas to be strictly controlled by rule. However, fringe areas of the urban settlements and vast closely scattered settled areas to be allowed for planned growth emphasizing upon construction of single storied buildings on stable slopes ii. Vegetation density in the urban zones and closely scattered settlement areas to be increased by social/agro forestry programmes by community efforts iii. Compulsory terrace cultivation for the extensive row crop cultivated lands to be adopted iv. Govt. Forest Departments should be insisted by the local people (by demonstration) to initiate Afforestation programme on the scattered tree field areas
Marginal sensitivity, moderately stable zones (Grade II, Moderate risk area)	i. Most tea garden and cinchona plantation areas ii. Wide areas of closely scattered settlements in higher altitude zones iii. Some scattered tree field areas iv. Large and small patches of forest cover	**Moderate** i. Within the gardens of tea and cinchona plantation if further expansion of colonies and quarters are necessary, it should be done following stipulated construction rules in well planned manner on the stable hill slopes strictly with single storied constructions ii. The closely scattered settlement areas on the high altitudes can be permitted for further planned growth preferably with single storied buildings on stable slopes with approval from concerned construction authority iii. Govt. forest departments should be insisted by the local people (by demonstration) to initiate Afforestation programme on the scattered tree field areas
Stable in sensitivity, stable zones (Grade I, Low to risk free area)	i. Most of the area under forest cover ii. Low altitude few tea garden areas iii. Little area under scattered tree fields in high altitude zone	**Minor** i. No immediate measures are necessary except the scattered tree fields which should be brought under intense afforestation programme as recommended for other zones

settlements (Fig. 17.4a), closely scattered concrete structures and buildings, erosion prone fallow or bare land surface, un-terraced row crop cultivation practice, sick tea gardens, and sparse vegetation cover made the entire zone fragile (Bhattacharya 2021) for which this landform has been graded as IV, i.e., **Very High Risk areas** from the viewpoint of terrain sensitivity and

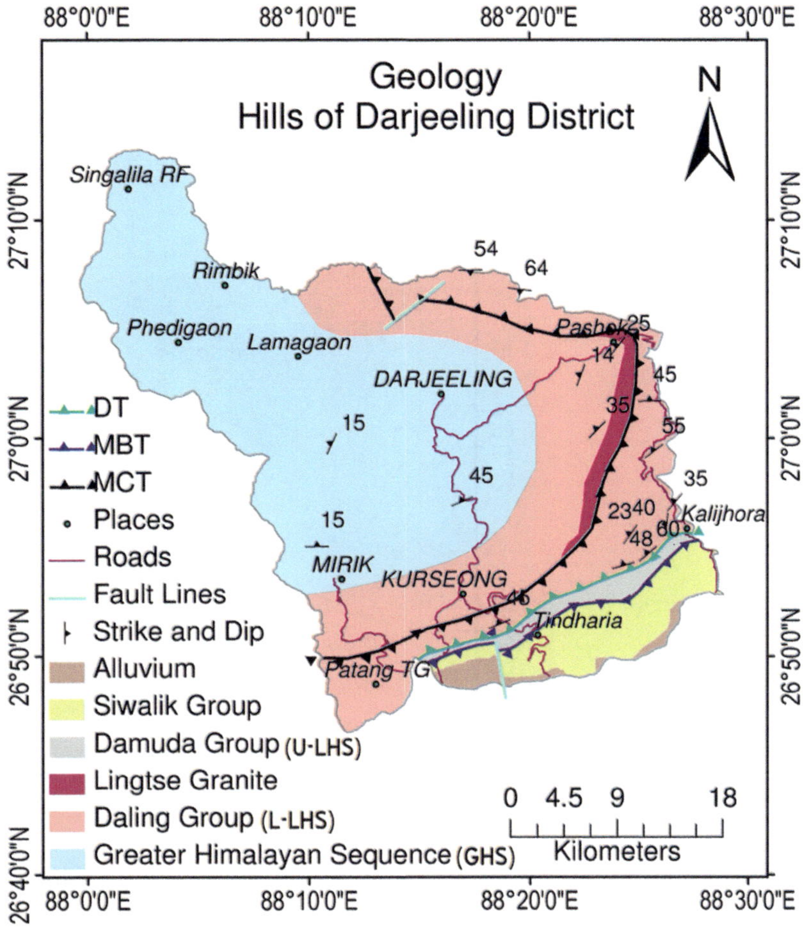

Fig. 17.6 Geological sequences of Darjeeling Himalaya

human perception. Another zone of same grade is identified over the northern part of the study area covering a region of geologically less disturbed and harder rock strata of Greater Himalayan Sequence (GHS) (Fig. 17.6) having Dharwarian Darjeeling gneissic rocks where very steady growth and expansion of urbanization in the town of Darjeeling and its fringe areas, mostly induced by flourishing hill tourism, have converted the area into a jungle of concretes and high rising buildings (Fig. 17.4b, c) on the lofty hills and on some densely settled areas developed due to rapid urban growth as well as on few congested settlements in the high altitude tea gardens. The huge pressure of settlements for urban growth and expansion along with poor condition in some tea gardens (Fig. 17.4d) on the high altitudes have converted this zone into high terrain sensitivity and people's perception accordingly marked this zone as environmentally degradable.

To restore these zones from further deterioration as well as to ensure the stability of the built environment of these zones a well thought management strategies have been recommended in the Table 17.3 where strict prohibition of further settlement growth and multi-storied building constructions have been proposed. The un-vegetated fallow or bare land surfaces in these zones have been recommended for intense root spreading and soil anchoring fodder and fuel wood plants. These plants, e.g., Chilawne (*Schima wallichii*), Gokul (*Ailanthus grandis*) and Utish (*Alnus nepalensis*) and shrubs like Guenyhlo (*Callicarpa arboria*), Amaro (*Spondius pinnata*) and Phirfira (*Firmiana colorata* R.

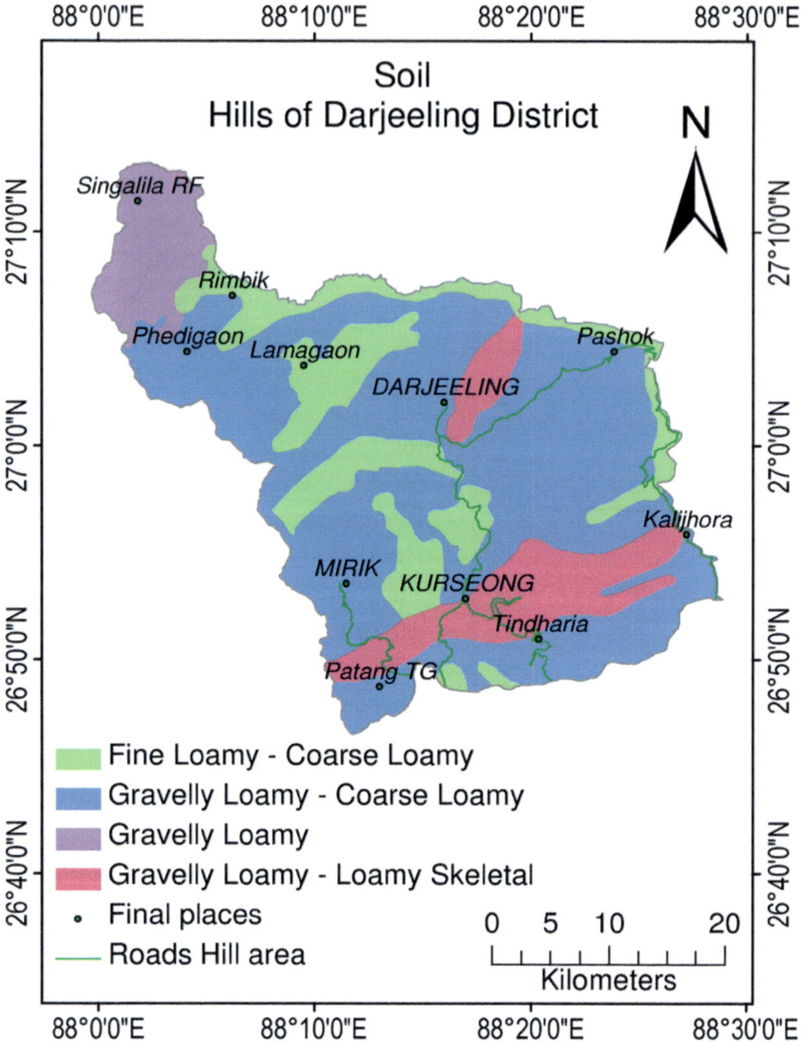

Fig. 17.7 Physical soil properties of the hills of Darjeeling district

Br.) easily thrive even on most difficult sites such as cliffs and joints, road cuts, hard and dry soils and stony surfaces along the hill slopes (Bhattacharya 2018) and in turn have the economic potential for giving return to the local people (Reddy 1983; Singh et al. 2006). Other than these plants, two important grass species Amlisho or Broom Grass (*Thysanolaena maxima*) and Vetiver (*Vetiveria zizanioides* L.) can be chosen for plantation on the bare soil surfaces. Amlisho or Broom Grass can grow almost everywhere on the hill slopes (Fig. 17.4h) and is economically very profitable since this grass stems are used for making domestic brooms (Bhattacharya 2013, 2018). Vetiver is a grass species which has been recommended by the agricultural scientists of World Bank as a soil and water conservation tool in a number of countries including Madagascar and India. Vetiver is a 'climax' plant and therefore if all plants are destroyed by draught, flood, pest, disease, fire or other adversities, vetiver will remain to protect the ground from the onslaught of the next rain (Sivamohan et al. 2010). Vetiver has economic potential too because it is a high value essential oil crop, used as fixative in perfumery and cosmetic industries (Yaseen et al. 2014). Compulsory terrace cultivation for the row crops should be ensured in association with increase of vegetation cover in the settled areas by community effort as well as improvement of the conditions of few tea gardens in these zones of terrain sensitivity.

The zones of moderate sensitivity perceived by the local people as deplorable zones (Figs. 17.2, 17.3 and Table 17.3) are again mostly confined within the both geological provinces of upper and lower Lesser Himalayan Sequence (LHS) and lower Siwalik rocks where thrust effect induced flakey nature of the rock strata coupled with Damuda and Siwalik groups of relatively softer rocks of medium hardness (Fig. 17.6) as well as Gravelly Loamy-Coarse Loamy to Gravelly Loamy-Loamy Skeletal soils (Figs. 17.4g and 17.7) are prominent. These altogether being affected by moderate to high dissection, moderately high to very high ruggedness and higher length of drainage and overburdened with congested urban and closely scattered settlements with inadequate vegetation cover on the land surface (Bhattacharya 2021) have moderately exceeded the limit of terrain tolerance. Therefore, these terrain zones have been identified as grade III, i.e., **High Risk areas**.

For proper planning and restoration of these zones and to bring back the entire region from deplorable to stable state, a high level of management has been proposed with the appropriate measures as in the Table 17.3 where emphasis has been given upon the restriction of further growth of the congested settled areas and only expansion of the settlements on the fringe areas of the built environments, increase of vegetation density in the settled areas, compulsory terrace cultivation and people's initiative to make Govt. forest departments aware to take up afforestation programme on the scattered tree fields. Increase of vegetation density in the settled areas can be done by introducing promising agro or social forestry species on the basis of suitability of plant species in the various altitudes of the study area (Chaturvedi et al. 2017).

The zones of marginal terrain sensitivity (Figs. 17.2, 17.3 and Table 17.3) which according to peoples' perception have been identified as moderately stable zones are extended over the lithological setup of Greater Himalayan Sequence, Upper Lesser Himalayan Sequence (U-LHS) and Siwalik formations (Fig. 17.6) having almost Fine Loamy-Coarse Loamy to Gravelly Loamy-Coarse Loamy soils (Fig. 17.7) over a wide range of terrain dissection and ruggedness where dominance of tea gardens and cinchona plantations along with large and small patches of forest cover and wide areas under closely scattered settlements in the higher altitudes of harder rocks (Bhattacharya 2021) have marginally exceeded the limit of physical terrain tolerance and thus, have been recognized as grade II i.e. **Moderate Risk areas**.

As these zones are relatively stable being the terrain sensitivity marginal and the peoples' perception about these zones as partially fragile some careful management strategies have been proposed in the Table 17.3 where growth and expansion of settlements in the tea gardens and cinchona plantation areas have been permitted for the prospect of these valuable plantation sectors (Bhattacharya 2021) with stipulated construction rules and adequate planning on the stable hill slopes strictly with single storied buildings. Similarly over closely scattered settled areas on high altitudes planned growth of the settlements have been proposed with single storied structures on stable slopes seeking approval from concerned construction authority. Regeneration of forest and vegetative growth on the scattered tree fields has mentioned here also like the grade III zones.

The zones which are stable in terrain sensitivity and perceived by the human as stable zones (Figs. 17.2, 17.3 and Table 17.3) are distributed over various geological setup, geomorphological setup and soil conditions and almost covered up by forest with few settlements (Bhattacharya 2021) provide a positive view of the importance of forest over the landscape for which terrain sensitivity of these zones did not exceed the limit of physical terrain tolerance and therefore this zones have been recognized as grade I, i.e., **Low to Risk Free areas**.

No immediate planning and measures are necessary for these zones. However, little areas under scattered tree fields are found in the high altitude zones which should be brought under afforestation programme as proposed for other grades of terrain in the hills of Darjeeling district.

17.6 Conclusion

In the terrain sensitivity assessment and gradation of different risk areas, it is highly contextual to consider in mind whether landform shapes the human habitation or human habitation shapes the landform. A great deal of dilemma lies between these two propositions. From geographical perspective 'landform shapes the human habitation' is quite accepted because landform determines what feasibilities of life are present over a landscape on the basis of which human habitation develops accordingly. From environmental perspective, 'human habitation shapes the landform' is more appropriate because any feasibility of life that facilitates rapid development of an area does not obey the rule of the tolerance of the landform. Therefore, the landform becomes subjected to human pressure leading to a kind of deterioration which converts the concerned area into an area of risk for safe human habitation.

From the terrain sensitivity assessment and perception study of the hills of Darjeeling district, it becomes quite clear that the landform especially the hilly terrain becomes degradable or fragile if not used judiciously. Judicious use of the landform can otherwise provide sustainability of the landform for future use without unacceptable risk of damage to the concerned landscape.

At the present time human plays a pivotal role in every sphere of development not because of the need only, but probably because of satisfying greed too. This emerges the inevitable consequences on the hill terrain. The hills of Darjeeling district is no exception from it for which the environmental planning and strategies proposed above, if implemented, can pave a way of sustainable human habitation in the coming years.

Acknowledgements This paper is a part of the North Bengal University Funded Research Project on "Terrain Sensitivity Assessment based on ASTER DEM and Field Investigation for Environmental Planning of the hills of Darjeeling district, West Bengal, India", for the period 2020–21. Author seeks to express his sincere thanks and gratitude to the Hon'ble Vice Chancellor of the University of North Bengal, West Bengal, India for providing such opportunity and financial assistance for carrying out this Research Project. During field work and data collection, many personnel of various offices and unnamed local inhabitants extended their helping hands and cooperation. Author is highly grateful to all of them.

References

Acharyya SK, Ghosh S, Mandal N, Bose S, Pande K (2017) Pre-Himalayan tectono-magmatic imprints in the Darjeeling-Sikkim Himalaya (DSH) constrained by 40Ar/39Ar dating of muscovite. J Asian Earth Sci 146:211–220

Bathrellos GD, Gaki-Papanastassiou K, Skilodimou HD, Papanastassiou D, Chousianitis KG (2012) Potential suitability for urban planning and industry development using natural hazard maps and geological–geomorphological parameters. Environ Earth Sci 66:537–548

Bathrellos GD, Gaki-Papanastassiou K, Skilodimou HD, Skianis GA, Chousianitis KG (2013) Assessment of rural community and agricultural development using geomorphological–geological factors and GIS in the Trikala prefecture (Central Greece). Stoch Environ Res Risk Assess 27:573–588

Bhattacharya SK (1993) A comprehensive study on the problems of management of the Rakti Basin in the Darjeeling Himalaya, West Bengal (pp. i-ii). Unpublished Ph.D. thesis of North Bengal University

Bhattacharya SK (1999) A constructive approach to landslides through susceptibility zoning and case study in the Rakti basin of the Eastern Himalaya. Trans Jpn Geomorphol Union Jpn 20(3):317–333

Bhattacharya SK (2002) Tindharia Cricket Colony landslide in Darjeeling hills—a comparative account, 1996 and 2001. In: Basu S (ed) Hanging environmental scenario of the Indian subcontinent. ACB Publication, Calcutta, pp 400–405

Bhattacharya SK (2009) Effect of deforestation on rainfall and runoff over the Terai and adjacent Hills in the Darjeeling district, West Bengal, India. Geogr Rev India Kolkata 71(2):130–136

Bhattacharya SK (2012a) Landslide disaster perception of the AILA cyclone in the Darjeeling town, West Bengal, India. Int J GeomatGeosci 3(1):14–29

Bhattacharya SK (2012b) The Depletion of forest cover, its impact and some strategies for conservation in the

Terai plains and adjacent hills in the Darjeeling district, West Bengal. J Geo-Environ Observer 1 (2):13–23

Bhattacharya SK (2013) Landslides and their control in the Darjeeling Hills of West Bengal. In: Geographical thoughts, N.B.U., Siliguri, 11, pp 49–57

Bhattacharya SK (2014) A calibration technique for better estimation of soil loss in the Terai and adjacent Hills of West Bengal, India. Res Forum Int J Soc Sci 2 (3):1–6

Bhattacharya SK (2016) A comparison of the predicted vulnerability zones with the data based on hazard zones of landslide in the Kurseong hill subdivision, Darjeeling district, West Bengal, India. Environ Earth Sci 75(10):1–10. https://doi.org/10.1007/s12665-016-5729-8

Bhattacharya SK (2018) Geomorphometric analysis and terrain evaluation for environmental management in the Kurseong hill subdivision of the Darjeeling district, West Bengal, India. Environ Dev Sustain 22:985–1016. https://doi.org/10.1007/s10668-018-0230-z

Bhattacharya SK (2021) Terrain sensitivity assessment based on ASTER DEM and field investigation for environmental planning of the hills of Darjeeling district, West Bengal, India. Unpublished Research Project Report submitted to University of North Bengal, West Bengal, India, pp 1–101

Chaturvedi OP, Handa AK, Uthappa AR, Sridhar KB, Kumar N, Chavan SB, Rizvi J (2017) Promising agroforestry tree species in India. Jointly published by Central Agroforestry Research Institute (CAFRI) and South Asia Regional Programme of the World Agroforestry Research Centre (ICRAF), 195p

Ghosh S (2011) Knowledge guided empirical prediction of landslide hazard. Ph.D. Dissertation, University of Twente, pp 7–9

Ghosh S et al (2009) GIS-based slope stability evaluation—case study from Paglajhora landslide complex, Darjeeling Himalaya, India. Engineering Geology Divisions, Geological Survey of India, Eastern Region, Kolkata, ppt. Slides, 1–29

Ghosh PK, Debajit D (2012) Coastal tourism and beach sustainability—an assessment of community perceptions in Kovalam, India. GEOGRAFIA OnlineTM Malaysia J Soc Space 8(7):75–87

Ghosh S, Bose S, Mandal N, Dasgupta S (2016) Dynamic recrystallization mechanisms and their transition in the Daling Thrust (DT) zone, Darjeeling-Sikkim Himalaya. Tectonophysics 674:166–181

Haughton G (2017) Environmental Planning. In: Richardson D, Castree N, Goodchild MF, Kobayashi A, Liu W, Marston RA (eds) The international encyclopaedia of Geography. John Wiley & Sons Ltd., pp 1–7. https://doi.org/10.1002/9781118786352.wbieg0668

Kaklauskas A, Gudauskas R (2016) Intelligent decision-support systems and the Internet of Things for the smart built environment. ScienceDirect, pp 1–27. https://www.sciencedirect.com/topics/engineering/built-environment. Accessed 6 Oct 2020

Kellett D, Grujic D, Mottram C, Mukul M (2014) Virtual field guide for the Darjeeling-Sikkim Himalaya, India. J Virtual Explorer:1–38. https://doi.org/10.3809/jvirtex.2014.00344

Mallet FR (1874) On the geology & mineral resources of the Darjeeling district & Western Duars. Memoirs Geol Surv India 11(1):1–72

McDonald SM (2012) Perception: a concept analysis. Int J Nurs Knowl 23(1):2–9. https://doi.org/10.1111/j.2047-3095.2011.01198.x

Papadopoulou-Vrynioti K, Bathrellos GD, Skilodimou HD, Kaviris G, Makropoulos K (2013) Karst collapse susceptibility mapping considering peak ground acceleration in a rapidly growing urban area. Eng Geol 158:77–88

Pawde MB, Saha SS (1982) Geology of the Darjeeling Himalayas. Geol Surv India Misc Publ 41:50–55

Reddy AKN (1983) Rural fuelwood: significant relationships. In: FAO (ed) Forestry for local community development programme—GCP/INT/365/SWE: wood fuel surveys. Food and Agriculture Organization of the United Nations, Rome, pp 1–21. http://www.fao.org/3/Q1085e/q1085e05.htm#2.%20rural%20fuelwood:%20significant%20relationships. Accessed 12 March 2021

Singh RK, Lama TD, Satapathy KK (2006) Treatment technologies for watershed development and management in North East hill region. ENVIS Bull: Himalayan Ecol 14(1):14–21

Sivamohan MVK, Scott CA, Walter MF (2010) Vetiver grass for soil and water conservation: prospects and problems. In: Pimentel D (ed) World soil erosion and conservation, chap 13. Cambridge University Press, pp 293–310. https://doi.org/10.1017/CBO9780511735394.015

Wikipedia (2021) Built environment, pp 1–10. https://en.wikipedia.org/wiki/Built_environment. Accessed on 15 Jan 2021

Yaseen M, Singh M, Ram D (2014) Growth, yield and economics of vetiver (*Vetiveria zizanioides* L. Nash) under intercropping system. Ind Crops Prod 61:417–421. https://doi.org/10.1016/j.indcrop.2014.07.033

Wastewater Treatment in India—A New Perspective

18

Debjani Mandal, Subhankar Mondal, Sayan Biswas, Souradip Seal, Sayan Das, Suparna Bagchi, Rudrajit Mandal, Sk. Fulchand, Atanu Mondal, and Abhishek Basu

Abstract

The increasing population of India has initiated a growing demand for freshwater. In the present scenario, scale of urbanization and industrial growth has generated unprecedented volume of wastewater, often dumped into the surface water resources. The technologies used for domestic and industrial wastewater treatment are not sufficient to handle this problem. The installed and operational capacities of Sewage Treatment Plants (STPs) are much lower compared to the total sewage generated per day in India. Under these circumstances, innovative biotechnological tools could be used for treatment of domestic and industrial wastewater in cost-effective and eco-friendly manner. Microbial population, plants and fungi could be used for sequestration and removal of toxic heavy metals, degradation of complex organic matter and bio-transformation and bio-volatilization of harmful substances in the wastewater. These techniques of bioremediation, phytoremediation and mycoremediation could be independently used or amalgamated with existing technologies for wastewater treatment. There are several instances throughout the world, where adsorption, accumulation, degradation, volatilization and transformation potential of plants, bacteria, fungi and protozoa have been used for wastewater treatment.

Keywords

Wastewater · Sewage treatment plant · Bioremediation · Phytoremediation · Mycoremediation · Case study

18.1 Introduction

The ever increasing population has initiated a growing demand for freshwater. Water is an essential commodity and access to clean water is vital in everyday life. Groundwater and surface water from rivers, lakes and streams are the chief sources of freshwater in both rural and urban parts of India. The growing consumption of water has led to a higher production of wastewater. Domestic and industrial wastewater is mostly dumped in water bodies such as rivers with or without proper treatment. The rivers passing through big cities are polluted by municipal wastes including sewage and oil-based pollutants. Our modern lifestyle has introduced various toxic components into the ecosystem, many of which

D. Mandal · S. Mondal · S. Biswas · S. Seal · S. Das · S. Bagchi · R. Mandal · Sk.Fulchand · A. Mondal · A. Basu (✉)
Department of Molecular Biology and Biotechnology, Sripat Singh College (Under University of Kalyani), Jiaganj, Murshidabad, West Bengal, India
e-mail: abhishek@mbbtsripatsinghcollege.in

J. Das and S. K. Bhattacharya (eds.), *Monitoring and Managing Multi-hazards*, GIScience and Geo-environmental Modelling, https://doi.org/10.1007/978-3-031-15377-8_18

are non-biodegradable in nature. Pollutants like sewages and chemical wastes from industries alter the biochemical balance of water bodies, if released in untreated form. Sewage and domestic wastewater contain detergents that increase alkalinity of the water and also promote cultural eutrophication. Under such condition, the dissolved oxygen of the water bodies decreases and the Biochemical Oxygen Demand (BOD) increases, which cause massive damage to the aquatic ecosystem. As the pollutants pile up over time, they not only affect the environment but also put strain on the availability of valuable and scarce resources. The addition of specific pollutants depends on the local usage pattern of substances, responsible for production of the pollutants. Rural regions using pesticides and fertilizers in high amounts show presence of these chemicals into the streams or these chemicals leach through the aquifer, thus contaminating the surface water and the groundwater. Municipal wastes contain a variety of pollutants including sewage, various chemicals, microplastics, hydrocarbons, etc. As of now the sewage water treatment in India is not very effective and exhaustive. An inefficient collection system of sewage and municipal wastes leads to their direct dumping in the rivers or in the ocean. Industrial wastes contain chemical pollutants that can cause damage to the biodiversity of the sink water body, if discharged without treatment (Rajput et al. 2017; Almuktar et al. 2018; Boelee et al. 2019; Wanhong et al. 2020).

The bioremediation process could complement existing water treatment methods in treating a vast range of pollutants. Suitable microorganisms can remove the pollutants by different biological processes. Some microorganisms can passively bind to the pollutants by a process called adsorption. In some cases the bacteria may uptake the pollutants into their system thereby removing it from the contaminated water body. Certain pollutants can also be broken down to simpler and less toxic forms, which can then be slowly incorporated in the environment by natural process of biodegradation. Plants and associated microbes are used to detoxify wastewater by the process of phytoremediation. Mycoremediation is a similar process that uses fungi. These biological agents can be used to remove heavy metals, pesticides, hydrocarbons, organic matters, phosphorus, ammonia and other xenobiotics. Traditional methods of wastewater management are effective up to a certain extent. The ever increasing volume of wastewater can over well the system. The use of microbes can provide a cost efficient and eco-friendly way to combat wastewater crisis with a low learning curve and help to preserve the ecosystem (Brar et al. 2017; Kapahi and Sachdeva 2017; Kumar et al. 2018; Masiá et al. 2020; Shah and Shah 2020).

18.2 Status of Domestic/Sewage/ Municipal Wastewater and Sewage Treatment Plants (STPs) in India

Sewage/domestic/municipal wastewater is produced from households. This wastewater is characterized by non-white colour, unpleasant odour, turbid nature, high BOD and COD, oil, fat, grease content, etc. Industrialization and urbanization have taken a toll on water resources. With the increasing demand for water, there is also an increase in scarcity of potable water and deterioration in quality of groundwater (CPCB 2021). India, a country with 1.39 billion people is failing to meet the need of water for domestic requirements, due to the lack of capacity of water reuse for non-potable purposes. 80% of supplied domestic water is discharged as domestic wastewater, which if treated and reused for non-potable purposes, could have fulfilled the demand for water to some extent. In most cases, domestic wastewater is discharged in rivers, lakes and other natural water bodies without any treatment. The impact of such discharge is deterioration of water quality of the sink water body. Domestic wastewater is mainly composed of small amount of organic and inorganic carbon, volatile and non-volatile dissolved and suspended solids like oil, soap, salts, metals, faeces, food particles, etc. (CPCB 2021). Such untreated wastewater leads to an imbalance in the natural floral and faunal system of the receiving water body. The sewage/municipal wastewater can be treated in sewage treatment plants (STPs) located throughout the country. But in most of the cases the number of operational STPs is far less than the total number of STPs installed. In 2021, Central Pollution Control Board estimated that in India minimum 72,368 mega litre of sewage is generated from urban centres per day. Total number of STPs (including proposed STPs) in India is 1631 with a total sewage treatment capacity of 36,668 mega litre per day. But out of 1631 STPs only 1093 STPs are operational (CPCB 2021). Some of the installed STPs are non-functional due to lack of proper maintenance and man power, and lack of sewage collection network. 274 STPs are under construction and 162 STPs have been proposed for construction. Maharashtra produces maximum amount of sewage (9107 mega litre per day), followed by Uttar Pradesh (8263 mega litre per day), Tamil Nadu (6421 mega litre per day), West Bengal (5457 mega litre per day), Gujrat (5013 mega litre per day) and so on (CPCB 2021). These states also have greater number of STPs compared to other states. Arunachal Pradesh, Andaman and Nicobar Islands, Assam, Lakshadweep, Manipur, Meghalaya and Nagaland do not have any STP installed. These states and union territories produce less sewage compared to other larger states and union territories. But unfortunately, all the sewage water from these states is added to the natural water bodies without any treatment. None of the state in India treats the entire sewage water generated and therefore, much of the sewage water is discharged without any treatment. Therefore, India having installed sewage treatment capacity of 31,841 mega litre per day, treats only 20,235 mega litre of sewage out of 72,368 mega litre of total sewage generated per day (CPCB 2021). Most of the STPs use technologies like Sequential Batch Reactor, Activated Sludge process, Upflow anaerobic sludge blanket, Moving bed biofilm reactor, etc. In some of the cases technologies like Suspended aeration, Fludized aerobic bed reactor, Oxidation pond, Trickling filter, Bio-tower, Electro-coagulation, etc. are also used in STPs to treat sewage water. As 72% of sewage water is left untreated, this water cannot be reused for non-potable and industrial purposes. Such utilization would have decreased the demand on freshwater sources and provide the chance of natural reclamation and conservation of natural resources (CPCB 2021).

18.3 Status of Industrial Wastewater Treatment in India

Industrial wastewater is highly toxic for our nature and affects the daily life of human. The industrial wastewater with alkaline pH, dark black or dark brown colour, unbearable odour, consists of a mixture of dissolved solids, organic and inorganic pollutants. Molasses-based distilleries are one of the most polluting industries with respect to industrial wastewater. Textile industries generate maximum chemical pollutants and have huge affect on potable water. Many unused chemical substances, complexes, dyes, heavy metals, additives, grease, waxes, etc., are disposed in the wastewater in streams around the textile, diary and glass industries. The wastewater is mostly acidic with high temperature, BOD and COD. Physical, chemical and biological treatment methods like coagulation, flocculation, filtration, reverse osmosis, activated sludge process, aerated lagoons, etc., can be applied for wastewater treatment of these industries (Figs. 18.1, 18.2 and 18.3). Direct disposal of toxic wastewater affects the ecological balance of the aquatic ecosystem by changing the temperature, pH and salinity. The wastewater released from chemical industry consists of abundant amount of chemicals like poly-aluminium chloride, poly-aluminium sulphate, poly-aluminium ferric chloride and poly-ferric chloride. These are the pre-hydrolysed coagulants. Industrial wastewater in semi-solid, fluid or semi-fluid forms pollute the nearby soil, surface water (lakes, pond, rivers and streams) and the underground water. If we take a look at the water

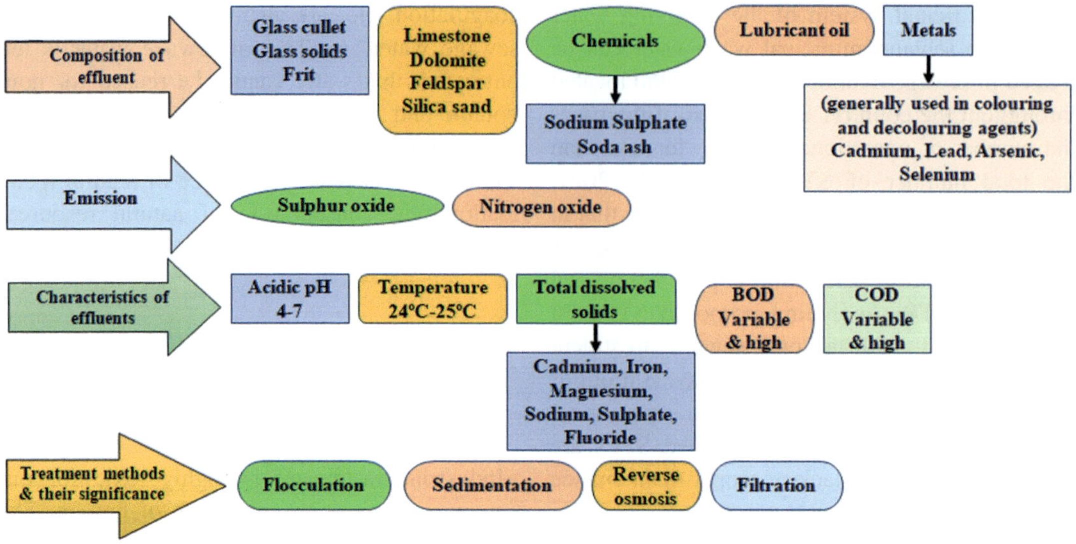

Fig. 18.1 Composition and characteristics of effluents from glass industry and their treatment methods

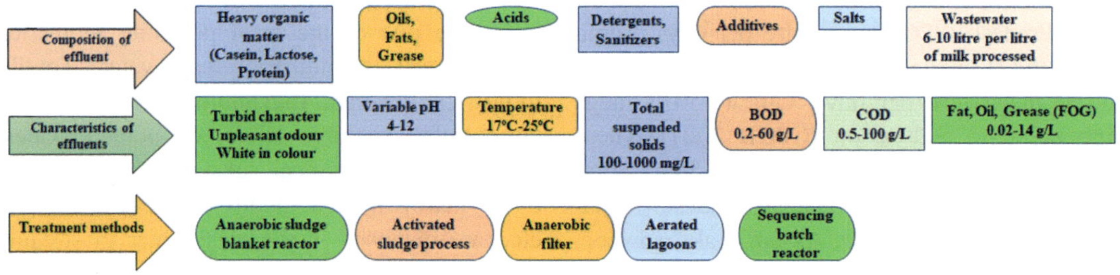

Fig. 18.2 Composition and characteristics of effluents from diary industry and their treatment methods

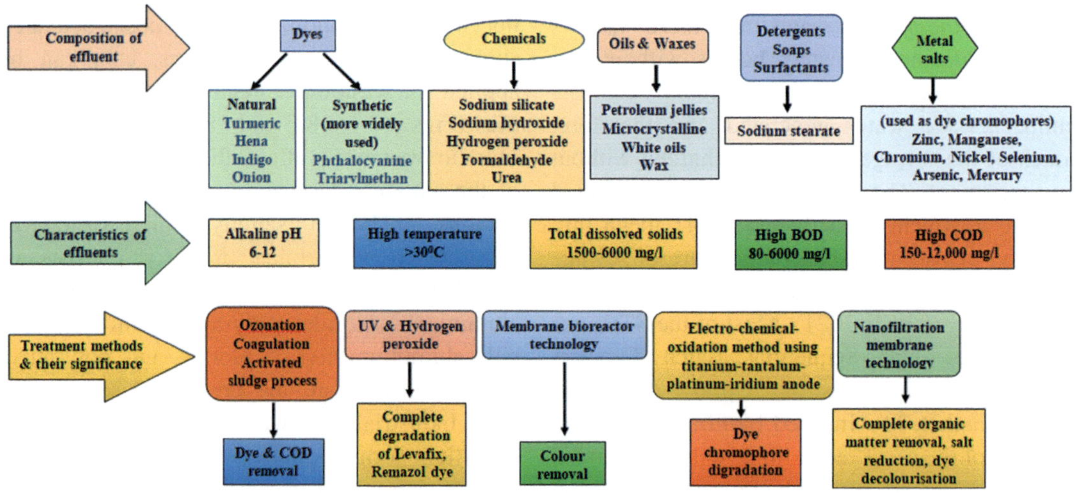

Fig. 18.3 Composition and characteristics of effluents from textile industry and their treatment methods

pollution caused by industrial wastewater, we see many problems in their nearby locality. This type of pollution of the water bodies decrease the dissolved oxygen level and increase the BOD. Industrial wastewater contaminates the soil of the locality and creates problem in farming. There is bio-magnification of the toxic industrial pollutants in the food chain due to their accumulation in local crops and vegetables. Hence, the pollutants can finally enter the human system and cause diseases like typhoid, dysentery, cholera, abdominal pain, etc. Battery manufacturing industries release by-products containing toxic heavy metals like cadmium, cobalt, copper, cyanide, chromium, iron, mercury, nickel, lead, zinc and silver. Removal of dissolved pollutants demands high end technology. At present, industries employ special methods and devices for the treatment of wastewater. Ultrafiltration is a pressure driven process that uses a membrane to remove oils, metal hydroxides, emulsions, dispersed materials, suspended solids and other high molecular weight materials from wastewater. Ultrafiltration system is designed to remove 98% oil from wastewater without the use of any chemical. Such system is capable of recycling wastewater and soap solution for reuse. Reverse osmosis removes dissolved solids and impurities from water by using semi-permeable membrane, which allows the water to pass, leaving behind the majority of the dissolved solids/salts and other contaminants. Reverse osmosis can remove up to 99.5% of dissolved contaminants. Reverse osmosis generally follows the ultrafiltration treatment of wastewater. Solid bowl centrifuge is used to separate solid from liquid in metal processing industries. Solid bowl centrifuge provide high performance separation of all types of solids and liquid particles (both metallic and non-metallic). It is available both in manually cleaned rotor style and in fully automated self-cleaning design. In tramp oil separation wastewater treatment process, contaminated fluid flows through a series of baffles and porous media bed and the mechanically dispersed oils are separated from the fluid (Satyawali and Balakrishnan 2008; Rajput et al. 2017; Saxena et al. 2017).

18.4 Present Methods of Wastewater Treatment in India

The different sewage treatment plants in India employs a conventional sequential process which involves physical, chemical and biological operation to remove solid particles, organic and inorganic substances, and nutrients from the wastewater. Then the treated wastewater is allowed to be either reused or released into the river. In the conventional process of wastewater treatment, as preliminary treatment, different type of physical operation is performed on the wastewater for removal of solids like plastics, rags, woods, etc. These solid wastes are used for landfill. Primary treatment involves separation and sedimentation of large contaminants using large-screen filters, membrane system of different pore size and sedimentation tank. The primary treatment produces primary sludge and primary filtrate. The primary filtrate is sent for subsequent treatment, whilst the primary sludge is often used for production of biogas. The purpose of secondary treatment is removal of organic or biodegradable substances of the wastewater. In this step, a Trickling Filter and/or an Activated Sludge system are used. Both these systems are conventional method of biological wastewater treatment which employs aerobic microorganisms for decomposition of organic materials in presence of oxygen. Trickling Filter is a solid material packed tank (rocks, gravel, shredded PVC bottles, or special pre-formed plastic) and colonies of microorganism are allowed to grow on its surface. Whereas in Activated Sludge system, the semi-solid sludge contain the colonies of microorganism. Effluent from secondary treatment is sent for tertiary treatment, where rest of the organic and inorganic (like nitrates and phosphates) compounds are removed using different anaerobic bacteria like denitrifying bacteria (biological process), chemical agents like ferric chloride, activated carbon and alum for sedimentation of different chemical ions. Lastly, the tertiary filtrate is disinfected by application of disinfecting agents (chlorination). Oxidation

Ponds as a means of secondary treatment of wastewater, are used in warmer climates. However, Oxidation Pond requires large area and considerable amount of time (2–3 weeks) for the treatment process to get completed (CPCB 2021).

At present, there has been a lot of improvement over the conventional wastewater treatment processes. State of the art methods are very efficient and are different from each other in their operational principle. Sequencing Batch Reactor (SBR), Moving Bed Biofilm Reactor (MBBR) and Membrane Bioreactor (MBR) are predominantly used in wastewater treatment plant in India. Other than these, Submerged Aerated Filter (SAF), Suspended Media Filters (SMF), Rotating Disc System (RDS), etc., are also used. The Sequencing Batch Reactor employs the principle of activated sludge process, where the wastewater is passed through a "Flow-through" system and in presence of oxygen and aerobic microorganisms, oxidizes and degrades almost all the organic substances inside the reactor. SBR also removes nitrogen, phosphorus and the treated effluent is reasonably good for application in lands or release to the surface water. Some of the SBR-based water treatment plants in India are Mundhwa Sewage Treatment Plant (Pune), Sewage Treatment Plant (SBR) Kalpataru Construction Overseas, Mumbai, etc. Membrane Bioreactor combines the process of Ultrafiltration (UF) and Microfiltration (MF) along with biological wastewater treatment process (Activated Sludge process). Initially, the membrane zone of MBR removes the contaminants (solid, organic and inorganic) and then the filtrate is allowed to go through clarifiers and filters used in activated sludge system. This process is mostly used in industrial and municipal waste water treatment. Delhi Games Village-Water treatment plant, India's first wastewater recycling plant IOCL Panipat, are MBR based water treatment plants. Moving Bed Biofilm Reactor is made up of an aeration tank with plastic (highly dense polyethylene)-based media carrier called biofilm. It is a biological method for removing organic contaminants with the help of different microorganisms grown in the biofilm. This process can give desired level of Mixed Liquor Suspended Solids (MLSS), which is critical for subsequent processes. This type of reactor is also suitable for retreatment of industrial and municipal wastewater (CPCB 2021).

One example of wastewater treatment plant is Zero Liquid Discharge (ZLD) plant which employs MBR, RO, evaporation or electrodialysis technology in order to remove liquid contaminants completely from industrial wastewater. Effluent Treatment Plant (ETP) is also designed for industrial wastewater, which employs activated sludge system and different chemical and physical processes to produce a reusable effluent from the industrial wastewater. Grey Water Treatment Plant (GTP) is mainly used for laundry and sanitary wastewater. This plant uses both biological and chemical processes for treatment and recycling of the wastewater.

18.5 Need for Biological Wastewater Treatment

The main constituents of the two type of wastewater (domestic and industrial wastewater) are carbon, nitrogen and phosphorus. However, the composition and percentage of trace elements vary based on the source of wastewater. This makes the wastewater treatment process versatile, complex and in many cases inefficient in complete removal of trace toxicants. Untreated or poorly treated wastewater discharged in rivers, streams or any water bodies, cause local and global pollution leading to ecosystem imbalance (Ahammad et al. 2013). Domestic wastewater rich in organic matter, depletes the oxygen level of the sink water bodies, reduces biodiversity and imparts a long term negative impact on the environment. Industrial effluents add non-recommended level of toxicants to the water bodies, which affect the floral and faunal diversity, and put a stress on the health of inhabitants surrounding the water bodies. Therefore, aim of wastewater treatment include reduced concentration of toxicants below the permissible standards, safe and secure disposal of sludge, efficient handling of wastewater, minimizing

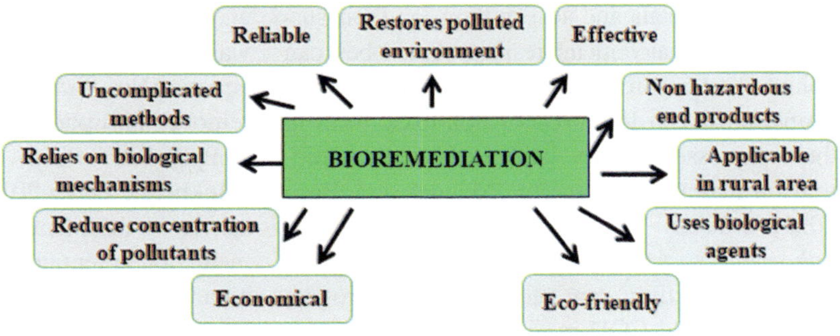

Fig. 18.4 Bioremediation as a tool to combat pollution and toxicity

negative impacts on the environment and recycling and recovery of valuable end products (Ahammad et al. 2013; Samer 2015). Optimal treatment process depends on the wastewater treatment goals, composition and relative percentage of constituents of wastewater. Wastewater treatment approaches can be physical, chemical or biological in nature. Physical approaches involve application of physical force in processes like filtration, flocculation, floatation, sedimentation, etc. Chemical reactions based adsorption, disinfection, precipitation, etc., are associated with chemical treatment of wastewater. In some cases, especially for industrial wastewater treatment, the physical and chemical processes are combined together to achieve the good quality of treated water. These processes generate high quantity of sludge which is difficult to dispose and hence, increase environmental stress (Ahammad et al. 2013). Further, indiscriminate use of chemical additives and physical agents turn these processes into an expensive one. Also, expensive instruments, large man power, high energy, etc., are some of the prerequisites of physical and chemical processes. Therefore, wastewater treatment is now diverting from application of physicochemical processes to biological treatment processes involving agents like plants, fungi, fish, microbes, etc. In a nutshell, application of biological mechanisms and processes in wastewater treatment, decrease the overall cost of the treatment and make the entire treatment process efficient, eco-friendly and simple in terms of design and operation. Such treatment processes also consume less energy in the long run and generate less sludge. In most of the cases sludge are well stabilized, easily handled and disposed (Fig. 18.4) (Ahammad et al. 2013).

18.6 Mechanisms Involved in Biological Wastewater Treatment

Implementation of biological agents in treatment process not only decreases the total cost of the treatment, but also make the process eco-friendly and suitable for recycling and recovery of valuable end products. Bioremediation is the removal, transformation, decomposition or degradation of toxic pollutants by microbes, fungi, plant, etc., into less toxic or non-toxic forms. These bioremediators (living or dead) have the potential to remove organic and heavy metallic contaminants from wastewater (Ahammad et al. 2013). Microbes decompose organic matters either by bio-oxidation or by biosynthesis. The end products of bio-oxidation remain in the solution and are often discharged along with the effluent. Dissolved organic matters of wastewater are transformed into new biomass by biosynthesis, which is further removed by sedimentation (Samer 2015). Heavy metallic contaminants are removed from wastewater by microbes mediated bio-accumulation, bio-adsorption, bio-volatilization and bio-transformation. Heavy metal resistant microbes accumulate heavy metals intracellularly in conjugated or free forms in a process called bio-accumulation. In bio-adsorption the components of microbial membrane associate in ionic and non-ionic interactions with heavy metals and adsorb them on their surface. Some heavy metal resistant microbes can volatilize or transform heavy metals into different forms, mostly in less toxic or non-toxic form in the processes bio-volatilization and bio-transformation, respectively (Mandal and Basu 2021; Mandal et al. 2021). For example, *Sphaerotilus natans* and *Heliscomenobacter hydrosis* together can reduce total nitrogen, phosphorus and COD in diary industry wastewater. *Lactobacillus casei* reduces lactose present in diary industry wastewater. *Stenotrophomonas* (K-T-1 strain) has nylon bioremediation property (Table 18.1) (Slavov 2017; Scalenghe 2018). Plants can also be used for wastewater treatment. Phytoremediation is the application of plants and plant-based mechanisms for removal of toxicants from contaminated site. Many plants can accumulate (phyto-accumulation), transform (phyto-transformation) or stabilize (phyto-stabilization) toxic pollutants like explosives, pesticides, hydrocarbons and heavy metals in their shoot system. In phyto-accumulation toxic pollutants are accumulated in the vacuole of the respective pollutant resistant plants. Plants can enzymatically transform toxic pollutants or degrade them into less toxic forms, in the processes phyto-transformation and phyto-degradation, respectively (Ahammad et al. 2013). Plant roots also play a vital role in extraction and stabilization of toxic pollutants in them. Plant roots can extract toxicants from wastewater by the process of phyto-extraction. Some contaminants are resistant to complete degradation. Such components of wastewater are partially degraded and stabilized within the plant roots. This is called phyto-stabilization, which can be applied in biological wastewater treatment (Mandal and Basu 2021). For example, *Eichhornia crassipes,* a water hyacinth weed, reduces approximately 99.5% of chromium from wastewater (Saha et al. 2017). *Eichhornia crassipes* also absorbs and thus, reduces the concentration of organic dyes (Sharma et al. 2021). The aquatic plant, Darkweed absorbs chloride and sulphate from wastewater. It also degrades and removes other toxic components of wastewater (Saha et al. 2015). *Trapa natans,* a watery plant absorbs heavy metals and chemicals, and reduce COD and BOD in municipal wastewater (Kumar and Chopra 2018). The aquatic plant *Pistia stratiotes* have the potential to remove phosphorous, ammonia and organic components from municipal wastewater (Mukherjee et al. 2015). Microbes at the rhizosphere of the plants often enhance the phytoremediation potential of plants by biostimulation or bioaugmentation of the phytoremedial mechanisms (Samer 2015). Mycoremediation can be defined as the application of fungal mycelia in bioremediation of toxic pollutants from contaminated setting. Mycelium, the vegetative part of a fungus, secretes variety of acids and enzymes that degrade and decompose toxic contaminants of wastewater. For instance, aromatic and chlorinated compounds are easily degraded by wood degrading fungi. Fungi can also filter out toxic pollutants from wastewater in a process called microfiltration (Samer 2015). *Pleurotus* sp. absorbs and removes heavy metals from wastewater (Kapahi and Sachdeva 2017). Various enzymes from *Giberella* and *Morticrella* react with hydrocarbons from petroleum industry wastes and break them into stable simpler forms (Horel and Schiewer 2020). *Phlebia acerina* (wood rotting fungi) degrades toxic contaminants of wastewater (Kumar et al. 2018). Earthworms are used for vermicomposting/worm composting/vermicasting/worm casting. Two common earthworms generally used for this purpose are *Eisenia fetida* (red wiggler) and *Lumbricus rubellus*. Earthworms can also be used for filtration of wastewater. They can decompose the organic matter in wastewater and change its composition. They can also decrease the carbon nitrogen ratio and increase the surface area of organic matter for microbial decomposition (Samer 2015). Increased surface area increases the microbial growth and metabolism leading to increased degradation of organic matter. The microbial diversity in such filtration unit is more compared to the microbial diversity in any conventional filtration unit. Application of earthworm in wastewater filtration is called vermifiltration, which is developing as a new biological treatment process (Samer 2015). The insect *Tenebrio molitor* L. can be used in polystyrene

Table 18.1 Significance of biological agents in treatment of different types of wastes

Biological agents	Significance
A. Significance of some biological agents in textile wastewater treatment	
Phragmites australis	Textile industry pollutant removal
Eichhornia crassipes	Reduction of concentration of organic dyes and Wastewater quality improvement
Lemna minor L.	Dye Acid Orange 7 (AO7) bioremediation COD removal
Myriophyllum spicatum L. and *Ceratophyllum demersum* L.	Dye Basic Blue 41 removal
Spirogyra spp.	Azo dye reactive yellow 22 (RY22) bioremediation
B. Significance of some biological agents in diary wastewater treatment	
Sphaerotilus natans and *Heliscomenobacter hydrosis*	COD reduction Total nitrogen reduction Total phosphorus reduction
Lactobacillus casei	Lactose reduction COD degradation
C. Significance of some biological agents in plastic waste treatment	
Fusarium	Low density polyethylene degradation
Tenebrio molitor L. *Aspergillus* *Penecillum* *Fusarium*	Polystyrene biodegradation
Stenotrophomonas (K-T-1 strain) *Fusarium* (K-T-2 strain)	Nylon bioremediation
Aspergillus tubingensis	Polyurethanes biodegradation
D. Significance of some biological agents in municipal wastewater treatment	
Trapa natans	Heavy metals and chemicals removal COD and BOD reduction
Pistia stratiotes	Removal of phosphorous, ammonia and organic components

biodegradation of microplastics in wastewater (Scalenghe 2018) (Table 18.1).

Biological processes and treatments are temperature and pH dependent. Microbial growth and metabolism rate varies with change in temperature and pH of the medium. Biological treatment processes are inefficient in handling shock loads and thus require uniform loading of wastewater. Therefore, biological wastewater treatment is sensitive to multiple abiotic and external factors. Appropriate bio-environmental condition is a prerequisite for effective biological wastewater treatment (Samer 2015). High temperature increases substrate removal but causes oxygen limitation in the medium as well. All the parameters like temperature, pH, air, moisture, uniform loading and unloading, etc., are maintained in the bioreactor, which provides appropriate bio-environmental conditions for growth and metabolism of the microbes. These bioreactors are cylindrical, stainless steel container with all the input and output devices attached to them. Majority of biological treatments are carried out at mesophilic temperature. The optimum temperature in these cases usually ranges from 20 to 40 °C (Samer 2015). Biological treatments are classified according to the dominant pathway that the microbes employ for wastewater bioremediation. Oxygen requirement during the wastewater treatment classifies the processes into aerobic and anaerobic processes (Ahammad et al. 2013).

18.7 Classification of Biological Wastewater Treatment Based on Oxygen Demand

18.7.1 Aerobic Processes

Aerobic processes are most widely used for wastewater treatment. Such processes use aerobic respiration and produce more biomass and residual solids compared to other processes. In such treatments microbes must be metabolically active for a long time. Microbes are supplied with constant oxygen by aeration. These processes involve microbial conversion of complex organic compounds into simple organic compounds, carbon dioxide and biomass in presence of oxygen. Aeration also reduces harmful gas emission like methane, nitrogen oxides, ammonia and increases the escape of dissolved gases and organic volatile compounds (Samer 2015). Aerobic treatments are carried out in oxidation ponds or aerated lagoons (Fig. 18.5). Oxidation ponds are usually 1.2–1.8 m deep. There are two zones in oxidation ponds-aerobic zone or the zone of bacterial activity, and anaerobic zone or the zone of solid settling, sludge formation and anaerobic degradation. Such ponds are inhabited by both aerobic and anaerobic bacteria. The ponds appear dark green in colour due to excessive algal growth, which feeds on inorganic compounds. Photosynthetic algae of the euphotic zone (sunlight zone of few centimetres) of the oxidation pond associate in a symbiotic relationship with aerobic heterotrophic bacteria (Samer 2015). Photosynthetic algae use sunlight for their growth and produce molecular oxygen for aerobic heterotrophic bacteria. Aerobic heterotrophic bacteria use this molecular oxygen and release inorganic molecules as feed for photosynthetic algae. The anaerobic zone of the pond is inhabited by anaerobic bacteria which are involved in sludge formation and degradation of the solid settling in absence of oxygen. Aerated lagoons are 3–4 m deep. Aeration is done by aerators, which provide molecular oxygen as well as proper suspension of microbes for enhanced microbial Activity. The hydraulic retention time is 3–8 days depending on treatment and waste type, temperature, etc. (Samer 2015). Plant mediated phytoremediation like phyto-degradation, phyto-accumulation and rhizofiltration of wastewater is carried out in wetlands. These wetlands appear similar to the aerated lagoons and oxidation ponds. Various types of wetlands like subsurface flow constructed wetlands, horizontal flow reed bed, free

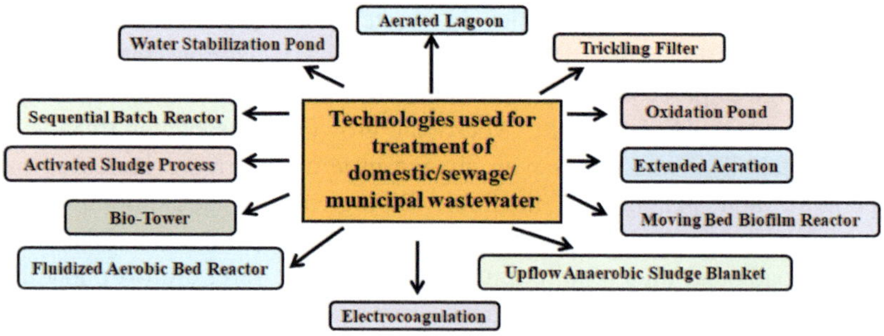

Fig. 18.5 Technologies used for treatment of domestic/sewage/municipal wastewater

water surface system, vegetated submerged bed system, etc., are nowadays used for wastewater treatment (Samer 2015).

18.7.2 Anaerobic Processes

These processes involve microbes mediated anaerobic, sequential degradation of organic matter, where the active groups of microbes participate in metabolic interactions with functional groups of organic matters. Such degradation is carried out in a complex multistep process, where the four major reaction stages are hydrolysis, acidogenesis, acetogenesis and methanogenesis. Hydrolysis involves microbes mediated enzymatic hydrolysis of organic polymers like protein, carbohydrates and lipids into simple organics (Ahammad et al. 2013). Simple organics like amino acids, fatty acids, sugars, etc., are made available for other bacteria. In the second stage of anaerobic digestion, acidogenesis, these simple organics are further converted to short-chain organics like ketones, formic acid, lactic acid, propanoic acid, succinic acid, carbon dioxide and hydrogen gas. Organic acids and alcohols formed in the second stage are further used by acetogenic bacteria (obligate hydrogen producing acetogens) in the third stage. (Ahammad et al. 2013). The products acetate, methanol and methylamine of the third stage, act as substrates for microbe mediated methanogenesis in the fourth stage of anaerobic digestion. The ultimate product is methane gas. Methanogenesis is the rate limiting step of anaerobic digestion. The acetogenic and methanogenic bacteria used in the digestion, participate in a symbiotic relationship with each other. Acetogenic bacteria produce hydrogen gas for methanogens and in lieu methanogens create a suitable environment for acetogenic conversion by utilizing the hydrogen gas (Ahammad et al. 2013). Anaerobic processes are complex multistep processes carried out at mesophilic or thermophilic temperatures. These processes involve anaerobic respiration in absence of free or combined oxygen. Such processes are time consuming in nature. But they are advantageous for treatment of wastewater which is excessively rich in organic wastes. High level of purification is achieved in these processes with less biomass and biosolid production (Samer 2015). Anaerobic processes produce effluents with soluble organic matters which are treated aerobically. Anaerobic treatment processes are carried out in an anaerobic digester or anaerobic lagoon. Anaerobic digesters can be classified as completely mixed anaerobic digesters, upflow anaerobic sludge blanket reactor, anaerobic fluidized bed reactor and expanded granular sludge bed reactor (a variant of upflow anaerobic sluge blanket reactor) (Fig. 18.5) (Samer 2015). The completely mix anaerobic digester mixes the anaerobic microorganisms with the wastewater whilst maintaining an anaerobic condition. Equal hydraulic and solid retention time is provided in this case, which usually ranges from 15 to 40 days. In an upflow anaerobic sludge blanket reactor, the influent travel upward through the sludge blanket with reduced upflow velocity and enhanced solid retention time. This is an attached self-immobilized cell system, where the bacterial aggregates form a flowing blanket and water flows upward through this blanket. It is the most common and widely used anaerobic digester. The microbes in the floating blanket convert complex organics into methane, hydrogen gas and carbon dioxide. Microbes are attached and immobilized on sand or on granular activated carbon in an anaerobic fluidized bed reactor. These carriers (sand and granular activated carbon) provide large surface area for microbial activity, high flow rate and less clogging due to their large pores. Expanded granular sludge bed reactor is a variant of upflow anaerobic sludge blanket reactor but with less upflow velocity and partial bed fluidization (Samer 2015; Ersahin et al. 2011). In all the three reactors methanogenic bacteria produce methane gas which can be used for industrial and domestic purposes. Anaerobic lagoons maintain an anaerobic condition for anaerobic bacteria mediated wastewater treatment. These lagoons are usually deeper than 2.5 m with no aeration, heating or mixing of wastewater. Deeper lagoons are more efficient in anaerobic treatment of wastewater. The anaerobic sludge layer accumulates at the bottom of the pond whereas the final effluent supernatant forms the upper layer (Samer 2015).

18.8 Classification of Biological Wastewater Treatment Based on Growth Condition of Bioremediators Within a Bioreactor

Biological wastewater treatment can be classified on the basis of the growth condition provided to the microbes inside a bioreactor. The treatment processes involving microbes attached or immobilized on a solid support are called attached growth processes. Suspended microbial cells are used in some wastewater treatment processes which fall under the category of suspended growth processes of wastewater treatment (Ahammad et al. 2013).

18.8.1 Attached Growth Processes

In an attached growth process microbes are either attached on the surface of packaging material or immobilized on a solid support (flocs or granules). Aerobic attached growth processes includes trickling filters, packed bed reactor, rotating biocontactors, roughing filters, etc. Anaerobic attached growth process includes method involving hybrid anaerobic reactor, upflow anaerobic sludge blanket reactor, anaerobic fluidized bed reactor, anaerobic biological contactors and anaerobic up and down flow reactor (Fig. 18.5) (Ahammad et al. 2013). In attached growth process, the substrate concentration available for microbes is less than the substrate concentration in bulk liquid. Trickling filter is one of the most commonly attached growth process of wastewater treatment. In these filters, microbial cells are attached on a support material with large pores and surface area. These large pores allow the unrestricted entry of water and air through the filters and high level of microbial activity. In some cases, submerged aerated filters are also used for wastewater treatment (Ahammad et al. 2013). Rotating biological contactors are also widely used in wastewater treatment. In rotating biological contactors aerobic and anaerobic bacteria form fixed biofilm on circular plastic discs made up of polystyrene and polyethylene. Large number of discs are connected together with a shaft and 40% submerged in wastewater. The rotation of discs at a speed of 1–1.6 RPM spread the wastewater over the disc for microbe mediated bioremediation (Samer 2015). Mycoremediation is performed in trickle bed bioreactor, rotating disc bioreactor or fluidized bed bioreactor. In these three cases, fungal mycelium is either attached or immobilized within the cubes, disc or column, respectively, and wastewater is passed through it. Fungal mycelia filter or convert the organic matters of the wastewater into simple forms. Biomass formed in this process is separated from the supernatant effluent (Samer 2015).

18.8.2 Suspended Growth Processes

In suspended growth process microbes are suspended in a liquid phase. The substrate concentration available for microbes and the concentration in the bulk liquid are equal in suspended growth process. Aerobic suspended growth processes include sequencing batch reactor, aerated lagoon and aerated sludge process, whereas anaerobic suspended processes include stirred tank reactor, bag digester, plug flow digester and baffled reactor with microbes (Fig. 18.5) (Ahammad et al. 2013). Activated sludge process is the predominant aerobic suspended growth process used in wastewater treatment. The process is associated with microbes mediated mineralization of complex organic compounds to generate new biomass and simple organic compounds. The solid part of the process consists of complex mixture of microbes, which forms the active part of the process as well. The liquid part is constituted by mixing liquor with wastewater and its intrinsic microbial population. The uncovered open aeration tank of activated sludge process uses mechanical aerators like surface and brush aerator to provide sufficient amount of oxygen to the microbes. The microbes in the aeration tank use organic matters to thrive (Ahammad et al. 2013). Diffusers are more efficient in transferring oxygen to deep water. However, both diffusers and aerators consume large amount of energy. Even after the use of aerators and diffusers there can be some dead zones (poorly mixed area) in the tank. These zones with less oxygen and liquor, and anoxic or anaerobic condition, decrease the settling efficiency and increase sludge bulking. The zone also generates unpleasant odour. The wastewater is then transferred to a secondary clarifier where, by gravity settling the solid biomass is separated from the supernatant liquid. The flow rate of the liquid is far less than the settling velocity of the solid which settles the solid at the bottom (Ahammad et al. 2013). The supernatant produced at this stage is then taken to the disinfection chamber following which the effluent is discharged. This process produces a large amount of sludge which is often returned to the bioreactor to maintain the microbial concentration. As the entire process requires unlimited nutrients and oxygen for microbial growth and metabolism, the process is applicable when huge running cost is available. There are various types of activated sludge plant like conventional plant, complete mix plant, contact stabilization plant and step aeration plant (Samer 2015).

18.9 Biological Removal of nitrogen and Phosphorus

Biological removal of nitrogen involves two slow, low yielding, microbe mediated processes-nitrification and denitrification. In the first step, nitrification, ammonia is oxidized to nitrate by chemoautotrophic bacteria under strict aerobic condition. The nitrification takes place in two stages- Ammonia oxidation, where ammonia is oxidized to nitrite and nitrite oxidation, where nitrite is oxidized to nitrate. Both the stages use molecular oxygen as electron acceptor and carbon dioxide as carbon source (Ahammad et al. 2013). The bacteria involved in ammonia oxidation are *Nitrosovibrio, Nitrospira, Nitrosomonas, Nitrosolobus, Nitrosococcus,* etc. *Nitrocystis, Nitrospina, Nitrobacter,* etc., perform nitrite oxidation. Denitrification is carried out by heterotrophic denitrifiers. These bacteria play a significant role in heterotrophic bioconversion or reduction of oxidized nitrogen compounds (like nitrite and nitrate) into nitrogen gas, under anoxic condition. The process uses nitrite and nitrate as electron acceptors and organic carbon as the energy source. *Hellobacterium, Paracoccus, Pseudomonas, Thiobacillus* and *Alcaligens* are some common denitrifiers used in denitrification process (Ahammad et al. 2013). Advance processes like Sharon process (single reactor high activity ammonia removal over nitrite), canon process (autotrophic nitrogen removal) and anammox process (anaerobic ammonia oxidation) are also used in biological nitrogen removal during wastewater treatment. Anammox process includes anaerobic ammonia oxidation involving lithoautotrophic bacteria (*Planctomycetes*) for biological conversion of ammonia to nitrogen. The bacteria use carbon dioxide as the carbon source and nitrite as the electron acceptor under anoxic condition. The autotrophic nitrogen removal method (canon process) involves combination of partial nitrification and anammox process. In this case, both the reaction takes place within a single aerated reactor using *Nitrosomonas* and *Planctomycetes. Nitrifers* consume oxygen and oxidize ammonia to nitrite. Therefore, the oxygen concentration decreases and an anaerobic condition is created for anammox process (Ahammad et al. 2013). Sharon is also a two-step reaction process involving autotrophic nitrification and heterotrophic denitrification in a single reactor. It is a partial nitrification process for wastewater treatment, which mainly serves the purpose of preliminary treatment (Ahammad et al. 2013; Milia et al. 2016). Biological phosphorus removal involves phosphate accumulating organisms like *Acinetobactor, Lampropedia, Aeromonas, Pseudomonas, Microlunatus, Phosphovorus,* etc. Phosphate accumulating bacteria accumulate phosphate and store them as polyphosphates within the cell. These bacteria can decrease the concentration of phosphorus from the wastewater in activated sludge process. Phosphate accumulating organisms can uptake volatile fatty acids (as carbon source) from wastewater under anaerobic condition (Ahammad et al. 2013).

18.10 Valuable End Products of Biological Wastewater Treatment

Biological wastewater treatment process has potential for recycling and recovery of valuable end products. Most of the plastics are easily biodegradable or recyclable from plastic wastes (Table 18.2). Methane gas is one of the valuable products of anaerobic digestion of wastewater, which can be used for industrial and domestic purposes. Vermifiltration of wastewater produces bio-fertilizers and soil conditioners as its end products (Samer 2015). Biological wastewater treatments can also generate electricity as an end product. In a microbial fuel cell microbial growth occurs on the anode by oxidizing organic matters present in the wastewater. This releases electrons from the anode. Air is sparked on the cathode and reaction between protons, oxygen and electrons completes the circuit to produce electrical energy. Therefore, biological wastewater treatment processes actually mimic the natural processes and deal with adequate amount of wastes along with production of valuable end product for commercial and industrial applications (Ahammad et al. 2013; Samer 2015). Wastewater is not actually a problem, but it is a valuable resource which if treated properly can give economically valuable products. We just need the appropriate treatment method which will be eco-friendly, cost-effective, simple to design, give minimized negative environmental impacts and maintain ecological balance. Biological wastewater treatment processes are time consuming processes and are sensitive to seasonal variation. Also, the treatment processes are applicable only for biodegradable substrates. Wastewater might contain some bio-resistant components that could not be degraded or decomposed by any biological agents. Therefore, in some cases biological wastewater treatment is combined with

Table 18.2 Type of plastics used in manufacturing of common products and their fate towards recycling

Type of plastic	Used in manufacturing of	Recyclable nature
Polytetrafluoroethylene	Gaskets, Non-stick frying pan coating, Gears, Seals, Plates	Non-recyclable
Polystyrene	Cafeteria trays, Plastic utensils	Non-recyclable
Polyethylene terephthalate	Cups, Beverage bottles, Packaging materials	Mostly/partially
High density polyethylene	Cups, Bottles, Milk jugs	Mostly/partially
Polyvinyl chloride	Siding, Flooring, Pipes	Mostly/partially
Low density polyethylene	Tubing, Plastic bags	Mostly/partially
Polypropylene	Food containers, Auto parts, Industrial fibres	Mostly/partially
Polycarbonate	Auto parts, Safety glasses, Phones, Bullet resistance materials, Laptops, CD, DVD	Totally recyclable
Polylactic acid	Plastic films, Bottles, Medical devices (likes Screws, Pins, Rods, Plates)	Totally recyclable
Polyurethane	Insulation panels, Elastic wheels, Tyres, Seals, Gaskets	Totally recyclable

physicochemical treatment for degradation of bio-resistant components (Ahammad et al. 2013; Samer 2015).

18.11 Instances of Innovative Methods of Wastewater Treatment

The domestic water needs in Qatar gets fulfilled by the groundwater and desalinized water. Another alternate source is the Industrial Wastewater (IWW). Here, the water is purified using phytoremediation by removal and degradation of contaminants, which is showing promising results. For phytoremediation, many native plants have been effectively used to treat wastewater and polluted water bodies. *Ricinus communis* is one of the plants in Qatar used for remediation of pesticides and hydrocarbons from wastewater. Other promising plants for phytoremediation are *Amaranthus* sp., *Typha domingensis, Nerium oleander*, etc. For heavy metal accumulation green areas are created around Doha, Qatar, where wastewater from different industries and domestic circuits are being treated. Inorganic and organic contaminants from polluted water and wastewater are removed using native flora and microorganisms. (Al-Thani and Yasseen 2020). In many countries around the world, freshwater shortage is a noticeable problem. Desalinization using photosynthetic organisms offers a potential opportunity for development of efficient technology in this regard. In brackish and sea water, cyanobacterial culture can produce a low salt reservoir (Amezaga et al. 2014). Some industries release heavy metal like mercury, which is a very toxic pollutant. This heavy metal can accumulate in the fish and aquatic animals, and gets bio-magnified in the food chain, thereby affecting the human health. Microbes have evolved a machinery to detoxify mercury using *mer operon*. Mercury hyper-tolerant microbes perform intracellular biotransformation of Hg^{2+} to Hg^0 (which is a relatively non-toxic form) by using mercury reductase enzyme and Hg^0 finally defuses from the cell. This process is functional under fluctuating mercury concentration. Removal or transformation of mercury using mercury resistance microbes is a very cost-effective, simple and eco-friendly process. Highly mercury resistance bacterium *Brevundimonas* sp. is often used for this process (Wagner-Döbler et al. 2003).

18.12 Conclusion

The aim of using bioremediation tools in water treatment is to maintain a good standard of the water for domestic use in a cost-effective and eco-friendly manner. Bioremediation is a passive process, which requires minimal energy over time and the components of the cleaning process doesn't add up any pollutant to the environment. This method could complement or supplement the existing physical and chemical methods of wastewater treatment. The requirement for sophisticated machinery, specialized protocols and special operational training is minimal. Therefore, bioremediation should develop as a sustainable alternative to the existing traditional techniques of wastewater treatment.

Acknowledgements The authors acknowledge Department of Science and Technology and Biotechnology, Government of West Bengal, India for funding the research by its R&D Project Scheme—'Gobeshonay Bangla' (Grant No. STBT-11012(15)/26/2019-ST SEC).

References

Ahammad SZ, Graham DW, Dolfing J (2013) Wastewater treatment: biological. In: Jorgensen SE (ed) Encyclopedia of environmental management, vol 4. Taylor & Francis, Bosa Roca, pp 2645–2655. https://doi.org/10.1081/E-EEM-120046063

Al-Thani RF, Yasseen BT (2020) Phytoremediation of polluted soils and waters by native Qatari plants: future perspectives. Environ Pollut 259:113694. https://doi.org/10.1016/j.envpol.2019.113694

Almuktar S, Abed SN, Scholz M (2018) Wetlands for wastewater treatment and subsequent recycling of treated effluent: a review. Environ Sci Pollut Res Int 25(24):23595–23623. https://doi.org/10.1007/s11356-018-2629-3

Amezaga JM, Amtmann A, Biggs CA, Bond T, Gandy CJ, Honsbein A, Karunakaran E, Lawton L, Madsen MA, Minas K, Templeton MR (2014)

Biodesalination: a case study for applications of photosynthetic bacteria in water treatment. Plant Physiol 164(4):1661–1676. https://doi.org/10.1104/pp.113.233973
Boelee E, Geerling G, van der Zaan B, Blauw A, Vethaak AD (2019) Water and health: from environmental pressures to integrated responses. Acta Trop 193:217–226. https://doi.org/10.1016/j.actatropica.2019.03.011
Brar A, Kumar M, Vivekanand V, Pareek N (2017) Photoautotrophic microorganisms and bioremediation of industrial effluents: current status and future prospects. 3 Biotech 7(1):18. https://doi.org/10.1007/s13205-017-0600-5
CPCB (2021) National inventory of sewage treatment plants. Central Pollution Control Board. Ministry of Environment, Forest and Climate Change, Government of India
Ersahin ME, Ozgun H, Dereli RK, Ozturk I (2011) Anaerobic treatment of industrial effluents: an overview of applications. In: Waste water—treatment and reutilization. https://doi.org/10.5772/16032
Horel A, Schiewer S (2020) Microbial degradation of different hydrocarbon fuels with mycoremediation of volatiles. Microorganisms 8(2):163. https://doi.org/10.3390/microorganisms8020163
Kapahi M, Sachdeva S (2017) Mycoremediation potential of Pleurotus species for heavy metals: a review. Bioresources Bioprocess 4(1):1–9. https://doi.org/10.1186/s40643-017-0162-8
Kumar V, Chopra AK (2018) Phytoremediation potential of water caltrop (*Trapa natans* L.) using municipal wastewater of the activated sludge process-based municipal wastewater treatment plant. Environ Technol 39(1):12–23. https://doi.org/10.1080/09593330.2017.1293165
Kumar R, Negi S, Sharma P, Prasher IB, Chaudhary S, Dhau JS, Umar A (2018) Wastewater cleanup using *Phlebia acerina* fungi: an insight into mycoremediation. J Environ Manage 228:130–139. https://doi.org/10.1016/j.jenvman.2018.07.091
Mandal D, Basu A (2021) Role of heavy-metal resistant bacteria isolated from rhizosphere in bioremediation and plant development. In: Rhizobiology: molecular physiology of plant roots. Springer, Cham, pp 411–435. https://doi.org/10.1007/978-3-030-84985-6_22
Mandal D, Sonar R, Saha I, Ahmed S, Basu A (2021) Isolation and identification of arsenic resistant bacteria: a tool for bioremediation of arsenic toxicity. Int J Environ Sci Technology:1–18.https://doi.org/10.1007/s13762-021-03673-9
Masiá P, Sol D, Ardura A, Laca A, Borrell YJ, Dopico E, Laca A, Machado-Schiaffino G, Díaz M, Garcia-Vazquez E (2020) Bioremediation as a promising strategy for microplastics removal in wastewater treatment plants. Mar Pollut Bull 156:111252. https://doi.org/10.1016/j.marpolbul.2020.111252
Milia S, Perra M, Cappai G, Carucci A (2016) SHARON process as preliminary treatment of refinery wastewater with high organic carbon-to-nitrogen ratio. Desalin Water Treat 57(38):17935–17943. https://doi.org/10.1080/19443994.2015.1087341
Mukherjee B, Majumdar M, Gangopadhyay A, Chakraborty S, Chaterjee D (2015) Phytoremediation of parboiled rice mill wastewater using water lettuce (*Pistia stratiotes*). Int J Phytorem 17(7):651–656. https://doi.org/10.1080/15226514.2014.950415
Rajput RS, Pandey S, Bhadauria S (2017) Status of water pollution in relation to industrialization in Rajasthan. Rev Environ Health 32(3):245–252. https://doi.org/10.1515/reveh-2016-0069
Saha P, Banerjee A, Sarkar S (2015) Phytoremediation potential of Duckweed (*Lemna minor* L.) on steel wastewater. Int J Phytoremediation 17(6):589–596. https://doi.org/10.1080/15226514.2014.950410
Saha P, Shinde O, Sarkar S (2017) Phytoremediation of industrial mines wastewater using water hyacinth. Int J Phytorem 19(1):87–96. https://doi.org/10.1080/15226514.2016.1216078
Samer M (2015) Biological and chemical wastewater treatment processes. In: Wastewater treatment engineering, p 150.https://doi.org/10.5772/61250
Satyawali Y, Balakrishnan M (2008) Wastewater treatment in molasses-based alcohol distilleries for COD and color removal: a review. J Environ Manage 86(3):481–497. https://doi.org/10.1016/j.jenvman.2006.12.024
Saxena G, Chandra R, Bharagava RN (2017) Environmental pollution, toxicity profile and treatment approaches for tannery wastewater and its chemical pollutants. Rev Environ Contam Toxicol 240:31–69. https://doi.org/10.1007/398_2015_5009
Scalenghe R (2018) Resource or waste? A perspective of plastics degradation in soil with a focus on end-of-life options. Heliyon 4(12):e00941. https://doi.org/10.1016/j.heliyon.2018.e00941
Shah A, Shah M (2020) Characterisation and bioremediation of wastewater: a review exploring bioremediation as a sustainable technique for pharmaceutical wastewater. Groundw Sustain Dev 11:100383. https://doi.org/10.1016/j.gsd.2020.100383
Sharma R, Saini H, Paul DR, Chaudhary S, Nehra SP (2021) Removal of organic dyes from wastewater using *Eichhornia crassipes*: a potential phytoremediation option. Environ Sci Poll Res 28(6):7116–7122. https://doi.org/10.1007/s11356-020-10940-8
Slavov AK (2017) General characteristics and treatment possibilities of dairy wastewater–a review. Food Technol Biotechnol 55(1):14–28. https://doi.org/10.17113/ftb.55.01.17.4520
Wagner-Döbler I (2003) Pilot plant for bioremediation of mercury-containing industrial wastewater. Appl Microbiol Biotechnol 62(2):124–133. https://doi.org/10.1007/s00253-003-1322-7
Wanhong L, Fang L, Fan W, Maiqi D, Tiansen L (2020) Industrial water pollution and transboundary eco-compensation: analyzing the case of Songhua river basin, China. Environ Sci Poll Res 27(28):34746–34759. https://doi.org/10.1007/s11356-019-07254-9

Adaptation to Climate Change in Agriculture at Teesta Basin in Bangladesh

19

Md. Abdullah Al Mamun, A. T. M. Sakiur Rahman, Most. Shayda Shamsea Aziz Shabee, Jayanta Das, Md. Aminul Islam, G. M. Monirul Alam, M. Mizanur Rahman, and Md. Kamruzzaman

Abstract

cTeesta river basin is one of the hazard-prone areas in Bangladesh. Flood, riverbank erosion and hailstorm are the most devastating natural hazards to agriculture in this area. Therefore, adaptation measures are obligatory to sustain the livelihood and food security of the households. This study has investigated the indigenous and institutional adaptation measures of the farming households using survey data collected from three districts of Bangladesh. The results reveal that more than half of the respondents have changed the crops' seeding or planting time, adopting resilient crop varieties to avoid the climatic hazards, whereas about 45% of respondents have adopted crop rotation to sustain soil fertility. They also adopt non-agricultural adaptation strategies such as a change in profession, taking credit from different sources and using agricultural land for non-agricultural purposes. Poor farmers receive relatively little support from the government to enhance their adaptation strategies. Institutional support such as technical and financial support is needed to promote farmers' adaptation process.

Keywords

Climate change · Adaptation · Agriculture · Teesta river basin · Bangladesh

19.1 Introduction

Climatic stress on natural resources (e.g., water resources) and world food production is likely to develop in the forthcoming decades (Yusuf et al.

Md. A. A. Mamun · Md. Kamruzzaman (✉)
Institute of Bangladesh Studies, University of Rajshahi, Rajshahi, Bangladesh
e-mail: mkzaman@ru.ac.bd

A. T. M. S. Rahman
Hydrology Lab, Department of Earth and Environmental Science, Kumamoto University, Kumamoto, Japan

Most. S. S. A. Shabee
Minatojima Nakamachi, Kobe City 650-0046, Hyogo, Japan

J. Das
Department of Geography, Rampurhat College, Birbhum, West Bengal 731224, India
e-mail: jayanta.daas@gmail.com

Md. A. Islam
Department of Media Studies and Journalism, University of Liberal Arts Bangladesh, Dhaka, Bangladesh
e-mail: aminul.islam@ulab.edu.bd

G. M. M. Alam
Department of Agribusiness, Bangabandhu Sheikh Mujibur Rahman Agricultural University, Ghazipur, Bangladesh
e-mail: gmalam@bsmrau.edu.bd

M. M. Rahman
Department of Geography and Environmental Studies, University of Rajshahi, Rajshahi, Bangladesh

J. Das and S. K. Bhattacharya (eds.), *Monitoring and Managing Multi-hazards*,
GIScience and Geo-environmental Modelling, https://doi.org/10.1007/978-3-031-15377-8_19

2009; Gosling and Arnell 2016; Fitton et al. 2019). Climatic hazards have affected farming and human livelihood by declining crop production and decreasing farmers' income (Kamruzzaman et al. 2022; Shikuku et al. 2017; Jamshidi et al. 2019). Agriculture in Bangladesh is also exposed to climate change. Crops, fisheries and livestock are hampered due to natural hazards, i.e., floods, salinity intrusion, drought, riverbank erosion and diseases (Alam et al. 2018). Therefore, adaptation is important to face climate change hazards. Adaptation is the process of adjustment to genuine or anticipated climatic variability and change, and adaptive capacity is the fruitful reaction including changes in human behavior, assets and technologies (Noble et al. 2014; Alam et al. 2017; Jamshidi et al. 2019).

In Bangladesh, the west-central part is highly drought-prone (Kamruzzaman et al. 2016b; Pal et al. 2021). Drought diminishes soil moisture and water-retaining capacity and dries out topsoil. The wind blows away the dried topsoil, and precipitation erodes topsoil too. It also drains the natural substance of soil that lessens microbial action and harms vegetation. The absence of organic materials affects both vegetative growth and the yield of crops (Habiba et al. 2011). A higher probability of rainfall increases the potential of flood in these areas results in damages to *aman* cultivation (Kamruzzaman et al. 2016a, 2021). Winter temperature was favorable for HYV *boro* and wheat but unfavorable for tomato production in this area (Azad et al. 2013). Deforestation leads to extensive soil erosion in the catchment area. It accelerates rapid siltation on the river bed that forms *chars* (sand bar in the river channel). *Chars* make obstacles to river flow, responsible for lengthy floods and riverbank erosion (Alam et al. 2018; Ahmed et al. 2021). Deep sand deposits make the *char* land unsuitable for cultivation (Raune et al. 2013; Alam et al. 2018). Due to the rapid glacier melting of Himalayan region, low-lying Bangladesh is very vulnerable to floods as it drains out 92% water of Ganges, Brahmaputra and Meghna rivers in the rainy season (Dastagir 2015). These floods often affect the Teesta basin that devastates infrastructures, crops, livestock, riverbank erosion and even human lives (Karim and Thiel 2017). It also imbalances the ecology along with disturbing the communication and essential services (Rakib et al. 2016). On the other hand, Teesta basin also faces severe drought in dry season characterized by low rainfall, high temperature, high rate of evaporation and significant groundwater depletion (Karim and Thiel 2017). As a result, income, rate of consumption of the people may fawhichhat also increases malnutrition (Rahman et al. 2007). Das et al. (2021) summarized that climate of Bangladesh has been changed within the study period (1966–2019), and the results exhibit a drastic change of precipitation over the whole study area except few pocket areas. Therefore, required measures need to be taken to address the above miseries of people of the Teesta basin.

Momtaz and Shameem (2016) revealed the issues regarding various adaptation measures, e.g., shifting to shrimp aquaculture from crop farming, the culture of mud crabs, mono-sex tilapia aquaculture in the coastal region of Bangladesh. Moreover, farmers cultivated transplanted *aman* at low-lying land in the rainy season when the salinity level was decreased. At medium high land, farmers cultivated traditional *aman* instead of HYV *aman*, because HYV *aman* is more sensitive to salinity. Baki et al. (2012) developed an inundation map of Shariatpur district by using hydrodynamic model HEC-RAS and HEC-GeoRAS. After considering an optimal flood depth, they calculated the volume of water above optimal flood depth. It revealed that 1.85 m flood depth from mean sea level was more suitable for ensuring the maximum aman and boro varieties production. Islam et al. (2014) noticed farmers' initiatives in diversifying crops in the case of floods and riverbank erosion. Instead of rice, farmers changed the cropping patterns by introducing maize, oil-seeds, groundnut, etc. Similarly, Alam et al. (2018) prioritized farmers' adaptation practices at riverbank erosion-prone Sirajganj and Tangail districts in Bangladesh. Changing plantation time, farming of pulses, spices and oilseeds, homestead gardening and tree plantation were the most common strategies. Sarker et al. (2019) studied climate-smart agriculture, which includes resource-conserving zero tillage or no-tillage farming practices to maintain agricultural production sustainability. Bhatta et al. (2017) studied the promising agricultural innovations in northern Bangladesh. Bangladesh Rice Research Institute and Bangladesh Institute of Nuclear Agriculture developed low-yielding rice varieties (BR-33 and BINA-7) for critical food scarce periods in this area. Alam et al. (2015) addressed the adaptation measures in the drought-prone Rajshahi district in Bangladesh. He identified groundwater irrigation as the most familiar adaptation practice in that area, influenced by schooling and household income, electricity and tenure status, institutional initiatives and awareness of climatic threats. Addressing peoples' responses to floods, Karim and Thiel (2017) told about the cultivation of water-tolerant crops in the Rangpur district of northern Bangladesh.

In the south Asian perspective, Manandhar et al. (2011) mentioned farmer's adaptation in response to severe climatic threats, i.e., erratic precipitation, flood and increased summer temperature in both upland and low land of Nepal. People shifted from local crop varieties to hybrid and early maturing crop varieties in low land. In the case of the coastal plain, Rautaray (2020) addressed the problems of excessive water in the wet season and scarcity of water in the dry season in southeastern Odisha in India. He suggested pond-based farming by constructing a pond in clay soil to protect harvested water from seepage and percolation. Moreover, dyke cropping ensures more supply of capillary water, which is beneficial for papaya, banana and vegetables along with fish production. Wide-spaced creeper vegetables (e.g., pumpkin, water melon, bitter ground, ridge ground and cucumber) are appropriate for saving water in the soil. In such cases, drip irrigation is essential to maximize the use of harvested water in ponds. Meldrum et al. (2018) addressed climate events in potato farming by changing cropping schedules and the shifting planting location in Bolivian Altiplano. In addition, farmers used to extend crop fields with fire. Hence, smoke scattered clouds that had deflected frost and hail. Moreover, spreading ashes on crop leaves was a widespread practice for pest control.

However, none of these studies looked at the adaptation measures to deep sedimentation due to flood and riverbank erosion. Moreover, there is a substantial gap in the literature about adaptation to cold waves and pests in the tropical zone. Thus, adaptation in these sectors needs more attention. Therefore, the objectives of this study are: (i) to assess the resilience of farmers by spotting different indigenous adaptation strategies; and (ii) to investigate the government and non-government assistance to boost the farmers' adaptive capacity.

19.2 Materials and Methods

19.2.1 Study Area

The Teesta basin in Bangladesh includes Lalmonirhat, Rangpur, Kurigram, Nilphamari and Gaibandha districts. This study has selected Lalmonirhat, Rangpur and Nilphamari districts from five districts by simple random aampling (SRS) using Microsoft Excel-2013. Similarly, the study has selected Lalmonirhat Sadar, Dimla and Kaunia upazilas from Lalmonirhat, Nilphamari and Rangpur districts, respectively, through SRS using Microsoft Excel-2013. The upazila is the third-order administrative unit in Bangladesh that comprises two or more unions. Likewise, the study has also selected Khaga Kharibari, Gokunda and Tepa Modhupur unions from Dimla, Lalmonirhat Sadar, Kaunia upazilas, respectively, through SRS. The union is the smallest unit of local government in rural Bangladesh that comprises villages. After that, the study has also selected Kismat Chatnai, Kursamari and Azam Khan villages, respectively, from Khaga Kharibari, Gokunda and Tepa Modhupur unions through SRS Microsoft Excel-2013 (without replacement).

It is well-known that the three forms of SRS, such as simple random sampling without replacement, simple random sampling with replacement and fixed cost simple random

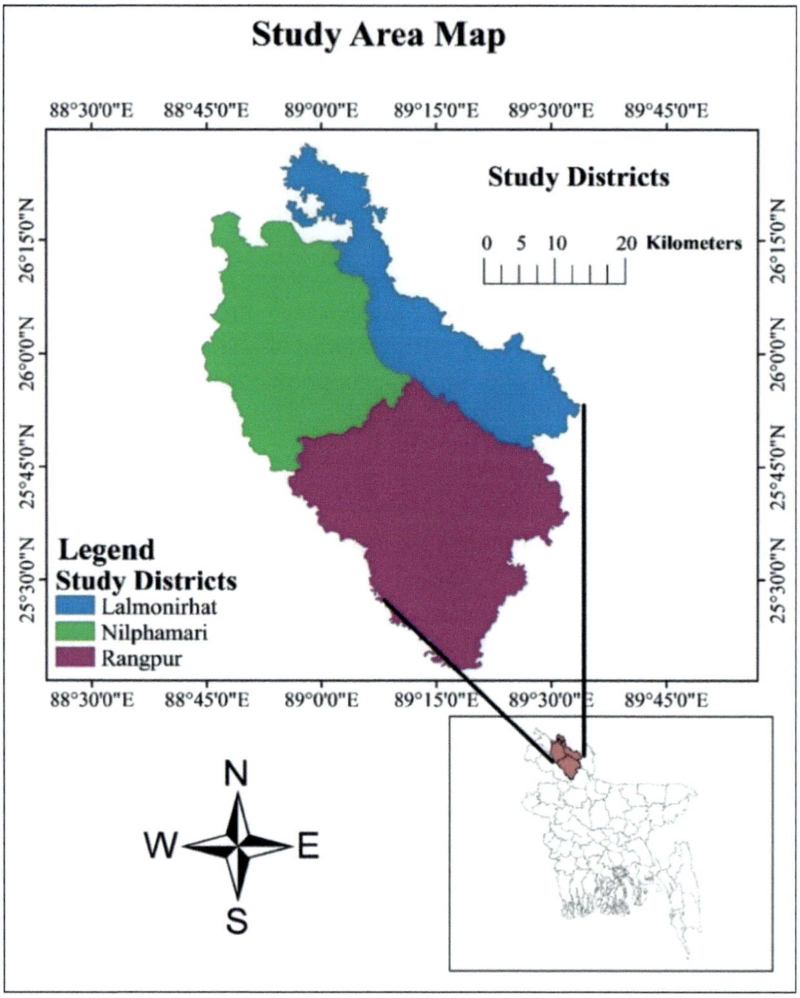

Fig. 19.1 Study area map with location of districts, upazilas and unions

sampling, constitute the sample mean both acceptable and unbiased (Mitra and Pathak 1984). Thus, this study has used the SRS without replacement technique. Selected districts from the north-eastern part of Bangladesh have been shown in Fig. 19.1.

19.2.2 Sample Size and Sampling Procedure

A questionnaire survey was conducted from September 2017 to November 2017 (3 months) to select sample households from three sample villages. Sample households have been selected from each village through SRS (without replacement) using Microsoft Excel-2013. The SRS technique of sample selection was also followed by Manandhar et al. (2011) and Kabir et al. (2016). Here, the list of households involved in farming has been used to select samples. The Agriculture Extension Offices (AEO) of respective upazilas have provided the list. Based on the farmers' list provided by the AEO, sample farmers (respondents) have been selected through SRS. In addition, the most experienced persons involved in farming and key decision-making have been selected from each household. This selection has also gone through SRS. Sample households for each village (n_i) have been calculated by the following formula,

$$n_i = \frac{N_i}{N} \times n$$

The size of the study population (N) is known. Total household sample size (n) has been determined by the following formula (Kothari 2004).

$$n = \frac{Z^2.p.q.N}{e^2(N-1) + Z^2.p.q}$$

Therefore, 276 sample households have been selected from three sample villages: 130 households from Azam Khan, 76 households from Kursamari village and 70 households from Kismat Chatnai village.

19.2.3 Data Collection Tools

Qualitative and quantitative data were collected for this study where the household head (the most experienced person in the family) was the respondent. Here, elderly farmers have been prioritized to tap into their long-term memory of extreme climatic events in farming. Quantitative data were collected through face-to-face interviews of the respondents. Qualitative data were also collected through in-depth interviews for cross-checking the data and information obtained from the survey. A semi-structured questionnaire was used to collect survey data. The duration of the interviews ranged from 30 to 40 min. The questionnaire contains information on demographic characteristics, effects of climate change on agricultural activities and peoples' adaptation practices to climate change. Secondary data were collected from relevant books, journals, research papers and research reports. Data were analyzed by using IBM SPSS (Statistical Package for Social Sciences) version 22. For primary data, inferential statistics have been used to explore the existing situation. Frequency distributions in LS (Likert Scale) have been used to express respondents' opinions about the adaptation processes to climate change in agriculture. This study has also explored both institutional and indigenous adaptation measures.

19.3 Results and Discussion

According to the field survey, more than half of the respondents are between the age of 30 to 49, and 37.3% of respondents are in the 50–69 age groups. It indicates respondents' high experience in farming. The average annual income of the respondent is BDT 67,178, whereas the annual per capita income of Bangladesh has raised to BDT 121,678 by 2017, which indicates a weak financial condition. About half of the respondents are illiterate, indicating their poor educational status. The state of higher education is also very fragile. Only four respondents have passed graduation or post-graduation. The average own land area of the people is 1.83 acres. Many farmers cultivate in leased land. The average leased land area is 1.27 acres.

19.3.1 Adaptation Measures to Climate Change

Adaptation in agriculture includes both technological and non-technological responses. Technological interventions include new crop varieties, better farming technologies, suitable crop calendars and adequate communication. On the other hand, non-technological actions include access to market and credit (Aryal et al. 2019). Both indigenous (based on knowledge and experience) and institutional measures/planned adaptation (supposed by government organizations and NGOs) are important to be successful in agricultural adaptation (Alam 2016; Alam et al. 2018). Drought, flood, riverbank erosion, sedimentation, severe cold and pests are the main climatic hazards in the study area. It hampers crop production extremely. Adaptation measures have been taken to address crop damage, soil erosion and diseases of the crops in the study area. It is also needed to retain soil fertility and reduce production costs. Farmers are more concerned about floods than other hazards. Change in seeding and planting time is a common adaptation measure to flood. Besides, crop rotation, new crop varieties and technologies are very crucial to adapt to flood.

Table 19.1 Assessment of risk from climatic hazards and adaptation actions

Climatic hazards	Assessment of risk	Adaptation actions
The flood	Damage to crops	Change seeding or planting time Maintain crop rotation Cultivate new crops varieties Introduce new technologies Provide financial and non-financial aids
Riverbank erosion	Damage to crops and erosion of soil	Develop infrastructure Change seeding or planting time
Sedimentation	Reduce of soil fertility	Provide supplementary irrigation Use sufficient fertilizer and insecticides Maintain crops rotation
Severe cold and pests	Increase in diseases of the crops Rise in production cost	Use sufficient fertilizer and insecticides Take credit Arrange training
Drought	Damage to crops	Provide supplementary irrigation Introduce new crops varieties

Source Field survey from September 2017 to November 2017

Irrigation is a vital adaptation to tackle all hazards except severe cold and pests. Farmers have no option but to provide fertilizer, insecticides and irrigation to uphold the fertility of soil in the case of deep sedimentation due to flash flood. Institutional adaptation like infrastructure development to reduce riverbank erosion and arrange farmers' training to prevent crops' diseases are vital in adaptation (Table 19.1). Major adaptation options in agricultural sectors in the Teesta basin are described in the following sections:

19.3.2 Indigenous Adaptation Measures

1. **Change of Seeding and Planting Time**
Farmers have taken different types of adaptation measures from their knowledge and experiences. Natural hazards often cause a disaster at some crops' seeding or planting time. *Transplanted aman* is very affected due to flood at the time of its seeding or planting. Total area of aman cultivation is 5.62 million hectares in Bangladesh (Salam et al. 2014). Severe cold, fog and rainfall during the winter season affect potato cultivation mainly in its harvesting time. Therefore, farmers do think about the change of seeding or planting time of these crops to avoid damages from flood and cold, respectively. Half of the respondents occasionally have changed seeding or planting time of their crops, and 22.8% of respondents have done it sometimes (Table 19.2). However, farmers in the Azam Khan village have practiced this on a larger scale. The traditional mindset of farmers about farming and lack of information regarding flood forecasting are the main reasons for maintaining the traditional seeding and planting timing.
2. **Maintain Crop Rotation**
Deep sedimentation by sand has reduced the fertility of the soil. Crop rotation can enhance soil fertility and ensure good production (Yusuf et al. 2009). About half (45.3%) of respondents occasionally have alternated crops, and 20.3% of respondents have done it sometimes (Table 19.2). More than one-fourth of respondents (26.3%) in this village sometimes have switched crops in fields. Moreover, in Kursamari village, more than half (56.6%) of the respondents have alternated crops occasionally (Table 19.3). In this context, People in Kursamari village have rotated crops more frequently than those of the other two villages.

Table 19.2 Overall percentage of indigenous adaptation strategies to climate change in agriculture

Adaptation strategies	Number of respondents in percent				
	Never	Occasionally	Sometimes	Often	At all times
Change in seeding or planting timing	19.2	50.0	22.8	5.8	2.2
Maintain crop rotation	31.5	45.3	20.3	2.5	0.4
Cultivate new crop varieties	39.1	49.6	9.1	2.2	0.0
Introduce new technologies	90.2	6.5	2.9	0.0	1
Supplementary irrigation	24.3	1.4	38.4	24.6	11.2
Supplementary fertilizer and insecticide	10.9	2.2	6.2	52.2	28.6
Use agricultural land in non-agricultural activities	40.2	50.0	8.3	1.1	0.4
Credit from rich people	38.8	39.5	17.0	4.3	0.4
Change in profession	48.6	41.3	7.6	2.5	0.0

Source Field survey from September to November 2017

3. **Cultivating New Crop Varieties**
   Diversification in farming generally demonstrates an expansion of the collective mixture of cultivating exercises inside a characterized range (Bradshaw et al. 2004). Farmers have followed diversified crop cultivation worldwide to cope with the adverse impact of climatic variability and change (Olesen et al. 2011; Maponya and Mpandeli 2012; Bhatta et al. 2017; Meldrum et al. 2018). Farmers in the study area are not very aware of the cultivation of new crop varieties. Almost forty percent (39.1%) of respondents have never adopted new varieties of crops but about half of the respondents (49.6%) have adopted new varieties of crops occasionally (Table 19.2). For example, many farmers in Kursamari village do not cultivate tobacco nowadays. To increase rice and maize production, farmers have to irrigate their crop fields frequently. Extensive irrigation has increased the moisture of the soil. Too much moisture in the soil is not suitable for tobacco cultivation. Besides, tobacco cultivation is time-consuming, which makes delayed starting of next crops. For the above reasons, rice, maize and vegetables have replaced tobacco cultivation. Similarly, respondents have replaced jute cultivation with the lady's finger, cabbage and cauliflower cultivation, because jute is very vulnerable to floods and hailstorms.
4. **Introduce New Technologies in Agriculture**
   Scientists and researchers are always trying to introduce new technologies to intensify agricultural production. The respondents in the study area do not use new technologies on a large scale. In the case of introducing new technologies, more than 90% of respondents have never chosen it (Table 19.2). The situation is little different in the Azam Khan village, where 13.1% of respondents have introduced new technologies in agriculture (Table 19.3).
5. **Provide Supplementary Irrigation**
   Rain-fed crops cultivation becomes very troublesome due to erratic precipitation. Hence, farmers are increasingly dependent on irrigation-based cultivation. Underground water has been very available for the farmers at a very low cost. Almost forty percent (38.4%) of respondents sometimes provide supplementary irrigation to their crop fields and 24.6% of respondents do it often (Table 19.2). The cultivation in the Kursamari village is not fully dependent on rainfall. Thus, the irrigation rate is relatively higher in this village. To produce plenty of crops in

Table 19.3 Percentage of indigenous adaptation strategies in agriculture on villages separately

Villages	Degree of effects	Change in seeding/planting timing	Change in crop rotation	Introduction of new varieties	Introduce new technologies	Supplementary irrigation	Use supplementary fertilizer and insecticide	Use agricultural land in non-agricultural activities	Credit from rich people	Change in profession
Kursamari	Never	19.7	13.2	32.9	94.7	15.8	11.8	31.6	27.6	36.8
	Occasionally	59.2	56.6	59.2	1.3	1.3	2.6	59.2	55.3	55.3
	Sometimes	15.8	26.3	6.6	3.9	56.6	5.3	7.9	7.9	5.3
	Often	5.3	2.6	1.3	0.0	23.7	65.8	1.3	9.2	2.6
	At all time	0.0	1.3	0.0	0.0	2.6	14.5	0.0	0.0	0.0
Azam Khan	Never	10.0	29.2	31.5	82.3	20.0	10.0	36.9	37.7	50.0
	Occasionally	51.5	43.8	51.5	13.1	2.3	3.1	49.2	31.5	34.6
	Sometimes	24.6	23.8	13.8	3.8	29.2	9.2	11.5	30.0	12.3
	Often	9.2	3.1	3.1	0.0	35.4	46.2	1.5	0.8	3.1
	At all time	4.6	0.0	0.0	0.8	13.1	31.5	0.8	0.0	0.0
Kismat Chatnai	Never	35.7	55.7	60.0	100	41.4	11.4	55.7	52.9	58.6
	Occasionally	37.1	35.7	35.7	0.0	0.0	0.0	41.4	37.1	38.6
	Sometimes	27.1	7.1	2.9	0.0	35.7	1.4	2.9	2.9	1.4
	Often	0.0	1.4	1.4	0.0	5.7	48.6	0.0	5.7	1.4
	At all time	0.0	0.0	0.0	0.0	17.1	38.6	0.0	1.4	0.0

Source: Field survey from September to November 2017

Kursamari village, 56.6% of respondents have provided supplementary irrigation during the winter and the summer (Table 19.3). Besides, the soil infiltration rate in this village is relatively low than that of the other two villages. Therefore, the cost of irrigation is less in this village. Farmers in Azam Khan village are also very concerned about this supplementary irrigation. HYD rice grows well in this village. Irrigation is mandatory for HYD rice cultivation. More than thirty-five percent (35.4%) of respondents in Azam Khan village provide supplementary irrigation in crop fields. On the other hand, 41.4% of respondents in Kismat Chatnai village never provided supplementary irrigation to their crop fields (Table 19.3). The soil of the village is very sandy. So, the infiltration rate is very high in this village. For this reason, irrigation-based HYD rice is not cultivated on a large scale in this village. Thus, only 24.6% of respondents irrigate in crops often in the study area (Table 19.2).

6. **Use of Fertilizer and Insecticides**
Fertilizer is essential for maximum utilization of soil and crop production. Farmers are well aware of using fertilizer. However, insects hinder the maximum production of crops. For example, different diseases affect potato cultivation in the study area. Therefore, it is needed to provide insecticides to maintain the expected production of potatoes. Thus, farmers are also aware of preventing crops from damage due to insects. Fertilizer and insecticide are the most frequent adaptation strategy in the study area (Fig. 19.1). Therefore, 28.6% of respondents have used both sufficient fertilizer and insecticides at all times to ensure expected productivity. More than fifty percent of respondents (52.2%) have often provided supplementary fertilizer and insecticide in their crop field (Table 19.2). In the Kursamari, Azam Khan and Kismat Chatnai villages, 14.5%, 31.5% and 38.6% of respondents have provided supplementary fertilizer and insecticide to their crop fields, respectively. Farmers in Azam Khan and Kismat Chatnai villages have provided fertilizer and insecticide more intensively than those of the Kursamari village. Azam Khan and Kismat Chatnai villages are closer to the Teesta River than the Kursamari village. That is why the sedimentation rate is relatively high in Azam Khan and Kismat Chatnai villages. The main component of the sediment is sand. Deep sandy sedimentation has made the land unproductive. So, it is required to supply adequate fertilizer to the crop fields in these villages.

7. **The Access to Credit**
The production cost of some crops, e. g., potatoes, is relatively high. Rich farmers can invest in potato cultivation, but marginal and poor farmers cannot invest. As a result, poor and marginal farmers have taken credit with very high interest from rich people of the society to continue their cultivation. Farmers have followed the old tradition of taking credit from rich people. From Table 19.2, 39.5% of respondents occasionally have taken credit from rich people, and 17% of respondents sometimes have got it. In the case of Azam Khan village, 30% of respondents sometimes have taken credit from rich people, whereas 7.9% and 2.9% of respondents in Kursamari and Kismat Chatnai villages got it, respectively (Table 19.3). The communication of Azam Khan village is not well with the nearest town. That is why NGOs do not show much interest in credit disbursement in this village. Moreover, government initiatives in credit disbursement are not satisfactory. Thus, farmers in Azam Khan village acquired more credit from rich people than those of other two villages. Sometimes, rich people have hired the land of marginal farmers for potato cultivation. Similar results were found by (Karim and Thiel 2017).

Besides, half of the respondents in the study area have occasionally converted agricultural land into non-agricultural land. Generally, people are using agricultural land for a brick building, making markets, poultry farms and building houses. Change in the profession is not a popular adaptation option for the inhabitants. About

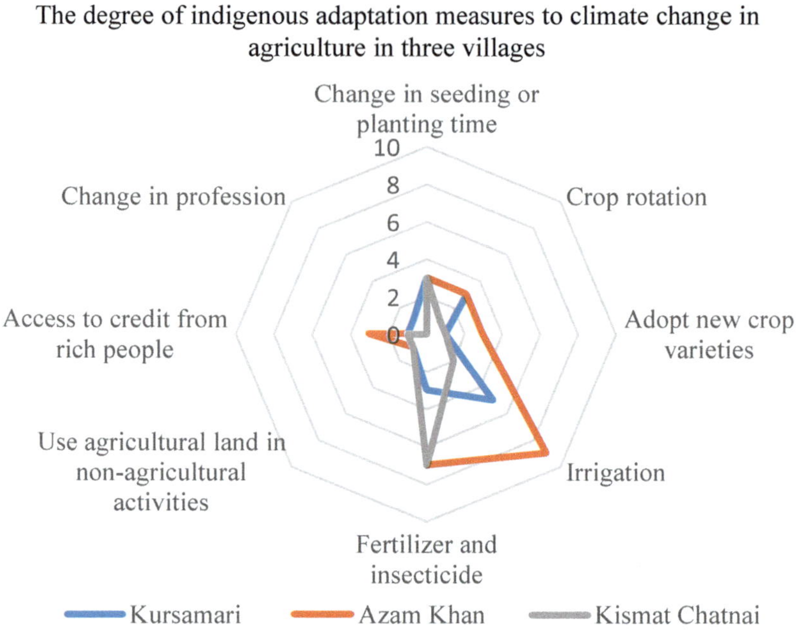

Fig. 19.2 Degree of indigenous adaptation measures to climate change in agriculture. *Source* Field survey from September to November 2017

41.3% of respondents have made it occasionally (Table 19.2).

Figure 19.2 shows that farmers of Azam Khan village are in a more advanced position than that of other villages in adopting different measures to address natural hazards. In this village, they have prioritized irrigation and fertilizer and insecticides to adapt to adverse climatic situations and get the expected agricultural yield. In response to adapt to natural hazards, farmers in Kismat Chatnai village also provide fertilizer and insecticides. However, farmers of this village do not have much interest in providing irrigation like the farmers of Azam Khan village. On the other hand, farmers in Kursamari village are very active in irrigating their crop fields. The cultivation of HYD rice varieties, dependent on irrigation, makes the difference between the villages. Farmers widely follow both changes in seeding and planting time and crop rotation in three villages. Interestingly, access to credit from rich people is only available to the farmers of Azam Khan village due to its limitations to get credit from the government and NGOs.

19.3.3 Institutional Interventions in Facilitating Adaptation to Climate Change

The government and NGOs are working on strengthening farmers' adaptive capacity to cope with different climatic hazards. These organizations have supported them by providing both financial and non-financial aid. Besides, they also have organized training for making the inhabitants skilled to reduce the calamity as well as make them capable of using different resources for rehabilitation. Infrastructure development is an important government intervention to promote the adaptation ability of the inhabitants.

Financial aid is one of the key incentives for farmers. However, the amount of this aid is not reasonable. About 66.7% of respondents have occasionally received government funds as aid (Table 19.4). All of the respondents have not received this aid at a satisfactory level. Azam Khan and Kismat Chatnai villages are flood and erosion-prone.

Table 19.4 Institutional adaptation strategies to climate change in agriculture

Adaptation strategies	Number of respondents in percent				
	Never	Occasionally	Sometimes	Often	At all times
Government funds as aid	28.3	66.7	4.3	0.7	0.0
Non-government funds as aid	19.6	72.5	7.6	0.4	0.0
Seed, fertilizer, insecticide, herbicide as govt. aid	90.6	2.9	6.5	0.0	0.0
Flood-tolerant crop varieties	88.8	5.8	5.1	0.4	0.0
Improve marketing system	89.1	2.5	8.0	0.4	0.0
Credit from govt. banks and financial organizations	88.0	6.9	4.7	0.4	0.0
Credit from non-govt. banks and financial organizations	17.8	63.8	17.8	0.7	0.0
Training from govt. organizations	25.0	64.5	9.8	0.4	0.4
Training from non-govt. organizations	31.2	58.7	7.2	2.5	0.4
Infrastructure development	15.9	59.4	15.6	8.3	0.7

Source: Field survey from September to November 2017

For this reason, these two villages have gotten more attention from the government. Therefore, 69.2%, 81.4% and 48.7% of respondents in Azam Khan, Kismat Chatnai and Kursamari villages, respectively, have received government financial aid occasionally (Table 19.5). In addition to this, NGOs are playing a significant role in enhancing the financial capacity of the people. About 72.5% of respondents have occasionally received non-government funds as aid (Table 19.4). According to respondents' opinions, some Non-Government Organizations (NGOs) such as Association for Social Advancement (ASA), Rangpur Dinajpur Rural Service (RDRS) and Grameen Bank are working in the study area. Farmers need not only financial support but also logistic support. However, the government has not substantially provided seed, fertilizer, insecticides and herbicides. Government and NGOs take least initiatives in providing flood-tolerant crop varieties to the farmers. About 14.5% of respondents in Kursamari village have occasionally got flood-tolerant crop varieties, and 17.1% of respondents have got it from the government, especially from the Bangladesh Agricultural Development Corporation (BADC) (Table 19.5).

On the other hand, the weak marketing system is a barrier to farmers' economic development. But this system is not taken into consideration for development so far. Only 8% of respondents have got benefit from the development of this marketing system (Table 19.4). However, the marketing system is getting better in the Kursamari village compared to other villages. Different non-government banks and financial organizations have provided credit to 63.8% of respondents, whereas government banks and financial organizations have occasionally provided credit to only 6.9% of respondents.

Moreover, 17.8% of respondents sometimes have got credit from non-government banks and financial organizations. So, compared to non-government initiatives, government initiatives in providing credit to farmers are very negligible. In the case of government credit, people have got credit from the Bangladesh Rural Development Board (BRDB) to increase agricultural production. Government and different NGOs have provided occasional training to 64.5% and 58.7% of respondents, respectively (Table 19.4). People in the Kismat Chatnai village have got more training from both government and NGOs compared to those of other villages. The status of

Table 19.5 Institutional adaptation strategies in agriculture in three villages in percent

Villages	Degree of effects	Government funds as aid	Non-governmentfunds as aid	Seed, fertilizer, insecticides, herbicides as government aid	Drought and flood tolerant crop varieties	Improve marketing System	Credit from government banks and financial organizations	Credit from non-government banks and financial organizations	Training from government banks and financial organizations	Training from non-government banks and financial organizations	Infrastructure development
Kursamari	Never	50.0	28.9	90.8	67.1	72.4	77.6	27.6	39.5	40.8	25.0
	Occasionally	48.7	64.5	5.3	14.5	6.6	11.8	52.6	53.9	44.7	48.7
	Sometimes	1.3	6.6	3.9	17.1	21.1	10.5	18.4	5.3	7.9	22.4
	Often	0.0	0.0	0.0	0.0	0.0	0.0	1.3	1.3	5.3	3.9
	At all time	0.0	0.0	0.0	1.3	0.0	0.0	0.0	0.0	1.3	0.0
Azam Khan	Never	27.7	16.9	86.9	96.2	93.8	88.5	11.5	23.8	32.3	13.8
	Occasionally	69.2	73.1	3.1	3.8	1.5	7.7	63.1	63.8	56.9	56.2
	Sometimes	3.1	10.0	10.0	0.0	3.8	3.1	24.6	11.5	9.2	14.6
	Often	0.0	0.0	0.0	0.0	0.0	0.8	0.8	0.0	1.5	14.6
	At all time	0.0	0.0	0.0	0.0	0.0	0.0	0.0	0.8	0.0	0.8
Kismat Chatnai	Never	5.7	14.3	97.1	98.6	98.6	98.6	18.6	11.4	18.6	10.0
	Occasionally	81.4	80.0	0.0	1.4	0.0	0.0	77.1	77.1	77.1	77.1
	Sometimes	10.0	4.3	2.9	0.0	1.4	1.4	4.3	11.4	2.9	10.0
	Often	2.9	1.4	0.0	0.0	0.0	0.0	0.0	0.0	1.4	1.4
	At all time	0.0	0.0	0.0	0.0	0.0	0.0	0.0	0.0	0.0	1.4

Source: Field survey from September to November 2017

infrastructure development is not satisfactory in the study villages.

19.4 Conclusion and Policy Implications

Adaptation is vital for agricultural sustainability in the face of climate change and variability. This study has investigated the adaptation strategies of the farmers using survey data. Due to climate change issues, farmers are found to adopt different adaptation measures such as changing seeding or planting time, maintaining crop rotation, intensive irrigation, use of fertilizers and insecticides to crop fields. Despite taking different adaptation measures, inefficient institutional measures and lack of access to finance might be a barrier in facing future climate change challenges. The government and NGOs have not launched new technologies on a large scale. The change of profession is not a popular option for them to adapt to hostile climatic conditions. Thus, strengthening the farmers' resilience to climate change will require a combination of actions, i.e., subsidy to seed, fertilizer, insecticides, fuel, easy access to credit, information, modern technologies and effective training facilities by the government and NGOs. Infrastructure development, i.e., development of cold storage and communication network, dams required more attention from the government. Moreover, long-term initiatives of the government, including river dredging, need to be emphasized, which can help to save the farmers from frequent flooding hazards.

References

Alam K (2015) Farmers' adaptation to water scarcity in drought-prone environments: a case study of Rajshahi District, Bangladesh. Agric Water Manag 148:196–206

Alam GMM (2016) An assessment of the livelihood vulnerability of the riverbank erosion hazard and its impact on food security for rural households in Bangladesh. Ph.D. thesis, University of Southern Queensland, Queensland, Australia

Alam GMM, Alam K, Shahbaz M (2017) Climate change perceptions and local adaptation strategies of hazard-prone rural households in Bangladesh. Clim Risk Manag 17:52–63

Alam GMM, Alam K, Mushtaq S, Khatun MN, Filho WL (2018) Strategies and barriers to adaptation of hazard-prone rural households in Bangladesh. In: Filho WL, Nalau J (eds) Limits to climate change adaptation. Springer International Publishing AG

Ahmed Z, Guha GS, Shew AM, Alam GMM (2021) Climate change risk perceptions and agricultural adaptation strategies in vulnerable riverine char islands of Bangladesh. Land Use Policy 103 (4):105295

Aryal JP, Sapkota TB, Khurana R, Khatri-Chhetri A, Rahut DB, Jat ML (2019) Climate change and agriculture in south Asia: adaptation options in smallholder production systems. Environ Dev Sustain

Azad MAK, Hossain MZ, Karmakar S (2013) Impact of climatic elements on winter crop productivity in Rangpur area of Bangladesh. J Environ Sci Nat Resour 6(2):149–155

Baky AA, Zaman AM, Khan AU (2012). Managing flood flows for crop production producing risk management with hydraulic and GIS modeling: case study of agricultural areas in Shariatpur. APCBEE Procedia, Part 8–324

Bhatta GD, Ojha HR, Aggarwal PK, Sulaiman VR, Sultana P, Thapa D, Mittal N, Dahal K, Thomson P, Ghimire L (2017) Agricultural innovation and adaptation to climate change: empirical evidence from diverse agro-ecologies in south Asia. Environ Dev Sustain 19:497–525

Bradshaw B, Dolan H, Smit B (2004) Farm-level adaptation to climatic variability and change: crop diversification in the Canadian Prairies. Clim Change 67:119–141

Das J, Mandal T, Rahman AS, Saha P (2021) Spatio-temporal characterization of rainfall in Bangladesh: an innovative trend and discrete wavelet transformation approaches. Theoret Appl Climatol 143(3):1557–1579. https://doi.org/10.1007/s00704-020-03508-6

Dastagir MR (2015) Modeling recent climate change induced extreme events in Bangladesh: a review. Weather Clim Extremes 7:49–60

Fitton N, Alexander P, Arnell N, Bajzelj B, Calvin K, Doelman J, Gerber JS, Havlik P, Hasegawa T, Herrero M, Krisztin T, Meijl HV, Powell T, Sands R, Stehfest E, West PC, Smith P (2019) The vulnerabilities of agricultural land and food production to future water scarcity. Glob Environ Chang 58:101944

Gosling SN, Arnell NW (2016) A global assessment of the impact of climate change on water scarcity. Clim Change 134:371–385

Habiba U, Shaw R, Takeuchi Y (2011) Socio-economic impact of droughts in Bangladesh. Commun Environ Disaster Risk Manage 8:25–48

Islam MS, Sultana S, Saifunnahar, Miah MA (2014) Adaptation of char livelihood in flood and river

erosion areas through indigenous practices: a study on Bhuapur riverine area in Tangail. J Environ Sci Nat Resour 7(1):13–19
Jamshidi O, Asadia A, Kalantaria K, Azadib H, Scheffran J (2019) Vulnerability to climate change of smallholder farmers in the Hamadan province, Iran. Clim Risk Manage 23:146–159
Kabir MI, Rahman MB, Smith W, Lusha MAF, Milton AH (2016) Climate change and health in Bangladesh: a baseline cross-sectional survey. Glob Health Action 9(1)
Kamruzzaman M, Rahman ATMS, Kabir ME, Jahan CS, Mazumder QH, Rahman MS (2016a) Spatio-temporal analysis of climatic variables in the western part of Bangladesh. Environ Dev Sustain 18(6)
Kamruzzaman M, Kabir ME, Rahman ATMS, Mazumder QH, Rahman MS, Jahan CS (2016b) Modeling of agricultural drought risk pattern using Markov Chain and GIS in the western part of Bangladesh. Environ Dev Sustain 18(6)
Kamruzzaman M, Mandal T, Rahman ATM, Khalek A, Alam GMM (2021) Climate modeling, drought risk assessment and adaptation strategies in the western part of Bangladesh. In: Alam et al (eds) Climate vulnerability and resilience in the global south, pp 103–129
Kamruzzaman M, Rahman ATMS, Basak A, Alam J, Das J (2022). Assessment of climate change impacts and adaptation strategies through the prism of farmers' perception: a case study. Int J Environ Sci Technol:1–20.https://doi.org/10.1007/s13762-022-04254-0
Karim M, Thiel A (2017) Role of community based local institution for climate change adaptation in the Teesta riverine area of Bangladesh. Clim Risk Manag 17:92–103
Kothari CR (2004) Research methodology: methods and techniques. New Age International (p) Limited, New Delhi
Manandhar S, Vogt DS, Perret SR, Kazama F (2011) Adapting cropping systems to climate change in Nepal: a cross-regional study of farmers' perception and practices. Reg Environ Change 11:335–348
Maponya P, Mpandeli S (2012) Climate change and agricultural production in south Africa: Impacts and adaptation options. J Agric Sci 4(10):48–60
Meldrum G, Mijatovic D, Rojas W, Flores J, Pinto M, Mamani G, Condori E, Hilaquita D, Gruberg H, Padulosi S (2018) Climate change and crop diversity: farmers' perceptions and adaptation on the Bolivian Altiplano. Environ Dev Sustain 20:703–730
Mitra SK, Pathak PK (1984) The nature of simple random sampling. Ann Stat 12(4):1536–1542
Momtaz S, Shameem MIM (2016) Experiencing climate change in Bangladesh: vulnerability and adaptation in coastal regions. Elsevier Inc
Noble IR, Huq S, Anokhin YA, Carmin J, Goudou D, Lansigan FP, Osman-Elasha B, Villamizar A (2014) Adaptation needs and options, in climate change 2014: impacts, adaptation, and vulnerability, part A: Global and sectoral aspects. 27 Contribution of working group II to the Fifth Assessment Report of the Intergovernmental Panel on Climate Change, pp 833–868
Olesen JE, Trnka M, Kersebaum KC, Skjelvåg AO, Seguin B, Peltonen-Sainio P, Rossi F, Kozyra J, Micale F (2011) Impacts and adaptation of European crop production systems to climate change. Eur J Agron 34:96–112
Pal S, Islam ARM, Patwary MA, Alam GMM (2021) Modeling household socio-economic vulnerability to natural disaster in Teesta basin, Bangladesh. In: Alam et al (eds) Climate vulnerability and resilience in the global south, pp 103–129
Rautaray SK (2020) Field design foe enhancing water productivity in waterlogged areas with efficient water harvesting and farming system. In: Agricultural research
Rahman AA, Alam M, Alam SS, Rabiuzzaman M, Rashid M, Rabbani G (2007) UNDP human development report 2007: background paper on risks, vulnerability and adaptation in Bangladesh. Bangladesh Centre for Advanced Studies
Rakib MA, Akter MS, Majumder K (2016) Climatic hazard and social crisis: a quick mitigation approach to accelerate sustainable development. J Soc Sci 12 (1):14–26
Ruane AC, Major DC, Yu WH, Alam M, Hussain SG, Khan AS, Hassan A, Rosenzweig C (2013) Multifactor impact analysis of agricultural production in Bangladesh with climate change. Glob Environ Chang 23:338–350
Salam MA, Iftekharuddaula KM, Siddique AB, Rashid MA, Rashid MH, Momin MSI, Kabir MS, Biswas JK (2014) Strategic plan for increasing Aus and Aman rice cultivation in Bangladesh. The Bangladesh Rice J
Sarker MNI, Wu M, Alam GMM, Islam MS (2019) Role of climate-smart agriculture in promoting sustainable agriculture: a systematic literature review. Int J Agric Resour Gov Ecol 15(4):323–337
Shikuku KM, Winowiecki L, Twyman J, Eitzinger A, Perez JG, Mwongera C, Läderach P (2017) Smallholder farmers' attitudes and determinants of adaptation to climate risks in east Africa. Clim Risk Manag 16:234–245
Yusuf AA, Iwuafor ENO, Abaidoo RC, Olufajo OO, Sanginga N (2009) Effect of crop rotation and nitrogen fertilization on yield and nitrogen efficiency in maize in the northern Guinea savanna of Nigeria. Afr J Agric Res 4(10):913–921

Land Use Land Cover Change Detection Through the Spatial Approach: A Case Study of the Badiadka Panchayath, Kerala

20

R. Nirmala and K. S. Harikrishna Naik

Abstract

Land use land cover (LULC) change is a key factor in managing strategies for monitoring natural resources and environmental changes. The main objective of the present study is to analyze the LULC change, using geospatial tool that took place in the Badiadka gram panchayath between 2013 and 2020. The changes of LULC pattern occurred in the study area is mainly influenced by huge plantation distribution plan held by the Krishi project, Kerala after 2017 which attracted people to farming in and around their home. The results shows that the agricultural land increased from 25 to 47% and settlement areas from 14 to 22% in 2020. The rate of land consumption and absorption coefficient between 2013 and 2020 reveal LULC changes of the study area is mainly due to immense agricultural and built-up land with rapid growth in population and decrease in vegetation and barren land. The present study assists in the selection and management of land use plans to fully meet basic human needs.

Keywords

Badiadka panchayath · LULC change detection · LCR · LAC

20.1 Introduction

The earth's surface is in some way threatening by the demands of increasing population change, and the use of land has an impact negatively on the environment, climate, and noticeable change pattern in LULC over time on a global and local scale (Lambin et al. 2000; Petit et al. 2001; Read and Lam 2002; Hema et al. 2012). Changes in land cover by land use do not inevitably suggest land degradation. But, many displaced land use forms are due to social causes that affect biodiversity (Lambin and Meyfroidt 2011; Akpoti et al. 2016; Yang et al. 2017; Siddaraju et al. 2018). Geospatial approach is an efficient tool for analyzing, modeling in LULC change detection, and advanced ecosystem management (Jensen and Cowen 1999; Saadat et al. 2011; Sekertekin et al., 2017; Praveen Kumar Mallupattu et al. 2013; Wu et al. 2018; Zhang 2018). The aim of the present work is to estimate the status of LULC of the Badiadka panchayath area between 2013 and 2020 in order to detect the land consumption rates.

R. Nirmala · K. S. H. Naik (✉)
Department of Marine Geology, Mangalore University, Mangaluru, Karnataka 574199, India
e-mail: munne8547@gmail.com

J. Das and S. K. Bhattacharya (eds.), *Monitoring and Managing Multi-hazards*,
GIScience and Geo-environmental Modelling, https://doi.org/10.1007/978-3-031-15377-8_20

20.2 Materials and Methods

20.2.1 Study Area

Badiadka also known as Perdala is a town and gram panchayat in the Kasaragod district, state of Kerala, India and belongs to North Kerala Division (Fig. 20.1) (District Urbanization report of Kasaragod 2020).

The study area belongs to the Low dissected plateau to the North (Fig. 20.2b) and pediment pediplain complex which covers most of the part and the Varada river passes through the middle of the area. It is broadly divisible into two geological belts of Khondalite and Peninsular Gneissic Complex and mainly comprising of Garnet Sillimanite, Granite Gneiss, and Pyribolite (Fig. 20.2a) (Chattopadhyay 1985).

The district receives an annual average rainfall of 3350 mm. In 2013, it experienced 2500 mm more rainfall, but in 2020 it recorded 1052 mm, which is drastically lower than 2013 rainfall amount. The mean maximum and minimum temperatures of the study area are 37 °C and 17 °C respectively (District Urbanization report of Kasaragod 2020).

20.2.2 Methods

The LANDSAT 8 OLI/TIRS C2 L1 satellite image for 2013 and 2019 has been used, and a classification pattern has been established for the study area, according to Anderson et al. classification (1976). LULC types per hectare are estimated for each study year to compare the results to identify the changes in LULC patterns (Fig. 20.3). The maximum likelihood classification and calculated land consumption rate (LCR) and land absorption coefficient (LAC) are followed with the help of the following formulas.

$$\text{LCR} = \frac{A(\text{Areal extent of the panchayath inhectares})}{P(\text{Population})}$$

$$\text{LAC} = \frac{A2 - A1}{P2 - P1}$$

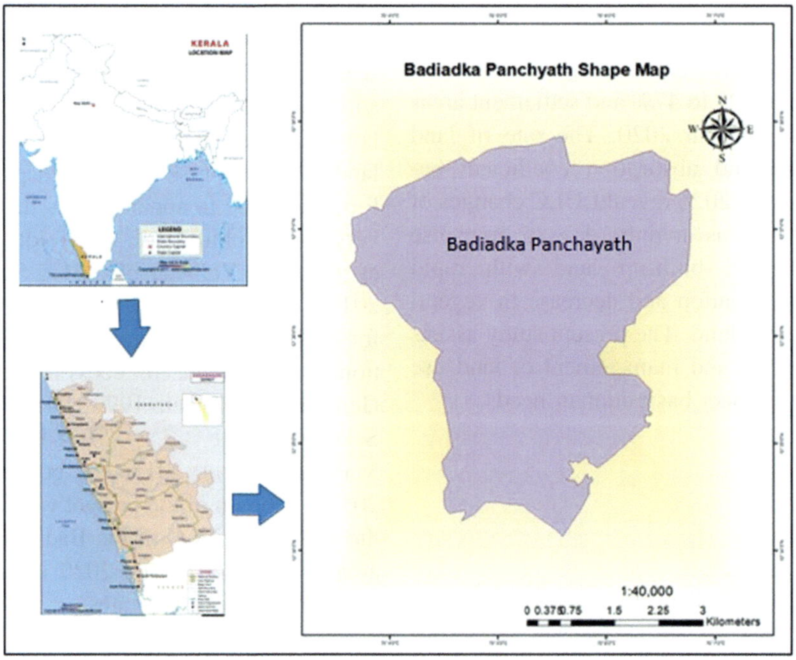

Fig. 20.1 Location map of Badiadka gram panchayath

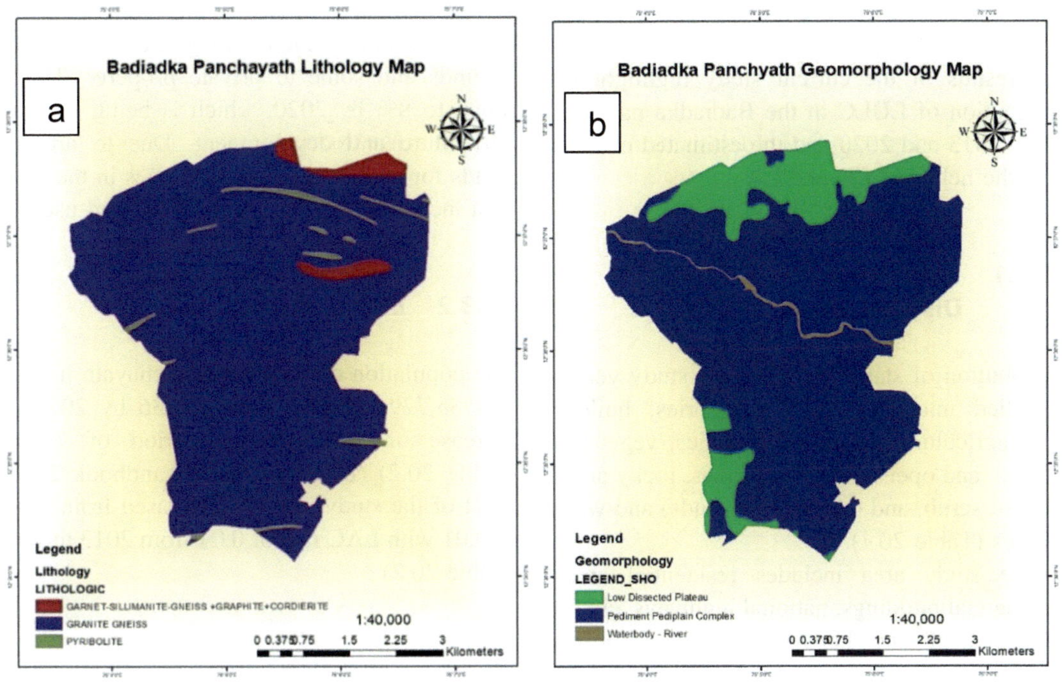

Fig. 20.2 **a** Lithology map, **b** geomorphology map of Badiadka Panchayath

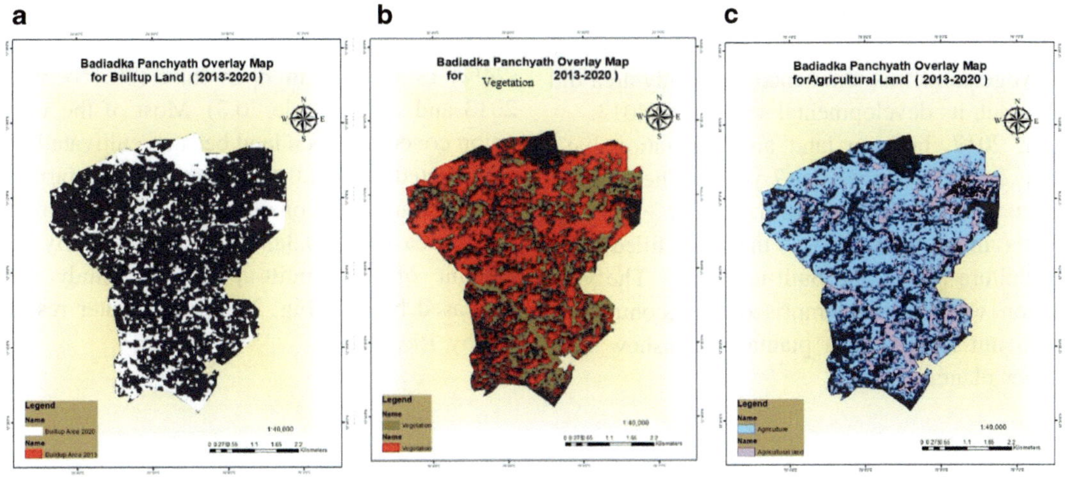

Fig. 20.3 Overlay map of 2013 and 2020: **a** built-up land, **b** vegetation, **c** agriculture

A = $A1$ and $A2$ are the areal extents (in hectares) for the early and later years.

P = $P1$ and $P2$ are population figures for the early and later years

$$\text{Change of percentage} = \frac{\text{Observed change}}{\text{Sum of change}} \times 100$$

20.3 Results and Discussion

The results of the current study highlight the distribution of LULC in the Badiadka panchayath in 2013 and 2020 and the estimated changes with the help of LCR and LAC.

20.3.1 Land Use Land Cover Distribution

Distribution of static LULC in the study year is classified into five major categories; built-up land, agricultural land, water bodies, vegetation (forest), and open land (sandy areas, rocky areas, grasses, scrub, and other barren lands) and water sources (Table 20.1).

The study area includes residential areas, commercial buildings, national highways, etc., in urban extent and houses, roads, institutions, health centers, etc., in rural areas. Mixed land use category (vegetation and settlements) and mining is also a part of the built-up area. The Badiadka covered more vegetation of 41% of the total land area in 2013, with 14% built-up land, 25% agricultural land and 15% open land, which means the vegetation was untouched and study area did not reach its developmental stage until 2013.

In 2017, built-up land and vegetation have seen drastic changes in 7 years. The rate of vegetation land declines from 41 to 17% in 2020. That's more than half of the area filled with agriculture (47%) and built-up (22%). The vegetation was mostly comprised of coconut tree, areca nut tree, banana plantation, cashew tree, rubber plantations, etc.

Approximately 15% of the study area had little open land, which belonged to some of the grounds and some to private property. This is down to 8% by 2020, which is being used for agriculture and development. Due to artificial ponds for irrigation, water resources in the study area increased from 5 to 6% of its land use.

20.3.2 LCR and LAC

The population of Badiadka Panchayath in 2013 was 35,729 and it will be 40,336 by 2020, an increase of 4607 over a period of 7 years (Table 20.2) (District Census handbook 2011). LCR of the study area are increased from 0.008 to 0.01 with LAC rate of 0.04 from 2013 to 2020 (Table 20.2).

20.3.3 Land Use Land Cover Change: Trend and Rate

The study indicates that noticeable changes have occurred in the Badiadka Panchayath with negative changes such as great loss in vegetation (24%) as well as in open land (7%) between 2013 and 2020 (Table 20.3). Most of the vegetation cover and open land became cultivated and constructed land in the Panchayath, and farming practices are also considered a great loss in the study area (Fig. 20.3a, b, c). Subsequently, the amount of land built-up by the study area increased by 9% (Fig. 20.4b) and water resources by 1%.

Table 20.1 LULC distribution during 2013 and 2020

LULC classes	2013		2020	
	Area (ha)	Area (%)	Area (ha)	Area (%)
Agriculture land	586.02	25	1106.05	47
Built-up land	314.48	14	523.13	22
Vegetation	982.05	41	406.28	17
Water bodies	124.10	5	133.56	6
Open land	362.78	15	200.07	8
Total	2369.45	100	2369.45	100

Table 20.2 Land Consumption Rate (LCR) and Land Absorption Coefficient (LAC)

Year	LCR	Population	Year	LAC
2013	0.008	35,729	2013–2020	0.04
2020	0.010	40,336		

Table 20.3 LULC cover change from 2013 to 2020

LULC categories	2020–2013 (area in ha)	Change in percentage	Annual rate of change (%) from 2013 to 2020
Agriculture land	520.03	21.9	3.14
Built-up land	208.65	8.8	1.28
Vegetation	−575.77	−24.3	−3.42
Water bodies	9.46	0.3	0.14
Open land	−162.71	−6.8	−1

The Badiadka Panchayath has witnessed a gradual decline in the vegetation cover from 41% in 2013 to 17% in 2020. A total down of 24% vegetation cover in the study area has been noticed, and this change is influenced by human activities mainly by massive plantation distribution project plan held by the Krishi bhavan, Kerala after 2017 to support agriculture (Fig. 20.4a, b). As a result of this project, most people were active in agriculture, which resulted in a gradual decrease in vegetation and open land, which was converted to agricultural land.

On the other hand, change under stability means there are no changes, loss or gains from

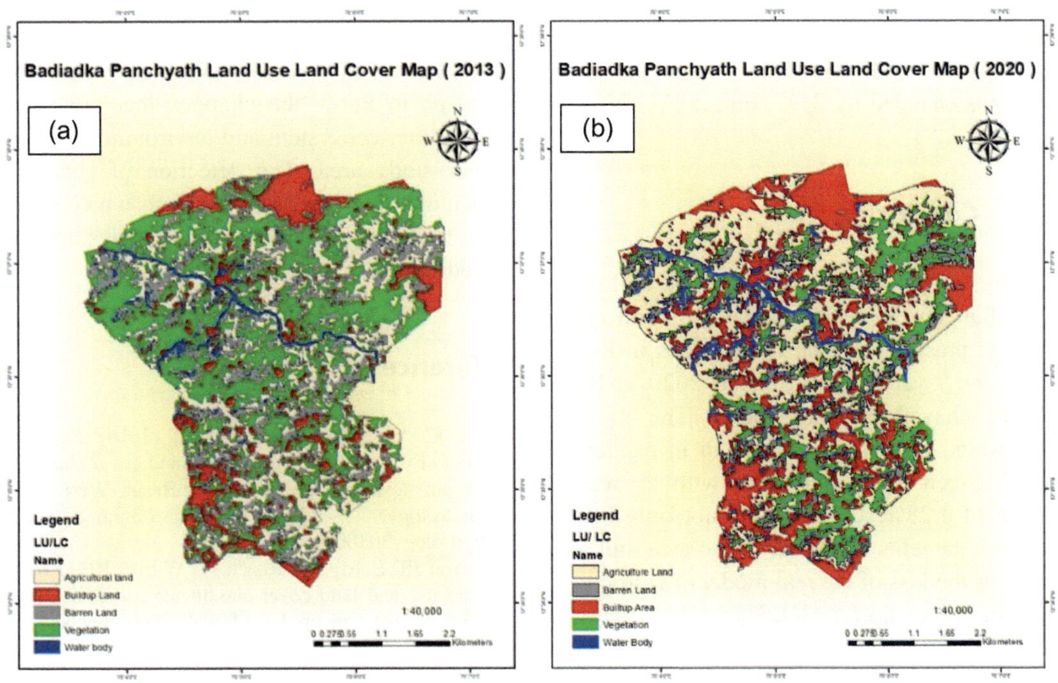

Fig. 20.4 LULC changes of Badiadka panchayath in **a** 2013 and **b** 2020

Fig. 20.5 Badiadka panchayath field photos

each class between 2013 and 2020 in study area. But, between 2013 and 2020 the water source is the only class with very little change and vegetation has changed by 24% (Fig. 20.5).

20.4 Conclusion

The present work demonstrates the ability of geospatial tool to detect spatial–temporal changes. The present work on the changes in LULC during the 7 years from 2013 to 2020 in Badiadka Panchayath is being studied. The observation results show a fastest growth in residential land between 2013 and 2020 with an annual change of 1.28%. Urbanization not only brings economic benefits or wealth to the area, but also increases the loss of the real model of land cover. Loss of vegetation (−3.42%) and agriculture gains were also observed in rural and urban areas (3.14%) due to higher agricultural activities. The pattern of inconsistent and random growth can lead to enormous problems, and a regional study reveals the need for a proper spatial planning to make the urban area less complex. This study helps us to know the changes that are taking place in our ecosystem and environment, mainly in our study area. The detection of change is shown in two different time periods which enable us to find areas where human activities are consumed more.

References

Akpoti K, Antwi EO, Kabo-Bah AT (2016) Impacts of rainfall variability, land use and land cover change on stream flow of the Black Volta Basin, West Africa. Hydrology, 3(3):26. https://doi.org/10.3390/hydrology3030026

Anderson JR, Hardy EE, Roach JT, Witmer RE (1976) A land use and land cover classification system for use with remote sensor data. USGS Professional Paper, 964, U.S. Geological Survey

Chattopadhyay S (1985) Deforestation in parts of Western Ghats region (Kerala), India. J Environ Manage 20:219–230

District Urbanisation report of Kasaragod, 2020

Government of India (2014–2015) District Census Handbook—Kasaragod (Part-A) 2011 (PDF). Directorate of Census Operations, Kerala

Hema HC, Givindaiah S, Jayaramaiah R, Shahinfar H (2012) Remote sensing and GIS application for land use land cover mapping in the Kanakapura watershed of ramanagar district, Karnataka, India. The Indian Mineralogist 2:41–57

Jensen JR, Cowen DC (1999) Remote sensing of urban/suburban infrastructure and socio-economic attributes. Photogramm Eng Remote Sens 65:611–622

Lambin EF, Meyfroidt P (2011) Global land use change, economic globalization, and the looming land scarcity. Proc Natl Acad Sci USA 108:3465–3472

Lambin EF, Rounsevell MDA, Geist HJ (2000) Are agricultural land-use models able to predict changes in land-use intensity? Agric Ecosyst. Environ 82(1–3):321–331. https://doi.org/10.1016/S0167-8809(00)00235-8

Mallupattu PK, Reddy J, Reddy S (2013). Analysis of land use/land cover changes using remote sensing data and GIS at an urban area, Tirupati, India. The Scientific World J 6. https://doi.org/10.1155/2013/268623

Petit C, Scudder T, Lambin E (2001) Quantifying processes of land-cover change by remote sensing: resettlement and rapid land-cover changes in south-eastern Zambia. Int J Remote Sens 22(17):34353456

Read JM, Lam NSN (2002) Spatial methods for characterising land cover and detecting land-cover changes for the tropics. Int J Remote Sens 23(12):2457–2474. https://doi.org/10.1080/01431160110106140

Saadat H, Adamowski J, Bonnell R, Sharifi F, Namdar M, Ale-Ebrahim S (2011) Land use and land cover classification over a large area in Iran based on single date analysis of satellite imagery. ISPRS J Photogrammetry Remote Sens 66:608–619

Sekertekin A, Marangoz AM, Akcin H (2017) Pixel based classification analysis of land use land cover using sentinel-2 and landsat-8 data. Int Arch Photogram Rem Sens Spatial Inf Sci, XLII-4/W6:91–93

Siddaraju K, Nagaraju D, Balasubramanian A (2018) Classification of land use and land cover in Hanur watershed area, Kollegal Taluk, Chamarajnagar district, Karnataka state, India, using remote sensing and GIS. Int J Recent Sci Res 9(2K):24574–24578

Wu T, Luo J, Fang J, Ma J, Song X (2018) Unsupervised object-based change detection via a Weibull mixture model-based binarization for high-resolution remote sensing images. IEEE Geosci Remote Sens Lett 15:63–67

Yang L, Zhu Z, Homer C (2017) A land cover change detection and classification protocol for updating Alaska NLCD 2001 to 2011. Remote Sens Environ 195:44–55

Zhang X (2018) Post-processing approach for refining raw land cover change detection of very high-resolution remote sensing images. Remote Sens 10:472

Delineation of Groundwater Potential Zones Through AHP: A Case Study from Tamil Nadu, India

21

Aneesah Rahaman, Brototi Biswas, Jonmenjoy Barman, V. Madha Suresh, Braj Kishor, and Jayanta Das

Abstract

Groundwater is one of the reliable and most valuable sources of freshwater worldwide. Due to various reasons such as overgrowth of population, urbanization, industrial development, and most significant change in climate play a significant role in respect of quality as well as quantity of the sources of groundwater. Unpredictable monsoons and surface water degradation tend to lower subterranean water levels. As a result, the primary goal of the current study is to detect and demarcate prospective groundwater regions in the study region. Seven parameters have been taken to fulfill the objective, including "slope, geology, geomorphology, lineament density, soil, drainage density, and rainfall distribution". Bhuvan portal and NBSS were used to gather geology, lineament, geomorphology, and soil map information. Drainage density and slope map have been prepared by using SRTM DEM (30 m resolution) satellite data from USGS earth explorer. Five years of average rainfall data (2015–2020) has been collected from IMD. Using ArcGIS (10.4) software, various themed maps were created and converted to raster format. These maps' weight, ranking, and reclassification were done in the ArcGIS platform using the Analytical Hierarchical Process (AHP). Then Groundwater Potential Zone (GPZ) map was made by overlaying the maps with five different zones, namely very high (0.533%), high (59.507%), moderate (17.843%), low (21.914%), and very low (0.202%) potential zone. The final output of the study showed that very high zone covered 17.798 km^2, high 1987.220 km^2, moderate 595.875 km^2, low 731.821 km^2, and very low zone covered 6.746 km^2, respectively. High drainage density (high rate of infiltration) and Low lineament density enhance the area marginally only in the very poor to poor prospective zones, according to the results. The groundwater potential zone map has been overlain with the research area's lineament density map for validation. In addition to well yields, groundwater level anomalies and rainfall data were used to corroborate the result. The yield values vary from 258 to 320 L per minute (LPM) in the high to very high potential zones, whereas 106 to 185 lpm within the low

A. Rahaman · V. M. Suresh
CNDHS, University of Madras, Chennai, India

B. Biswas (✉) · J. Barman
Department of Geography, Mizoram University, Aizawl, India
e-mail: brototibiswas@gmail.com

B. Kishor
St. John's College, Agra, India

J. Das
Department of Geography, Rampurhat College, PO-Rampurhat, Birbhum 731224, India
e-mail: jayanta.daas@gmail.com

J. Das and S. K. Bhattacharya (eds.), *Monitoring and Managing Multi-hazards*,
GIScience and Geo-environmental Modelling, https://doi.org/10.1007/978-3-031-15377-8_21

to very low potential zones. Furthermore, the results were authenticated by combining well output information which was combined with pre-monsoon and post-monsoon water level information and the result showed a strong association between the parameters.

Keywords

Groundwater · Geographical information system · Analytical hierarchical process · Weighted overlay · Remote sensing

21.1 Introduction

Subsurface water, often known as groundwater, is a natural resource that both urban and rural residents rely on for their everyday lives as well as economic purposes. As a result, groundwater plays a vital part in aquatic, terrestrial, and human well-being ecosystems (Mandal et al. 2021). Groundwater depth maps are used by planners and policymakers all around the world to enable proper managing of groundwater assets (Ahmed and Mansor 2018). The absence of proper groundwater management may lead to a host of environmental problems (Chen et al. 2016; Kumar et al. 2021). Many of the states in India are on the brink of a serious water shortage. If the necessary actions are not done in a timely manner, India would face an acute water issue in the future (Jasrotia et al. 2016). Groundwater resources contribute to about 85% needs of rural Indians, almost 50% of the urban domestic water 92% of irrigational water (Saha and Ray 2019). As the surface rinsing a large volume of rainwater, the problem of groundwater is deteriorating because of imbalances between groundwater and exploitation (CGWB 2014; Rani et al. 2019; Biswas et al. 2018). Discovering potential groundwater locations in order to meet the world's growing demand for freshwater in both home and agricultural arenas is most imperative. The use of remote sensing and GIS tools to identify the groundwater potential zone is a crucial role. The utilization of satellite data, together with standard maps and certified ground truth data, has simplified the processes for determining the groundwater potential zone this year (Pradhan 2009; Nigussie et al. 2019; Kumar et al. 2021). It is a powerful tool aiding through the collection and integration of data from a wide variety of earth science fields. RS has a wide range of utilities from providing a wide range of spatio-temporal information to saving resources and time. It is easy to find the lineament structure by using satellite data. The exploration of groundwater depends on the various influencing factors namely lineament, density of stream, geology, geomorphology, land-use, slope, texture of soil, and intensity of rainfall (Pande et al. 2021; Das and Pardesi 2018; Thapa et al. 2017). The widespread use of remote sensing data in recent years has been able to offer basic information on prospective groundwater zones (Kumar et al. 2021). Several researchers have reported the application of GIS in the monitoring and management of groundwater such as GPZ delineation (Pourtaghi and Pourghasemi 2014; Rahmati et al. 2015; Pande et al. 2017). Groundwater monitoring, assessment, and management have all benefited from the use of RS, as well as delivering time, space, and spectrum data for vast and remote areas in a short amount of time (Jackson 2002). In recent days, several researchers are taking advanced approaches to groundwater research including the FR method (Saravanan et al. 2021; Manap et al. 2014; Ozdemir 2011), LR method (Sangchini et al. 2016; Kalantar et al. 2018), fuzzy logic (Pourghasemi et al. 2012), Dempster-Shafer model (Mogaji et al. 2015), weights of evidence model (Madani and Niyazi 2015; Al Abadi 2015), artificial neural network (Mohanty et al. 2013; Chitsazan et al. 2015), maximum entropy model (Rahmati et al. 2016), and decision tree model (Lee and Lee 2015; Jeihouni et al. 2020) have been implemented successfully. Many academics around the world have created a GPZ map using the AHP model. (Pourghasemi et al. 2012; Pande et al. 2021; Ajay et al. 2020). Nowadays groundwater stress zonation also becoming popular (Basak et al. 2021).

As the global demand for drinkable water grows, innovative approaches for determining the spatial dissemination and appropriateness of groundwater assets are becoming more important (Arulbalaji et al. 2019). India being a semi-arid, heavily populated nation has a considerable portion of the population relying on groundwater for their water needs. As a result, it's crucial that groundwater in this nation is well-characterized in terms of geographical distribution and availability. A variety of hydro-geological, geophysical, geological, and photogeological techniques have historically been utilized to determine the availability of groundwater in a region (Fashae et al. 2014).

21.2 Aim and Objective

The primary goal and objective of this research is to define the potential groundwater zone in the Gummindipoondy area.

21.3 Study Area

The research area is Gummindipoondy Taluk, which is located in the coastal district of Thiruvallur in Tamil Nadu, India (Fig. 21.1). Geographical location of the study area is extending from 13° 21′0″ North to 13° 30′0″ North latitude and 79° 57′ 0″ East to 80° 15′0″ East longitude. The study area covers a total area of 425.47 km^2. The taluk is surrounded by another two taluks such as in the eastern side Ponneri taluk, in the south Utthukottai taluk, and in the north and west side of Andhra Pradesh state is situated. Topographically the study area is under flat terrain. The elevation of the taluk is all about 34 m above Mean Sea Level (MSL). Lithologically majority of the area is covered with lateritic rock and alluvial soil. Lineament can be found in the NW part of the region. Basically, it's a coastal area. The average rainfall of the taluk ranges between 800 to 1000 mm and rainfall receives both from north-east and south-west monsoon. In the summer, the average temperature is 37.9° C and the average temperature in winter is 18.5 °C of the region. The study area has a total population of 32,665 people, with 52% of men and 48% of women, according to the 2011 census. Net Annual Groundwater Availability in the review region is 1620.88 (ha.m). Precipitation is the fundamental wellspring of groundwater. According to State Ground and Surface Water Resources Data Center, 2019 during the period of pre-monsoon, the water level commonly in declining pattern goes from ground level to 15 m. The profundity of well subterranean level 12.0 m becomes dry during hot summer season in May, June, and July. At the time of post-monsoon, the water level is commonly in vertical pattern because of precipitation and it might arrive at the ground level.

21.4 Data Collection

In the present study, seven different types of parameters have been used such as "slope, geology, lineament density, geomorphology, soil, and rainfall and drainage density".Each and every parameter has been collected from individual sources and has been preceded in the GIS platform (Table 21.1). The slope map and drainage density map have been derived from the SRTM DEM data of 30 m resolution, downloaded from the earth explorer of the USGS. The hydrological tool (ArcGIS 10.4) was utilized for drainage density mapping. Geomorphology, geological and lineament maps have been derived from the NRSC (National Remote Sensing Center) Bhuvan portal from thematic section which was geo-referenced and digitized in the GIS platform. The soil map was received in 1:50,000 scales from NBSS (National Bureau of Soil Survey and Land Use Planning, Kolkata), while rainfall data for the last five years (2015–2020) was taken from the "India Meteorological Department (IMD, Chennai)". For rainfall distribution mapping,"IDW (Inverse Distance Weighting) technique" in ArcGIS spatial analyst tool has been applied. All the thematic layers have been projected to UTM zone 44 N in the GIS platform. AHP (Analytical Hierarchical Process) has been applied to assign the weight and rank of each parameter. All weighted layers were converted to raster and resampled to a

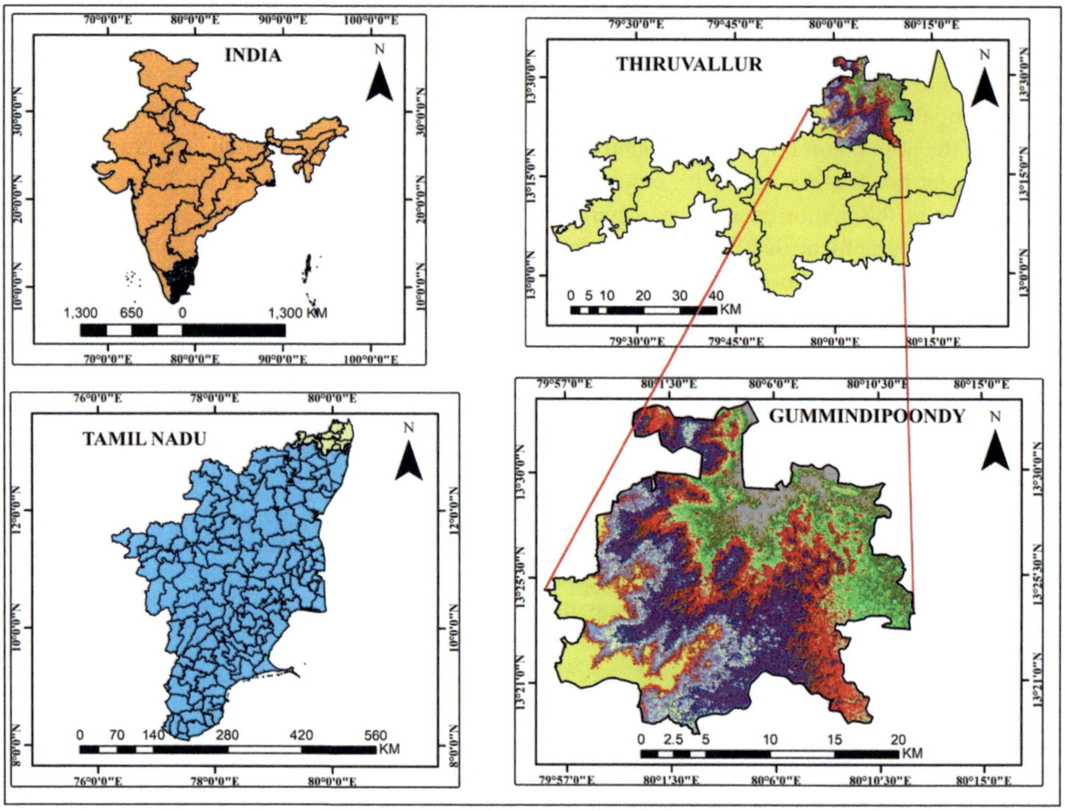

Fig. 21.1 Location map of the study area

Table 21.1 Data used in the present study

Data	Source of data	Data derived
SRTMDEM (30 m resolution)	USGS earth explorer	Slope, drainage density map
Geomorphology	NRSC Bhuvan	Landform features map
Geology	NRSC Bhuvan	Lithology, lineament density map
Soil	NBSS	Soil map
Rainfall (2015–2020)	IMD	Rainfall distribution map

30 * 30 m cell size. For Groundwater Potential Zone (GPZ) mapping, weighted overlay method has been used.

21.5 Methodology

The study methodology includes collection and preparation with ArcGIS software of all seven thematic layers. Weights calculated by the AHP were assigned to the geo-referenced maps. Standardized layer weights were achieved by the consistent index and CR value of the built-up matrix. If their CR value is below 0.1, the matrix created thus should be consistent. After the determined weights were assigned, the layers were subject to weighted overlay analysis. Five classes of groundwater potential zone were obtained on the basis of the indices calculated using Eq. 21.1.Well output and groundwater variation information from both the pre-monsoon and post-monsoon periods were used to cross-verify the study's conclusions.

$$\mathrm{GWPI} = D_r D_w + G_r G_w + Sl_r Sl_w + L_r L_w + S_r S_w + Ge_r Ge_w + R_r R_w \tag{21.1}$$

In this methodology, where "GWPI is groundwater potential Index, G is geomorphology, D is drainage density, L is lineament density, Sl is slope, Ge is geology, S is soil and R is rainfall". The suffix "r" and "w" represents the rank and weight of each layer.

21.5.1 Weights and Rank Determination Using AHP

Analytical Hierarchical Process (AHP) is an organized strategy utilized for investigating and sorting out complex choices utilizing the guideline of brain research and science. It was created by Thomas L. Saaty in 1970's. Instead of giving the right answer for the main thing in need of attention, it emerges with the best arrangement in light of the client's need. AHP manages stepwise tasks to give out an answer. It deals with the guideline of Multi-Criteria Decision Making (MCDM) and is valuable in dissecting this present reality issues which include numerous standards. In this cycle, data is divided into bundles of choices and models. This data is additionally incorporated to determine the overall positioning of choices. A last weight got gives the best result.

The analytical hierarchy is a systematic decision-making technique that leverages domain expertise to establish rank and weight by generating a pair-wise comparison matrix of its own value. This strategy is effective for making judgments in a problem with multiple variables that influence the outcome. It involves creating a pair-wise array that calculates the weights of each parameter while taking all other parameters' relative relevance into account (Saaty 2008). The elements impacting that result were constructed as a hierarchical hierarchy, and a pair-wise matrix with sequence and column requirements was designed. The column was referred to as B, while the row was referred to as A. The matrix's creator, Saaty, devised a scale of 1–9 as a measure of relative importance (Table 21.2). By analysing the criteria and ranking the matrix is completed according to their relative relevance. The diagonal of this matrix was given 1, which is equally important compared to criteria A and B. Criteria that have extreme influence on the other have been given a rank of nine. The relative importance of the Saaty scale is shown in Table 21.2, and the pair-wise matrix for this research is shown in Table 21.3. The consistency index and consistency ratio shall be calculated by means of Eqs. 21.2 and 21.3. The consistency matrix was then examined, and a consistency ratio of less than 0.1 was utilized as the analytical criterion.

This method offers a wide range of choices for the decision-maker (Das et al. 2017). The subjectivity to determine weights can also be evaluated using this method by examining their consistency. The matrix with various options can be reconstructed if the condition is not fulfilled (Chandio et al. 2013). Saaty's matrix scale (1980) is based on 9-point scales where 1 represents the equality between two factors and 9 represents the extremely important of one factor as compared to other factors. Table 21.3 displays the matrix of all the influencing factors. For the delineation of the groundwater potential zone, a pair-wise matrix was applied to each of the thematic layers.

The most common way of accepting loads is done subjectively as currently expressed. The weights are given in light of computation and it is confirmed with the assistance of specific terms. The terms incorporate Final Weightage for every parameter, Consistency Index, Random Consistency Index, and Consistency ratio.

$$CI = \frac{\lambda_{\max - n}}{n} \tag{21.2}$$

Table 21.2 Satty's scale of relative importance

"Intensity of importance"	Definition	Explanation
1	"Equally valuable"	There are two acts that both contribute to the goal
3	One has a lower priority than the other	Based on experience and judgment, one activity has a little edge over another
5	Importance that is essential or significant	Based on experience and judgment, one activity has a little edge over others
7	Importance that has been demonstrated	An activity is greatly favored, with evidence of its domination in practice
9	"Absolute significance"	The indication that favors one action more than another
2,4,6,8	Between two neighboring judgments, there are some values that are in the middle	When need to make a compromise

Table 21.3 Delineation of groundwater potential zone using the AHP matrix

Factors	Geom	LD	Litho	Slope	Soil	LULC	DD	Weight
Geomorphology	7	6	5	4	3	2	1	0.38
Lineament density	7/2	6/2	5/2	4/2	3/2	2/2	1/2	0.19
Geology	7/3	6/3	5/3	4/3	3/3	2/3	1/3	0.12
Slope	7/4	6/4	5/4	4/4	3/4	2/4	1/4	0.10
Soil	7/5	6/5	5/5	4/5	3/5	2/5	1/5	0.08
Rainfall	7/6	6/6	5/6	4/6	3/6	2/6	1/6	0.066
Drainage density	7/7	6/7	5/7	4/7	3/7	2/7	1/7	0.064

$$CR = \frac{CI}{CR} \tag{21.3}$$

21.6 Results and Discussion

Ranking of parameters into classes

The groundwater existence and flow can be controlled by various factors such as "geomorphology, geology, soil, slope, rainfall, and drainage system and the interrelationship among these factors of the particular place" (Jaiswal et al. 2003).

Geomorphology

It examines the shape and structure of landform and is related to the inner process of the "groundwater probable zone" (Swain 2015). Geomorphology has a significant role for the presence of groundwater potential zones. For the present work, the geomorphological map (Fig. 21.2) has been categorized into—coastal, denudation, and fluvial (alluvial) plain. Among the three classes, flood plain or fluvial origin has the highest water level surface and it means that it is the suitable landform for the groundwater potential zone. Hence, rank 38 has been given to geomorphology.

Lineament Density

Lineament is a part of geological features. It is the combination of joints, fractures, and faults which express the indirect relationship with groundwater potential zone and also provide a way to percolate water under the surface (Pinto et al. 2017). Groundwater potentiality can be determined through lineament density and the

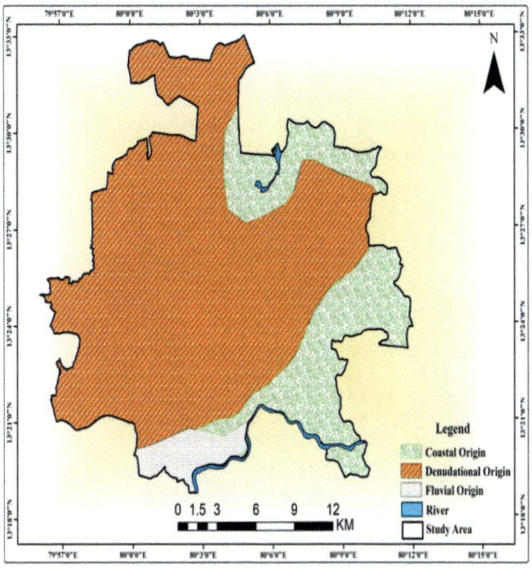

Fig. 21.2 Geomorphology of the study area

existence of lineament shows the permeable zone (Sitender 2010). Lineament density (Fig. 21.3) of the study area has been divided into five categories—"very low, Low, moderate, high, and very high". The maximum prospective for groundwater recharge is indicated by a high lineament density and the low density denotes the low potential recharge zone. Hence rank value of 19 has been assigned to lineament density.

Lithology

The various types of rock formation influence the subsurface water properties in the field of occurrence and movement of subsurface water (Saha et al. 2022). The lithology map (Fig. 21.4) is classified into three main categories such as unconsolidated sediment (alluvial), lateritic and marine sandstone. Unconsolidated sedimentary rock is most suitable for the groundwater potential zone. Unconsolidated sedimentary rocks consist of sands, gravel, and silts which can make good aquifers. A rank value of 12, according to AHP scheme, has been assigned to lithology parameter.

Slope

The slope map (Fig. 21.5) has been derived from the SRTM DEM satellite image with 30 * 30 m resolution. Runoff and infiltration can be affected by the slope. According to Gupta et al. 2018, gradient of the slope may increase the runoff process. So, steeper slope more runoff. The slope

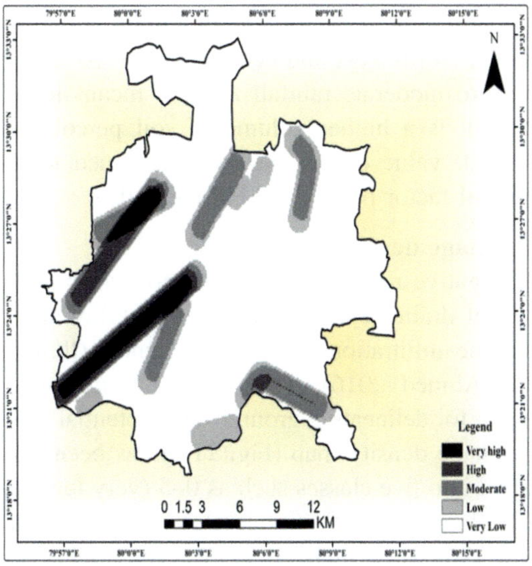

Fig. 21.3 Lineamnet density of the study area

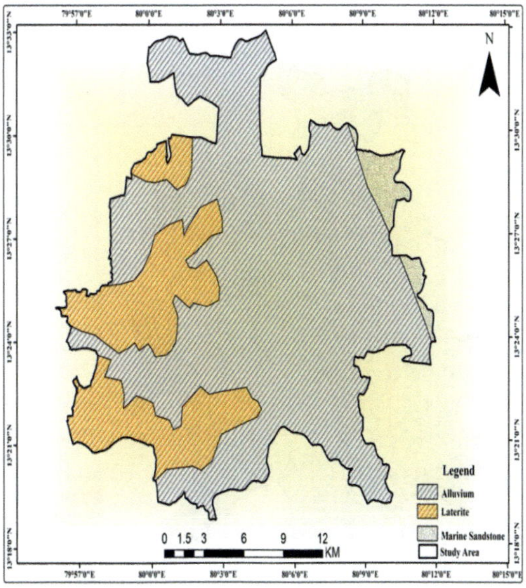

Fig. 21.4 Lithology of the study area

map of the research area has been divided into five categories such as 0–5° (very low), 6–10° (low), 11–15° (moderate), 16–20° (high) and <20° (very high) (Fig. 21.5). About more than 80% of area is having the flat terrain means low to very low slope. Based on the groundwater influencing factors, low to very low terrain or flat terrain is very much suitable. Based on the AHP method a rank value of 10 has been given to slope factor in this study.

Soil map

The top layer of the earth acts as a conduit for water to percolate into the soil profile. Based on the soil's porosity and water storage ability, the infiltration rate is dependent (Gupta et al. 2018). The soil map (Fig. 21.6) has been classified into three major classes' namely alluvial, black, and lateritic soil. About 30% the area is under alluvial soil, which is very much suitable for the groundwater potential zone. A rank value of eight values has been assigned to soil factor based on AHP method.

Rainfall

It is a key source of groundwater recharge and one of the most vital elements for examining groundwater potential zone. The research area's rainfall is influenced by the north-east and south-west monsoons. Rainfall duration and intensity have a substantial impact on percolation and rainfall amount (Abuzied et al. 2015). In this study, five years of average rainfall has been taken from 2015 to 2020. The rainfall map (Fig. 21.7) has been classified into five classes —"very low, Low, moderate, high, and very high". North-west and central portion have very high to moderate rainfall zone. It means higher rainfall is a higher volume of soil percolation. A rank value of 6.6 has been assigned to the rainfall factor based on AHP method.

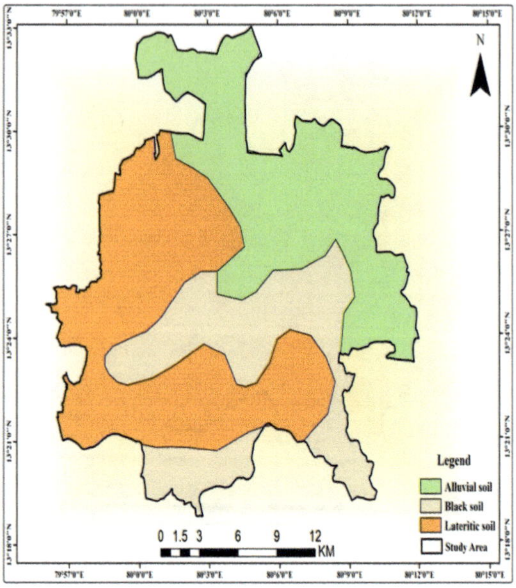

Fig. 21.6 Soil of the study area

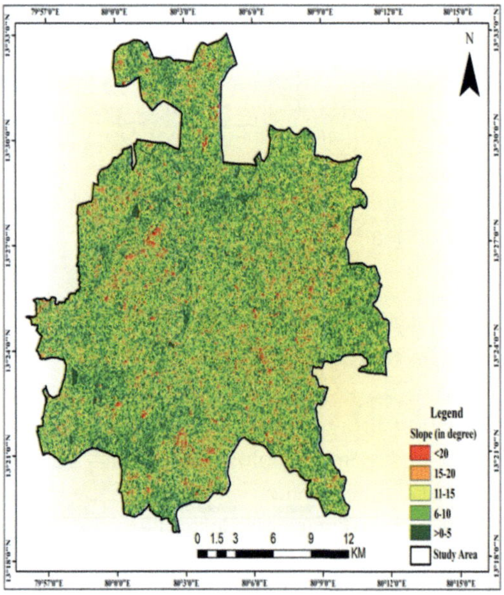

Fig. 21.5 Slope of the study area

Drainage density

A negative relationship exists between the density of drainage and surface permeability affecting the infiltration volume and runoff (Ibrahim and Ahmed 2016). Thus, it is an influencing factor for delineating groundwater potential zone. Drainage density map (Fig. 21.8) has been classified into five classes such as 0–5 (very low), 6–10 (low), 11–15 (moderate), 16–20 (high) and <20 (very high) km/km^2. The higher the density, is higher the chances of runoff. And lower the density increase the high percolation rate. About 70% of area is under low to very low

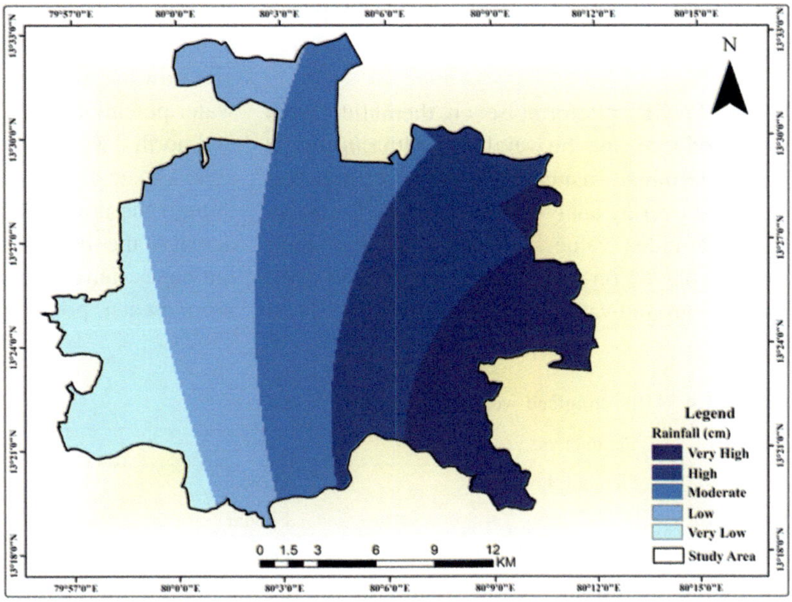

Fig. 21.7 Rainfall of the study area

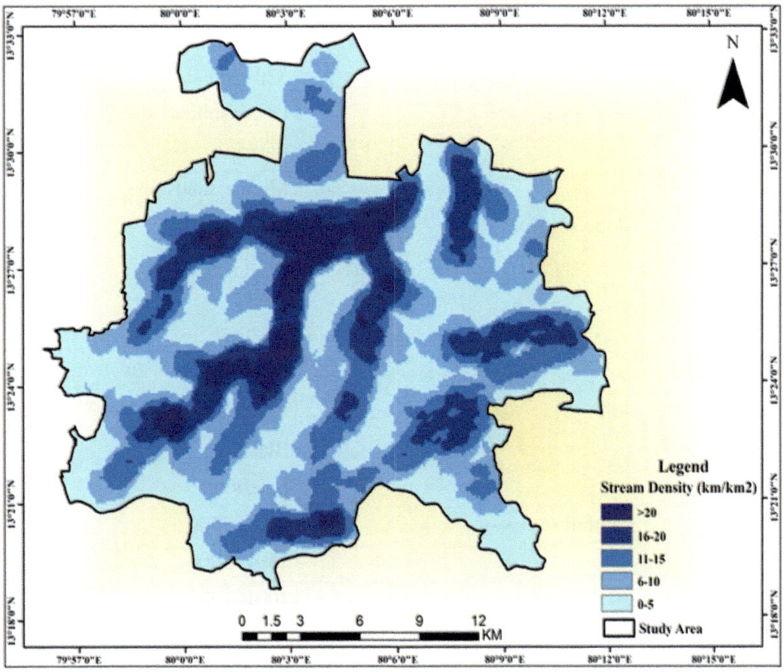

Fig. 21.8 Drainage density of the study area

drainage density due to flat terrain of the study area. Central and western portions of the study area is having the very high to high drainage density. Remain part is under a very low to moderate drainage density zone. A rank value of 6.4 has been assigned to drainage density parameter based on AHP method.

21.7 Groundwater Potential Zone (GWPZ)

In the ArcGIS platform, seven thematic layers were produced, and the weightage with rank value was determined using the AHP method for weighted overlay analysis (Table 21.4). Based on the AHP index value, the results of this study were displayed on a thematic map and grouped into five groundwater potential zones (Fig. 21.9) ranging from "very low to very high". As the terrain of the study area is flat, so it is ideal for groundwater recharging. A very high "groundwater potential zone" has been found in the north and north-east portion with 45.29 km^2, where the area is basically alluvial deposition covered. Even though the northeast region has a low lineament density, the high intensity of rainfall and low drainage density make the region a highly groundwater potential zone. Likewise, the east

Table 21.4 AHP rating and weight of each sub parameters

Sl. No.	Parameters	Classes	Rating	Weight
1	Geomorphology	Coastal	13	38
		Fluvial	11	
		Denudation	14	
2	Lineament density (km/km^2)	0–80	1.2	19
		81–160	2.2	
		161–240	3.5	
		241–320	4.6	
		320–400	7.5	
3	Geology	Unconsolidated sediment	8	12
		Lateritic	3	
		Marine sandstone	1	
4	Slope (degree)	0–5	5	10
		6–10	4	
		11–15	0.5	
		16–20	0.3	
		20–25	0.2	
5	Soil	Alluvial	6	8
		Black	2	
		Lateritic	0	
6	Rainfall (mm)	2001–2500	3	6.6
		1501–2000	1.8	
		1001–1500	1.1	
		501–1000	0.09	
		0–500	0.02	
7	Drainage density (km/km^2)	0–5	3	6.4
		6–10	2.6	
		11–15	0.5	
		16–20	0.2	
		21–25	0.1	

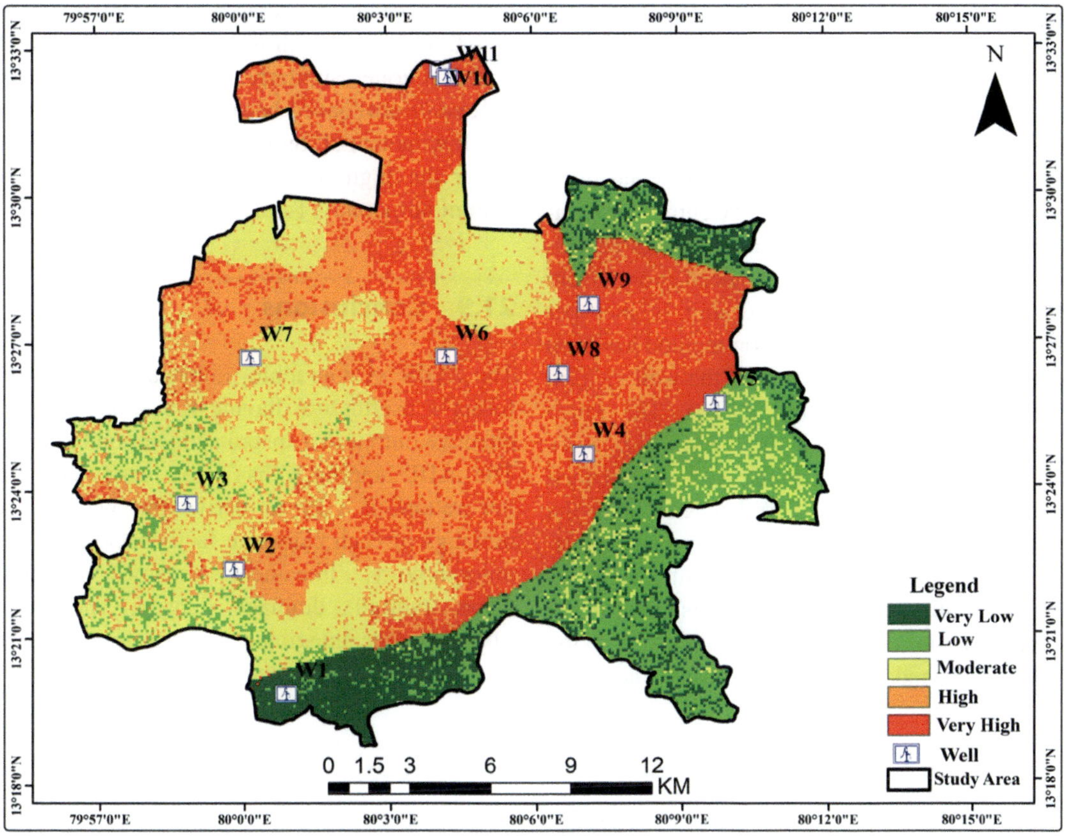

Fig. 21.9 Groundwater potential zone of the study area

and southern parts of the area fall under the low to very low potential zone with 225.37 km^2. Though the area has a high concentration of rainfall, due to the high drainage density, the infiltration rate is very high and there is no water for recharge to groundwater. High (63.2 km^2) and moderate (91.55 km^2) zones is found in the west and central portion of the study area. Though the lateritic rock is present, very high lineament density, moderate rainfall, alluvial soil, and low to very low slope have made the region high to moderate groundwater potential zone (Table 21.5).

Table 21.5 Categories of "groundwater potential zone" with area

Sl No	Category	Map Color	Area (km^2)	Area (%)
1	Very low		102.05	23.99
2	Low		123.32	28.99
3	Moderate		91.55	21.52
4	High		63.2	14.86
5	Very high		45.29	10.65

21.8 Groundwater Potential Zone Validation Using Well Yield (GWPZ)

The GWPZhas been validated with the well yield data and the water depth data of pre- and post-monsoon acquired from the "Central Groundwater Board" for the preceding 10 years (2010–2020). A total of 11 wells have been chosen from the study area and the yield unit is liter per minute (LPM). An interpolation method has been applied to find out the ranges of the well yield in the ArcGIS platform (Fig. 21.10). The range varies between 106 and 400 LPM (Fig. 21.4). Figure 21.10 shows that the results obtained using this methodology correlate well with the real-time data. Furthermore, pre- and post-monsoon groundwater measurements were utilized to confirm the level of replenishment in the selected regions. Figure 21.11 shows that the W8, W10, and W11 wells during the post-monsoon show an increase in water levels and that this water is falling within this very high potential zone. Similarly, W1 and W5 wells are at least or not loaded because they are located in the low potential region, remaining has fallen under the high to moderate potential zone of Gummindipoondy Taluk (Table 21.6).

21.9 Conclusions

The "GIS (Geographic Information System)" forms a useful tool having the ability to give effective output in a spatial and temporal context. The AHP approach has been used with the integration of GIS to derive the weight and rankings of each causative factor including "geomorphology, lineament density, lithology, slope, rainfall, drainage density, and soil". Therefore, the integrated tool has been used to identify the groundwater potential zone (GWPZ)

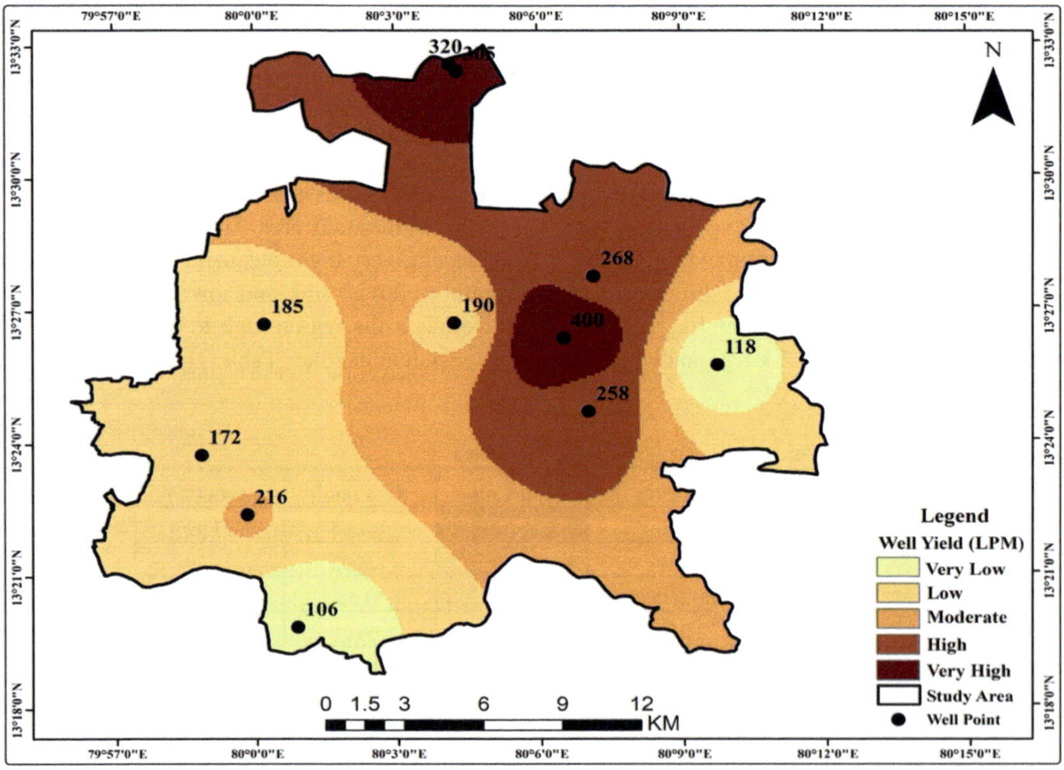

Fig. 21.10 Well yield of the study area

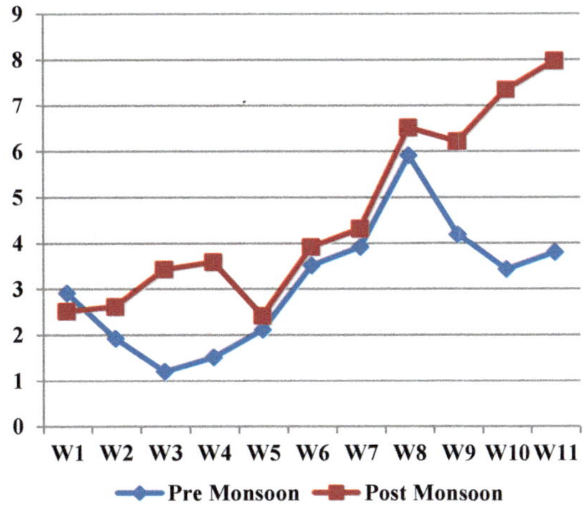

Fig. 21.11 Pre- and post-monsoon water level

Table 21.6 Well details for validation

Well No	Latitude	Longitude	Yield (LPM)
W1	13°19′58”	80°00′12”	106
W2	13°22′23”	79°54′57”	216
W3	13°23′14”	79°58′42”	172
W4	13°24′08”	80°07′34”	258
W5	13°25′40”	80°09′45”	118
W6	13°26′42”	80°04′45”	190
W7	13°26′49”	80°00′30”	185
W8	13°26′58”	80°06′04”	400
W9	13°28′28”	80°08′33”	268
W10	13°32′16”	80°03′49”	320
W11	13°32′23”	80°04′17”	305

for the Gummindipoondy Taluk with five potential zones such as “very high, high, moderate, low, and very low zone”. Very high potential zone occupied 10.65%, high zone is 14.86%, moderate zone is 21.52%, low is 28.99% and very low zone is 23.99%. The most critical characteristics controlling groundwater recharge in the study region were shown to be geomorphology, lineament density and lithology. As a consequence of this research, the study region is classified as having a high to moderate groundwater potential. Due to the over-exploitation of groundwater use for agriculture, industrial purposes, and mostly for household needs, the groundwater level is decreasing day by day. For future purposes, artificial recharge techniques can be done To increase the groundwater in the very low to moderate zone.

References

Abuzied SM, Yuan M, Ibrahim SK et al (2015) Delineation of groundwater potential zones in Nuweiba Area (Egypt) using remote sensing and GIS techniques. Int J Signal Process Syst 4:203–214. https://doi.org/10.12720/ijsps.4.2.109-117

Ahmed JB II, Mansor S (2018) Overview of the application of geospatial technology to groundwater potential mapping in Nigeria. Arab J Geosci 11(17): 1–16

Ajay Kumar V, Mondal NC, Ahmed S (2020) Identification of groundwater potential zones using RS, GIS and AHP techniques: a case study in a part of Deccan volcanic province (DVP), Maharashtra India. J Indian Soc Remote Sens 48:497–511. https://doi.org/10.1007/s12524-019-01086-3

Al-Abadi AM (2015) Groundwater potential mapping at northeastern Wasit and Missan governorates, Iraq using a data-driven weights of evidence technique in framework of GIS. Environ Earth Sci 74, 1109–1124. https://doi.org/10.1007/s12665-015-4097-0

Arulbalaji P, Padmalal D, Sreelash K (2019) GIS and AHP techniques based delineation of groundwater potential zones: a case study from Southern Western Ghats. India Sci Rep. https://doi.org/10.1038/s41598-019-38567-x

Basak A, Das J, Rahman ATM, Pham QB (2021) An integrated approach for delineating and characterizing groundwater depletion hotspots in a Coastal State of India. J Geol Soc India 97(11):1429–1440. https://doi.org/10.1007/s12594-021-1883-z

Biswas B, Jain S, Rawat S (2018) Spatio-temporal analysis of groundwater levels and projection of future trend of Agra city, Uttar Pradesh, India. Arab J Geosci 11:278. https://doi.org/10.1007/s12517-018-3577-4

Central Ground Water Board (CGWB) (2014) Ministry of Water Resources, G. o. I., Ground water scenario in India, p 50

Chandio IA, Matori ANB, WanYusof KB et al (2013) GIS-based analytic hierarchy process as a multicriteria decision analysis instrument: a review. Arab J Geosci 6:3059–3066. https://doi.org/10.1007/s1251.7-012-0568-8

Chen J, Wu H, Qian H (2016) Groundwater nitrate contamination and associated health risk for the rural communities in an agricultural area of Ningxia, northwest China. Exposure Health 8(3):349–359

Chitsazan M, Rahmani G, Neyamadpour A (2015) Forecasting groundwater level by artificial neural networks as an alternative approach to groundwater modeling. J Geol Soc India 85(1):98–106

Das J, Gayen A, Saha S, Bhattacharya SK (2017) Modelling of alternative crops suitability to tobacco based on Analytical hierarchy process in Dinhata subdivision of Koch Bihar district, West Bengal. Model Earth Syst Environ 3(4):1571–1587. https://doi.org/10.1007/s40808-017-0392-y

Das S, Pardeshi SD (2018) Integration of different influencing factors in GIS to delineate groundwater potential areas using IF and FR techniques: a study of Pravara basin, Maharashtra, India. Appl Water Sci 8:197. https://doi.org/10.1007/s13201-018-0848-x

Fashae OA, Tijani MN, Talabi AO, Adedeji OI (2014) Delineation of groundwater potential zones in the crystalline basement terrain of SW-Nigeria: an integrated GIS and remote sensing approach. Appl Water Sci 4(1):19–38

Gupta D, Yadav S, Tyagi D, Tomar L (2018) Multi-criteria decision analysis for identifying of groundwater potential sites in Haridwar 3:9–15

Ibrahim-Bathis K, Ahmed SA (2016) Geospatial technology for delineating groundwater potential zones in Doddahalla watershed of Chitradurga district, India. Egypt J Remote Sens Space Sci 19(2):223–234

Jackson TJ (2002) Remote sensing of soil moisture: implications for groundwater recharge. Hydrogeol J 10(1):40–51

Jaiswal RK, Mukherjee S, Krishnamurthy J, Saxena R (2003) Role of remote sensing and GIS techniques for generation of groundwater prospect zones towards rural development–an approach. Int J Remote Sens 24 (5):993–1008

Jasrotia AS, Kumar A, Singh R (2016) Integrated remote sensing and GIS approach for delineation of groundwater potential zones using aquifer parameters in Devak and Rui watershed of Jammu and Kashmir, India. Arab J Geosci 9(4):304

Jeihouni M, Toomanian A, Mansourian A (2020) Decision tree-based data mining and rule induction for identifying high quality groundwater zones to water supply management: a novel hybrid use of data mining and GIS. Water Resour Manage 34:139–154. https://doi.org/10.1007/s11269-019-02447-w

Kalantar B, Pradhan B, Naghibi SA, Motevalli A, Mansor S (2018) Assessment of the effects of training data selection on the landslide susceptibility mapping: a comparison between support vector machine (SVM), logistic regression (LR) and artificial neural networks (ANN). Geomatics Nat Hazards Risk 9(1):49–69

Kumar J, Biswas B, Verghese S (2021) Assessment of groundwater quality for drinking and irrigation purpose using geospatial and statistical techniques in a semi-arid region of Rajasthan, India. J Geol Soc India 97:416–427. https://doi.org/10.1007/s12594-021-1699-x

Lee S, Lee CW (2015) Application of decision-tree model to groundwater productivity-potential mapping. Sustainability 7:13416–13432. https://doi.org/10.3390/su71013416

Madani A, Niyazi B (2015) Groundwater potential mapping using remote sensing techniques and weights of evidence GIS model: a case study from Wadi Yalamlam basin, Makkah Province, Western Saudi Arabia. Environ Earth Sci 74:5129–5142

Mandal T, Saha S, Das J, Sarkar A (2021) Groundwater depletion susceptibility zonation using TOPSIS model in Bhagirathi river basin, India. Model Earth Syst Environ 1–21. https://doi.org/10.1007/s40808-021-01176-7

Manap MA, Nampak H, Pradhan B, Lee S, Sulaiman WNA, Ramli MF (2014) Application of probabilistic-based frequency ratio model in groundwater potential mapping using remote sensing data and GIS. Arab J Geosci 7(2):711–724

Mogaji KA, Lim HS, Abdullah K (2015) Regional prediction of groundwater potential mapping in a multifaceted geology terrain using GIS-based Dempster-Shafer model. Arab J Geosci 8(5):3235–3258

Mohanty S, Jha MK, Kumar A, Panda DK (2013) Comparative evaluation of numerical model and artificial neural network for simulating groundwater flow in Kathajodi-Surua Inter-basin of Odisha, India. J Hydrol 495:38–51

Nigussie W, Hailu BT, Azagegn T (2019) Mapping of groundwater potential zones using sentinel satellites (−1 SAR and-2A MSI) images and analytical hierarchy process in Ketar watershed, Main Ethiopian Rift. J Afr Earth Sc 160:103632

Ozdemir A (2011) GIS-based groundwater spring potential mapping in the Sultan Mountains (Konya, Turkey) using frequency ratio, weights of evidence and logistic regression methods and their comparison. J Hydrol 411(3–4):290–308

Pande CB, Khadri SFR, Moharir KN, Patode RS (2017) Assessment of groundwater potential zonation of Mahesh river basin Akola and Buldhana district, Maharashtra, India using remote sensing and GIS techniques. Sustain Water ResourManag

Pande CB, Moharir KN, Panneerselvam B et al (2021) Delineation of groundwater potential zones for sustainable development and planning using analytical hierarchy process (AHP), and MIF techniques. Appl Water Sci 11:186. https://doi.org/10.1007/s13201-021-01522-1

Pinto D, Shrestha S, Babel MS, Ninsawat S (2017) Delineation of groundwater potential zones in the Comoro watershed, Timor Leste using GIS, remote sensing and analytic hierarchy process (AHP) technique. Appl Water Sci 7(1):503–519

Pourghasemi HR, Pradhan B, Gokceoglu C (2012) Application of fuzzy logic and analytical hierarchy process (ahp) to landslide susceptibility mapping at Haraz watershed, Iran. Nat Hazards 63(2):965–996. https://doi.org/10.1007/s11069-012-0217-2

Pourtaghi ZS, Pourghasemi HR (2014) GIS-based groundwater spring potential assessment and mapping in the Birjand Township, southern Khorasan Province Iran. Hydrogeol J

Pradhan B (2009) Groundwater potential zonation for basaltic watersheds using satellite remote sensing data and GIS techniques. Cent Eur J Geosci 1:120–129. https://doi.org/10.2478/v10085-009-0008-5

Rahmati O, Samani AN, Mahdavi M, Pourghasemi HR, Zeinivand H (2015) Groundwater potential mapping at Kurdistan region of Iran using analytic hierarchy process and GIS. Arab J Geosci 8(9):7059–7071

Rahmati O, Pourghasemi HR, Melesse AM (2016) Application of GIS-based data driven random forest and maximum entropy models for groundwater potential mapping: a case study at Mehran Region, Iran. Catena 137:360–372

Rani M, Joshi H, Kumar K, Pande A, Rawat DS (2019) Development of recharge and conservation site suitability model for groundwater retrieval and evaluation of artificial recharge potential in a complex hydrogeological spring-fed river basin. Arab J Geosci 12 (18):1–11

Saaty TL (2008) Decision making with the analytic hierarchy process. Int J Serv Sci 1(1):83

Saha D, Ray RK (2019) Groundwater resources of India: potential, challenges and management. In: Groundwater development and management. Springer, Cham, pp 19–42

Saha S, Das J, Mandal T (2022) Investigation of the watershed hydro-morphologic characteristics through the morphometric analysis: a study on Rayeng basin in Darjeeling Himalaya. Environ Chall 100463. https://doi.org/10.1016/j.envc.2022.100463

Sangchini EK, Emami SN, Tahmasebipour N, Pourghasemi HR, Naghibi SA, Arami SA, Pradhan B (2016) Assessment and comparison of combined bivariate and AHP models with logistic regression for landslide susceptibility mapping in the Chaharmahal-e-Bakhtiari Province, Iran. Arab J Geosci 9(3):201

Saravanan S, Saranya T, Abijith D (2021) Application of frequency ratio, analytical hierarchy process, and multi-influencing factor methods for delineating groundwater potential zones. Int J Environ Sci Technol. https://doi.org/10.1007/s13762-021-03794-1

Sitender R (2010) Delineation of groundwater potential zones in Mewat District, Haryana, India. Int J Geomatics Geosci 2(1):270–281

Swain AK (2015) Delineation of groundwater potential zones in Coimbatore district, Tamil Nadu, using remote sensing and GIS techniques. Int J Eng Res Gen Sci 3(6):203–214

Thapa R, Gupta S, Guin S et al (2017) Assessment of groundwater potential zones using multi-influencing factor (MIF) and GIS: a case study from Birbhum district, West Bengal. Appl Water Sci 7:4117–4131. https://doi.org/10.1007/s13201-017-0571-z

Expected Climate-Induced Alterations in Sugarcane Yield and Its Agronomic Adaptation Strategies

22

P. Dhanya, A. Ramachandran, R. Jagannathan, Rajalakshmi Dhandapani, and E. Krishnaveni Prabhu

Abstract

The present study tries to evaluate climate change impact and its vulnerability on one of the significant crops, sugarcane using the DSSAT crop simulation model. It was also an endeavor to understand the current status of major sugarcane production and agricultural practices in the study area and evolve a synergistic agronomic adaptation framework involving all stakeholders. A dynamical downscaling technique employing the RegCM, the regional climate model, was employed. The period of study was from 1971 to 2098 (128 years). This simulation output was used to predict the crop productivity changes in the Kancheepuram area using crop simulation model DSSAT v 4.7. The future yield projections showed a decline in yield for sugarcane. The likely yield changes for sugarcane were ascertained for the near, mid, and end centuries. The alterations in productivity for sugarcane were just −1.8, −2.6, and −2.8 percent for the near, mid, and end of the twenty-first century, respectively. The effect of CO_2 fertilization (+520 ppm) could be noticed till the mid-century, thereafter yield was shown to fall when that threshold level was reached. Agronomic adaptation was simulated using the model, adopting different sowing windows that are prevalent in the study area. According to the findings, sugarcane planted on March 15 can augment yield rates by 3.5% in the near-century under the RCP 4.5. Success and implementation of planned climate change adaptation lie in the hands of the stakeholders and can reduce the weather and climate risks to a great extent.

Keywords

Climatic changes · Impacts · Sugarcane yield · Agronomic adaptation strategies · Diurnal temperature range · Groundwater management

22.1 Introduction

Despite the fact that climate change is a global phenomenon, its consequences are regional and local (Das et al. 2019). Its impacts are likely to have a considerable regional variation, as it affects crop productivity, negatively and positively across the globe. Research studies piloted around the world show that detrimental effects are more prevalent in the tropics, whereas

P. Dhanya (✉) · R. Jagannathan · R. Dhandapani
Tamil Nadu Agricultural University, Coimbatore, Tamil Nadu, India
e-mail: dhanya.eptri@gmail.com

A. Ramachandran · E. Krishnaveni Prabhu
Center for Climate Change and Adaptation Research, Anna University, Guindy, Chennai, Tamil Nadu, India

J. Das and S. K. Bhattacharya (eds.), *Monitoring and Managing Multi-hazards*,
GIScience and Geo-environmental Modelling, https://doi.org/10.1007/978-3-031-15377-8_22

favorable effects have been recorded for temperate regions (Lobell 2009). Under this purview, food security is a critical concern because it has the potential to exacerbate the existing challenges of food insecurity in a developing country like India, which has a high population growth rate (FAO 2008). This FAO report also states that India's arid and semiarid regions have the highest prevalence of food insecurity. It is critical to plan to cater to the rising food demand sustainably, especially in light of climate change, which exacerbates already existing strains on available natural resources, particularly water. Coastal areas are the hardest hit by natural catastrophes and are particularly vulnerable to climate change and other environmental issues. Sea-level rise associated with global warming and related topics salinity, soil fertility, waterlogging, flooding, cyclones, and storm surges are all factors to deliberate (Ranjan et al. 2006). The coastal areas in India have become increasingly vulnerable to the socio-economic effects of extreme weather events such as floods, droughts, cyclones, and heatwaves. Furthermore, Indian coastal regions account for 17% of the country's population, making them increasingly vulnerable (Khan et al. 2012). According to Patnaik and Narayanan (2005), India's eastern coast is more vulnerable than its western shore. The present study area comes on the east coast of India.

An increase in daytime extreme temperatures, several hot days, and dryness have been predicted for south Asian regions as a consequence of global warming (IPCC 2013). It is very essential to understand these possible changes and critically evaluate them spatially to understand the extent and degree of their impacts and prepare and plan suitable regional and local adaptation strategies (Ramachandran et al. 2015). According to the FAO (2017), developing nations will account for the majority of global population growth until 2050, ensuing intensified water and food demands. Cities and towns in lower-income nations will account for the majority of global population growth by 2050. It is essential to have regional and local-specific research studies to get the pulse of a region. Multiple hazards such as droughts, dry spells, heavy rainfall damages, floods, cyclones, and earthquake-induced tsunamis are frequent in Tamil Nadu due to its physical setting. Many academics have recommended that specific local-level studies be conducted to quantify the effects and to better understand agricultural susceptibility and society's responses to climate change.

Geospatial technologies play a significant role to capture the regional variations of the impact of climate change on available water, agriculture yield, etc. These tools support post-processing the modeled climate data to visualize spatial comparison (Quach and Jenny 2020). In the tropics, crops are particularly susceptible to not only hazards but even the minor changes in their microclimatic circumstances; hence, modeling and forecasts of crop productivity at the micro-scale would help to mitigate the potential repercussions. Many authors have modeled and forecasted the impacts on key crop yields using agroecosystem simulation for India (Patel et al. 2013). This study is also an endeavor to understand the anticipated warming of sugarcane productivity in the northeast agroclimatic zones of Tamil Nadu. This work deserves special focus as it is carried out in a comprehensive way combining both top-down and bottom-up approaches using the latest state of art methodologies and field surveys. Disseminating the climate change knowledge will help the scientific community and decision-makers in further research and policy formulations.

22.2 Materials and Methods

22.2.1 Profile of the Study Area

The present area of interest covers both Kancheepuram and Thiruvallur districts. Both districts share boundaries with the Bay of Bengal on the east (Fig. 22.1). The shaded areas in red in the study area map show the possible risk of inundation in coastal areas if the sea level rises by 0.50 cms. Building the resilience of coastal communities is a matter of concern in these areas as these areas are frequently hit by cyclones and storm surges. Palar and Cheyyar are the major

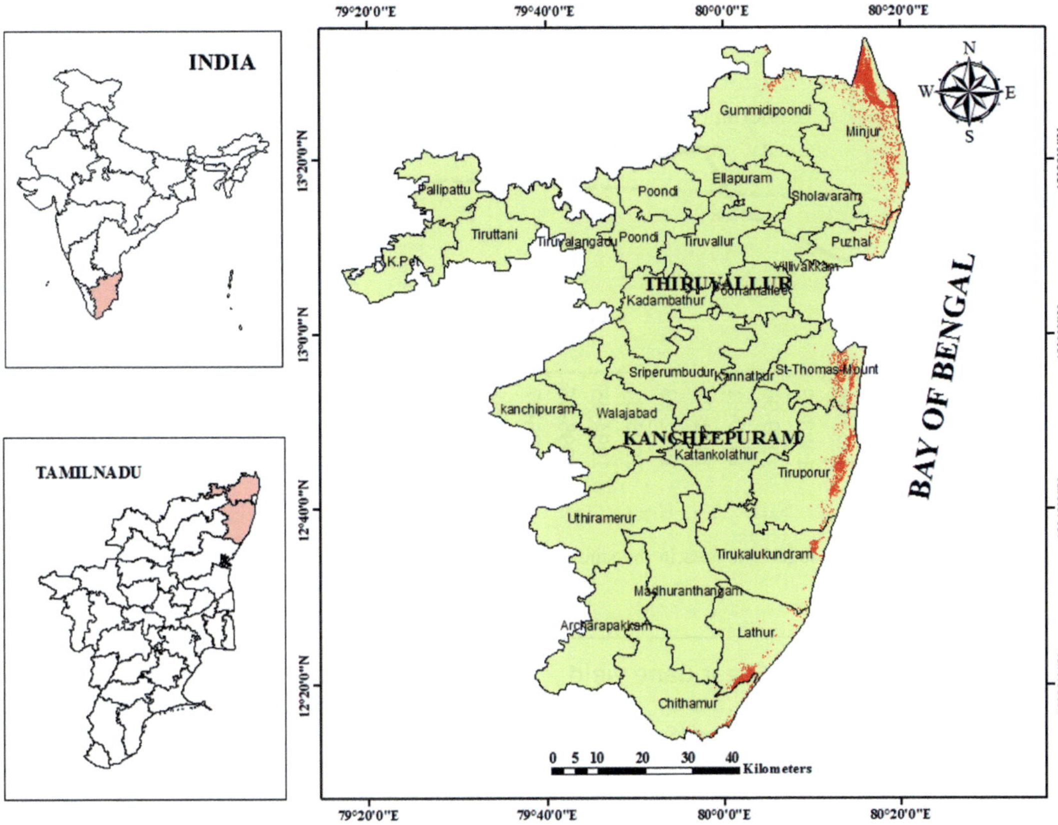

Fig. 22.1 Index map showing the low-lying areas in the study area

drainage sources. The major soil types found in this district are clay soil, red loams, and alluvial soil.

Field observations were carried out to comprehend the existing agronomical practices and management details of the key crops grown in the study area Kancheepuram and Thiruvallur. Information with regards to sugarcane production, phenology, the key ruling varieties as well as the results of experimental trials was gathered from research centers of Tamil Nadu Agricultural University for modeling purposes. Department of Agriculture Economics and Statistics, Chennai was the source for secondary data on crop coverage and production. Farmers follow three separate seasons when it comes to sugarcane farming procedures. The primary season commences from December to January. The dominant varieties cultivated in this season are CO-658, CO-62174, and CO-62198, and the mid-season commences from February/March, and the end season starts from April/ May. Secondary data shows that area under sugarcane has increased during both 1995–1996 and 2000–2001 and reduced during the period 2005–2006 (Seasonal Crop Reports, Fig. 22.2). Sugarcane yield for the Kancheepuram district was reduced due to droughts in 2002–2003 and 1990–91 (Fig. 22.3).

22.2.2 Model-Based Crop Yield Impact Assessment

Agronomic modifications can boost agricultural output. Forecast-based actions can support crop production systems that face multidimensional risks such as delays in monsoon onset dates, dry

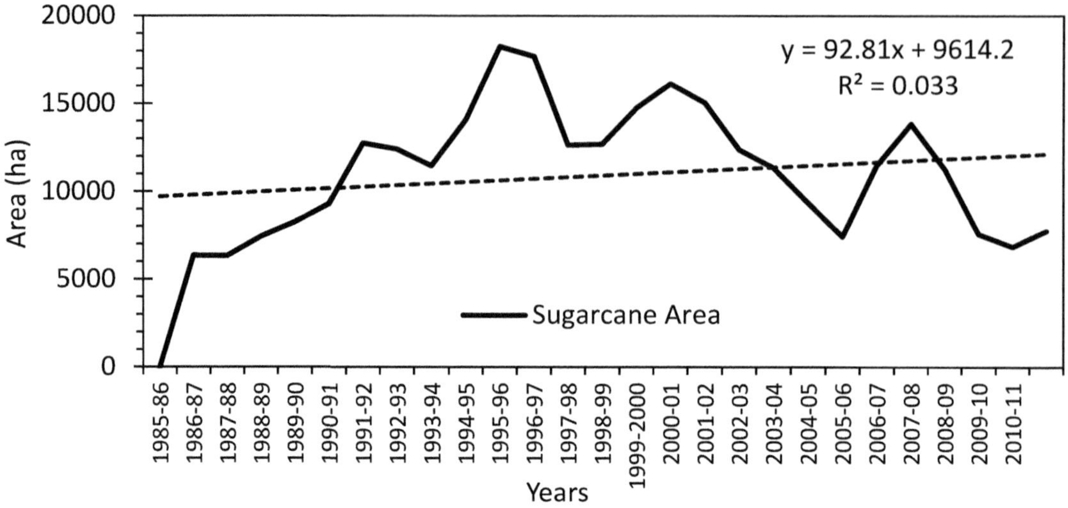

Fig. 22.2 Linear trends in sugarcane area in the study area from 1985 to 2012

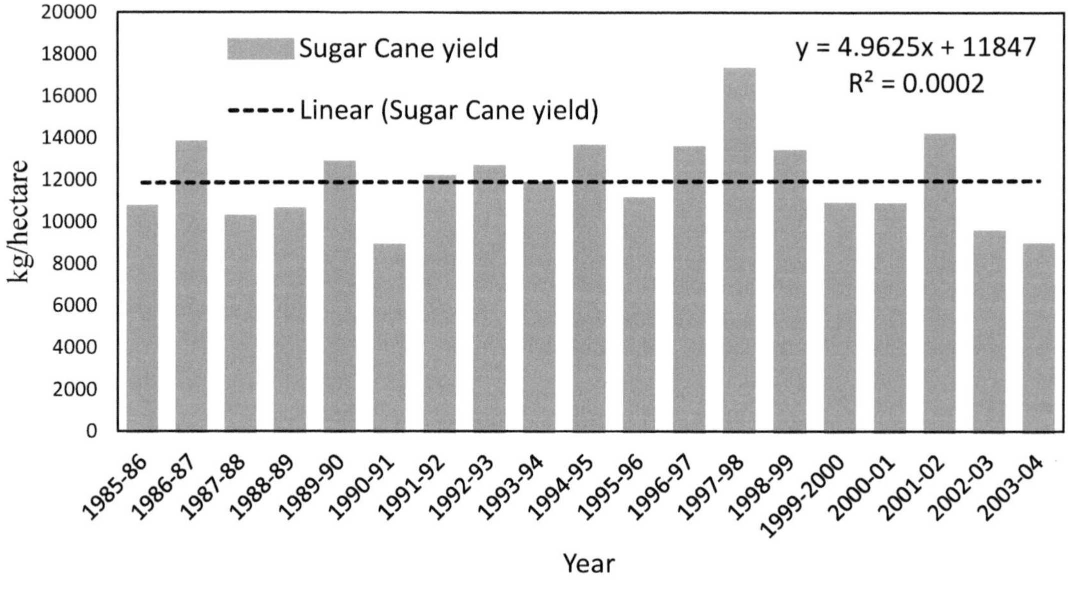

Fig. 22.3 Linear trends in sugarcane yield in the study area from 1985 to 2003

spells, acute water shortage, soil moisture stress, and insect pest attacks on crops and livestock (Naresh et al. 2012). The complex task of yield predictions can be carried out using crop simulation models (CSMs).

In this context, the goal of this research is to gain a better understanding of how agronomical adaptations such as adjustments in sowing dates can reduce the likelihood of crop output losses due to climatic changes. Decision Support System for Agrotechnology Transfer (DSSAT-V4.5) is a process-based CSM, (Hoogenboom et al. 2012). Under this, a sub-segment, CANE-GRO under sugar/energy modules was chosen to simulate the future likelihood of sugarcane productivity based on varying sowing adjustments

(Jones et al 2003). Daily weather, soil, and crop management files were generated for the simulation purpose.

DSSAT-CERES module has been used and validated by Ramachandran et al. (2017) and Rajalakshmi et al. (2013) for south India. Daily simulated weather output was transformed to monthly values and further averaged over 30-year period time slices such as reference period (1971–2000), near (2010–2040), mid (2041–2070), and end century (2071–2100).

The following equation was used to calculate the expected change in yield.

$$\text{Anticipated Yield Change} = \left(\frac{\text{Simulated Yield} - \text{Observed yield}}{\text{Observed Yield}}\right) * 100 \quad (22.1)$$

22.2.3 Agronomic Adaptation Simulation Using Changes in Sowing Date Using DSSAT

With the knowledge gained from the impact assessment using CSM, it was obvious that main crop yields are expected to decline by the end of the twenty-first century. This study also investigated the potential agronomic adaptation strategies, such as sowing date adjustment, to mitigate the projected yield change. Crop calendars gathered from field visits and a review of the literature were helpful to determine distinctive sowing windows for sugarcane.

22.3 Results and Discussion

The RegCM 4.4 model simulations based on the IPCC's RCP 4.5 pathway were used for the analysis. RCP4.5 is a stabilization scenario, which assumes a global economic framework that stabilizes the radiative forcing around 550 ppm CO_2-equivalent by 2100.

Previously published researches also reveal that a rise in day and night time temperature by 2.3 and 2.4 °C may be expected in the study area by mid-century (Dhanya et al. 2018). As a result, significant reduction strategies to mitigate carbon emissions in all economic sectors are required as the solution to mitigate the negative impacts. The simulated output was verified to assess the uncertainty and bias for the reference period (1970–2000). This model has performed with a cold bias for temperature projections. The annual maximum and minimum temperature was underestimated by −1.45 and −3.79 °C, respectively. A significant warm bias of 532 mm was noticed for annual rainfall projections. By the end century (2070–2098), daytime temperatures may be amplified by 2.3 and 2.5 °C, respectively. Precipitation is projected to drop by 16% during mid-century, but thereafter it may increase by 24% toward the end century. Relative humidity, wind speed, and solar radiation show the slightest decrease but are negligible.

Diurnal temperature ranges (DTR), one of the chief indicators of climate change, are affected by the differential heating and cooling of land. The mean annual DTR predictions indicated a declining trend for the study region. Similar outcomes have been briefed by various scientists (Lobell and Field 2007). A research study on Douro valley of Portugal reveals a significant dropping in DTR for the period 2020–2080, Jones and Alves (2012). Analysis shows that DTR for coastal zones may likely be ranging from 10 °C to 11 °C during the end of this century (Fig. 22.4). DTR of 11.5 °C to 12.5 °C may be experienced in the extreme northern, southern, and interior parts toward the end of the twenty-first century. The decline in DTR may be higher in and around Tambaram, Chengalpattu, and Cheyyur taluks.

The future yield forecasts using DSST CSM suggest that the sugarcane yields may drop due to warming. The anticipated productivity decline for sugarcane is −1.8, −2.6, and −2.8% for the 2040s, 2070s, and 2100, respectively, for the Kancheepuram district. The maximum decline was noted during the 2081–90 (−3.53%)

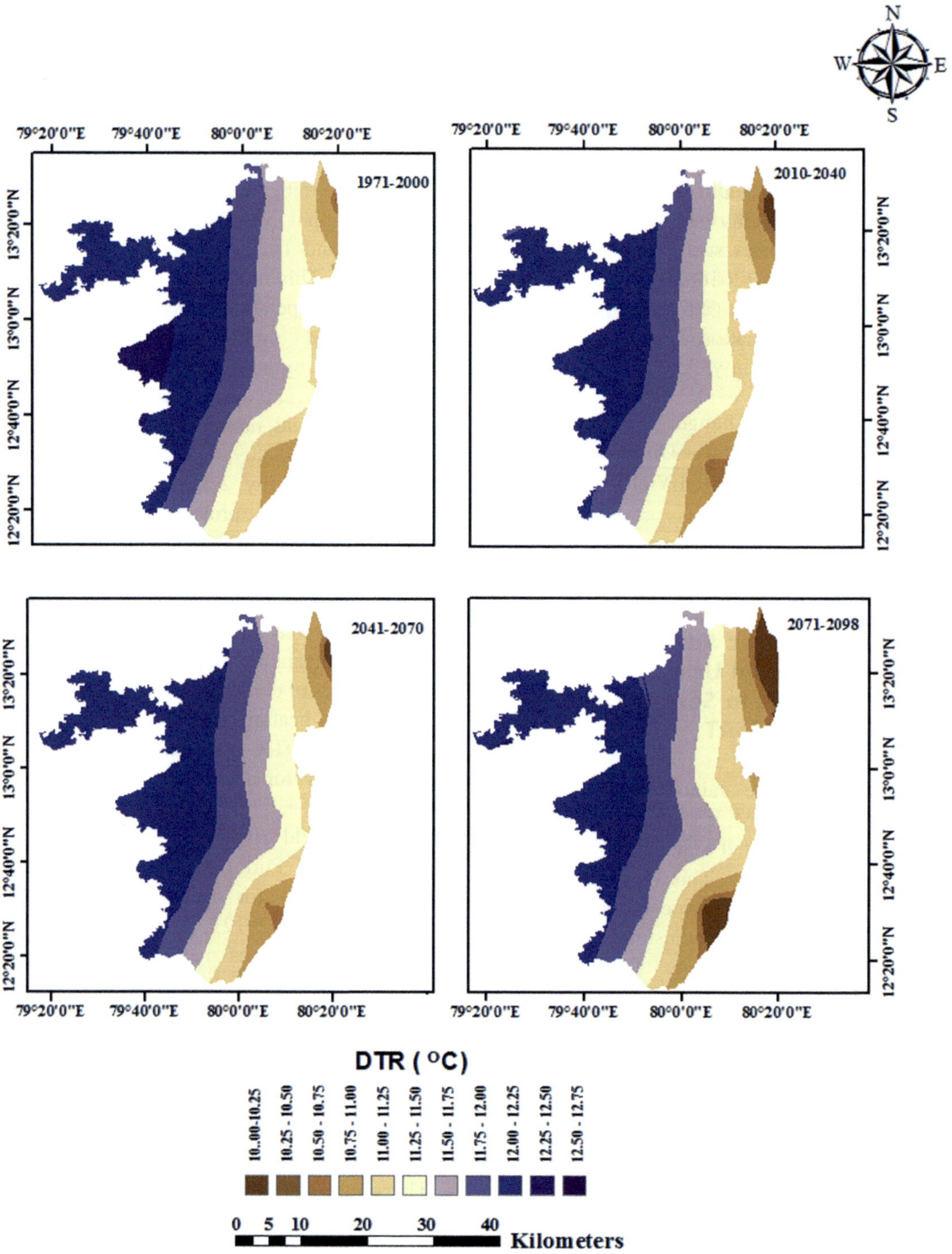

Fig. 22.4 Spatial variations in projected diurnal temperature range (DTR) during BL, NC, MC and EC

followed by 2071–80 (−2.56%) and 2041–50 (−2.4%). The CO_2 fertilization (+520 ppm) effect was visible till mid-century after that, and productivity was seen to be declining (Fig. 22.5).

Advancing or delaying sowing dates can provide better yield under future warming. Findings suggest that altering the sowing dates can result in higher yield rates in the future. The experimental planting dates selected were the 15 of December, January, February, March, April, and May. In the research location, the planting season lasts for six months. Thus wide span of sowing windows from December to May is employed for the experimental simulation. According to the results, sugarcane planted on March 15 can enhance yield rates by 3.5% in the 2040s under the RCP 4.5 future climate pathway. (Fig. 22.6). On the other hand, it yielded 5.1 percent and 6.9 percent higher yields comparatively in the mid and end century, respectively.

It has been discovered that changing planting dates and cultivars can enable improved production and better-coping capacity (Smit and Wandel 2006 and Sneh Gangwar 2013, Dettori 2011). Tingem et al (2009), on the other hand, observed that genotype adaptation was more efficient in mitigating unfavorable effects. The need to adapt to climate change is now widely recognized as greenhouse gas emissions continue to be unabated (Wilson et al. 2013). There is a wide gap between substantial rates of implementation of adaptation actions on the ground despite substantial investments in adaptation science. Hence, viable options such as adjustments in the form of adaptation are the need of the hour. Successful adaptation at the local-level necessitates a well-planned and integrated set of measures with both immediate and long-term adaptation goals in mind (Baas and Ramaswamy 2008). Breeding salt, heat and drought-tolerant crops, modifying the time of farm operations, changing cropping patterns, adopting water-saving techniques, and availing of farm risk insurances are all possibilities to be explored (Cote and Nightingale 2012).

It is necessary to enhance research and developments in agriculture as proactive changes in farm practices can have profound effects on negating climate risks. The best outcomes from the simulation output can be used for field-level assessments and farm-level implementation. The main challenge is to disseminate this information for the benefit of the farmers during their farming operations. Climate change poses a serious challenge to crop production systems, compromising a country's food and livelihood security (Chaturvedi et al. 2012). Groundwater is also significant to sustain sugarcane production in these areas. However, many of the blocks have groundwater that is either underutilized or in the critical category (Fig. 22.7).

Impact assessments are essential to understand future plausible climate change and its impacts on agriculture productivity will facilitate

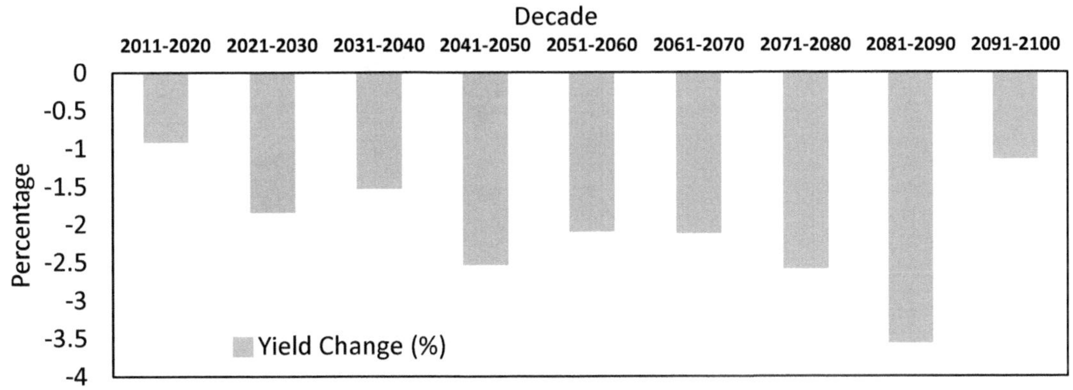

Fig. 22.5 Decadal changes in the sugarcane yield projected by the DSSAT model for Kancheepuram areas

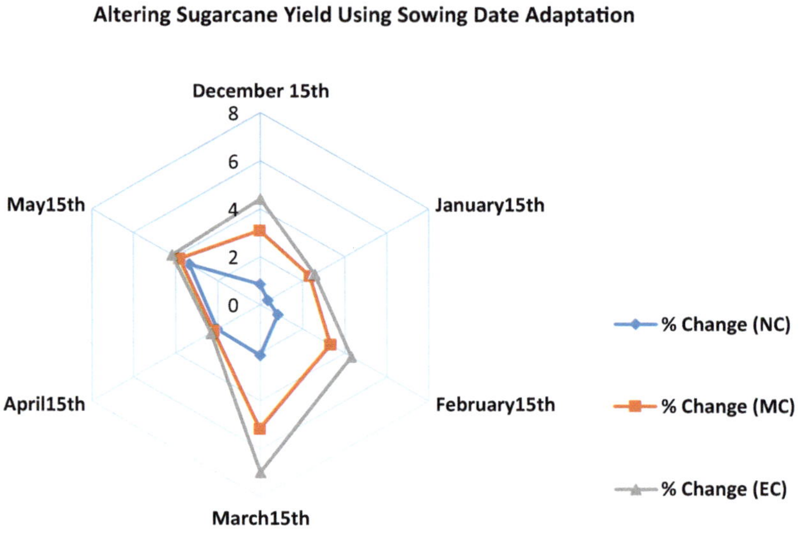

Fig. 22.6 Variations in the sugarcane yield under sowing date adaptation (*NC* near century; *MC* mid century and *EC* end century)

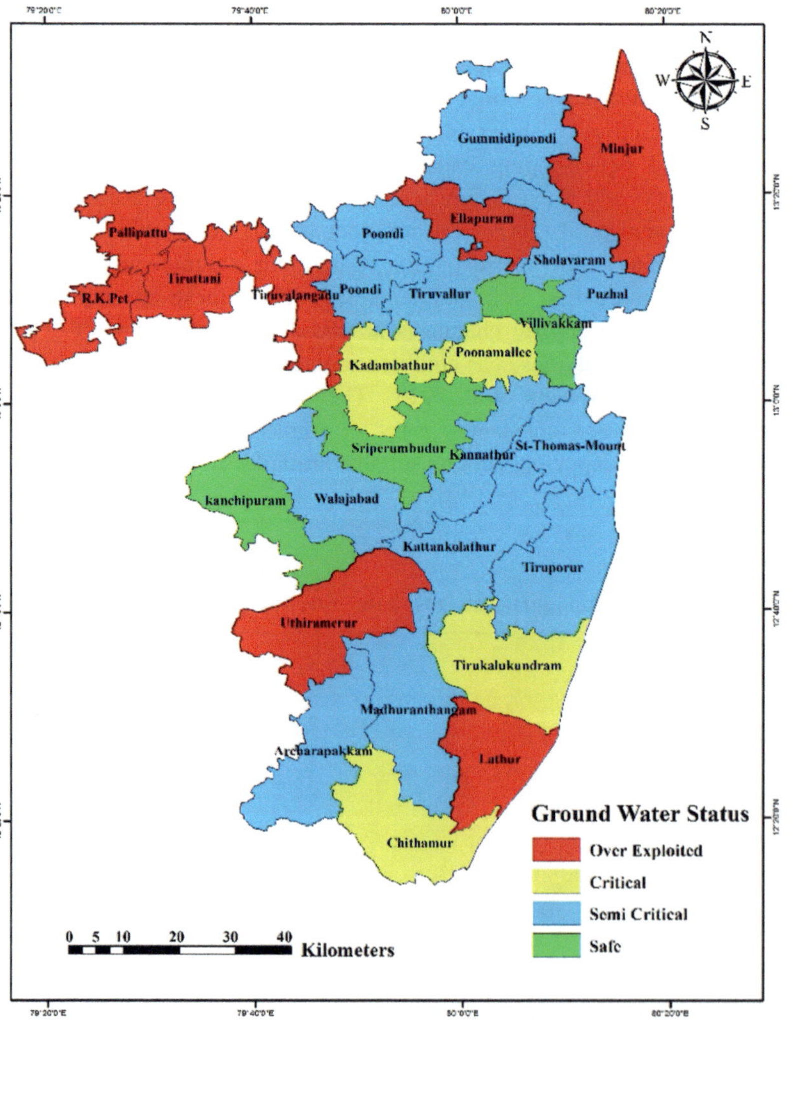

Fig. 22.7 Groundwater status in Kancheepuram and Thiruvallur districts

implementing site specific adaptation options. Moreover, this paper has tried to explore impacts of various agronomic practices as climate change adaptation strategies. The current study used an integrated approach that included both science and practice to analyze the effects of climate change on main crop yields in Chengalpattu. This detailed study at the sub-district level helped to identify the future possible climate variations under the RCP 4.5 emission trajectory and its likely impacts on the productivity of major crops. Concomitantly, this knowledge on impacts was acquired and disseminated among the stakeholders to provide awareness and to frame a suitable adaptation framework. This can be incorporated into local-level policy decisions and may be scaled up to state and national levels.

22.4 Conclusions

This study was an integration of both top-down and bottom-up approaches. It tried to incorporate both scientific and social perspectives on an environmental challenges like climate change. The model RegCM was successfully employed to project and assess the future likely impacts on major crop yields of Chengalpattu areas. Using DSSAT, the productivity of the major crops was found to be decreasing in the century. In the scenario of sea-level rise and increased warming combined with heavy groundwater tapping, saline intrusion may create severe problems to crop production. Hence to maintain food security, we need to have adaptations by changing the agricultural practices and improvised water management practices. Taking into consideration of scientific outcomes from the model simulated projections of impacts and stakeholders' decisions, changes in the timing of farming operations were a potential option to adjust the climate variability, and shifting to crop varieties that are heat, drought, or salt-tolerant were chosen as the best and popular adaptation options to reduce the risks from climate change. Adoption of planned adaptation lies in the hands of the stakeholders. Farmers should be willing to adapt and take advantage of the support provided through agro advisories Krishiwani by IMD or agricultural universities, etc. Science and technological research advancements should be capable of disseminating reliable advisories to the grass-root level.

Acknowledgements The authors sincerely acknowledge Anna University for providing Centenary Research Fellowship for the financial support and Tamil Nadu Agriculture University for providing guidance. The first author would like to thank the support provided by all concerned institutions and especially Dr. Saleem Khan, Center for Climate Change and Adaptation research, Anna University for all the guidance in sea-level rise.

References

Baas S, Ramasamy S (2008) Community based adaptation in action. Food and Agricultural Organization of the United Nations (FAO), Rome

Chaturvedi RK, Joshi J, Jayaraman M, Bala G, Ravindranath NH (2012) Multi-model climate change projections for India under representative concentration pathways. Curr Sci 7(103):791–802

Cote M, Nightingale AJ (2012) Resilience thinking meets social theory: situating social change in socio-ecological systems (SES) research. Prog Hum Geogr 36(4):475–489. https://doi.org/10.1177/0309132511425708

Das J, Mandal T, Saha P (2019) Spatio-temporal trend and change point detection of winter temperature of North Bengal, India. Spatial Inf Res 27(4):411–424. https://doi.org/10.1007/s41324-019-00241-9

Dettori M, Cesaraccio C, Motroni A, Spano D, Duce P (2011) Using CERES-Wheat to simulate durum wheat production and phenology in Southern Sardinia, Italy. Field Crop Res 120:179–188

Dhanya P, Ramachandran A, Palanivelu K (2018) Constructing local sea level rise scenarios for assessing possible impacts and adaptation needs: insights from coasts of India, Sea Level Rise and Coastal Infrastructure; IntechOpen. In: Zhang Y, Hou Y, Yang X (eds). https://doi.org/10.5772/intechopen.74325, URL https://doi.org/10.5772/intechopen.74325

FAO (2008) Report on the state of food insecurity in Rural India. Accessed from https://www.fao.org/3/i0291e/i0291e00.htm

FAO (2017) The future of food and agriculture—trends and challenges. Rome

Hoogenboom G, Jones JW, Wilkens PW, Porter CH, Boote KJ, Hunt LA, Singh U, Lizaso JL, White JW, Uryasev O, Royce FS, Ogoshi R, Gijsman AJ,

Tsuji GY, Koo J (2012) Decision support system for agrotechnology transfer (DSSAT) version 4.5. The University of Hawaii, Honolulu, Hawaii
IPCC (2013) Climate change 2013: the physical science basis, fifth assessment report of ar5/wg1/, viewed on May 20, 2014
Jones GV, Alves F (2012) Impact of climate change on wine production: a global overview and regional assessment in the Douro Valley of Portugal. Int J Glob Warm 4(3–4):383–406
Jones JW, Hoogenboom G, Porter CH, Boote KJ, Batchelor WD, Hunt LA, Wilkens PW, Singh U, Gijsman AJ, Ritchie JT (2003) The DSSAT cropping system model. Eur J Agron 18:235–265
Khan AS, Ramachandran A, Usha N, Punitha P, Selvam V (2012) Predicted impact of sea-level rise at vellar-coleroon estuarine region of Tamil Nadu Coast, in India: mainstreaming adaptation as a coastal zone management option. Ocean Coast Manag 69:327–339
Lobell DB, Field CB (2007) Global-scale climate-crop yield relationships and the impacts of recent warming. Environ Res Lett 2:014002
Lobell DB (2009) Crop yield gaps: their importance, magnitudes, and causes. Annu Rev Environ Resour 34:179–204
Naresh KS, Singh AK, Aggarwal PK, Rao VUM, Venkateswaru B (2012) Climate change and Indian agriculture salient achievements from ICAR network project, Indian Agricultural Research Institute publications, 2012, New Delhi
Patel HR, Lunagaria MM, Karande BI, Vyas P, Yadav SB, Shah AV, Rao VUM, Nareshkumar S (2013) Impact of projected climate change on groundnut in Gujarat. J Agrometeorol 15(I):41–44
Patnaik U, Narayanan K (2005) Vulnerability and climate change: an analysis of the eastern coastal districts of India. In: Presented at the GECHS international workshop human security and climate change, 21–23 June, Oslo, Norway. Available at. http://ideas.repec.org/p/pra/mprapa/22062.html. Accessed online on Feb 2013
Quach Q, Jenny B (2020) Immersive visualization with bar graphics. Cartography Geographic Inf Sci 47 (6):471–480. https://doi.org/10.1080/15230406.2020.1771771
Rajalakshmi D, Jagannathan R, Geethalakshmi V (2013) Uncertainty in seasonal climate projection over Tamil Nadu for 21st century. Afr J Agric Res 8(32):4334–4344. https://doi.org/10.5897/AJAR2013.7618.9
Ramachandran A, Dhanya P, Jaganathan R, Palanivelu K (2015) Projected and observed aridity and climate change on the east coast of South India under RCP 4.5. The Scien.World J 2015(169761):1–11. https://doi.org/10.1155/2015/169761
Ramachnadran A, Dhanya P, Jaganathan R, Rajalakshmi D, Palanivelu K (2017) Spatiotemporal analysis of projected impacts of climate change on the major C3 and C4 crop yield under representative concentration pathway 4.5: insight from the coasts of Tamil Nadu, South India. PLoS ONE 12(7):e0180706. https://doi.org/10.1371/journal.pone.0180706
Ranjan P, Kazama S, Sawamoto M (2006) Effects of climate change on coastal fresh groundwater resources. Glob Environ Chang 16:388–399
Seasonal Crop Report (2009–2010) Tamil Nadu2009–10, Fasli 1419, Department of Economics and Statistic, Tamil Nadu
Smit B, Wandel J (2006) Adaptation, adaptive capacity and vulnerability. Glob Environ Chang 16(3):282–292. https://doi.org/10.1016/j.gloenvcha.2006.03.008
Gangwar S (2013) Climate change vulnerability and risk assessment: focusing on coastal India. Int J Environ Eng Manag 4(6):605–612
Tingem M, Rivington M, Bellocchi G (2009) Adaptation assessments for crop production in response to climate change in Cameroon. Agron Sustain Dev 29:247–256. https://doi.org/10.1051/agro:2008053
Wilson S, Pearson LJ, Kashima Y, Lusher D, Pearson C (2013) Separating adaptive maintenance (resilience) and transformative capacity of social-ecological systems. Ecol Soc 18(1):22. https://doi.org/10.5751/es-05100-180122

Index

J. Das and S. K. Bhattacharya (eds.), *Monitoring and Managing Multi-hazards*, GIScience and Geo-environmental Modelling, https://doi.org/10.1007/978-3-031-15377-8

GPSR Compliance

The European Union's (EU) General Product Safety Regulation (GPSR) is a set of rules that requires consumer products to be safe and our obligations to ensure this.

If you have any concerns about our products, you can contact us on ProductSafety@springernature.com

In case Publisher is established outside the EU, the EU authorized representative is:

Springer Nature Customer Service Center GmbH
Europaplatz 3
69115 Heidelberg, Germany

Batch number: 10732807

Printed by Printforce, the Netherlands